D0609493

NANO-CMOS CIRCUIT AND PHYSICAL DESIGN

NANO-CMOS CIRCUIT AND PHYSICAL DESIGN

Ban P. Wong
NVIDIA

Anurag Mittal
Virage Logic, Inc.

Yu Cao
University of California–Berkeley

Greg Starr
Xilinx

WILEY-INTERSCIENCE
A JOHN WILEY & SONS, INC., PUBLICATION

Published by John Wiley & Sons, Inc., Hoboken, New Jersey.
Published simultaneously in Canada.

For general information on our other products and services please contact our Customer Care Department within the U.S. at 877-762-2974, outside the U.S. at 317-572-3993 or fax 317-572-4002.

Wiley also publishes its books in a variety of electronic formats. Some content that appears in print, however, may not be available in electronic format.

Library of Congress Cataloging-in-Publication Data:

Nano-CMOS circuit and physical design / Ban P. Wong ... [et al.].
p. cm.
Includes bibliographical references and index.
ISBN 0-471-46610-7 (cloth)
1. Metal oxide semiconductors, Complementary–Design and construction. 2. Integrated circuits–Design and construction. I. Wong, Ban P., 1953–
TK7871.99.M44N36 2004
621.39′732–dc22 2004002212

Printed in the United States of America.

10 9 8 7 6 5 4 3 2 1

CONTENTS

PART I PROCESS TECHNOLOGY AND SUBWAVELENGTH OPTICAL LITHOGRAPHY: PHYSICS, THEORY OF OPERATION, ISSUES, AND SOLUTIONS

FOREWORD

Relentless assaults on the frontiers of CMOS technology over several decades have produced a marvel of a technology. The world we live in has been changed by complex integrated circuits now containing a billion transistors with line widths of less than 100 nm, fabricated in plants costing several billion dollars. This microelectronics revolution was made possible only through the dedication and ingenuity of many specialized experts with detailed knowledge of their crafts.

Yet IC designers, device integrators, and process engineers have always recognized the benefits of a broad understanding of different aspects of IC technology and have combated the compartmentalization of knowledge through continuing learning. For IC designers, a good understanding of the underlying physical constraints of device, interconnect, and manufacturing is crucial for fully achieving the product values attainable. For technology developers, knowing the impact of technology on advanced designs provides the necessary foundation for making sound technological decisions.

While the need to acquire knowledge in the neighboring field has always existed, it has grown in recent years for several reasons. The pace of new technology introduction and the rate of rise of circuit speed increased significantly beyond the historical rates of the previous two decades. This accelerated pace may or may not be sustained for long; nevertheless, there is now a larger body of new knowledge that awaits engineers to learn and use than before. A second reason is that as technology scaling becomes more difficult, trade-offs such as those between leakage and performance and between line width and variability must, more than ever, be made judiciously with careful consideration of design, device, and manufacturing. Finally, a large and increasing number of engineers work for companies that specialize in either design or manufacturing (i.e., companies without fabrication facilities or silicon foundries). These engineers face greater challenges in seeing the complete picture than do those working for integrated IC companies.

There are many books devoted to either silicon process technology or IC design, but few that give a comprehensive view of the current status of both. It is in this area of integration of nanometer processes, device manufacturability, advanced circuit design, and related physical implementation that this book adds the most value. It starts with a section of three chapters on recent and future trends in devices and processing and continues through a second section of six chapters describing design issues, with special attention paid to the interactions between technology and design, such as signal integrity and interconnects as well as practical solutions. The third and final section addresses the impact of design on yield or design for manufacturability.

This book is for both IC designers and technologists who want a convenient and up-to-date reference written by expert practitioners of the industry. In IC technology there are still many more new territories to be pioneered and new vistas to be discovered. This book is a good addition to our travel bags!

CHENMING HU

Taiwan Semiconductor Manufacturing Company
and the University of California–Berkeley
January 2004

PREFACE

In 1965, Gordon Moore formulated his now famous Moore's law, which became the catalyst for advancements in the semiconductor industry. The semiconductor industry has brought us the sub-100-nm era with all the advancements we see today. With these advancements come difficulties in process control and subsequent challenges to circuit and physical design. As a result, the degrees of freedom in design methodology are fast shrinking and will require a revolutionary change in the way we put together chips that are not only functional but also meet the design objectives and are high yielding.

However, the explosive growth of semiconductor models developed in the absence of fabrication facilities has resulted in the isolation of process/device engineers from circuit design engineers, leading to some lack of understanding of the impact of their designs upon manufacturability, yield, and performance, due to the fundamental limitations of technology and device physics. As we enter the nano-CMOS era, knowing how to traverse these issues is critical to the success of products and companies. These communities of engineers must work together to fill each other's knowledge gaps, which are ever widening as we travel down the road of dimensional scaling. Only by doing this can goals be realized.

While faced with these issues during the course of our duties, we could find no book that addresses them in a single bound volume. The information exists in bits and pieces and mostly locked up in the minds of experts, some of whom we have consulted in the course of our jobs. This book is an attempt to provide a seamless entity that talks about these interactions and their impact on manufacturability, yield, and performance. It provides practical guidelines to help designers avoid some of the pitfalls inherent in advanced semiconductor processes as well as the strongly needed bridge from physical and circuit design to fabrication processing, manufacturability, and yield. The concepts we present in this text are extremely significant, especially as technology moves into the nano-CMOS feature sizes.

The book is organized into three parts. In the first part we provide detailed descriptions of the deep-submicron processes to help designers understand the issues associated with them and to provide more insight into the limitations brought about by dimensional scaling. In the second part we provide an overview of the impact of process scaling on circuit design and physical implementation. In the final part we cover issues concerning manufacturability and yield and provide guidance to ensure that a part is manufacturable and meets the yield and performance targets.

Chapter 1 provides an overview of the issues designers face in the deep-submicron processes. This chapter provides a framework for the rest of the book. Part I contains Chapters 2 and 3. In Chapter 2 we review the current status and possible future solutions of FEOL and BEOL processing systems for 90 nm and below. The FEOL section deals with gate dielectric and strain engineering developments, including related equipment issues. It also provides an in-depth discussion of CMOS scaling issues such as gate tunneling and NBTI. In the BEOL section we discuss local and global interconnect scaling, copper wire development, and low-κ interlayer dielectric challenges along with integration schemes such as dual damascene. Chapter 3 is a tutorial on optical lithography which encompasses the physics and theory of operation, including issues associated with advanced processes and corresponding solutions.

Part II consists of Chapters 4 through 9. In Chapter 4 we provide a brief overview of design issues facing mixed-signal circuits and guidance for avoiding some of the pitfalls associated with designing circuits for advanced processes. In Chapter 5 we provide an overview of the ESD issues designers face in the creation of complex systems on a chip. Issues such as multiple supply protection are covered in detail to equip designers in the evaluation of specific ESD requirements. The latest SCR structures are also included as yet another option for developing an ESD protection strategy. Chapter 6 outlines the current trends in I/O buffer design. An overview of the various I/O specifications is provided along with current trends for implementing designs. Power busing issues and simultaneous switching noise issues are discussed at length to illustrate the importance of developing the I/O power bus scheme up front. On-die decoupling is also discussed at length, as this is becoming a key feature required to meet high-speed interface specifications. Chapter 7 takes the reader through the basics of DRAM design and then goes into the techniques to successfully scale the storage capacitor, access transistor, and sense amplifier into nano-CMOS processes. Chapter 8 focuses on signal integrity analysis and design solutions for on-chip interconnects. First, efficient parasitics extraction techniques are presented, with particular emphasis on inductance issues. Then analytical approaches for signal timing, crosstalk noise, and waveform integrity analysis are discussed. In the last part of the chapter we investigate physical and circuit design solutions to improve signal integrity in high-speed signaling. Chapter 9 provides a comprehensive overview of existing design- and run-time low-power design techniques on different levels of a system design, with a focus on circuit-level logic and memory design approaches. The perspective of ultralow power design techniques for future technology nodes beyond 90 nm is discussed at the end of the chapter.

Part III comprises Chapters 10 and 11. Chapter 10 provides guidelines for achieving a manufacturable design. Numerous examples, including post-OPC simulations, are shown of potential issues with the physical layout of circuits along with methods for improvements. In Chapter 11 we cover the design principles for robust and high-performance circuits despite process variation. The chapter begins with a discussion of the sources of process and other variations, and their impact on circuit functionality and performance. Three principal design areas (clocks, SRAM, and selected digital circuits) were chosen as case studies to illustrate these principles. The chapter also includes some guidelines for a DFM-friendly design. The chapter concludes with a brief overview of the need for statistical device modeling for nano-CMOS designs, followed by a brief description of the new features incorporated in the BSIM4 model.

ACKNOWLEDGMENTS

We would like to acknowledge the many people who contributed to the completion of this book. First we thank the subject experts who wrote some of the chapters or sections. We thank the technologists at Applied Materials, Inc.—Reza Arghavani, Faran Nouri, and Gary Miner—for their contributions to the section on equipment requirements for front-end processing. We are indebted to Khaled Ahmad of Applied Materials, Inc. for providing the oxide characteristic data used in the front-end processing section of Chapter 2. We thank Qiang Lu, a technologist at the University of California at Berkeley, currently with Advanced Micro Devices, Inc., for his contributions to the FEOL section. We also thank lithography expert Franz Zach of IBM Microelectronics for the excellent tutorial on optical lithography for the nano-CMOS regime, included as Chapter 3. For Chapter 5 we thank Professor Ming-Dou Ker of the National University of Taiwan, who is a recognized authority on the subject. We thank Martin Brox, the memory guru of Infineon, for Chapter 7. We acknowledge Xuejue Huang of Rambus for her excellent contributions to Chapter 8, and Huifang Qin of UC–Berkeley for writing most of Chapter 9 and combining the work of the authors into this excellent chapter.

We also thank Altera Corporation for supporting this effort, especially Wanli Chang, William Hwang, Kang Wei Lai, Richard Chang, Leon Zheng, Mian Smith, and Howard Kahn for the simulations they ran. We thank Cynthia P. Tran for the physical layouts used as illustrations in this book as well as providing input to the lithographic simulation. We thank John Madok and Michael Smayling for help finding experts within Applied Materials to help write sections of the book as well as acting as consultants.

We gratefully acknowledge Shuji Ikeda of Trecenti/Hitachi; Ryuichi Hashishita, Yashushi Yamagata, and Toshiaki Hoshi of NEC; Richard Klein and Qiang Lu of Advanced Micro Devices for supplying technical data and numerous SE and TE micrographs used in the book. We thank Fung Chen, Armin Liebchen, and Sabita Roy of ASML Masktools for their help with the lithographic simulations as well as supplying the simulation tool used in generating the simulated aerial view of the resist profile used as illustrations.

We thank Professor Mark Greenstreet of the University of British Columbia for reviewing the initial table of contents and for the many valuable suggestions. Last but not least, we express our gratitude toward Professor Chenming Hu for his insightful suggestions and the affirming foreword he has written for the book.

CHAPTER 1

NANO-CMOS SCALING PROBLEMS AND IMPLICATIONS

1.1 DESIGN METHODOLOGY IN THE NANO-CMOS ERA

As process technology scales beyond 100-nm feature sizes, for functional and high-yielding silicon the traditional design approach needs to be modified to cope with the increased process variation, interconnect processing difficulties, and other newly exacerbated physical effects. The scaling of gate oxide (Figure 1.1) in the nano-CMOS regime results in a significant increase in gate direct tunneling current. Subthreshold leakage and gate direct tunneling current (Figure 1.2) are no longer second-order effects [1,15]. The effect of gate-induced drain leakage (GIDL) will be felt in designs, such as DRAM (Chapter 7) and low-power SRAM (Chapter 9), where the gate voltage is driven negative with respect to the source [15]. If these effects are not taken care of, the result will be a nonfunctional SRAM, DRAM, or any other circuit that uses this technique to reduce subthreshold leakage. In some cases even wide muxes and flip-flops may be affected.

Subthreshold leakage and gate current are not the only issues that we have to deal with at a functional level, but also the power management of chips for high-performance circuits such as microprocessors, digital signal processors, and graphics processing units. Power management is also a challenge in mobile applications.

Furthermore, optical lithography will be stretched to the limit even when enhanced resolution extension technologies (RETs) are employed. These techniques

Nano-CMOS Circuit and Physical Design, by Ban P. Wong, Anurag Mittal, Yu Cao, and Greg Starr
ISBN 0-471-46610-7

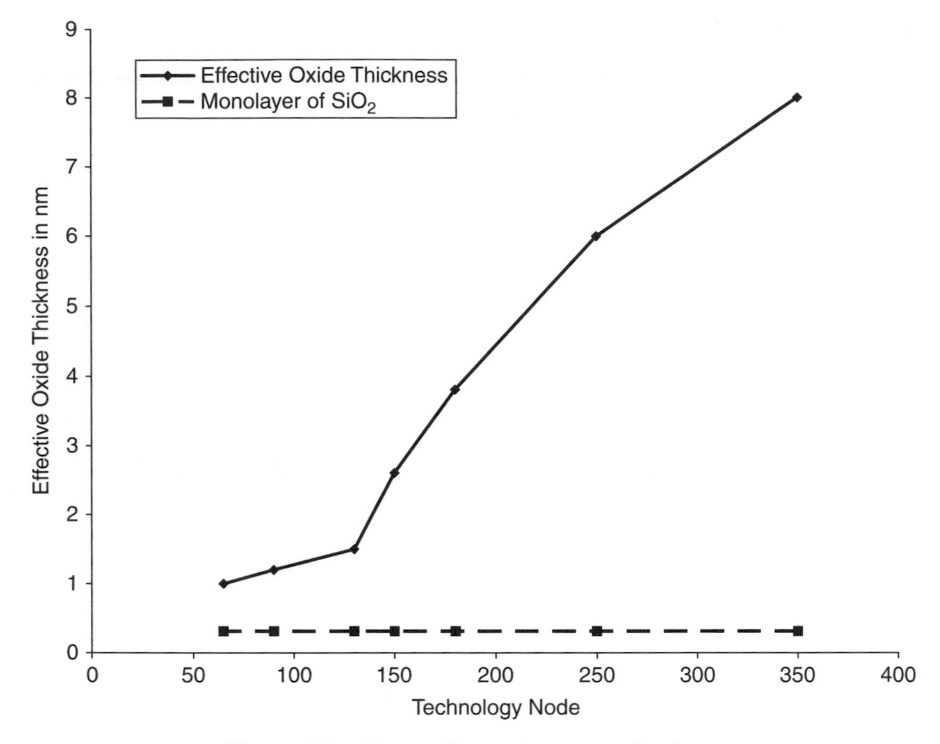

Figure 1.1 Gate oxide trend versus technology.

result in increased cost of the mask and longer fabrication turnaround time. It is no longer cost-effective to respin the design several times to get to a production-worthy design. In the past, processor designers would tape out their design when the verification confidence level was around 98%. Debug continued on silicon, which is usually several orders of magnitude faster and would result in getting a product to market sooner. Now, due to the increased mask cost and longer fabrication turnaround time, the trade-off to arrive at the most cost-effective product and shortest time to market will certainly be different [28].

Since design rules do not all shrink at the same rate, legacy designs must be reworked completely for the next node unless one anticipates the shifting rules and sacrifices density at previous nodes so that the design is scalable without redesign of the physical layout. There is still a need to resimulate the critical circuits, and that, too, can be minimized if one uses scaling-friendly circuit techniques. This will require prior thought and design rule trade-offs to achieve a scalable design, so that a faster and smaller chip for a cost-effective midlife performance boost can be realized through process scaling with a minimum, if any, rework. The key in foreseeing the changing trend in design rules is a good understanding of the process difficulties and tooling limitations, which are covered in detail in subsequent chapters.

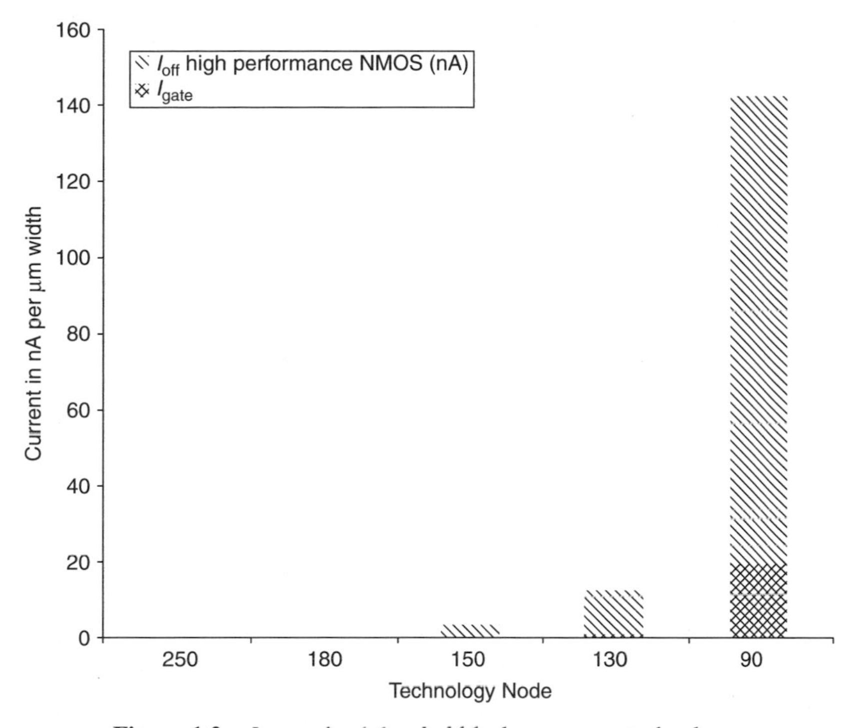

Figure 1.2 I_{gate} and subthreshold leakage versus technology.

1.2 INNOVATIONS NEEDED TO CONTINUE PERFORMANCE SCALING

The transistor figure of merit (FOM) is now deviating from the reciprocal of the gate length. As can be seen in Figure 1.3, the fanout-of-4 delay is tailing off with advancing technology. Furthermore, global wiring is not scaling, whereas wire resistance below 0.1 μm is increasing exponentially. This is due primarily to surface scattering and grain-size limitations in a narrow trench, resulting in carrier scattering and mobility degradation [2]. The gate dielectric thickness is approaching atomic dimensions and at 1.2 nm in the 90-nm node [22] is about five atomic layers of oxide. Figure 1.1 shows that gate oxide scaling is slowing as it approaches the limit, which is one atomic layer thick [26]. Source–drain extension resistance (RSD) is getting to be a larger proportion of the transistor "on" resistance. Source–drain extension doping has been increased significantly for the 130-nm node, and the ability to reduce this resistance has to be traded off with other short-channel effects, such as hot-carrier injections (HCIs) and leakage current due to band-to-band tunneling. Source–drain diffusions are getting so thin that implants are at the saturation level and resistance can no longer be reduced unless additional dopants can be activated [21].

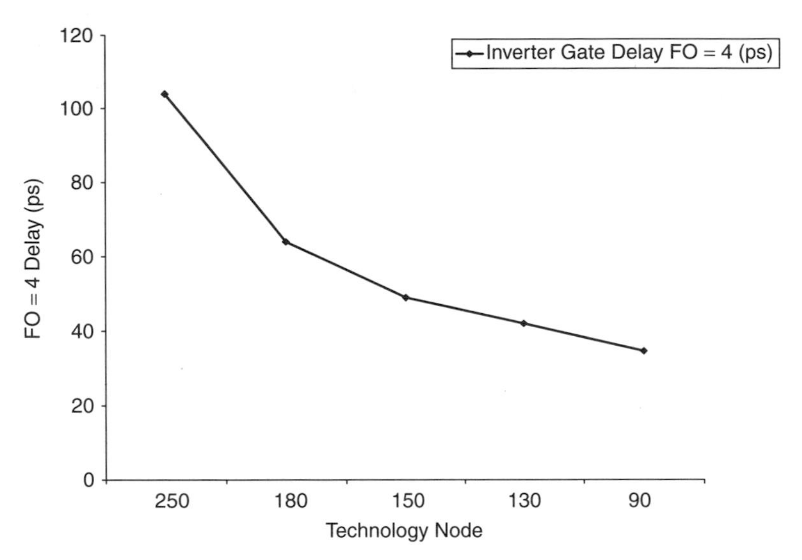

Figure 1.3 Gate delay versus technology.

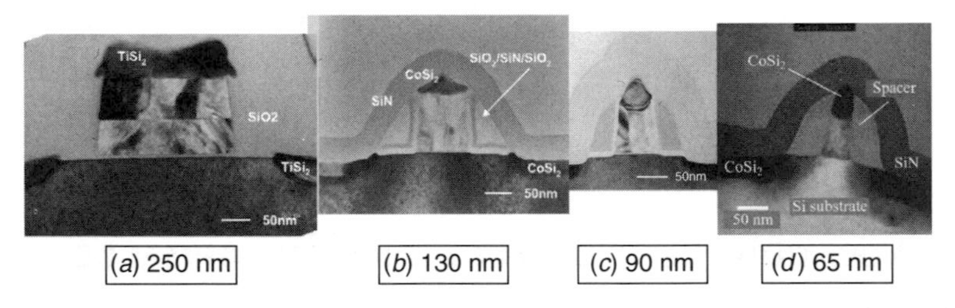

Figure 1.4 Transistor TEM. [Parts (*a*), (*b*), and (*d*) courtesy of NEC and Trecenti/Hitachi; part (*c*) © Advanced Micro Devices, Inc., reprinted with permission.]

Poly lines are getting to be quite narrow, between 70 and 90 nm for the 130-nm node and 50 nm for the 90-nm node (see Figure 1.4). This requires a trade-off between poly sheet resistance and source–drain leakage. To lower the narrow poly line resistance would require more silicidation of the poly. Since the silicidation process is common between poly and source–drain diffusion, increasing silicidation of the poly would result in higher silicide consumption of source and drain diffusions. Due to the extreme shallow junctions at the source and drain, this can result in punch-through as a result of silicide consumption of the source–drain diffusion. Research is ongoing to bring raised source–drain technology online to mitigate this effect for the 65-nm node and possibly for the 90-nm node as well. Some manufacturers might be able to bring this technique online by the later part of the 90-nm node.

Starting at the 180-nm technology node, the critical feature size (poly) is already subwavelength compared to the ultraviolet (UV) wavelength used in

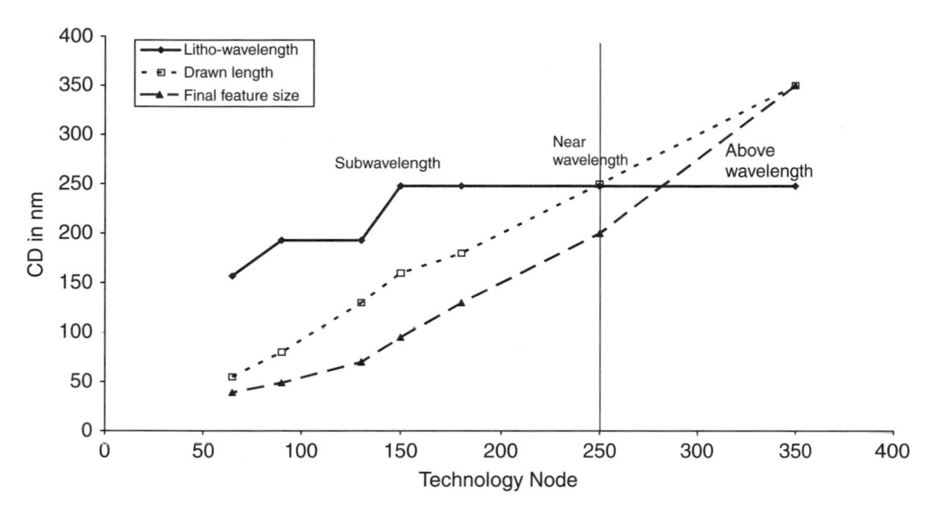

Figure 1.5 Poly CD versus lithographic UV wavelength at each technology node.

lithography. The gap is increasing at each subsequent technology node (see Figure 1.5). At the 65-nm technology node, even with aggressive RET, 193-nm lithography will run out of gas. To extend the resolution of 193-nm scanners, research is ongoing to increase the numerical aperture (NA) of the lithography system, including immersion lithography. More details on the challenges of lithography are presented in Chapter 3. The challenges of 157-nm and extreme UV (EUV) lithography are monumental and will increase tooling and mask costs and fabrication turnaround time. If 157-nm lithography is not brought online by the 65-nm technology node, we will see the subwavelength gap widen further. Circuit and physical designers can no longer design simply by technology design rules and expect a functional, let alone a scalable design that also meets varied design goals, such as high performance and low-power mobile applications from a single mask set. Designers must know when to use more relaxed rules and not simply relax the rules on the entire design, which negates physical scaling.

Combinations of materials and processes used to fabricate new structures create integration complexities that require design and layout solutions [20]. Process engineers and technology developers will not be able to resolve all the issues that arise as a result of sub-100-nm scaling, which includes integration complexities and fabrication and process control difficulties. We will suggest techniques that circuit and physical designers can employ to mitigate the challenges of working with sub-100-nm technologies, and provide some understanding of the process technology with which they are designing. Similarly, it is important for process engineers to understand the basis of physical design so that the technology can be tailored for a robust and scalable design that can continue with both physical and performance scaling.

It will require some innovation on the part of technology developers to bring new processes online, and will necessitate the development of new materials as well. It is an undisputed fact that performance scaling derived from mere physical scaling has already reached an inflection point and is no longer providing much, if any, gain in performance. To continue performance scaling we have already witnessed some innovations at work and more are under development. Silicon-on-insulator (SOI) technology has been shown to improve transistor performance by about 20 to 30%, depending on the source of the data. Some microprocessors have already adopted SOI as the technology of choice. Strained silicon using relaxed silicon–germanium substrates has been demonstrated to offer up to 30% improvement in carrier mobility. Since these substrates are expensive and are prone to dislocation defects, they are not as widely accepted.

An innovation that demonstrates yet another method of achieving strain in silicon for carrier mobility improvement is use of a nitride capping layer. Such a layer generates strain due to the compressive stresses on source–drain diffusion, thus creating strain in the transistor channel as the source–drain diffusions are pulled apart. This works only at 90-nm node and below because of the need for the channel to be in close proximity to source–drain stress. A longer-channel device will see less gain. Even at the 90-nm nodes transistors with drawn length longer than minimum will have diminished gain. Unfortunately, at the 130-nm node, this option for performance improvement is limited. This technique will be the preferred method to create strain since it requires no special substrates, and no dislocation has been seen so far. Best of all, it requires no extra steps, just a recipe change.

The switch to copper interconnects gave short-term relief on pressure to continue performance scaling in the near-limit regime. This is an example of an innovation that required a material change. Many other out-of-the-box innovations are in the pipeline, including raised source–drain (SD) diffusion, dual-gate FET, FinFET, high-κ gate dielectrics, and metal gates [4]. Whether they will pan out depends on the risks versus the benefits, as well as the cost, integration and fabrication complexity and turnaround time.

1.3 OVERVIEW OF SUB-100-NM SCALING CHALLENGES AND SUBWAVELENGTH OPTICAL LITHOGRAPHY

1.3.1 Back-End-of-Line Challenges (Metallization)

Metal Resistance Line width below 0.1 μm is accompanied by an exponential increase in resistivity. The higher-resistivity barrier material is becoming a larger proportion of the conductor cross-sectional area for narrower lines. Reduced electron mobility due to surface scattering plays a part in the increased resistivity [2]. Narrow lines result in smaller grains, which cannot be recrystallized into larger grains while encased in a narrow groove thus increasing the resistivity further.

Furthermore, variations in critical dimensions (CDs) of the barrier material and groove (line width) result in larger resistance variation. These, along with chemical–mechanical planarization (CMP) dishing and erosion, as well as

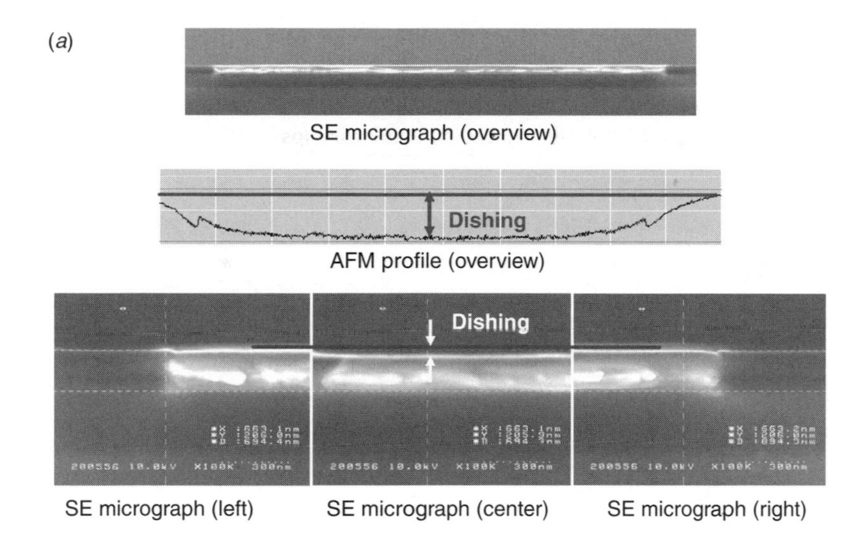

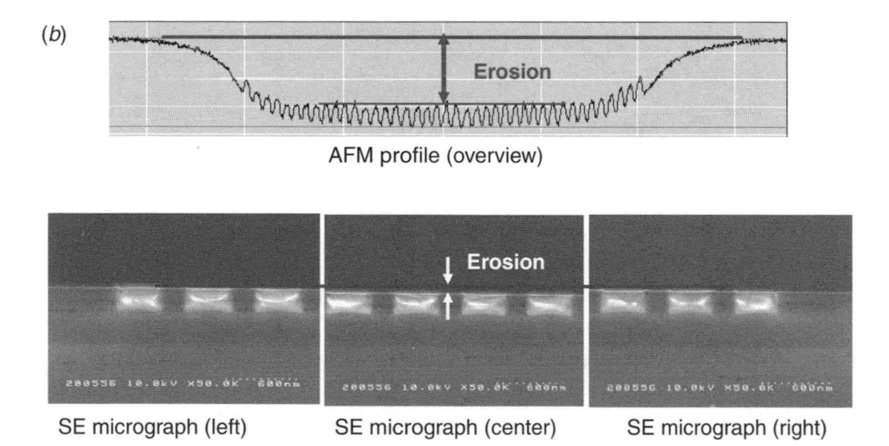

Figure 1.6 (*a*) Interconnect dishing: wider line area. (*b*) Interconnect erosion: line and space area. (Micrographs courtesy of Trecenti/Hitachi.)

lithographic and etch distortions, cause further variation in the line resistance [19] (Figure 1.6).

Interconnect *RC* values are increasing at the 130-nm node and getting worse for both local and global wiring beyond the 130-nm node. As explained above, resistivity is increasing (see Figure 2.25) while the scaled capacitance is not decreasing, leading to increased delay for local wiring even though the length of local wires is getting shorter (Figures 1.7 to 1.9). The length of global wires is not reduced since chip size is not being reduced as more functionality is added

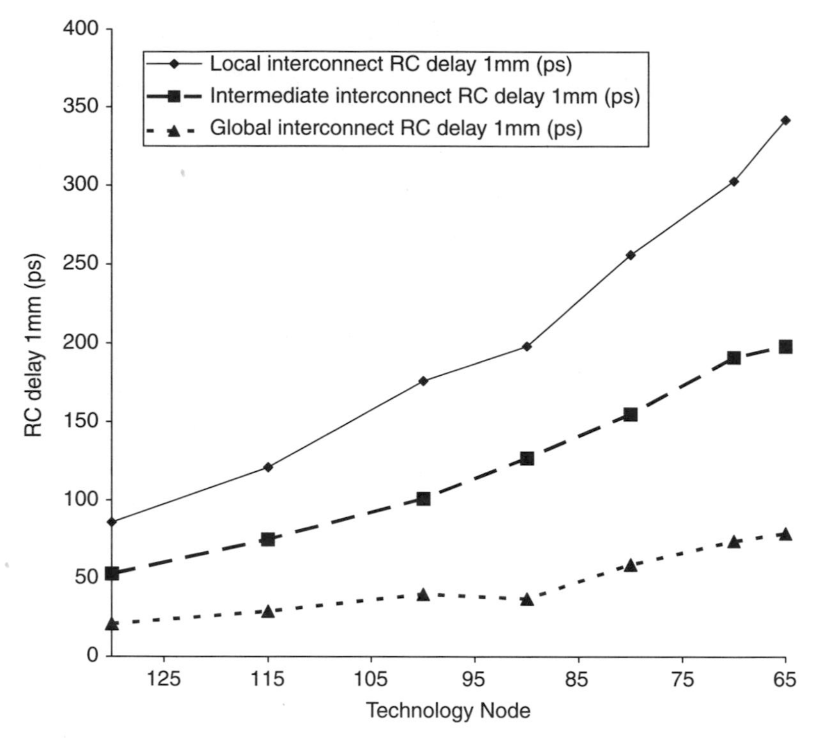

Figure 1.7 Interconnect delay versus technology node.

to new designs. For example, the Pentium 4 Willamette core in the 180-nm process had 42 million transistors; for the Northwood core in the 130-nm process, the number of transistors increased to 55 million. This is because the L2 cache increased from 256 kB to 512 kB for the Northwood core. The fraction of reachable area in a clock cycle is diminishing as the technology scales. This is further exacerbated for designs in the advanced technology nodes by the increase in clock frequency while the die size is not decreasing.

Interconnect Dielectric Constant Low-κ dielectric enables wire scaling in the nano-CMOS regime but is getting harder to implement as width and space are decreasing. Low-κ dielectric also poses potential leakage and reliability hazards, due to time-dependent dielectric breakdown (TDDB) in narrowly spaced lines. Packaging difficulties dictate the need to form a "hard crust" to provide a mechanically sound die against the stresses imposed on a chip by the packaging processes. This crust means that higher-dielectric-constant material is needed for the upper layers of the metal stack, somewhat reducing the effectiveness of the low-κ metal technology. Low-κ dielectric will be limited to four or five layers of metallization in eight- or nine-layer metal technology. The mitigating factors

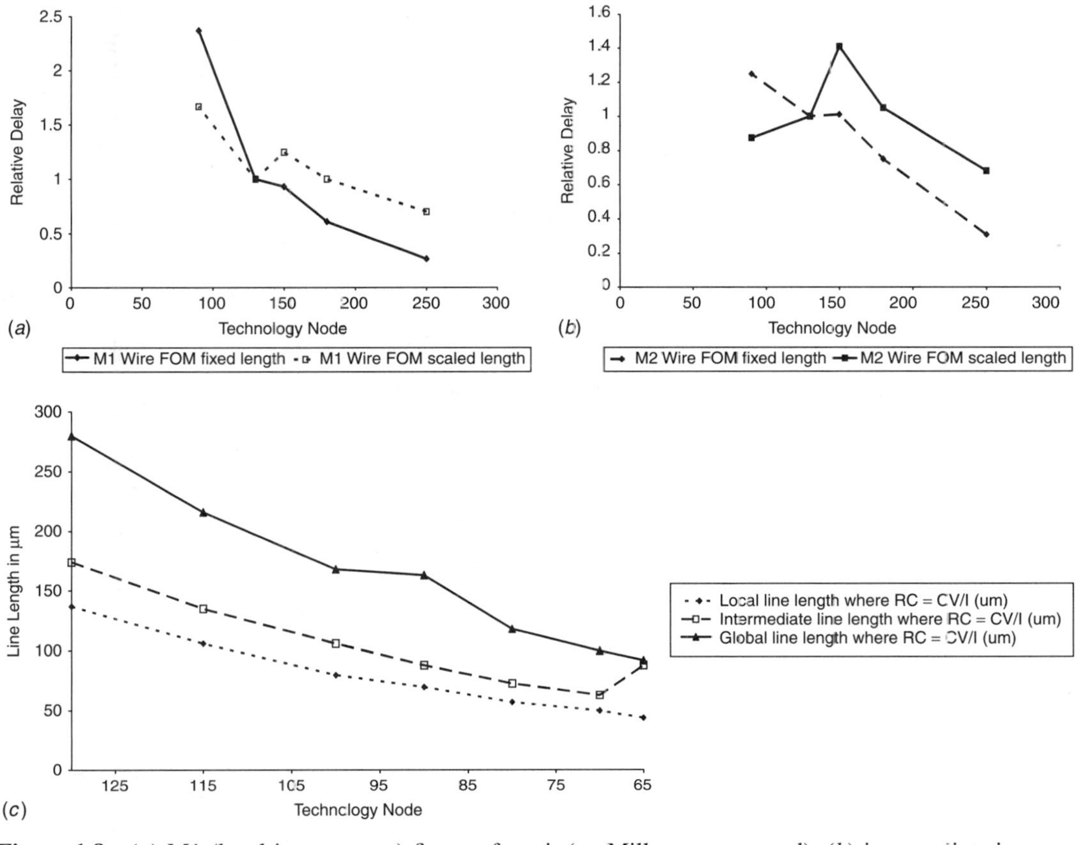

Figure 1.8 (*a*) M1 (local interconnect) figure of merit (no Miller, nonrepeated); (*b*) intermediate interconnect figure of merit (no Miller, nonrepeated); (*c*) line length equivalent to NMOS CV/I versus technology.

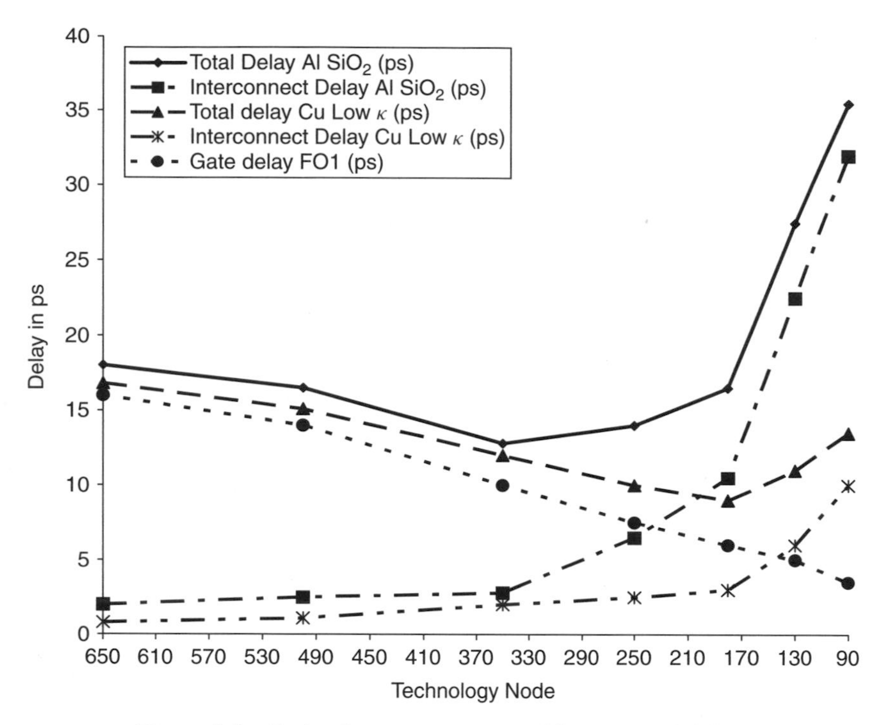

Figure 1.9 Technology versus gate and interconnect delay.

are the way the upper metals are used. Normally, the upper layer metals are used for power distribution. In most designs they are also used as clock distribution layers, thus increasing the power of the clock network and also requiring more stages to buffer up from the PLL, resulting in higher skew as well.

Low-κ Interconnect Roll-out Lagging Significantly The lag in the introduction of low-κ technology is due to problems with copper barrier material, mechanical integrity against bumping force during packaging, and a host of fabrication process issues. This has resulted in several manufacturers reverting to fluoro-silicate glass (FSG) dielectric.

Low-κ dielectric is like jelly and very porous, and thus it is susceptible to moisture and contaminant absorption and outgassing. Since the material is soft, it suffers from CMP ripouts, causing yield loss and erosion, affecting wire resistivity as well. Low-κ dielectric is also a poor conductor of heat, thus degrading the electromigration (EM) property of the interconnect, negating to some extent the good EM property of copper.

Interconnect Figure of Merit The unscaled interconnect FOM has been decreasing at every technology node (see Figures 1.7 to 1.9). In the past, transistor performance was lagging. We have arrived at a point where the interconnect performance will be the chip performance limiter. Local interconnect performance will not scale, while global wiring is getting really slow, especially if wire length does not scale due to additional functions [12–14]. Chip size invariably stays at the same size as in previous designs, despite technology scaling, due to increased functionality of newer designs. In other cases, as in microprocessors, chip size actually increases despite technology scaling. As the chip grows larger despite scaling, we need global wires to ship signals between blocks.

It has been predicted that the fraction of the total chip area reachable in one cycle will diminish as we scale the technology, while the clock frequency increases [13]. This will force designers to insert more repeaters on global wires, and in some cases pipelining of the global signals may be necessary, so that interconnect-dominated paths can scale better and will not be frequency-limiting paths. However, this will increase chip area, power consumption, and clock load [14], as well as increasing the complexity for full-chip timing. The result of higher clock load translates into higher clock skew as well. Then there is also an increase in signal latency due to the pipelining, which has other microarchitectural impacts as well.

These issues force designers to back off on interconnect pitch to improve global wire performance as well as signal integrity. Increasing wire pitch will reduce line-to-line coupling, but the capacitance will reach an asymptote where it is not reduced with further increase in line space (see Figure 1.10). The space where minimum capacitance is achieved also depends on the interlayer dielectric thickness. Further scaling beyond the 130-nm technology node will do little to improve wiring density, due to performance issues and signal integrity problems, which will require shields for some wires and spacing for others. This leads to the need for more metal layers to be able to route a complex chip.

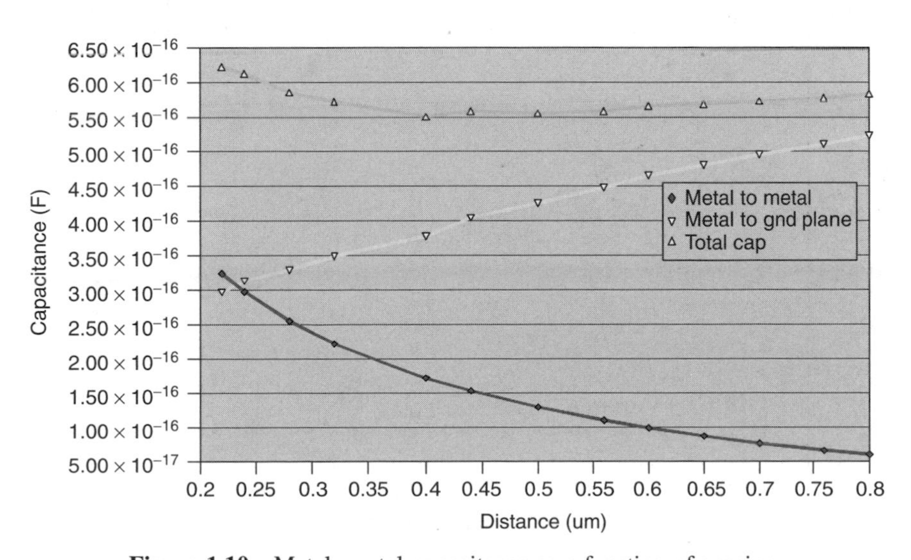

Figure 1.10 Metal–metal capacitance as a function of spacing.

Contact and Via Not Scaling Well Contacts for most 130-nm technologies are already at 0.16 μm and vias are at 0.2 μm. It will be difficult to scale them by much in future nodes. Certainly, they will not scale at the same rate as other features. Another limiter is the contact and via resistance, which will go up as they scale.

At the 130-nm nodes these two layers already require optical proximity correction (OPC) and phase-shift lithography. Mask data preparation and mask making for these layers take almost twice as long as for other layers that do not yet require OPC and/or phase shift [5].

1.3.2 Front-End-of-Line Challenges (Transistors)

Transistor Performance The transistor figure of merit is now deviating from being proportional to the reciprocal of gate length. Some of the main contributing factors are:

- $V_{gs} - V_{\mathrm{th}}$ is diminishing and V_{t}/V_{dd} is getting larger (Figure 1.11).
- RSD as a proportion of total transistor "on" resistance is getting to be significant, determined partly by the spacing of contact to polygate and the RSD.
- Thin junctions drive dopant levels to saturation. No further reduction in RSD is possible; at the same time, junction capacitance is increasing.

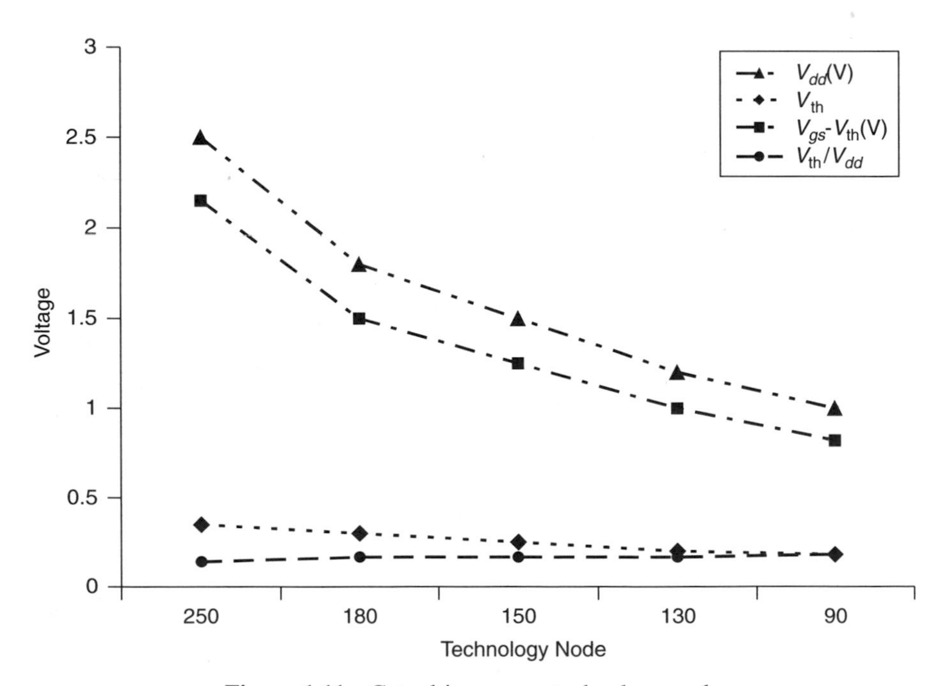

Figure 1.11 Gate drive versus technology node.

- Thinner source and drain diffusion increases RSD further, due to current crowding.
- Shallow trench isolation (STI) stress-induced mobility degradation is more pronounced, negatively affecting the NMOS transistor while the PMOS transistor improves slightly with STI stress [10,11].
- ΔW is becoming significant as well, even with STI for the smaller transistors.
- Drain capacitance reduction now proceeds at a slower pace than area reduction.
- Dopant loss and statistical dopant fluctuation on small-geometry devices increase device variability: input/output, analog, and memory designs are especially sensitive.
- Increasing channel doping concentration to control drain-induced barrier lowering (DIBL) reduces carrier mobility while increasing body effect.
- Thin gate oxide results in dopant penetration, which affects PMOS drive current [6].
- Gate oxide scaling is also slowing as it approaches the monolayer thickness of SiO_2 (see Figure 1.1).

Leakage Problems Subthreshold leakage is increasing at a rate that will eventually be equal to the dynamic power of the chip (Figure 1.12), especially for high-performance microprocessors within a couple of technology generations if design methodology is not changed to mitigate this increase. Gate current (Figure 1.13) has been seen to increase 2.5 times for each 1-Å reduction in oxide

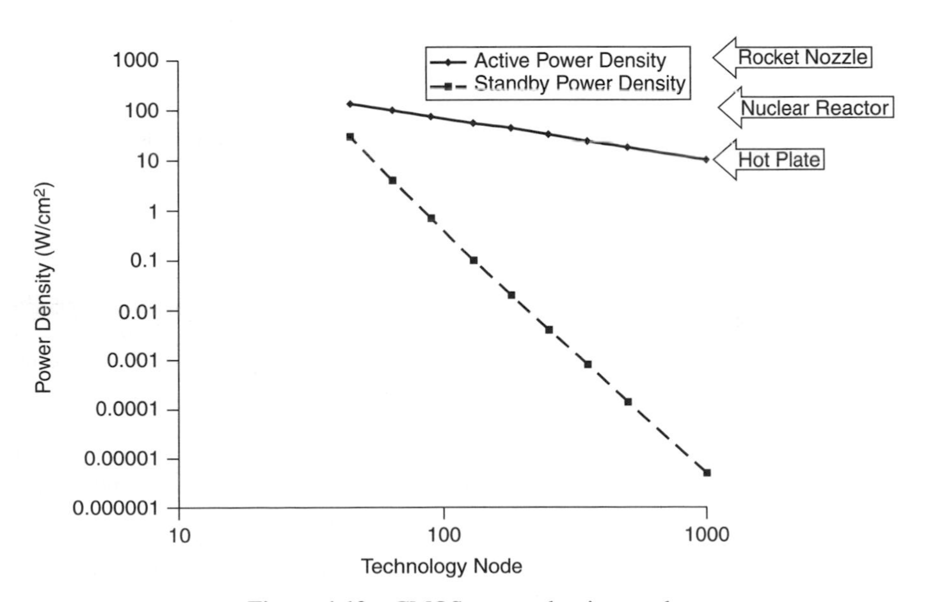

Figure 1.12 CMOS power density trend.

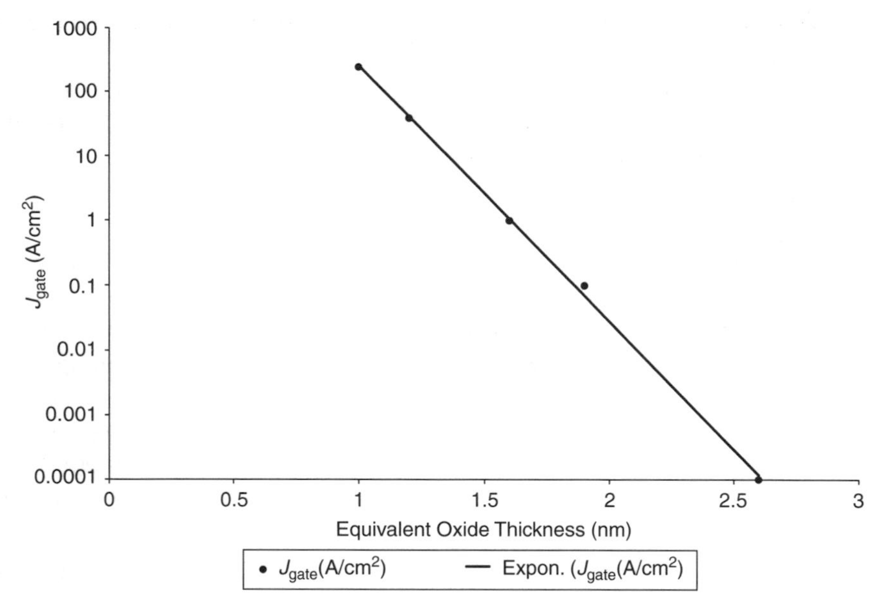

Figure 1.13 J_{gate} (A/cm^2) for NMOS versus equivalent oxide thickness. (Data courtesy of NEC.)

thickness, which is about two orders of magnitude in each generation from the 130-nm node.

Gate resistance is also increasing as feature size shrinks, and SD resistance will increase with an ever-thinning junction. This will require careful trading of SD resistance for junction leakage until a raised source–drain junction is a manufacturing reality. Figure 1.4 shows that the gate poly thickness has changed very little, scaling from 250 through 65 nm. The only obvious change is the gate length or the width of the poly. Thus, resistance to the channel increases with scaling and will need to be considered in transistor models.

To continue transistor $I_{d\ sat}$ improvement, serious research and development efforts are being poured into strained silicon channel transistors, where 10 to 20% improvements have been reported using SiGe strained silicon [22]. Less drastic strained technology relies on nitride capping film to provide the strained channel but offers modest $I_{d\ sat}$ improvements. Raised source–drain technology is also being developed, requiring selective epitaxial processing, a difficult manufacturing process. Many new materials are being introduced to make high-κ gate oxide a reality, and NiSi is being introduced to replace CoSi [22]. High-κ gate oxide comes with a major integration challenge, as it seems to be incompatible with silicon but works well with metal gates [4]. Metal gates have a distinct advantage over polysilicon gates since they are not depleted, so that process engineers do not need to use thinner gate oxide for the same capacitance effective oxide thickness (CET) [4]. Therefore, for a given oxide CET, metal gate technology would theoretically have lower accumulation mode gate leakage. Since metal gates are not self-aligning, innovation will be required for implementation. As this book is being prepared, predoped polysilicon is being used to reduce poly depletion at the expense of etching problems. Some manufacturers already have a handle on such problems, due to the use of predoped polysilicon. When the industry began, the materials count was about five, but it has risen to about 20 at present [23].

Performance derived from physical scaling is near the limit, but dimension scaling is expected to continue to grow as predicted by Moore's law. Performance is now improved through innovations such as new transistor designs and the introduction of new materials and processes, including high-κ gate dielectric, FinFET, SOI, strained-silicon, and isotopically pure silicon substrates, to mention just a few of the recent developments.

1.4 PROCESS CONTROL AND RELIABILITY

Absolute physical variation of gate-length critical dimensions (CDs) is not scaling with the technology, thus for future technology generations CD variation as a percentage of gate length will be higher [7]. On top of that, as the gate length goes below 100 nm, line-edge roughness (LER) is becoming an increasing concern, affecting several transistor parameters. LER control is critical in sub-100-nm technologies, since its effect is more significant for devices with shorter poly length as we scale. It is an artifact of the lithographic and etching steps that can only be improved by better process control. The adverse effect of large LER is the higher overlap capacitance C_{gd}, especially for the PMOS. The other device parameters affected include DIBL and threshold voltages, since the effective channel length reduces with LER after the anneal cycle, especially for the PMOS (Figure 1.14). As the $L_{\mathrm{effective}}$ value of the transistor is reduced due to the LER effect, the V_{th} and punch-through voltage of the PMOS will be affected adversely.

V_{th} variation is influenced by random dopant fluctuations and gate CD variation. Thin gate oxide in conjunction with dopant channeling causes dopant variation in the channel, depending on the morphology of the gate polysilicon (Figure 11.7). These effects make V_{th} control more difficult, and transistor V_{th} matching is even more difficult, especially for small-geometry devices. It can be seen in Figure 11.37 that V_{th} variation is largest for the smallest devices but reaches an asymptote. It would be prudent to stay away from minimum-width transistors unless the V_{th} variation does not cause circuit failures.

Negative bias temperature instability (NBTI) is an effect that surfaced as gate oxide thickness was scaled. Gate oxide thickness for the 130-nm technology node has already resulted in sensitivity to NBTI [18]. Any processing step that causes bond breaking will exacerbate NBTI. Plasma or reactive ion etch, in particular, is a process that can cause bond breaking, thus exacerbating NBTI. At the 65-nm technology node, gate oxide thickness is projected to be at or below 10 Å thickness. At this thickness, interface control will be critical. Poly depletion

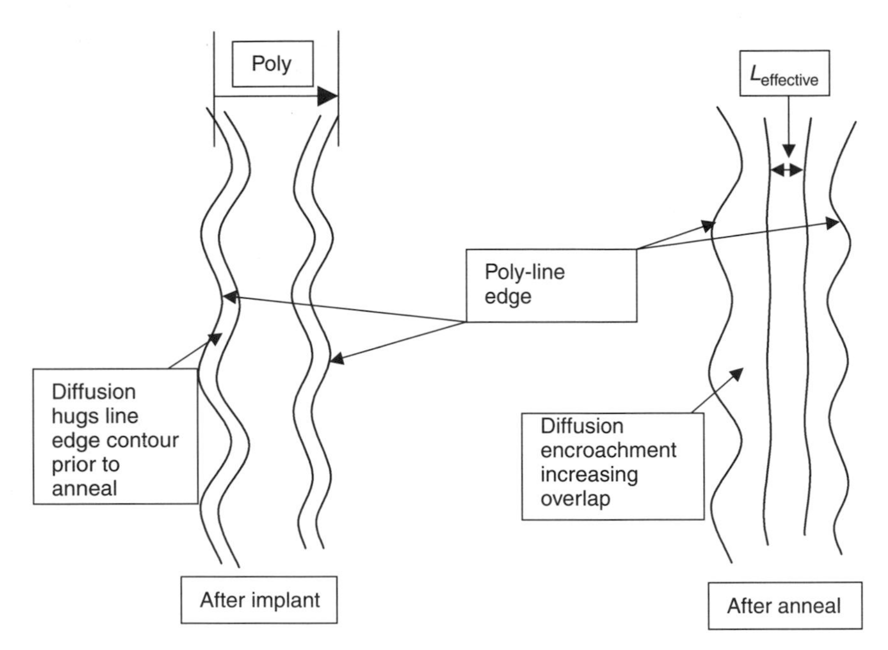

Figure 1.14 LER increases overlap capacitance and reduces channel length.

will be a limiter for further performance scaling and will require nondepleting gate material. Gate oxide thickness control at the 90-nm node and below will be critical to maintain a predictable, low gate current. Gate current increases about 2.5-fold for every 1 Å of gate oxide thickness reduction (see Figure 1.13).

1.5 LITHOGRAPHIC ISSUES AND MASK DATA EXPLOSION

Beginning with the 180-nm node, we crossed over to the subwavelength regime. The subwavelength gap for optical lithography is widening (see Figure 1.5) because of numerous obstacles that have to be surmounted to bring a new lithographic generation online. Changes in the physical design would therefore be required in sub-100-nm nodes so that the design will print reliably without needing next-generation lithography. Below the 90-nm node, aggressive OPC would be necessary and lithographically friendly physical designs, mandatory. Resolution extension technology results in mask data explosion after fracture, which will increase mask cost [8]. As a result of the widening subwavelength gap, mask and lithographic costs will increase exponentially in subsequent generations; hence, only the best-funded manufacturing facilities can afford to deploy leading-edge lithographic equipment. For others, the degrees of freedom in the physical design would have to be limited along with increased numerical aperture values and aggressive OPC to extend the resolution of 193-nm lithography [9]. A detailed tutorial on this subject is presented in Chapter 3.

1.6 NEW BREED OF CIRCUIT AND PHYSICAL DESIGN ENGINEERS

CMOS technology scaling is at a point where traditional assumptions that allowed total decoupling between circuit and physical design from process development are falling apart. This therefore demands a paradigm shift in the way we implement circuits [20]. Even the Application-Specific Integrated Circuit (ASIC) design methodology, which pushes the performance envelope, must adapt to this shift if the design is to be functional and scalable beyond the 100-nm drawn feature sizes.

High-performance design, in particular, will require significantly different approaches. This demands a new breed of circuit and physical design engineers who understand the difficulties so that they can be a part of the solution by creating lithographically friendly physical design to enable a robust, scalable, and high-yielding design. The design for future technology nodes must tolerate a lot of leakage, both subthreshold (including GIDL) and gate leakage. Variation tolerance is another requirement of designs for future nodes.

Many processing steps are affected by layout styles. Most notably, polygon density has a major impact on interlayer dielectric thickness. Diffusion density has a significant impact on the fabrication yield of the final product. Other layout styles may mitigate dopant fluctuation and poly-CD variation in circuits where device matching is important for circuit functionality.

The new breed of circuit and physical designers must understand the proximity effects on circuits and design circuits accordingly, so that silicon behavior is as predicted during simulations. Proximity effects can arise as a result of placing a transistor next to a well or in poly-dense or poly-sparse areas. Having transistors next to another structure can cause dopant fluctuations during the implant step which can deflect dopants onto a transistor next to the resist mask. As long as every transistor has a similar neighbor, the proximity effect is consistent. If not, the proximity effect can cause device V_{th} variation. Other proximity effects include poly-CD variations due to photolithographic and etch proximity effects as a result of suboptimal layout styles due to etch microloading and optical proximity effects. Many of the systematic proximity effects can be avoided through good layout style and by means of photolithographic techniques and biases. But designers must understand the limitations and apply design techniques to mitigate these effects. We cover these techniques in more detail later in the book to provide a background that will enable circuit and physical engineers to better deal with these effects through physical design.

1.7 MODELING CHALLENGES

Continued physical scaling increases electrical parameter tolerance and will be a modeling challenge. Prior to BSIM4, gate current was not modeled and designers had to fend for themselves. Statistical dopant variation in the channels is difficult to model and affects the small-geometry transistors used in bit cells, which can least afford poor modeling [3]. Proximity effects and STI stress mobility degradation will be difficult to model since they are very layout dependent [10,11].

TABLE 1.1 Summary of Device Modeling Challenges for Sub-100-nm Processes

Parameter	Reason for Effect	Synopsis of Effect
RSC	Halo implants (technology, physical device effect)	Reverse short-channel effect due to lateral nonuniform doping; when channel length varies, V_{th} varies
DITS	Halo implants (technology, physical device effect)	Drain induced threshold-voltage shift, due to change in DIBL for long-channel-length devices when the halo implant's influence on the channel diminishes
Early voltage and output resistance [17]	Halo implants (technology, physical device effect)	Change in DIBL for long-channel device similar to above
Poly depletion [25]	Ultrathin gate oxide (technology, physical device effect)	Poly depletion is getting significant for ultrathin gate oxide, which accounts for about a 8-nm increase in equivalent oxide thickness (EOT) for most devices, less for predoped poly
Gate tunnel current	Ultrathin gate oxide (technology, physical device effect)	Direct tunneling from gate to channel occurs due to ultrathin gate oxide
Mobility-dopant dependence	Halo implants (technology, physical device effect)	Mobility improves with reduction in dopants
Linear proximity effects	Dense, isolated	Partly due to lithographic effects and partly to etch microloading effects, also due to dopant scattering from the poly, causing systematic dopant variation as a function of poly-line space of the design
Nonlinear proximity effects	Optical proximity correction (OPC)	Subwavelength lithography requires resolution extension
GIDL	Band-to-band tunneling	High field in the drain to gate causes band-to-band tunneling, due to high junction doping and abrupt junctions of sub-100-nm devices
Diffusion and poly flaring	Technology and layout effects	Subwavelength lithography causes flaring of diffusion and poly, causing device variations of small-geometry devices and proximity of poly contact pads to diffusion edge
Well proximity	Devices at the edge of the well	The lateral scattering of well implant atoms out of the resist, which leads to threshold voltage increase for devices close to the well edge; typically, 50 and 20 mV for NMOS and PMOS, respectively
STI stress	Proximity effect of STI to device channel	STI stress reduces electron mobility but increases hole mobility, thus affecting $I_{d\ sat}$

Some new tools have become available that offer some assistance in this area through layout extraction. The best work-around is to comprehend the effects, then create the physical design that minimizes these effects on the circuits. We go over these effects in detail in Chapter 2.

Analog modeling of logic processes with halo implants leads to inaccuracies due to the anomalous behavior with channel length, unless analog transistors are available for use by mixed-signal engineers [17]. This adds to the cost and may sometimes be unavailable. Unless you are working with a foundry that has the capability of modeling halo effects on the DIBL, V_{th}, and early voltage versus the channel length of the transistor, it may be wise to rely on analog transistors. As can be seen in a paper [16] published at IEDM 2002, such a model is not impossible but may not be available at every foundry. If for whatever reason you have to use halo-processed transistors for analog design and your SPICE models do not take into account the halo effects (reverse short channel and drain-induced threshold voltage shift) and the output resistance and early voltage variations, it is very important to select transistor sizes where they intercept the points at which the models are fitted, to avoid inaccuracies due to nonlinear transistor characteristic changes with respect to channel length.

New physical effects that must be modeled include halo implant effects on V_{th} [16] versus transistor poly length [reverse short-channel (RSC) effect] [16], gate-induced drain leakage (GIDL), drain-induced threshold voltage shift (DITS) [24], output resistance and early voltage variations, and gate current [15]. Some of these new effects are only modeled beginning with BSIM4 [24]. For technologies beyond the 130-nm node, it is highly recommended that BSIM4 be used in all simulations, including digital circuit simulations.

Statistical modeling is necessary to work around some of these problems, due to process variations in implants and critical dimensions as described above. Unless one judiciously picks the model combinations that make sense for the particular circuit, reliance on corner models will result in unrealistic process combinations and overdesign of the circuit at the expense of speed, power, and area. On the other hand, a critical corner may not be modeled by the traditional five corners methodology, hence the worse-case corner for the particular circuit may not be exercised. The modeling challenges are summarized in Table 1.1.

1.8 NEED FOR DESIGN METHODOLOGY CHANGES

In the past, capacitive noise analysis would suffice, but now, signal integrity has been extended to inductive noise. Whereas timing used to be the main concern, we now have to worry about functionality as well. There is a need to develop noise-tolerant circuits to reduce the long analysis and modeling of on- and off-chip signal integrity issues. There is also a need to develop correct-by-construction signal-integrity-proof signal transmission methodology. This could be the way that repeaters are placed, spreading out wires where space allows. In some places, shields may be needed. A robust power distribution system that also doubles as an inductive shield and return path for large, wide buses is needed as well.

Power integrity has recently surfaced as a result of higher clock frequencies coupled with voltage scaling along with device scaling. The power consumption continued on an upward trend despite the scaling, due to an increase in functionality to satisfy the demand for ever-increasing chip performance. With the increase in power as power supply voltages drop, the supply current is on the rise and so is di/dt and resistive voltage drop. $L(di/dt)$ is getting to be a major performance limiter. To deal with this issue, the design methodology must now extend the design of the power distribution of the chip to the package and system board to ensure a total system solution. Otherwise, it will not be possible to achieve the supply impedance desired to mitigate the high resistive and $L(di/dt)$ drop.

Variations in the process, whether device or interconnect variations, will be a major issue for nano-CMOS designs. For the design to survive the much larger variations, the methodology must have provisions to deal with variations. The traditional five-process corner methodology becomes increasingly meaningless and will lead to costly overdesign at the expense of chip area and power in some cases and in other cases missing an important worse-case condition entirely.

The number of degrees of freedom in the design methodology is shrinking. Future designs will see the need to align critical poly. This will also dictate a change in the design of bitcells, the design at present having the pass transistor poly orthogonal to the pull-down and pull-up transistors. New bit-cell designs address this issue and have all the poly lines aligned. The reason for having all poly lines in the same direction is due to the angled halo implants. Positioning gates orthogonal to each other will result in variations due to the different times at which each edge of the poly gate receives the halo implant. Hence, implants received by the horizontal gate will receive half the dose at a different time and can contribute to V_{th} variation. There is also a higher CD variation for poly lines drawn orthogonal to each other, due to lithographic effects and due to the mask. See Chapter 11 for further details.

Leakage (subthreshold, GIDL, and gate) is the next nemesis that we have to address in the new design methodology. Memories must be designed to tolerate more leakage than before, yet should not significantly decrease array efficiency. In large arrays such as the L2 and L3 caches, the higher leakage is not merely a performance and functionality issue but an area and power issue as well. It may be necessary to design L2 and L3 caches for more than one-cycle access, since they can tolerate higher latency. This is needed to compensate for the slower access time, due to the need for longer channel length and higher V_{th} implants to reduce leakage power at the expense of access time. Some speed can be recovered from the fact that longer channel length provides better matching of the bit-cell transistor and allows for a more aggressive pull-down/pass transistor ratio.

Wide domino gates are no longer a feasible design style in the nano-CMOS era, due to the difficulty of trading among functionality, noise tolerance, and speed. A functional wide domino circuit will no longer be faster than one implemented in two stages. Ratioed logic will also be abandoned. Device and leakage variations will cause a well-designed ratio logic to go off its optimum operating point, in some cases rendering it nonfunctional.

The trade-off between power consumption, performance, and process complexity is getting more difficult and requires designers to ask judiciously for the optimum number of V_{th} implants available to the transistors and must be weighed against the cost. When lower-V_{th} transistors are applied judiciously to the design, one can improve chip performance without an enormous increase in standby power.

The switch to copper interconnects has provided a boost to electromigration (EM) and interconnect performance in the 130-nm generation technology. However, as chip size increases, designers demand more interconnect performance, which we have seen has not been improving with each subsequent node after the 130-nm technology node. Process engineers are trying to implement low-κ dielectric in an attempt to scale wire performance. Since low-κ dielectric has lower thermal conductivity, EM has resurfaced as an issue in nano-CMOS technologies. This, coupled with the higher signal speed, pushing a higher current pulse through the wire, further exacerbates the EM problem.

1.9 SUMMARY

We have covered most of the issues brought about by scaling beyond 100-nm feature sizes and how they can become challenges if we continue with the design methodology developed for previous-generation technology nodes. It should be clear that we need a paradigm shift to continue to take advantage of technology scaling in future designs to continue tracking Moore's law [27].

Although we have seen that performance scaling as a result of device dimension scaling is tailing off, performance scaling can continue as device and process engineers invent new processes and materials that work around problems that limit performance scaling due to physical limitations [23].

Nonetheless, now is the time when circuit and physical design engineers must understand the effects brought about through aggressive dimension scaling to take advantage of such a technology and to ensure functional and robust designs. As mask costs increase, it is even more urgent that designers understand these effects so as to avoid the pitfalls and achieve a functional design on first silicon.

REFERENCES

[1] *IBM J. Res. Dev.*, Vol. 46, No. 2/3, 2002.

[2] P. Kapur, Performance challenges of the future on chip metal interconnects and possible alternatives, Stanford University, May 23, 2002.

[3] Near limit scaling, workshop, Solid State Circuits Technology Committee, 2003.

[4] The future of semiconductor manufacturing, short course, *IEEE International Electron Devices Meeting*, 2002.

[5] S. Schulze, Mentor Graphics Corp., Wilsonville, OR, Effecting mask costs by solving the data explosion bottleneck in mask data preparation, *Semiconduct. Int.*, July 1, 2003.

[6] H. S. Momose, S. Nakamura, T. Ohguro, T. Yoshitomi, E. Morifuji, T. Morimoto, Y. Katsumata, and H. Iwai, Study of the manufacturing feasibility of 1.5 nm direct-tunnelling gate oxide MOSFETs: uniformity, reliability, and dopant penetration of the gate-oxide, *IEEE Trans. Electron Devices*, Vol. 45, No. 3, Mar. 1998.

[7] A. Allan, D. Edenfeld, W. Joyner, A. Kahng, M. Rodgers, and Y. Zorian, International Technology Roadmap for Semiconductors, *IEEE Comput.*, Jan. 2002.

[8] S. Schulze, Effecting mask cost by solving the data explosion bottleneck in mask data preparation, *Semiconductor Int.,* July 1, 2003.

[9] Y. Pati, Sub-wavelength lithography, Tutorial, *Design Automation Conference*, 1999.

[10] C. Diaz, M. Chang, T. Ong, and J. Sun, Process and circuit design interlock for application-dependent scaling tradeoffs and optimization in the SoC era, *IEEE J. Solid State Circuits*, Vol. 38, No. 3, Mar. 2003.

[11] G. Scott, J. Lutze, M. Rubin, F. Nouri, and M. Manley, NMOS drive current reduction caused by transistor layout and trench isolation induced stress, *IEEE International Electron Devices Meeting*, 1999.

[12] M. Horowitz, R. Ho, and K. Mai, The future of wires, *Semiconductor Research Corporation Workshop on Interconnects for Systems on a Chip*, May 1999.

[13] V. Agarwal, M. Hrishikesh, S. Keckler, and D. Burger, Clock rate vs. IPC: the end of the road for conventional microarchitectures, *27th Annual International Symposium on Computer Architecture*, June 2000.

[14] T. Sakurai, Issues of current LSI technology and an expectation for new system-level integration, *International Conference on Solid State Devices and Materials*, pp. 36–37, Sept. 2001.

[15] K. Osada, Y. Saitoh, E. Ibe, and K. Ishibashi, 16.7fA cell tunnel-leakage-suppressed 16 Mb SRAM for handling cosmic-ray-induced multi-errors, Session 17.2, *International Solid-State Conference*, 2003.

[16] R. Rios, W. K. Shih, A. Shah, S. Mudanai, P. Packan, T. Sandford, and K. Mistry, A three-transistor threshold voltage model for halo processes, *IEEE International Electron Devices Meeting*, Dec. 2002.

[17] K. Cao, W. Liu, X. Jin, K. Vasanth, K. Green, J. Krick, T. Vrotsos, and C. Hu, Modeling of pocket implanted MOSFETs for anomalous analog behavior, *IEEE International Electron Devices Meeting*, 1999.

[18] C. Liu, M. Lee, C. Lin, J. Chen, Y. Loh, F. Liou, K. Schruefer, A. Katsetos, Z. Yang, N. Rovedo, T. Hook, C. Wann, and T. Chen, Mechanism of threshold voltage shift (ΔV_{th}) caused by negative bias temperature (NBTI) instability in deep sub-micron pMOSFETs, *Jpn. J. Appl. Phys.*, Vol. 41, Pt. 1, No. 4B, pp. 2424–2425, Apr. 2002.

[19] A. Stamper, Interconnection scaling to 1 GHz and beyond, *MicroNews*, Vol. 4, No. 2, first quarter 1998.

[20] International Technology Roadmap for Semiconductors, *http://public.itrs.net.*

[21] P. Ranade, H. Takeuchi, W. Lee, V. Subramanian, and T. King, Application of silicon–germanium in the fabrication of ultra-shallow extension junctions for sub-100 nm PMOSFTs, *IEEE Trans. Electron Devices*, Vol. 49, No. 8, Aug. 2002.

[22] S. Thompson et al., A 90 nm logic technology featuring 50 nm strained silicon channel transistors, 7 layers of Cu interconnects, low κ ILD, and 1 μm^2 SRAM cell, *IEEE International Electron Devices Meeting*, 2002.

[23] A. Grove, Changing vectors of Moore's law, *IEEE International Electron Devices Meeting*, 2002.

[24] J. Assenmacher, *BSIM4 Modeling and Parameter Extraction*, CL TD SIM, Infineon Technologies, Workshop Analog Integrated Circuits, Berlin, Germany, Mar. 19, 2003.

[25] C. Choi, Modeling of nanoscale MOSFETs, Ph.D. dissertation, Stanford University, 2002.

[26] G. Brown, The tyranny of roadmap: new CMOS gate dielectrics with reliability promises and challenges, ISMT Reliability Engineering Working Group, Dec. 12, 2001.

[27] G. Moore, Cramming more components onto integrated circuits, *Electronics*, Vol. 38, No. 8, Apr. 19, 1965.

[28] G. Moore, No exponential is forever…, keynote, *IEEE International Solid-State Circuits Conference*, 2003.

CHAPTER 2

CMOS DEVICE AND PROCESS TECHNOLOGY

2.1 EQUIPMENT REQUIREMENTS FOR FRONT-END PROCESSING

The past decade has seen significant breakthroughs in the field of integrated circuit (IC) technology. In the back end of the line, *RC* improvements are due to migration to copper and low-κ interlayer dielectrics with unique integration schemes such as dual damascene. In the front end of the line, only a few atomic monolayers reliable gate oxynitride is used routinely in high-performance devices. Strain engineering, combined with significant progress in ultrashallow junction technology, enables routine sub-130-nm production in both high- and low-power devices. In this chapter we review the current status and possible future trends of front-end-of-line processing systems for sub-130-nm technology.

2.1.1 Technical Background

Over the past 40 years, the semiconductor industry has continued its rapid pace of development, offering more compact electronic products with more speed and functionality at lower cost. This rapid growth has been fueled by the industry's ability to scale the MOSFET (metal-oxide-semiconductor field-effect transistor), the most commonly used building block for integrated circuits [1]. Despite all the challenges involved, Moore's law continues to set the guideline for transistor scaling in IC technology. Traditionally, gate length and gate oxide scaling have been two key elements of transistor scaling. Significant progress has been made in

Nano-CMOS Circuit and Physical Design, by Ban P. Wong, Anurag Mittal, Yu Cao, and Greg Starr
ISBN 0-471-46610-7

scaling of the gate length to less than 130 nm in production and less than 30 nm in showcase research transistors. Yet the anticipated performance improvement from this scaling has been limited by fundamental quantum-mechanical tunneling in ultrathin gate oxide as well as proper control of short-channel effects and off-state currents. As such, new dimensions have been added to the traditional MOS architecture. Strain engineering to enhance channel mobility by a variety of techniques, such as introduction of SiGe, is one example. Another approach has been a move away from bulk planar transistors and the introduction of silicon-on-insulator (SOI) and finFET three-dimensional devices.

Typical processing of a silicon-based integrated circuit starts with the fabrication of isolation structures. Both shallow and deep trenches are used in volatile (e.g., SRAM) and nonvolatile [e.g., flash; Figure 2.1(*a*)] device processing. The

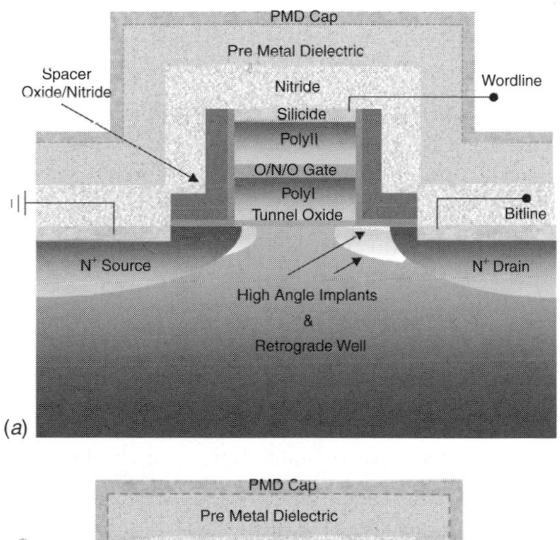

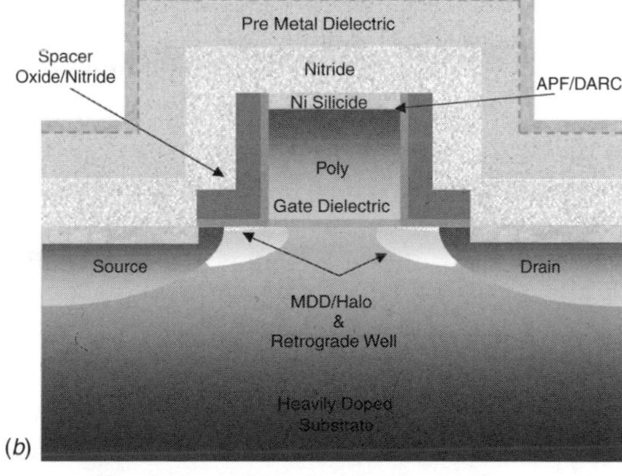

Figure 2.1 Typical NMOS flash (*a*) and MOSFET (*b*) cell. Although most processing steps are identical between the two cells, fundamental unique challenges face the design and processing of each. For example, whereas the MOSFET gate is scaled to the sub-2-nm regime, the flash cell (due to its intolerance to gate leakage) is in the sub-10-nm regime.

trench etch has significant challenges, such as attaining the correct sidewall profile. Just as critical is the filling of the trench. The correct selection of dielectric material has been key in eliminating voids, parasitic junctions, and unwanted stress on the silicon device channel. Generally, electromechanical polishing is used to remove excess dielectric material after trench fill. Various wet and dry cleaning processes are then used to prepare the silicon surface for well implants and then gate dielectric processing. Variations of low-energy angled implants, offset spacers, and short spike annealing are used for the creation of ultrashallow junctions after poly etch processing. Low-thermal-budget spacer formation, followed by source–drain implant and activation, are next, followed by low-thermal-budget salicidation. Nitride layers are used as a contact etch stop to allow an offset for contact to land on both trench oxide and source–drain contacts. Finally, a deposited film such as borophosphosilicate glass (BPSG) or high-density-plasma (HDP) forms the first interlayer dielectric layer, which completes the front end of the processing. Similar processing exists for flash cell integration. A tunnel oxide on the order of 10 nm is used to allow for channel hot-electron injection into poly I, the floating gate. Fowler–Nordheim tunneling then allows the cell to be erased. Asymmetric source–drain is sometimes used in a Flash cell. An oxide–nitride–oxide stack is used as the second gate dielectric between a floating poly I gate and a control poly II gate in a flash cell [2,3]. With concentration on gate stack, strain engineering, and rapid thermal processing, in this chapter we review front-end processing and key enabling equipment used in sub-130-nm nodes.

2.1.2 Gate Dielectric Scaling

The scaling of MOSFETs [Figure 2.1(*b*)] requires an increase in the dielectric capacitance and hence a decrease in gate dielectric thickness. In this section we review the scaling trends for gate dielectrics as the industry faces the challenge of possibly replacing SiO_2 as the gate dielectric [1]. The gate stack consists of the gate dielectric (SiO_2 or SiON) followed by the highly doped N+ (for NMOS) and P+ (for PMOS) polysilicon gate electrode. The scaling trend has required scaling of the gate dielectric for improved performance, increased density, and better control of the short-channel effects. The industry faced new challenges when gate oxide was first scaled below 4.0 nm. Examples of such challenges were boron penetration through the highly doped P+ polysilicon electrode for PMOS, increased leakage, and increased reliability concerns. The silicon oxynitrides (SiON) formed by thermal nitridation (N_2O, NH_3, or NO) were introduced to block boron penetration through the oxide as well as to enhance the hot-carrier immunity. Plasma oxynitrides were later introduced as the dielectric was scaled below 2.0 nm to incorporate higher levels of nitrogen in the dielectric and for better control of the nitrogen profile [4]. Despite early concerns, aggressive voltage scaling has allowed ultrathin oxides to continue to meet the reliability requirements. The gate leakage current through ultrathin oxides may, however, become the limiting factor for further scaling of the dielectric, as it can lead to excessive standby power consumption and degrade the dielectric integrity and reliability.

Carrier Transport in Gate Dielectrics The large SiO_2 energy bandgap of 9 eV and its large barrier height allow silicon dioxide to approach an ideal insulator under moderate bias conditions and at a thickness greater than 4.0 nm. This is in contrast to films such as Si_3N_4 or higher-κ dielectrics, where conduction may be characterized by bulk-limited mechanisms such as Frenkel–Poole emission [5,6]. The energy required to bring an electron from the Fermi level to vacuum is the work function ϕ_m of the electrode. Under applied bias $V_{ox} = E_{ox}t_{ox}$, the electrons have a finite probability of tunneling through the Si–SiO_2 potential barrier from the Si conduction band to the SiO_2 conduction band. Conduction through the triangular barrier is characterized by Fowler–Nordheim tunneling, and the current density measured can be described by [7–9]

$$J_{FN} = AE_{ox}^2 \exp\left(\frac{-B}{E_{ox}}\right) \tag{2.1.1}$$

where A is a constant related to the Si–SiO_2 barrier height, ϕ_b, and B is a constant related to the electron effective mass m^* and ϕ_b. As oxide thickness is scaled and V_{ox} drops, electrons no longer enter the conduction band and tunnel directly through the trapezoidal barrier. Direct-tunneling current density for V_{ox} smaller than the barrier height ϕ_b can be characterized by equation (2.1.2) [10,11].

For dielectrics below 3.0 nm, direct tunneling is the dominant current conduction mechanism. Since the direct-tunneling current is exponentially dependent on the oxide thickness, scaling the dielectric to the 1.0 nm range can result in unacceptably high leakage currents, leading to high standby power consumption and possible reliability and dielectric integrity concerns. The NMOS leakage current is expected to be the limiting factor in scaling the gate dielectric. The tunneling gate current in PMOS is roughly 10 times smaller than that of NMOS, due to its higher barrier for hole tunneling [11]:

$$J_n = AC(V_g, V_{ox}, t_{ox}, \phi_b) \exp\left\{\frac{-B\left[1 - (1 - V_{ox}/\phi_b)^{3/2}\right]}{E_{ox}}\right\} \tag{2.1.2}$$

$C(V_g, V_{ox}, t_{ox}, \phi_b)$ is a correction function, related to V_g, V_{ox}, t_{ox}, and ϕ_b, and is developed by empirical fitting [11].

Capacitance–Voltage and Equivalent Oxide Thickness The capacitance–voltage (CV) measurement at low and high frequencies is commonly used to extract metal-insulator-semiconductor (MIS) characteristics such as dielectric thickness, flat-band voltage, fixed charge, and interface state density. For thin oxides, particularly in the below 2.0 nm, range both measurement and interpretation of CV data become complicated. Tunneling current through thin

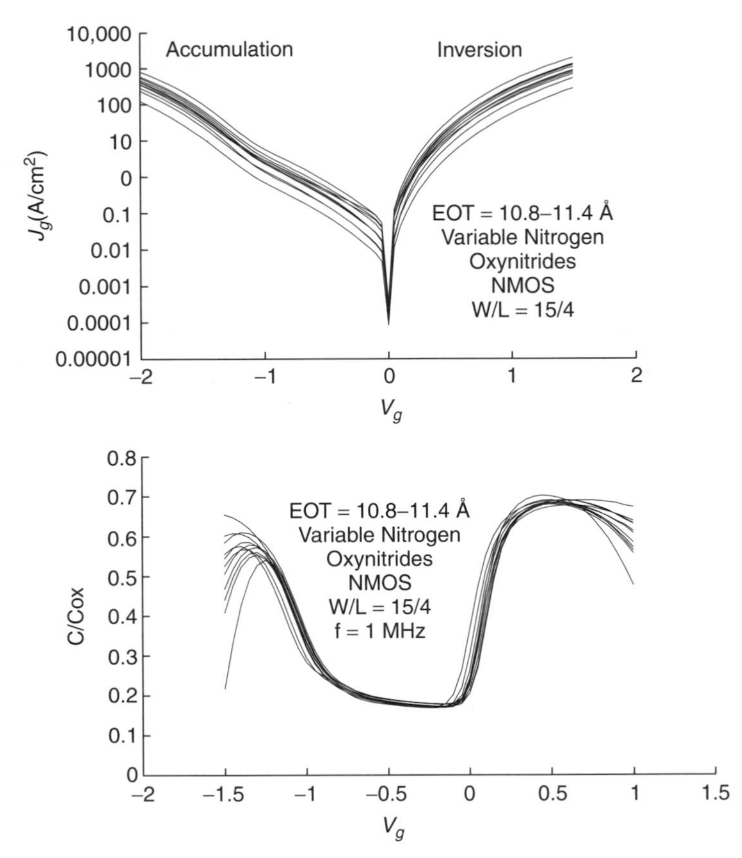

Figure 2.2 $J_g - V_g$ and CV curves for thin oxynitrides. Increasing nitrogen content reduces the tunneling leakage. Reducing the tunneling leakage reduces the capacitance attenuation in both inversion and accumulation. (Data courtesy of Applied Materials, Inc. [13].)

dielectrics, which increases exponentially with decreasing thickness (about 10-fold per 0.2 nm of physical oxide thickness), leads to voltage drops along the series resistances in the gate electrode and the substrate (Figure 2.2). The gate dielectric can be modeled as a voltage-dependent resistor in parallel with a capacitor. The gate electrode and substrate act as distributed series resistances [12]. Capacitance attenuation due to channel resistance may also become dominant in strong inversion, setting a limit on the channel lengths used when measuring MOSFETs [13].

Significant work in recent years has focused on accurate measurement, extraction, and interpretation of capacitance–voltage curves, as documented in the references in this section. The electrical thickness of a dielectric is the distance between the centroid of charge in the gate and the substrate [6]. Depletion of mobile charge carriers in polysilicon near the gate dielectric interface, particularly in inversion, results in a shifting of the charge centroid away from the interface by more than 0.3 nm. This effect can be modeled as an additional capacitance in series with the oxide capacitance [5], resulting in a dielectric that is electrically thicker than expected. Similarly, in an inversion or accumulation layer in the substrate, carriers are confined in a narrow potential well near the surface, and their motion in the direction normal to the surface must be treated quantum mechanically. A simplified, closed-form analytical treatment is not adequate, and correct treatment requires solving the coupled effective-mass Schrödinger and Poisson equations self-consistently [14]. The quantum-mechanical (QM) treatment of the inversion layer results in a shift of the inversion charge centroid away from the interface by more than 0.3 nm. In ultrathin dielectrics, the increased electrical thickness due to poly depletion and QM effects becomes increasingly significant [5–15]. This creates a large discrepancy between the expected capacitance of the dielectric and the measured dielectric capacitance.

Capacitance effective thickness (CET) is the electrical thickness of a dielectric and can be described by [12]

$$\mathrm{CET}(V) = \frac{\varepsilon_0 \varepsilon_{\mathrm{SiO_2}} A_{\mathrm{gate}}}{C(V)} \tag{2.1.3}$$

where ε_0 is the permittivity of free space, $\varepsilon_{\mathrm{SiO_2}}$ the permittivity of SiO_2, and A_{gate} the gate area. $C(V)$ is the capacitance at a given voltage V, which includes the series capacitances due to poly depletion and the QM effects in the substrate. CET will hence depend on the type of electrode, the electrode work function, and depletion in the electrode as well as substrate doping and gate voltage [12].

By contrast, the equivalent oxide thickness (EOT) of a dielectric is not dependent on the electrode properties or the substrate doping. The EOT is the thickness of SiO_2 that would produce the same CV curve as that of an alternative dielectric and is defined as [13]

$$\mathrm{EOT} = \frac{\varepsilon_{\mathrm{SiO_2}}}{\varepsilon_{\mathrm{high}\ \kappa}} \bullet t_{\mathrm{high}\ \kappa} \tag{2.1.4}$$

where $t_{\mathrm{high}\ \kappa}$ is the physical thickness of the high-κ dielectric and $\varepsilon_{\mathrm{high}\ \kappa}$ is the permittivity of the dielectric. Since the dielectric constant of SiON or other mid- and high-κ dielectrics are typically not known, the EOT must be determined by capacitance measurements as described above [12]. Once the CV is measured, the challenging task of correcting and interpreting the data remains. Different models have been proposed to account for poly depletion and quantum effects and to extract EOT. Variations in the models and algorithms can therefore lead to variations in the EOT extracted, and one must be careful when comparing the EOTs of dielectrics extracted by different methods [12,13,16,17].

Scaling Limit for SiO_2 and Alternative Dielectrics As described earlier, the gate dielectric has been scaled to improve device performance and to suppress short-channel effects. Several fundamental limits threaten further scaling of SiO_2 dielectrics to an EOT of less than 1.0 nm. The thickness at each interface required to achieve a full SiO_2 bandgap is shown to be about 0.35 to 0.40 nm, resulting in a total thickness of about 0.7 to 0.8 nm for both interfaces [6]. This sets an absolute physical limit of 0.7 nm for SiO_2 scaling. Other practical limits may be reached earlier, however, including excessive leakage and limited or zero performance gain with decreasing oxide thickness. As shown in equation (2.1.2), tunneling current increases exponentially with decreasing physical thickness of the dielectric. In addition, as the dielectric thickness is scaled, the relative significance of the silicon channel and poly electrode interfaces on EOT and channel mobility increase [5,6]. Larger mobility degradation reported for thinner oxides results in smaller than expected gains in $I_{d\ \mathrm{sat}}$ with decreasing dielectric thickness [62].

Silicon oxynitrides can be formed by thermal nitridation or annealing of SiO_2 in NO, N_2O, or NH_3 or by plasma nitridation of SiO_2 (Figure 2.3). The addition of nitrogen changes the material properties in several important ways (Figure 2.4). Nitrogen in the oxide provides a barrier to boron penetration, which can cause large V_{th} shifts in PMOS and degrade the dielectric reliability. The refractive index of SiO_2 increases with increased nitrogen content, from $\eta_{\mathrm{SiO_2}} = 1.46$ to $\eta_{\mathrm{Si_3N_4}} = 2.0$. In addition, the relative dielectric constant ($\kappa_{\mathrm{SiON}} = \varepsilon_{\mathrm{SiON}}/\varepsilon_0$) increases linearly with increasing nitrogen, from $\kappa_{\mathrm{SiO_2}} = 3.9$ to $\kappa_{\mathrm{Si_3N_4}} = 7.5$. The increased κ value allows the use of physically thicker films for the same EOT, as seen in equation (2.1.4), resulting in a smaller tunneling current [5]. However, the addition of nitrogen to SiO_2 decreases the bandgap and therefore the barrier height (ϕ_b) for electron and hole tunneling [5,8,19,20]. This means that the reduced direct tunneling due to the larger physical thickness of SiON is partially offset by the smaller effective barrier height [18,20,21]. Most commonly,

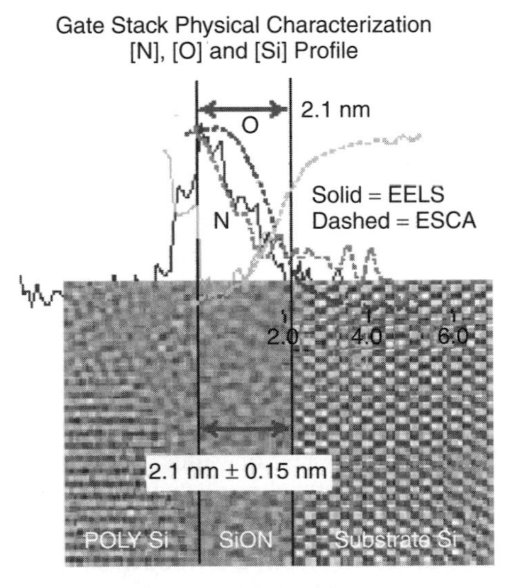

Figure 2.3 Plasma nitridation incorporates nitrogen at the polysilicon–oxynitride interface, as measured by both electron energy-loss spectroscopy (EELS) and electron spectroscopy for chemical analysis (ESCA). (From Ref. 4.)

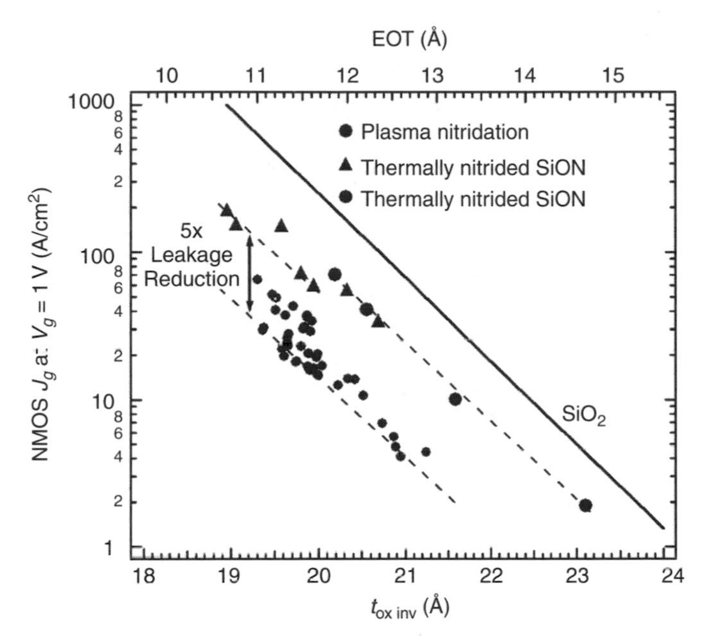

Figure 2.4 Typical J_g versus EOT plot for thermally nitrided and plasma-nitrided SiON. As in Figure 2.2, adding nitrogen decreases the tunneling current. (Data provided by Applied Materials, Inc.)

oxynitrides are grown or annealed in nitric oxide (NO). Nitrogen incorporation in NO nitrided oxides is limited, and nitrogen typically piles at the interface. For ultrathin oxides, a higher level of nitrogen (5 to 20%) is necessary to reduce leakage further and to prevent boron penetration [4]. Plasma nitridation is used for sub-1.5-nm oxides to better control the percentage and placement of nitrogen in the dielectric [4,24,27–29].

Nitrogen in the dielectric affects the mobility of both N- and PMOS devices. For PMOS devices, hole mobility decreases for all electric fields with increasing nitrogen. For NMOS, at low nitrogen levels, the peak electron mobility degrades with increasing nitrogen, but the high-field electron mobility fall-off improves with increasing nitrogen concentration (Figure 2.5) [23]. Larger amounts of nitrogen in the film can create traps at the interface or act as scattering centers for carriers in the channel resulting in large mobility degradation [10]. The impact of nitrogen on carrier mobility can be modulated by the nitrogen profile and the proximity of nitrogen to the channel [22].

An intense search for an alternative dielectric with higher permittivity has been under way to limit the gate leakage current and continue scaling of the dielectric. A material with a higher dielectric constant than SiON will be physically thicker than an SiON film of the same EOT by κ/κ_{SiON}, hence suppressing the tunneling current according to equation (2.1.2). Silicon nitride, aluminum oxide, zirconium oxide, and hafnium oxide and their silicates are just a few of the higher-κ

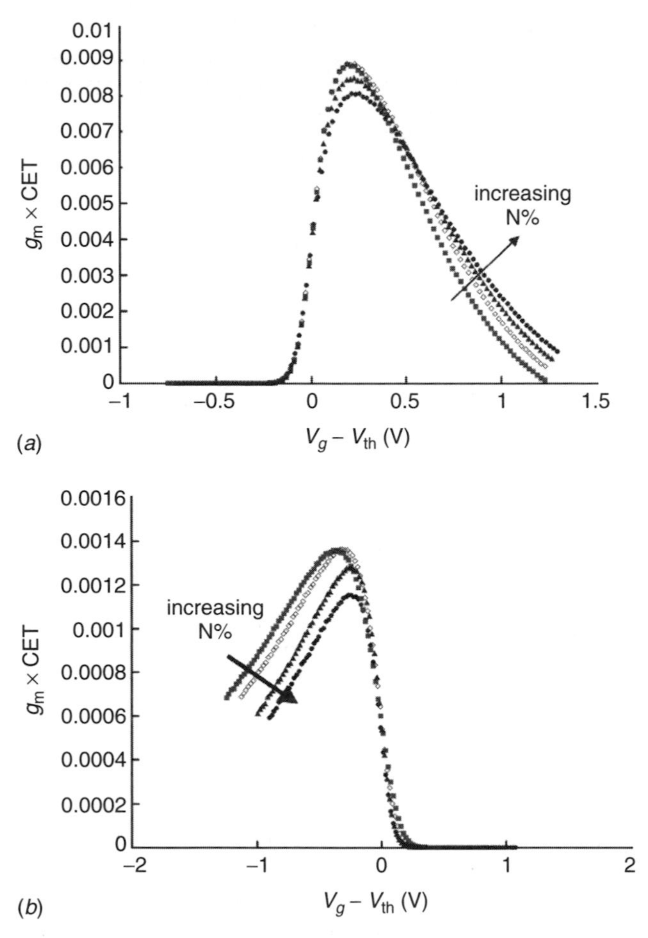

Figure 2.5 Normalized transconductance of plasma-nitrided long-channel (*a*) NMOS and (*b*) PMOS devices. High-field transconductance improves with increasing nitrogen in NMOS but degrades with increasing nitrogen for PMOS. (Data provided by Applied Materials, Inc.)

dielectrics studied. Important characteristics of an alternative dielectric include its permittivity, bandgap, band alignment to silicon, thermodynamic stability, interface quality, film morphology, reliability, compatibility with the gate electrode, and CMOS processing [5,6]. Significant progress has been made to improve the mobility degradation generally associated with higher-κ gate dielectric materials especially when metal gates are used. Promising results for HfSiON have been reported [25]. However, other possibly fundamental characteristics, such as Fermi-level pinning at the polysilicon–metal oxide interface (which results in large shifts in the threshold voltage) have delayed the adoption of high-κ dielectrics [26]. It is likely that high-κ dielectrics will be first introduced in low power applications where leakage requirements are more stringent.

2.1.3 Strain Engineering

It has been nearly six decades since the first preparation of homogeneous SiGe alloys by Stohr and Klemm [30] and Wang and Alexander [31]. The pioneering work of Johnson and Christian [32] and a series of classic papers by Braunstein et al. on single-crystal and polycrystalline SiGe alloys set the foundations of today's introduction of SiGe in advanced CMOS devices [33–36]. This work measured the variation of lattice constant and bandgap as the percent mole fraction of germanium in silicon is varied. Their work shows a nearly linear change (a quadratic fit based on later results) in the lattice constant from 5.43 for silicon to 5.66 for germanium. This nearly 4.2% lattice mismatch between germanium and silicon single-crystal lattice structures leads to a significant electronic band structure variation in the alloys of SiGe. Unlike the nearly linear change of lattice constant over the entire compositional range, the bandgap of Ge_xSi_{1-x} alloys first decreases linearly with a smaller slope and then switches to a steeper slope near the 85% germanium fraction in silicon. The change in bandgap is due to a switch from the silicon-like ($E_g = 1.14$ eV) conduction band structure to that of germanium ($E_g = 0.67$ eV) at the critical value when the percent mole fraction of silicon in germanium falls below 15%. The valence band structure remains virtually the same throughout the alloy compositional change, with its maximum at the center $k(000)$. The conduction band of the alloy is first silicon-like, with a minimum along [100] at 0.8×. But at an 85% germanium fraction in silicon, the band minimum switches from silicon-like to germanium-like.

The pseudomorphic deposition of Ge_xSi_{1-x} on silicon thus requires significant adjustment in lattice constant in the growth direction ([100] Si; Figure 2.6). Parallel to the growth direction the lattice constant must remain the same as that of silicon throughout the compositional change of germanium in silicon. The diamond structure of silicon or germanium lattice is now changed into a tetragonal structure with significant compressive strain parallel to the growth condition. The degree of the strain is thus related to the percent mole fraction of germanium in silicon. This strain, caused by commensurate deposition of Ge_xSi_{1-x} on silicon, modifies the conduction and valence band structure significantly by splitting the

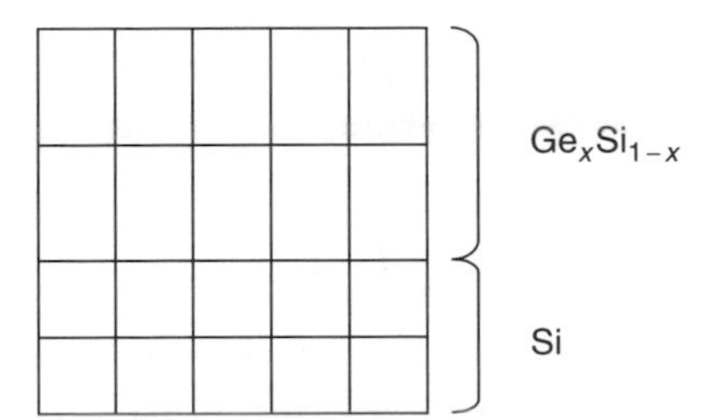

Figure 2.6 Compressive strain in the lattice of SiGe alloys allows for commensurate deposition. $Ge_{0.2}Si_{0.8}$ has an approximately 1% larger lattice constant than that of silicon, and up to a critical thickness of a few hundred angstroms can be grown pseudomorphically on silicon.

bands. The valence band of the strained alloy, for example, is split into two bands of heavy and light holes. Since the bandgap is a measure of energy difference between the maximum of the valence and the minimum of the conduction band, the overall bandgap of strained Ge_xSi_{1-x} can now be significantly lower than that of unstrained bulk alloys at a given fraction of germanium in silicon [38–41].

The consequence of a lower bandgap in SiGe alloys is a band offset between silicon and the alloy in heterostructures of Si/Ge–Si. In a type I band alignment, where alloys of Ge–Si are deposited on silicon, the offset occurs in the valence band and the conduction band is mostly aligned. In a type II band alignment, on the other hand, where silicon is grown pseudomorphically on SiGe, the offset occurs in both the conduction and valence bands. Another fundamental characteristic of the SiGe alloys is the higher hole mobility compared to silicon, due to the higher hole mobility of germanium. Furthermore, since mobility is a function of both scattering and effective mass, in a strained SiGe alloy the mobility is enhanced even further than the unstrained alloys. The lower effective mass and less scattering is due to the lifting of degeneracy in the energy band diagram [42–47].

Molecular beam epitaxy has been the onset of pseudomorphic growth of SiGe on silicon and the fundamental study of this family of heterostructures. In manufacturing, however, chemical vapor deposition (CVD) is the method of choice for SiGe deposition. The design of a typical epitaxial CVD system includes both atmospheric and reduced pressure processes. Atomically clean surfaces are the key to selective deposition, and the surface of silicon generally is precleaned in diluted hydrofluoric acid solutions. Prior to deposition, an in situ bake at high temperature in ambient hydrogen removes the native oxide. The deposition itself is at a lower temperature, depending on the chemistry used. Typically, silane or dichlorosilane is used as the silicon source and germane as the germanium source. To enhance selectivity to oxide and nitride, HCl gas is mixed with silane and germane [48]. At temperatures of $<800^\circ C$ a nearly ideal selective deposition of SiGe can be achieved on silicon. A combination of chamber design and lamp heating with individual zone temperature control can provide a uniform deposition by precise control of temperature and gas flow. Dopant incorporation as high as 10^{21} can be achieved. Optimization of gas mixtures, temperature, and chamber design, together with independent flow of dopant and germanium gases, can achieve less than 1% 1σ uniformities on both germanium and dopant concentration uniformity across the wafer.

2.1.4 Rapid Thermal Processing Technology

Rapid thermal processing (RTP) has been an important semiconductor manufacturing technology for over 10 years, with its initial development dating back to the 1960s [49]. RTP began replacing batch furnace thermal processing for two primary reasons: superior ambient control and reduced thermal budget. Ambient control was the primary motivation for the first widespread implementation of RTP. The process application was titanium disilicide ($TiSi_2$) formation, and its requirements were a good match for RTP capabilities. The nitrogen gas ambient must be strictly controlled, with background oxygen levels below 10 ppm. Such low oxygen levels are easily achieved in a low-volume single-wafer RTP chamber but are difficult to maintain reliably in large-volume batch furnace tubes. $TiSi_2$ formation is not highly sensitive to temperature control and repeatability, which was a weakness of early RTP systems, but it does benefit from the short process times of RTP in reducing thermal degradation and lateral overgrowth or bridging.

The second motivation for moving process steps from batch furnaces to RTP was to reduce thermal exposure. When performing high-temperature oxidations or anneals, batch furnaces expose wafers to excess thermal treatment as the temperature is slowly ramped up and down (typically, 10°C/min). This additional time leads to unnecessary dopant and defect diffusion. As shown in Figure 2.7, RTP accelerates the temperature ramp up and down to typically greater than 75°C/s, eliminating this excess thermal exposure and the unwanted diffusion.

RTP Technology Rapid thermal processing demands unique capabilities in equipment technology: precise ambient control; temperature ramp-up at up to 250°C/s; spike anneal temperature uniformity of 3°C, 3σ; temperature control range from 300 to 1200°C; and temperature measurement independent of wafer emissivity. Precise ambient control is required in most RTP processing applications, including metal silicidation, implant annealing, and thermal oxidation. Early RTP equipment relied on extensive purging for ambient control, because the chamber was opened to the factory atmosphere to load and unload each wafer. This limited the throughput and ultimate ambient control that could be achieved. One approach is to mount the RTP chamber on a cluster tool that employs vacuum load locks to exchange air quickly for a pure nitrogen environment. The cluster also allows for single- or multiple-chamber processing. Advances in gas flow modeling have also enabled non-load-locked systems to

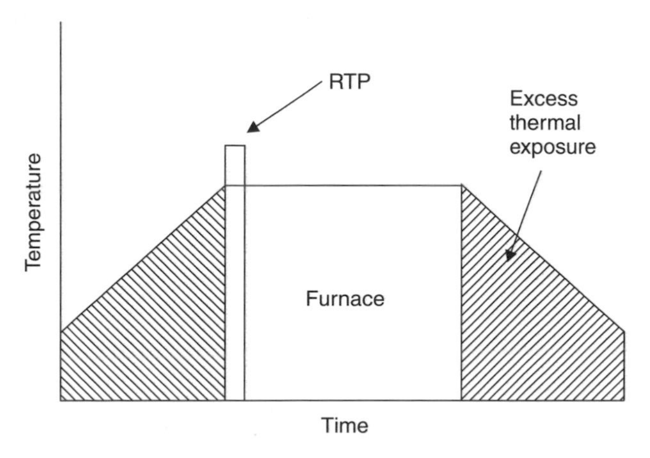

Figure 2.7 Thermal exposure comparison between RTP and batch furnace.

reach sub-part-per-million gas purity levels with high throughput. These systems typically employ high-velocity gas purges during wafer transfer to minimize oxygen incursion into the chamber while the chamber door is open.

To achieve the ramp rates required for implant annealing, RTP requires the minimum thermal mass to be heated, typically consisting of the wafer and its direct support structure. Most RTP equipment isolates the wafer thermally from the chamber walls and uses lamp radiation to supply the burst of energy required to ramp the temperature quickly. Another approach is to introduce the wafer rapidly to an environment with a large thermal mass relative to the wafer. The wafer causes only a small perturbation of the environment, and the wafer temperature asymptotically approaches that of the environment.

To maintain temperature uniformity of 3°C, 3σ (three times the standard deviation of all measured points within a wafer and from multiple wafers) over the broad range of temperature required for RTP, active uniformity control is required. Due to the difference in the ratio of surface area to volume at the edge and center of a wafer, different energy distributions are required at different parts of an RTP temperature cycle. As shown in Figure 2.8 during ramp-up, the wafer edge temperature tends to lead the center, while during steady state and ramp-down, the edge tends to trail the center. In lamp-based RTP equipment, this is addressed by grouping the lamps into different zones for center and edge. The lamp powers can then be changed during the thermal cycle to maintain temperature uniformity at all times. The most advanced system provides high-speed, multiple-point measurement and multiple-zone control for active real-time temperature uniformity control.

The most challenging process for uniformity control is the high-temperature spike anneal used for activation of implanted ions in the most advanced devices. Spike anneals do not hold at a peak temperature but ramp down as soon as the desired temperature is reached, resulting in a triangular or spike-shaped temperature-versus-time profile. Implant activation is a highly temperature sensitive process. To meet device requirements for the shallow junctions in the source–drain area of the transistors requires 3°C, 3σ temperature control.

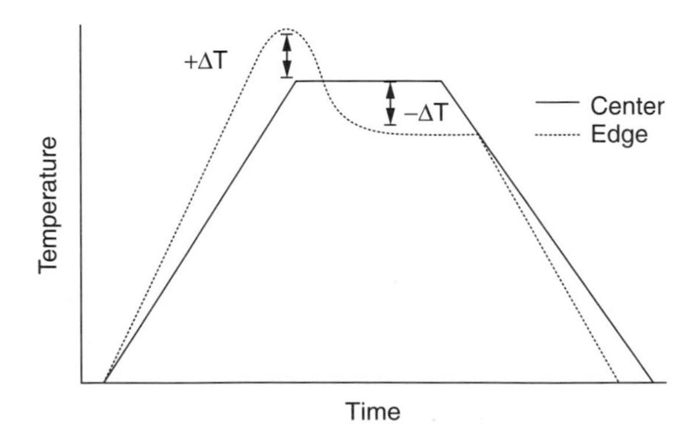

Figure 2.8 Center versus edge temperature profile without active uniformity control.

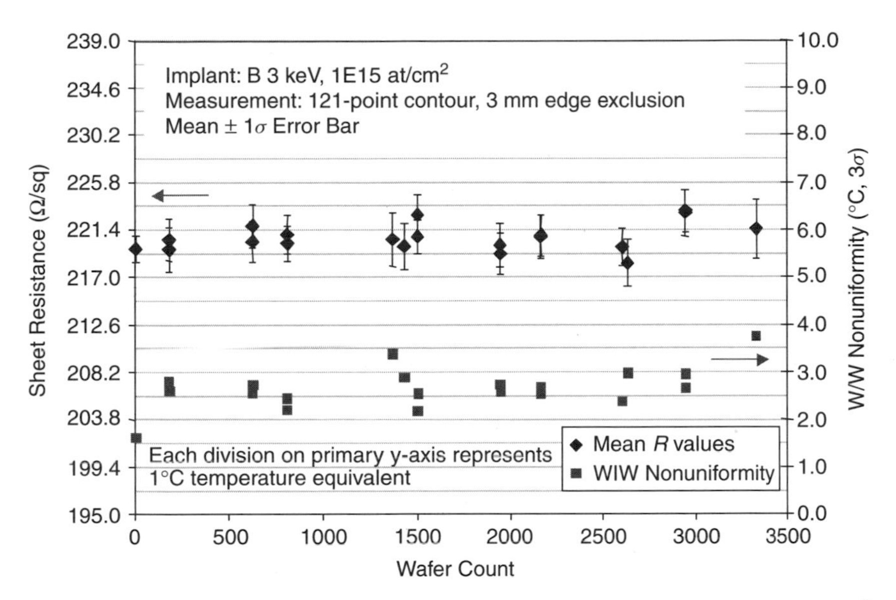

Figure 2.9 Sheet resistance uniformity/repeatability for RTP spike anneal with 3.2°C, 3σ control. (From Ref. 50.)

Figure 2.9 shows this level of temperature uniformity demonstrated on 300-mm wafers measured to within 3 mm of the wafer edge.

RTP equipment relies on optical pyrometry for temperature measurement. Pyrometry measures the gray-body radiation emitted by the wafer itself to determine its temperature. Direct contact with thermocouple detectors was employed in older-generation RTP systems, but this approach was abandoned due to inconsistent thermal contact and local thermal perturbation leading to temperature errors and nonuniformity. Accurate pyrometric measurement requires knowledge of the wafer emissivity (the ratio of the energy emitted by the wafer and the energy emitted by an ideal blackbody at the same temperature) at the wavelength of measurement. Two primary approaches are used to address these challenges. One is to create a high-reflectivity environment underneath the wafer to approach the properties of an ideal blackbody cavity. This enhances the effective emissivity of the wafer and reduces the potential for measurement errors. In addition, a separate probe measures the emissivity in real time and corrects for any remaining error [51]. A second approach employs a real-time reflectance measurement of a periodic lamp power variation to calculate the emissivity assuming no wafer transmission (emissivity = 1 − reflectance). The measured emissivity is then used to correct the pyrometer reading [52].

These technologies—ambient control, low thermal mass, uniformity control, and temperature measurement—together form the core of RTP equipment and enable RTP to cover a broad range of applications over a broad temperature range. For future device nodes at 65 down to 45 nm, even the limited thermal budget of RTP annealing may be too much for some process steps. One approach is to reduce the time exposure to the millisecond range by heating only the wafer surface. This is called *thermal flux annealing* and can be performed using a flashlamp or laser [53,54].

RTP Applications As RTP process equipment has evolved and matured, it is now capable of performing all thermal annealing and oxidation processes in typical CMOS and DRAM fabrication. Primary anneal applications include refractory metal silicide formation (e.g., $TiSi_2$, $CoSi_2$, NiSi), ion implant annealing, BPSG stabilization and reflow, and contact metal annealing (e.g., TiN). Primary oxidation applications include gate oxidation and nitridation, trench isolation liner oxidation, and sacrificial oxidation.

Self-aligned titanium silicide (salicide) was introduced to CMOS technology in the 1980s to reduce sheet resistance in the source–drain and gate regions and to provide good ohmic contact between the transistor and metal interconnects. The process is self-aligned, because following a blanket metal sputter, the silicide forms only where the metal is in direct contact with silicon in the source–drain and gate areas and does not form on sidewall spacers and isolation structures. To enable device scaling, the preferred metal has changed in recent years from titanium to cobalt to nickel. Salicide is typically formed using two RTP steps, as illustrated in Figure 2.10. Following a preclean and blanket metal film deposition by sputtering or physical vapor deposition (PVD), the first RTP anneal (commonly referred to as RTP1) forms the silicide. The RTP1 temperature must be high enough to initiate the metal–silicon reaction and to form sufficient thickness and the desired phase of silicide but low enough to prevent lateral silicide growth onto the gate sidewall spacers that could lead to gate-to-source/drain shorts. Ambient control is critical for RTP1, because the metal films will rapidly be consumed by oxidation if even a few parts per million of oxygen are present in the gas ambient.

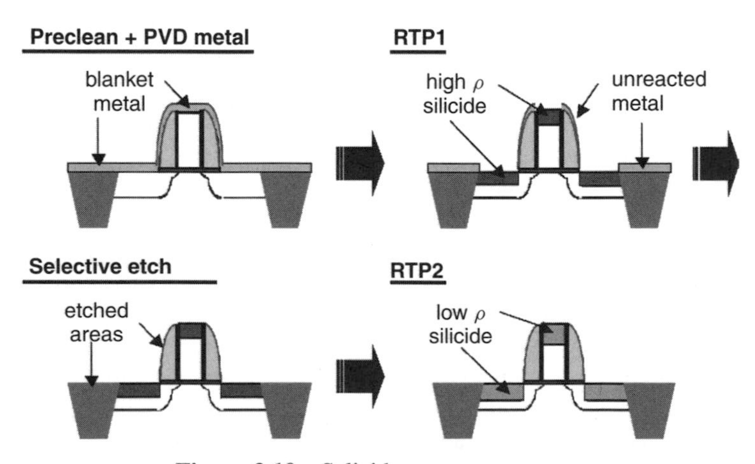

Figure 2.10 Salicide process sequence.

Following a selective etch to remove the unreacted metal from spacer and isolation areas, the second anneal (RTP2) is then used to reduce the sheet resistance by transforming the phase or compound of silicide. RTP2 requires a temperature typically several hundred degrees higher than RTP1, which is too high to allow the silicide to be formed in a single step. The maximum temperature for RTP2 is typically limited by morphological degradation, called *agglomeration*, or by resistance degradation caused by transformation of the phase or composition of the silicide.

Titanium salicide was the first material to be used broadly for logic applications. The C49 phase of $TiSi_2$ is typically formed with an RTP1 of 650°C for 30 s. RTP2 transforms the silicide phase from C49 to C54, achieving a low resistivity of 13 to 16 $\mu\Omega \cdot cm$, and is typically performed at 850°C for 20 s or less. Scaling issues made it difficult to extend titanium salicide below the 0.25-μm node. As line widths and silicide thickness are reduced, it becomes increasingly difficult to nucleate the C49- to-C54 phase transformation without agglomeration of the silicide [55]. These limitations led to increasing use of shorter times at high temperatures where the process window is broadest.

Cobalt silicide was selected as the successor to $TiSi_2$ for its improved scaling to narrow line widths. In the cobalt salicide process, RTP1 is typically 500 to 550°C for 30 to 60 s and forms CoSi. RTP2 is typically 750 to 800°C for 30 s and transforms the CoSi to $CoSi_2$ with a low resistivity of 14 to 18 $\mu\Omega \cdot cm$. Cobalt is typically capped with titanium or TiN during the blanket metal deposition to protect it from oxidation. The cap is then removed during the selective etch between RTP1 and RTP2.

Scaling limitations of cobalt silicide are seen below 50-nm line widths and are driving the industry to shift to nickel silicide. NiSi maintains low sheet resistance below 30-nm line widths [56]. NiSi can be formed in a single RTP step, typically between 400 and 500°C, rather than the two-RTP-step sequence shown in Figure 2.10. However, this results in excessive silicidation of narrow lines, indicated by a decrease in sheet resistance versus gate length below 100 nm. Two-step RTP for NiSi formation controls excessive silicidation and resulting problems with increased poly depletion and junction leakage. The first step, RTP1, forms a nickel-rich phase, Ni_2Si, and RTP2 completes the reaction to the low-resistivity phase, NiSi. Two-step NiSi formation requires advanced capabilities of RTP equipment. First, the RTP1 step requires significantly lower process temperatures of 250 to 350°C. This is below the typical range of temperature measurement using pyrometry. Second, the RTP1 step requires precise temperature uniformity. The silicide thickness is determined by the formation temperature rather than the deposited nickel thickness, and the formation reaction is extremely temperature sensitive [57]. Finally, as with cobalt and titanium before it, NiSi requires precise ambient control with oxygen levels at or below 1 ppm.

CMOS processing relies primarily on ion implantation and RTP annealing to position dopant profiles that determine device performance. Thermal annealing is required to repair damage resulting from implantation and to position dopant atoms in substitutional sites where they become electrically active. RTP replaced batch furnace processing for annealing implants, due to its precise control of thermal budget, enabling scaling of both lateral and vertical transistor dimensions. The elimination of excess thermal exposure, shown in Figure 2.7, enables RTP to produce shallower junctions with equivalent sheet resistance than with batch anneals.

RTP thermal budgets have continued to scale, moving from 30- to 60-s soak anneals between 950 to 1050°C and 0-s spike anneals from 1050 to 1100°C. Further reduction in RTP thermal budget has been enabled by increasing temperature ramp rates from 50°C/s to greater than 250°C/s. Further thermal budget scaling is needed to meet the International Technology Roadmap for Semiconductors (ITRS) requirements for source–drain and extension junctions. The ITRS specifies junction depth and sheet resistance targets for these junctions as shown in Figure 2.11. Also shown is the typical performance of low-energy implants coupled with RTP spike anneals.

While the 90-nm requirements can be met, 65-nm requirements require further reduction in junction depth and sheet resistance. Several technologies are being considered to enable this scaling, including thermal flux annealing, solid-phase epitaxial regrowth of implanted regions, and coimplantation of ions that can limit diffusion during RTP annealing.

Several factors have broadened RTP applications to include all thermal anneals and oxidations used in CMOS processing. First, thermal budget control throughout the process flow is essential to enable scaling of critical device dimensions. Second, all oxide thickness and anneal steps have scaled to within a practical range for single-wafer processing. Third, novel processes, such as in-situ steam generation (ISSG), have enabled improved device performance [59,60]. Finally, single-wafer processing has reduced cycle time and reduced work in progress that enable improve fabrication efficiency and product time to yield [61].

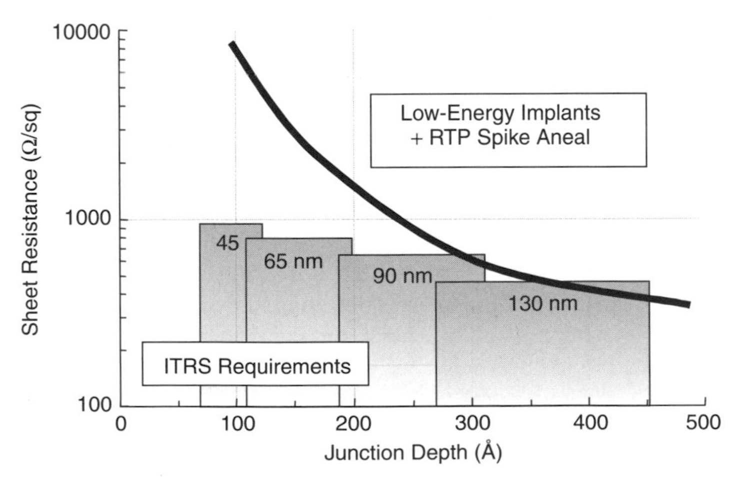

Figure 2.11 Sheet resistance versus junction depth with RTP capability versus ITRS requirements (From Ref. 58.)

In summary, RTP is a critical and enabling technology in semiconductor manufacturing. Extended capabilities for low-temperature processing and millisecond thermal exposures will continue to enable device scaling. Economic efficiencies of single-wafer processing will continue to drive transition of thermal processes from batch to RTP equipment for years to come.

2.2 FRONT-END-DEVICE PROBLEMS IN CMOS SCALING

The continued scaling of CMOS technology into the sub-100-nm regime gives rise to a number of new device physics and process issues that were not significant in the past. Consequently, it is necessary to introduce new materials and new device structures to the CMOS technology to maintain the pace of technology scaling as projected by the International Technology Roadmap of Semiconductors. In this section we survey some of the important front-end-related physical effects in aggressively scaled CMOS devices and discuss the implications of some new process technologies from a device and design point of view.

2.2.1 CMOS Scaling Challenges

The conventional bulk-silicon MOSFET structure has been scaled successfully for about three decades. In the nanometer regime, however, continued scaling is faced with both fundamental physical obstacles and economical constrains. Following is a brief description of the main issues and factors affecting MOSFET scaling.

Short-Channel Effects In an ideal MOSFET, the gate voltage has full control over the channel conduction state. As the gate length gets smaller, however, the drain region has more influence on the channel potential. The threshold voltage V_{th}, of devices with very short channel length can be reduced significantly. As a result, the off-state leakage increases, and the variation of V_{th} also becomes much larger in the roll-off region. To suppress the short-channel effect, it is necessary to reduce coupling between the drain and the channel and to increase coupling between the gate and the channel. Reduction of drain–channel coupling requires a smaller depletion region width in the channel and drain region, which can be achieved by doping profile engineering in the substrate. The retrograde channel doping profile (i.e., low doping at the surface and high doping deeper into the substrate) and halo doping profile are effective methods of reducing the depletion region width. The source and drain junction depths also need to be reduced accordingly. As the halo doping has a very high concentration, it brings the side effect of high field at the drain–substrate junction and consequently, high junction leakage due to the band-to-band tunneling. To increase channel-to-gate coupling, thinner electrical thickness of the gate oxide is needed.

Equivalent Gate Oxide Thickness Scaling Thinner gate oxide, or larger per unit area gate capacitance, is necessary for improving both the off- and on-state device characteristics. In the state-of-the-art technologies, the gate oxide thickness has been scaled to below 20 Å, with the nitridation process for better resistance to the dopant penetration. For such thin SiO_2, the direct-tunneling gate leakage is no longer negligible, even with reduced power supply voltage. It creates serious problems of off-state power consumption. High-κ gate dielectrics, which can achieve a given equivalent oxide thickness (EOT) with thicker physical thickness than SiO_2, is a potential solution to this problem. Currently, some issues of the high-κ gate dielectrics are still being investigated, such as the mobility degradation and flat-band voltage shift. Given the challenges of process technologies, it is also necessary to develop techniques on circuit and systems levels to overcome the gate leakage current problem. Differentiating the application types (i.e., defining high performance, low operating power, and low standby power applications) also helps better match the process technology to the design needs.

Related to the gate dielectric EOT scaling is the polysilicon gate depletion effect. For a typical polysilicon gate doping concentration, a thin (about 1 nm) depletion layer exists in the polysilicon when the device is turned on. This depletion layer weakens the gate-to-channel capacitive coupling, or equivalently, reduces the effective gate overdrive voltage. When the gate dielectric EOT approaches 1 nm, the addition from the gate depletion layer is significant. Although increasing the active doping concentration in the gate is helpful, a complete solution to this problem requires the use of metallic gate electrodes.

The quantum-mechanical effect in the channel also contributes an additional capacitance in series with the gate oxide capacitance. When the vertical electric field is sufficiently high, a condition satisfied in the current and future CMOS devices, the vertical (perpendicular to the substrate surface) movement of the carriers in the channel is confined in a potential well. The energy states related to this movement consequently become discrete instead of being a continuum in the classical scenario. One effect is that the peak of the carrier distribution is a small distance (about 1 nm) from the substrate–oxide interface, which means another few angstroms' addition to the gate dielectric EOT. Like the polysilicon gate depletion effect, this is also a seemingly small effect that was not a concern until the sub-100-nm technology generations. The quantum effect depends on several factors, such as the vertical field, the substrate doping, and the silicon body thickness in the case of ultrathin body MOSFETs. This is a fundamental physical limit which cannot be solved simply by process improvements.

Channel Carrier Mobility The channel carrier mobility affects the device and circuit performance directly. With device scaling, mobility is adversely affected by a few factors. The generally higher channel doping concentration, which is related to the need to control the short-channel effect, results in more impurity scattering and consequently, lower mobility. Due to nonscalable factors such as thermal voltage and the silicon bandgap (the bandgap affects the setting of the threshold voltages but is not regarded as a fundamental barrier to scaling), the power supply voltage cannot be scaled at the same rate as the device dimensions. Therefore, the average vertical electric field experienced by the carriers in the channel gradually increases. The higher vertical field results in lower mobility, as described by the universal mobility model. The undesirable stress in the channel due to the STI process also contributes to the mobility degradation of n-MOSFETs. In addition, most candidate high-κ gate dielectrics that are currently being evaluated have worse interface quality than the thermal SiO_2 gate dielectric. The poor interface also degrades the channel carrier mobility and is responsible for the generally degraded performance of MOSFETs using a high-κ gate dielectric.

A few possible solutions exist to improve the carrier mobility. As discussed earlier, the introduction of channel strain engineering techniques is projected by the ITRS. In some novel device structures, such as ultrathin body silicon on insulator (SOI) or the FinFET, the channel is very lightly doped. This also helps improve the carrier mobility.

Process Variations The process variation is a very important concern for design. With device scaling, variations in key device parameters such as the threshold voltage V_{th} will increase. Major sources of variations include dopant fluctuation, oxide thickness variation, critical dimension (CD) variation, and line edge roughness, among others. For smaller device dimensions, the total number of dopants in the channel decreases, so the statistical fluctuation increases by $1/N^{1/2}$, where N is the number of dopants. Although the resulting V_{th} variation depends on the doping profile, this general trend is unfavorable for device scaling. Some novel device structures that require very low doping concentration in the channel, such as the double-gate MOSFET, can be employed to solve this problem. But those devices may be sensitive to other sources of variations (e.g., the silicon body thickness). With smaller device dimensions, tolerance for CD and film thickness variations keeps shrinking. More robust design methodologies are therefore necessary in addition to process improvements to achieve better yield and performance.

Novel Device Structures Although bulk-silicon MOSFETs have been demonstrated with a gate length down to 15 nm [64], the ultimate CMOS scaling will probably require novel device structures. In recent years, the FinFET and its variations with multiple gates have been demonstrated successfully with very aggressively scaled dimensions by industrial and academic researchers. Ultrathin-body SOI MOSFETs are also a promising candidate. A feasible solution to the scaling problem must offer significant benefits in the form of improvement of performance or reduced cost per function. This will require integration of the various novel process and device innovations in an economical way.

2.2.2 Quantum Effects Model

In the past, gate oxide thickness was the dominant factor in gate-to-channel capacitance. With the gate oxide thickness scaled to below 20 Å, however, two other factors become important. One is related to the quantum confinement effect of inversion carriers in the channel. The other is the capacitance due to the gate depletion region under the inversion gate bias.

The thermal voltage, which affects the off-state leakage directly, is not scalable. Consequently, the threshold voltages need to be sufficiently large so as to keep the off-state leakage acceptable. Combined with the drive current requirement, this imposes a constraint on the scaling of the power supply voltage. As a result, the vertical electrical field in the channel and the oxide field increase with device scaling. The high vertical field at the substrate–oxide interface forms a potential well, which confines the movement of the inversion carriers along the vertical direction. The carriers become a two-dimensional gas which has discrete eigen-energy levels associated with the vertical movement. In addition, peak distribution of the inversion carriers, which depends on the wave function of all carriers in different energy bands, can be 1 to 2 nm away from the substrate–oxide interface, resulting in extra capacitance under the gate electrode (Figure 2.12). This capacitance is no longer negligible when the gate dielectric EOT is scaled to below 20 Å.

A strict treatment of the inversion charge distribution requires self-consistent solution of the Poisson and Schrödinger equations in the oxide and substrate regions, which can only be done using numerical methods. The inversion carrier distribution depends on the gate voltage and on device parameters such as the channel doping concentration and gate oxide thickness. Based on numerical simulation results and experimental data, an empirical model has been proposed to provide a simple estimation of the inversion charge ac centroid with good accuracy [65]. It was shown that in the case of thin oxide and a typical channel doping level, a universal relationship exists between the ac charge centroid X_{ac}, the gate voltage V_g, the threshold voltage V_{th}, and the gate oxide

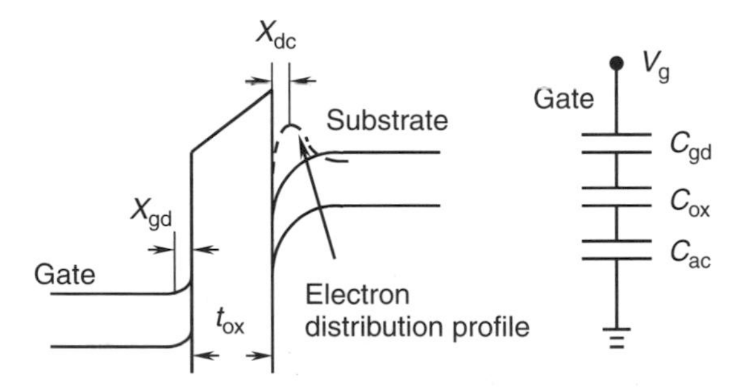

Figure 2.12 *Left:* Band diagram of an n-channel MOSFET with n^+ polysilicon gate biased in inversion. The channel electrons populate the discrete energy states in the potential well near the surface of the substrate. The charge centroid is located at X_{dc} (dc charge centroid) from the surface of the substrate. The polysilicon gate has a depletion width of X_{gd} (discussed in Section 2.2.3). *Right:* Ac equivalent circuit of the gate capacitance stack. C_{gd}, C_{ox}, and C_{ac} are contributed by the gate depletion region, gate oxide, and the ac centroid of the inversion charge, respectively.

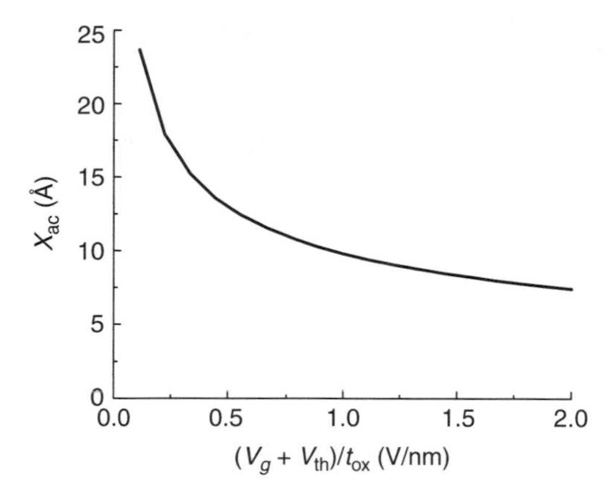

Figure 2.13 A universal relation exists between the ac charge centroid X_{ac} and $(V_g + V_{th})/t_{ox}$.

EOT t_{ox}:

$$X_{ac} = 6.20 \times 10^{-5}\left(\frac{V_g + V_{th}}{t_{ox}}\right)^{-0.4} \tag{2.2.1}$$

where V_g and V_{th} are in volts, and X_{ac} and t_{ox} are in centimeters. Figure 2.13 shows relation (2.2.1). Beyond the 90-nm node, the X_{ac} is roughly in the range 8 to 10 Å, which adds about 3 Å to the inversion capacitance equivalent thickness (CET). For a gate dielectric with about 10 Å EOT, this obviously causes a significant decrease in the gate capacitance.

The model above was developed based on a conventional bulk-silicon MOSFET structure. In the case of some novel device structures, such as the double-gate MOSFET or ultrathin-body SOI MOSFET, the numerical formula may be different, due to the different boundary conditions.

2.2.3 Polysilicon Gate Depletion Effects

When an electric field exists in the gate oxide, the electrostatic boundary conditions necessitate band bending in the polysilicon gate near the oxide interface. The depletion region contributes capacitance between the gate and the channel, with a noticeable effect when the gate oxide is sufficiently thin. Figure 2.12 shows the equivalent circuit of the components contributing to the gate capacitance in the inversion region. The equivalent gate capacitance (i.e., the CET) is related to these components by

$$\begin{aligned}\frac{1}{C_g} &= \frac{1}{C_{gd}} + \frac{1}{C_{ox}} + \frac{1}{C_{ac}}\\ &= \frac{X_{gd}}{\varepsilon_0\varepsilon_{Si}} + \frac{t_{ox}}{\varepsilon_0\varepsilon_{ox}} + \frac{X_{ac}}{\varepsilon_0\varepsilon_{si}}\end{aligned} \tag{2.2.2}$$

where the definitions of the variables follow those given in Figure 2.12. $\varepsilon_0 = 8.85 \times 10^{-12}$ F/m is the dielectric constant of vacuum, and $\varepsilon_{Si} = 11.7$ is the relative dielectric constant of silicon. In equation (2.2.2) an arbitrary gate dielectric with physical thickness t_{ox} and relative dielectric constant ε_{ox} is assumed. The EOT of the gate dielectric is defined by

$$\mathrm{EOT} = t_{ox}\frac{\varepsilon_{SiO_2}}{\varepsilon_{ox}} \tag{2.2.3}$$

where $\varepsilon_{SiO_2} = 3.9$ is the relative dielectric constant of SiO_2. So in terms of the capacitance, a gate dielectric of thickness t_{ox} is equivalent to a SiO_2 film with a thickness given by the EOT value. The quantity used to describe the overall effective gate capacitance is CET:

$$\mathrm{CET} = \frac{\varepsilon_0\varepsilon_{SiO_2}}{C_G} \approx \mathrm{EOT} + \frac{X_{gd} + X_{ac}}{3} \tag{2.2.4}$$

Both X_{gd} and X_{ac} are bias dependent, so the CET also depends on the gate bias V_G. Under typical device operation, the gate depletion and quantum effect each contribute a few angstroms to the CET, so in the case of an ultrathin gate dielectric, the CET can be significantly larger (by percentage) than the EOT, which is determined by the gate dielectric. A thinner CET and higher gate voltage translate to higher inversion charge density and better device characteristics; therefore, with reduced power supply voltage for each technology generation, the gate stack scaling requires not only the thinning of gate dielectric (EOT) but also reduction of the gate depletion effect (X_{gd}).

A simple estimation of the gate depletion effect and the CET can be made using electrostatics. Consider an n^+ polysilicon gated n-MOSFET biased in the inversion region (Figure 2.14); the flat-band voltage (V_{FB}) and the voltage drop

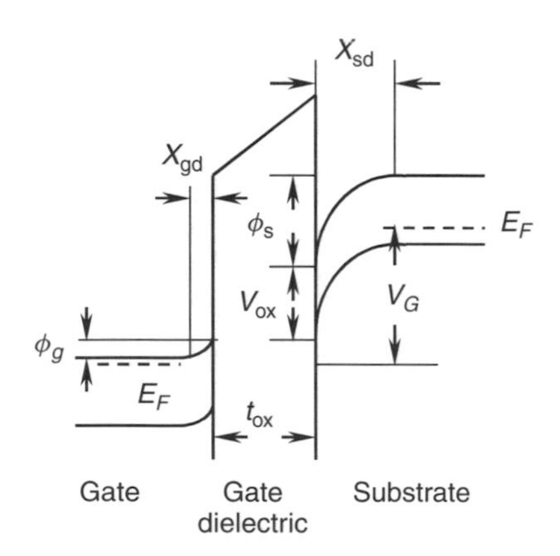

Figure 2.14 Band diagram for an n-MOSFET under positive gate bias. A depletion layer with finite thickness exists in the polysilicon gate at the oxide interface.

in the substrate (φ_s), in the gate (φ_g), and across the gate oxide (V_{ox}) add up to the gate voltage V_g.

$$V_g = V_{FB} + V_{ox} + \varphi_s + \varphi_g \tag{2.2.5}$$

The doping concentration of the gate is usually much higher than that of the channel. So in the subthreshold region, the continuity of the electric displacement across the oxide interfaces implies that the band bending of the gate is much smaller than that of the substrate. Under strong inversion, the boundary condition at the upper interface of the gate oxide is

$$\varepsilon_{ox}\frac{V_{ox}}{t_{ox}} = \varepsilon_{Si}E_{gate} \tag{2.2.6}$$

Using the depletion approximation and assuming a uniform doping concentration in the gate (N_{gate}), the electric field E_z in the gate can be solved using Gauss's law:

$$\frac{dE_z}{dz} = \frac{qN_{gate}}{\varepsilon_0\varepsilon_{Si}} \tag{2.2.7}$$

so the voltage drop on the gate can be solved:

$$V_{ox} = \gamma_G\sqrt{\varphi_g} \tag{2.2.8}$$

where $\gamma_G \equiv \sqrt{2\varepsilon_0\varepsilon_{Si}qN_{gate}}/C_{ox}$ and $C_{ox} = \varepsilon_{ox}\varepsilon_0/t_{ox}$.

Beyond the strong inversion onset, substrate band bending φ_S saturates at roughly $2\varphi_b = 2(k_BT/q)\ln(N_{sub}/n_i)$. Note that

$$V_{th} = V_{FB} + 2\varphi_b + \gamma_S\sqrt{2\varphi_b} \tag{2.2.9}$$

where the definition of γ_S is similar to that of γ_g with N_{gate} replaced by N_{sub}. So equation (2.2.5) becomes

$$\varphi_g + V_{ox} - (V_g + \gamma_S\sqrt{2\varphi_b} - V_{th}) = 0 \tag{2.2.10}$$

The gate depletion width X_{gd} for $V_g > V_{th}$ can be solved from (2.2.6–2.2.10):

$$\begin{aligned} X_{gd} &= \frac{\varepsilon_{Si}\varepsilon_0}{C_{ox}}\left[\sqrt{1 + \frac{2C_{ox}^2}{\varepsilon_{si}\varepsilon_0 qN_{gate}}(V_g - V_{th} + \gamma_S\sqrt{2\varphi_b})} - 1\right] \\ &\approx \frac{C_{ox}(V_g - V_{th} + \gamma_S\sqrt{2\varphi_S})}{qN_{gate}} \end{aligned} \tag{2.2.11}$$

Combining the charge centroid model, the CET can be estimated from equations (2.2.4) and (2.2.11) for given gate voltage, oxide thickness, gate doping concentration, and threshold voltage. For example, assuming that EOT = 12 Å, $V_G = 1.1$ V, and $V_{th} = 0.3$ V, we can obtain the ac charge centroid

$X_{ac} = 9.2$ Å from equation (2.2.1). Further assuming that $N_{sub} = 3 \times 10^{17}$ cm^{-3} and $N_{gate} = 1 \times 10^{20}$ cm^{-3}, we can estimate the gate depletion width $X_{gd} =$ 15.7 Å. Therefore, the quantum effect and the gate depletion effect combined contribute 8.3 Å to the CET, and the gate dielectric accounts for the remaining 12 Å. Apparently, with very thin EOT, further gate dielectric scaling becomes less effective in reducing CET, as polysilicon gate depletion and the quantum effect account for a higher percentage of the CET. Given that the quantum effect cannot be eliminated, it is critical to reduce the polysilicon gate depletion effect by process improvements or by using novel gate electrode materials, such as metallic gate electrodes. The International Technology Roadmap for Semiconductors (ITRS) projects an 8-Å EOT addition from the quantum effect and the gate depletion effect for a few years to come; and a 5-Å addition, which requires the use of metal gates, is needed at the introduction of the 65 nm node in 2007 [66].

2.2.4 Metal Gate Electrodes

The polysilicon gate depletion problem will be a bottleneck for device performance improvements in a couple of generations. As already discussed, the polysilicon gate cannot meet the latest ITRS requirement. The solution is to use metallic gate electrodes. An extra benefit of it is that metal gates generally have lower gate resistance than that of the silicided polysilicon gate, which helps reduce the gate *RC* delay.

Using metal gate electrodes creates many processing and integration challenges. From a device and design point of view, the gate work function is a primary concern. Proper threshold voltages of n- and p-MOSFETs are readily achieved using n^+/p^+ polysilicon gate electrodes with appropriate channel doping. For bulk-silicon CMOS, the same gate work-function requirements are difficult to satisfy using metal gates. In general, metals with high work functions (p^+ silicon-like) are not reactive, therefore difficult to etch, and those with low work functions tend to be too reactive, causing thermal stability problems in contact with the gate dielectric. Using a single metal with a midgap work function on bulk-silicon CMOS results in undesirably high threshold voltages for both n- and p-FETs. It is possible, however, to use one metal on both types of devices with the gate work functions tailored separately to optimize the threshold voltages for both. Currently, an acceptable metal gate solution is still on search. In addition to obtaining the appropriate work functions, a tight control of the work-function distribution is also of great importance to metal gate technology.

The use of high-κ gate dielectrics also affects the precise setting of threshold voltages. Many high-κ dielectrics are known to have high interface trap density and fixed charges. These translate to noticeable V_{th} shifts and larger V_{th} variations. In addition, an important physical effect occurs when metal gates are used with high-κ gate dielectrics. It was first observed experimentally that p-MOSFETs with molybdenum gate and different gate dielectrics exhibited different gate work-function values (Figure 2.15). This was explained by the different screening effects of the interfacial dipole layers of those dielectrics [68]. The

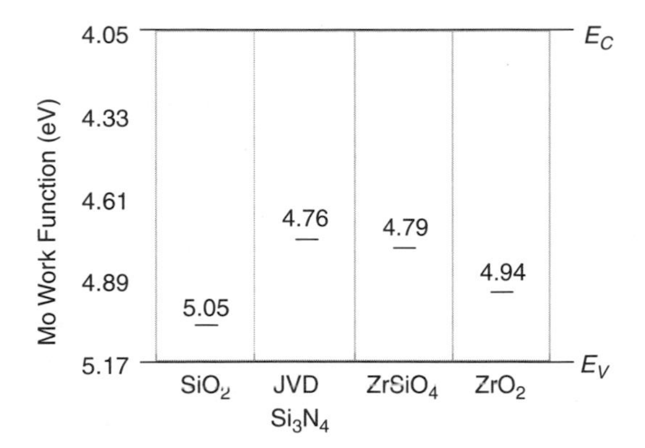

Figure 2.15 The apparent gate work-function values of molybdenum on different gate dielectrics, measured on p-MOSFETs. The work function varies with the underlying gate dielectric and is generally different from the vacuum value. (From Ref. 67.)

polysilicon gate is much less susceptible to this effect because of the negligible density of states in the bandgap. The theoretical model predicts that to achieve n^+/p^+ silicon-like work functions on high-κ gate dielectrics, a significantly larger range of metal work functions is needed, which poses to the candidate metal electrodes an even more stringent requirement.

For devices with very low channel doping concentration, such as the FinFET or ultrathin-body SOI MOSFETs, the gate work functions required for proper V_{th} are closer to the silicon midgap. Therefore, the selection of candidate materials could be slightly easier. It is projected by simulation that for the FinFETs, the gate work functions of ± 0.2 eV from silicon midgap is suitable for p- and n-channel devices [69]. Several techniques could be used to achieve the relatively small work-function range using metal gates, such as the doped nickel-silicide gate, the implantation of nitrogen into molybdenum, and metal intermixing and alloying. A complete solution, however, remains to be found.

2.2.5 Direct-Tunneling Gate Leakage

For ultrathin gate oxide, significant gate leakage currents can exist due to the direct-tunneling process even under low gate voltages. With aggressive device scaling, the gate leakage current has become a more and more serious problem for the power consumption. At the 65-nm technology node, the EOT will be close to 1 nm, and the gate leakage needs to be several orders of magnitude lower than that of 1-nm SiO_2 (requirements differ by applications). Therefore, there has been intense research on high-κ gate dielectrics as the replacement of the SiO_2 gate oxide.

Below 20 Å, the conventional thermal silicon dioxide can no longer serve as an adequate barrier for electrons and holes, thereby causing unacceptably high gate leakage. Unlike the Fowler–Nordheim tunneling mechanism, there is not a simple analytical equation for the direct-tunneling current. In addition, the quantum confinement effect of the channel carriers cannot be ignored in calculation of the tunneling current. Consequently, a strict physical model of the tunneling current involves the numerical routines that solve the Schrödinger and Poisson equations self-consistently and compute the tunneling contributions of the carriers in different quantum states [70]. A relatively simple closed-form model, on the other hand, is very useful for circuit simulation and quick estimation of the gate leakage. Lee et al. proposed a semiempirical direct-tunneling model for SiO_2 [71]. In general, three components of direct tunneling are needed to model the total gate leakage current: electron conduction band (ECB) tunneling, hole valence band (HVB) tunneling, and electron valence band (EVB) tunneling (Figure 2.16). Depending on specific bias conditions, some mechanisms can be negligible. For example, the EVB tunneling is forbidden when the oxide voltage $|V_{ox}|$ is smaller than the silicon bandgap voltage, 1.12 V. Under low V_G (<1.5 V), the dominant contribution to the inversion bias gate leakage comes from the ECB tunneling for n-FETs and the HVB for p-FETs, respectively.

For each tunneling mechanism, the direct-tunneling current is formulated as

$$J = \frac{q^3}{8\pi h \phi_b \varepsilon_{ox}} C(V_G, V_{ox}, t_{ox}, \phi_b) \exp\left\{ -\frac{8\pi\sqrt{2m_{ox}\phi_b^3}}{3hq|E_{ox}|}\left[1 - \left(1 - \frac{V_{ox}q}{\phi_b}\right)^{3/2}\right]\right\} \tag{2.2.12}$$

where the exponential term represents the WKB approximation of the tunneling probability of a carrier at the conduction band edge, and C is basically an empirical shape factor to obtain the correct $J_g - V_g$ curvature at a low gate voltage range:

$$C(V_g, V_{ox}, t_{ox}, \phi_b) = \exp\left[\frac{20}{\phi_b}\left(\frac{|V_{ox}| - \phi_b}{\phi_{b0}} + 1\right)^{\alpha}\left(1 - \frac{|V_{ox}|}{\phi_b}\right)\right]\frac{V_g}{t_{ox}}N \tag{2.2.13}$$

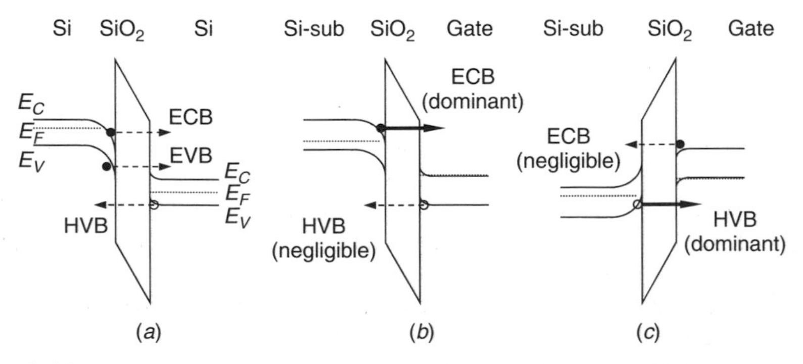

Figure 2.16 (*a*) In general, three components are needed to model direct gate tunneling current through thin gate oxides. Under low gate voltage, the dominant contribution is (*b*) ECB tunneling for n-FETs and (*c*) HVB tunneling for p-FETs. (From Refs. 71 and 72.)

N in equation (2.2.13) is related to the density of tunneling carriers. It has the following general form for the inversion and accumulation region:

$$N = \frac{\varepsilon_{ox}}{t_{ox}} \left\{ S \ln \left[1 + \exp \left(\frac{V_{ge} - V_{th}}{S} \right) \right] + v_t \ln \left[1 + \exp \left(- \frac{V_g - V_{fb}}{v_t} \right) \right] \right\} \tag{2.2.14}$$

In equation (2.2.14), S is the subthreshold swing, v_t the thermal voltage, and V_{th} the threshold voltage of the MOSFET.

The fitting parameters in this model are the effect mass m_{ox}, the barrier heights ϕ_b and ϕ_{b0}, and α. By using appropriate parameters, the model can fit the tunneling currents for different gate bias and gate and substrate doping types. The model parameters for each of the aforementioned tunneling components are different. Figure 2.17 illustrates the accuracy of this model. In the figure the solid lines are results by numerical simulation [70], the dashed lines by this closed-form model, and scattered symbols are the data measured on real devices. Good agreements among the three can be seen. In addition to SiO_2, the model above was also demonstrated to apply to single-layer non-SiO_2 gate dielectrics [72]. It was shown that direct-tunneling leakage through jet vapor–deposited (JVD) silicon nitride can also be very well fitted using this model with a single set of fitting parameters. Table 2.1 summarizes the parameters for the major tunneling mechanisms for n- and p-FETs with SiO_2 and JVD silicon nitride gate dielectrics. The barrier heights can be obtained independently by other methods, so essentially only the effective mass and α need to be determined from experimental gate leakage data.

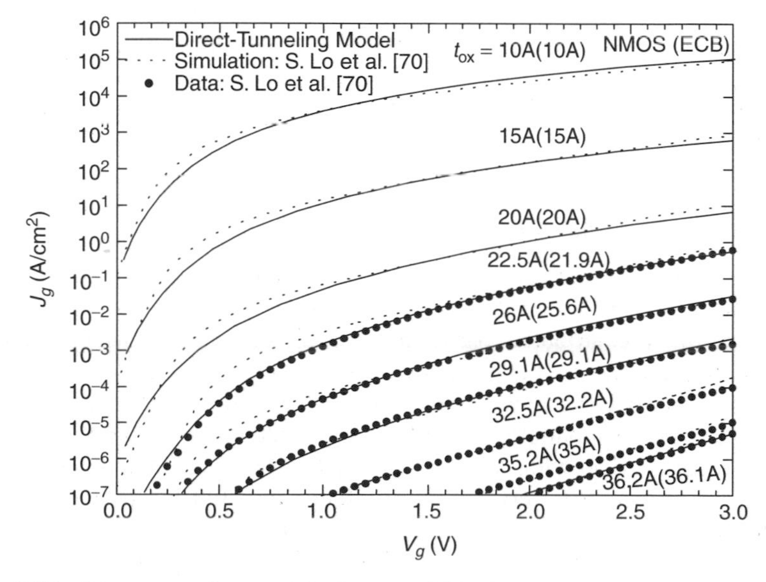

Figure 2.17 Direct-tunneling model of Lee and Hu [71] agrees well with experimental data and numerical simulations.

TABLE 2.1 Direct-Tunneling Model Parameters for SiO_2 and JVD Silicon Nitride Gate Dielectrics

	SiO_2		JVD Silicon Nitride	
	ECB	HVB	ECB	HVB
ϕ_b	3.10	4.50	2.10	1.90
ϕ_{b0}	3.10	4.50	2.10	1.90
m_{ox}	0.40	0.32	0.50	0.41
α	0.6	0.4	0.4	1.0

Source: Ref. 72.

The concepts of this model have been incorporated into the BISM4 gate leakage model, with modifications for more efficient numerical implementations [73].

It should be noted that in the case of some high-κ gate dielectrics there exists a thin interfacial layer between the high-κ gate dielectrics and the silicon substrate, which may either be intentionally introduced for better film quality or may result from undesirable chemical reactions between the high-κ layer and the substrate. So the gate insulator actually consists of two or even more layers of different dielectrics. Based on a single-layer concept, the model above is not directly applicable to a multilayer dielectric structure. A recent work extended the single-layer tunneling model in BSIM4 to the multilayer case [74]. It was shown that with appropriately constructed tunneling probability through each of the multiple layers, which is expressed in the tunneling parameters in a BSIM model, the BSIM direct-tunneling model also fits well the experimental data of multilayer gate dielectrics. Currently, there is a very limited amount of experimental gate leakage data for multiple-layer tunneling with an accurately known layer structure. Further refinement of the models will be needed as high-κ gate dielectric process technologies are improved.

2.2.6 Parasitic Capacitance

The gate-to-source/drain capacitance is an important factor affecting the CMOS device and circuit performance. The coupling between the gate and the source–drain is contributed partly by the direct overlap, which cannot be scaled in proportion with the general technology scaling. Therefore, the parasitic capacitance can become more serious for devices with shorter gate length. The spacer dielectric material affects the fringing capacitance. In addition, replacing the SiO_2 gate oxide with high-κ gate dielectrics also affects this parasitic component. Although a very accurate calculation of the capacitance requires two-dimensional numerical methods, a simple physical model can be used to understand those mechanisms [75]. As shown in Figure 2.18, the gate-to-drain (or source) capacitance consists of three coupling paths, the two fringing components and the gate-to-drain direct overlap, as indicated by the three arrows. The fringing

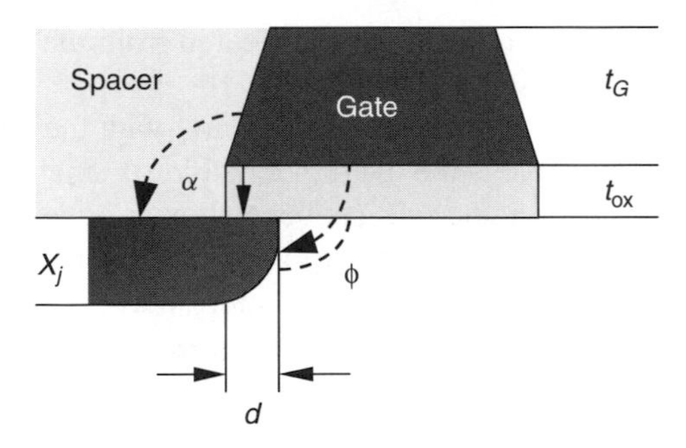

Figure 2.18 Simple illustration of the coupling paths between the gate and the drain of a MOSFET. The three coupling paths are indicated by the arrows. The device width W is perpendicular to the page. (From Ref. 75.)

capacitance through the spacer region is

$$C_1 = \frac{\varepsilon_{\text{SPCR}} W}{\alpha} \ln\left(1 + \frac{t_G}{t_{\text{ox}}}\right) \tag{2.2.15}$$

W is the width of the device, t_G and t_{ox} are the physical thicknesses of the gate dielectric and the gate, α is the angle as shown in Figure 2.18, and $\varepsilon_{\text{SPCR}}$ is the permittivity of the spacer dielectric.

The direct overlap capacitance is

$$C_2 = \frac{\varepsilon_{\text{ox}} W(d + \Delta)}{t_{\text{ox}}} \tag{2.2.16}$$

where d is the overlap length and Δ is zero when α is $\pi/2$. The coupling capacitance through the side of the junction is

$$C_3 = \frac{\varepsilon_{\text{ox}} W}{\beta} \ln\left(1 + \frac{x_j \sin\beta}{t_{\text{ox}}}\right) \tag{2.2.17}$$

with $\beta = \pi\varepsilon_{\text{ox}}/2\varepsilon_{\text{Si}}$. It should be noted that this coupling path is screened when there are free carriers under the gate, so this component does not exist in strong inversion. In the equations above, all permittivity values are in F·cm.

C_1 has a linear dependence on the device width and the permittivity of the spacer dielectric and a weaker dependence on the thickness ratio between the gate and the gate oxide. Assuming that $\alpha = \pi/2$, $\varepsilon_{\text{SPCR}} = 3.9\varepsilon_0$ (SiO_2), $t_{\text{ox}} = 12$ Å, and $x_g = 1000$ Å, we obtain $C_1 = 0.097$ fF/μm. Thicker physical gate dielectric can reduce C_1 slightly. If a high-κ gate dielectric is used such that $t_{\text{ox}} = 48$ Å (EOT remains 12 Å) and everything else is the same as in the example above, C_1 is reduced to 0.068 fF/μm. The key is to ensure that the spacer dielectric has low permittivity. If a nitride spacer ($\varepsilon_{\text{SiN}} \approx 2\varepsilon_{\text{SiO}_2}$) is used instead of SiO_2, the fringing capacitance increases roughly by a factor of 2. If during gate stack etching the high-κ gate dielectric cannot be cleared from the bottom of the spacer region, C_1 may also increase visibly.

The direct overlap capacitance C_2 scales linearly with the gate–channel capacitance C_{ox}. The main variable is the overlap length d. A smaller d is preferable, as it reduces the overlap capacitance. It has been shown, however, that a minimum overlap length of 15 to 20 nm is necessary to avoid the drive current degradation in short-channel MOSFETs [76]. Overly small source–drain extension (SDE) depth or SDE-to-gate overlap add high extrinsic resistance to the channel, causing the drain current to degrade. Figure 2.19 shows the trade-off between the reduction in the Miller capacitance ($C_M = C_1 + C_2$) and the improvement in the drive current $I_{d\ sat}$. To avoid the drive current degradation, the minimum overlap requirement implies that with device scaling, the overlap capacitance will be larger relative to the gate capacitance C_{ox}. Assuming that the gate dielectric EOT = 12 Å, a minimum overlap of 15 nm translates to 0.43 fF/μm, while for a channel length of 60 nm, the gate–channel capacitance is 1.73 fF/μm.

Another source of parasitic capacitance is the source and drain junction capacitance. A depletion region exists at the p-n junctions. When the junction bias voltage varies, the depletion width varies in response, contributing to the capacitive load. In MOSFETs, these junctions are typically much more heavily doped on one side, so the depletion width is determined primarily by the more lightly doped side:

$$C_j = \frac{\varepsilon_{Si}\varepsilon_0}{X_D} \approx \sqrt{\frac{\varepsilon_{Si}\varepsilon_0 q N_{sub}}{2(\varphi_{bi} + V_R)}} \tag{2.2.18}$$

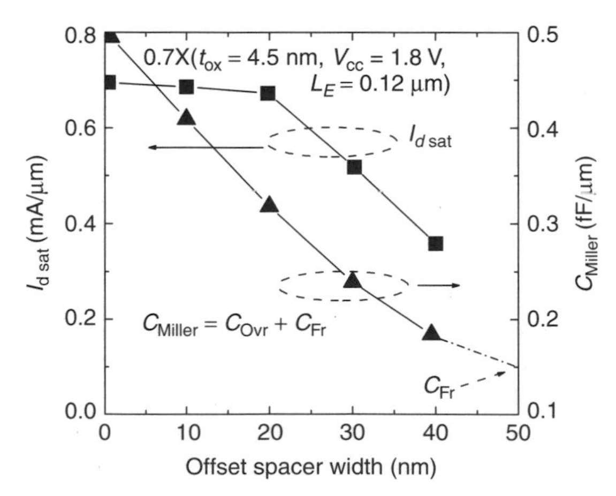

Figure 2.19 Effects of the offset spacer thickness on the Miller capacitance (overlap capacitance plus fringing capacitance) and the saturation drain current for a given process technology [76]. Experimentally, the width of the offset spacer, which was used to control the spacing between the SDE implant and the gate edge, was varied from 0 to 40 nm to modulate the SDE-to-gate overlap from approximately 40 to 0 nm. The degradation of the drain current occurs when the overlap length is below a minimum of 15 to 20 nm.

where X_D is the depletion region width, N_{sub} the substrate doping concentration, V_R the reverse bias voltage across the junction, and ϕ_{bi} the built-in potential of the p-n junction. Equation (2.2.18) gives the junction capacitance per unit area. To minimize the total junction capacitance, it is important to keep the substrate doping concentration under the junction region as low as possible. In addition, the junction areas should be minimized in the circuit layout.

The use of high-κ gate dielectrics may add one more parasitic coupling path that is not important in the case of SiO_2. As illustrated in Figure 2.20(*b*) when the gate dielectric has a very high permittivity, its physical thickness is much larger than needed for SiO_2 to achieve a given EOT. The fringing fields from the source and the drain affect the channel through the sides of the thick gate dielectric. The additional coupling degrades the short-channel characteristics of the devices. Simulations were performed to compare the V_{th} roll-off characteristics among the three cases shown in Figure 2.20. For the same EOT, the gate dielectric is set to be SiO_2 only (*a*), high-κ dielectric only (*b*), and a high-κ dielectric on top of a thin interfacial layer (*c*) [77]. Assuming the same device parameters, the EOT of 1.5 nm is achieved by using either 1.5-nm SiO_2 or a dielectric with $\kappa = 200$, or by using a dielectric with $\kappa = 200$ and EOT $= 0.5$ nm on top of 1.0-nm SiO_2. The threshold voltages as a function of the gate length are shown in Figure 2.21. It can be seen that the V_{th} roll-off characteristic is significantly degraded when a single layer of dielectric with a very high κ value is used. Having a thin interfacial layer with low κ value can effectively avoid the undesirable behavior. The simulations also found that the same physical t_{ox}-to-gate length ratio leads to the same amount of short-channel performance degradation even with different EOT. It should be noted that currently the most hopeful high-κ dielectrics have much lower κ values than that assumed in the simulation. Therefore, this effect will not be significant until the gate length becomes smaller than 40 to 50 nm, where the t_{ox}/L_G ratio becomes sufficiently large. Essentially, from a short-channel effect point of view, a very high κ value is not desirable even if it is available.

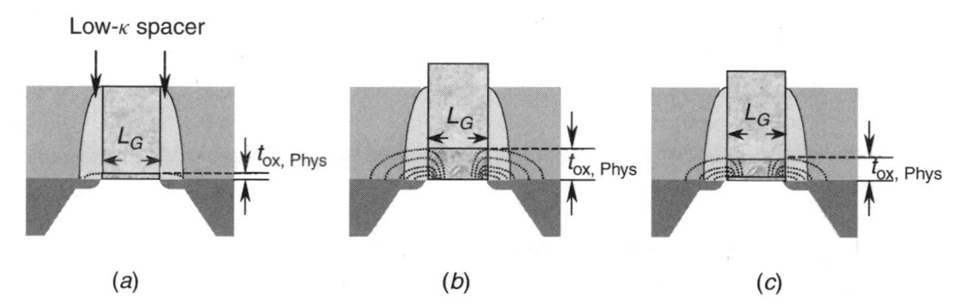

Figure 2.20 For a given EOT value, the single-layer high-κ gate dielectric (*b*) has a larger physical thickness than the SiO_2 (*a*). A more realistic structure of the gate dielectrics is a high-κ layer on top of a thin low-κ (e.g., SiO_2) interfacial layer (*c*). The fringing field between the source–drain and the channel is negligible for ultrathin SiO_2 but is enhanced by the use of high-κ gate dielectrics.

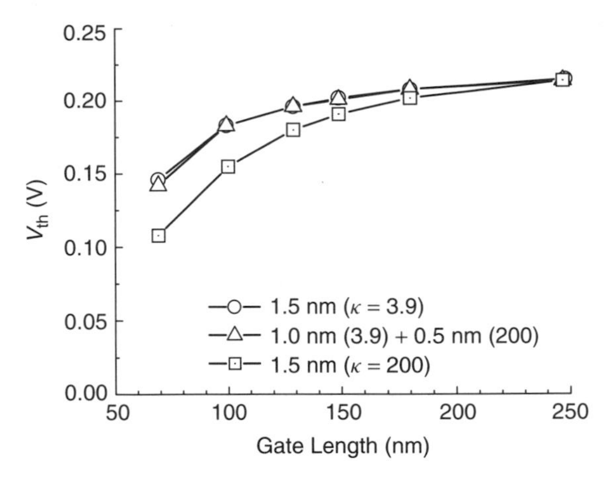

Figure 2.21 V_{th} roll-off characteristics of the three gate stacks shown in Figure 2.20, all having an EOT of 1.5 nm. Clear degradation is seen for a single-layer gate dielectric with very high κ value due to the enhanced drain-to-channel coupling through the side of the gate dielectric. Using a thin low-κ interfacial layer is effective to avoid the degradation. (From Ref. 77.)

From a device point of view, the only benefit of using high-κ gate dielectrics is the reduction of direct-tunneling gate leakage. Currently, there are a number of serious device and processing problems for high-κ dielectrics, such as the mobility degradation, flat-band voltage shift, and so on. Regardless of the availability of a manufacturable high-κ solution, the power consumption problem will have to be tackled on multiple fronts, from device and process to circuit and system design.

2.2.7 Reliability Concerns

The device reliability is a critical aspect of the CMOS technology. The smaller device dimensions and thc introduction of new materials, such as copper/low-κ dielectric, high-κ gate dielectrics, and metal gate electrodes, have a significant impact on CMOS device reliability. The increased scale of integration poses more stringent reliability requirements for each element in the circuit. The cost and time of testing complicated low-failure rate circuits also poses serious challenges. Below is a short overview of some major front-end-related CMOS reliability issues.

Time-Dependent Dielectric Breakdown The mechanism of the time-dependent dielectric breakdown (TDDB) of SiO_2 has been a very active research area for decades, and a unanimously accepted conclusion has yet to be reached. With operating voltage and thickness scaling, new failure modes such as soft breakdown also become important. In recent years, one major topic has been the correct procedures that can be used to predict low-voltage-operation lifetime based on high-voltage and high-temperature accelerated tests. A related question is: What is the reliability limit for the scaling of the SiO_2 gate dielectric? Some researchers believe that with improved microscopic and macroscopic gate oxide uniformity, the SiO_2 gate dielectric can be scaled to below 20 Å, which agrees with recent manufacturing experiences. In a study by Weir et al. it was demonstrated that it is possible to operate 16-Å oxides at 1.2 V and 70°C [78]. Based on present knowledge and the ITRS gate leakage requirements, scaling of the SiO_2 is probably limited by direct-tunneling leakage rather than TDDB reliability.

The TDDB reliability of high-κ gate dielectrics is much less well studied than is SiO_2. In general, many high-κ dielectrics have significantly smaller bandgaps and higher interface and bulk trap densities, which have adverse effects on their TDDB reliability. There have been studies on the TDDB reliability of some hafnium-based gate dielectrics, which are regarded as one of the most promising candidates. Some techniques, such as that of nitridation of the substrate/high-κ interface, were shown to improve the breakdown voltage of the HfO_2 gate dielectrics [79]. Other undesirable effects of the high-κ gate dielectrics, however, such as mobility degradation, are currently among the most important factors affecting its adoption.

Hot-Carrier Effects During normal operation, the high lateral electric field in the MOSFETs can generate hot carriers near the drain side which damage the gate dielectric and cause degradation of device characteristics such as g_m degradation and V_{th} shift. As the channel length is reduced, the peak electric field in the channel will increase for a given power supply voltage. The scaling of supply voltage can mitigate this problem. The doping profiles can also be adjusted to improve hot-carrier reliability. The power supply voltage in state-of-the-art CMOS technologies has been scaled to close to 1 V; therefore, the energy gain of the hot carriers is lower than the minimum energies required for thermal emission and impact ionization. However, no abrupt change in the hot-carrier degradation was observed with the power supply voltage reduced to below those critical energy levels [80]. Alternative hot-carrier degradation mechanisms have been proposed to explain the foregoing experimental observation. The practical implication is that hot-carrier reliability will remain an issue that requires appropriate attention, although it is not expected to be an obstacle to technology scaling.

Negative Bias Temperature Instability Negative bias temperature instability (NBTI) is a significant source of threshold voltage shift in p-MOSFETs. Under NBTI stress, the source, drain, and substrate are grounded, and the gate is biased at a negative stress voltage. At high temperatures such a bias enhances positive charge generation in the gate oxide as well as interface state generation. Consequently, g_m degradation and V_{th} shift occur. In some analog or mixed signal applications where extremely stable threshold voltages are required, the NBTI can be a limiting factor to circuit lifetime. The impact of NBTI on state-of the-art digital circuits was reported in a study by Reddy et al. [81], where it was demonstrated that the dominant degradation mode for static CMOS operation was the p-MOSFET NBTI. The relative frequency degradation of NBTI-stressed

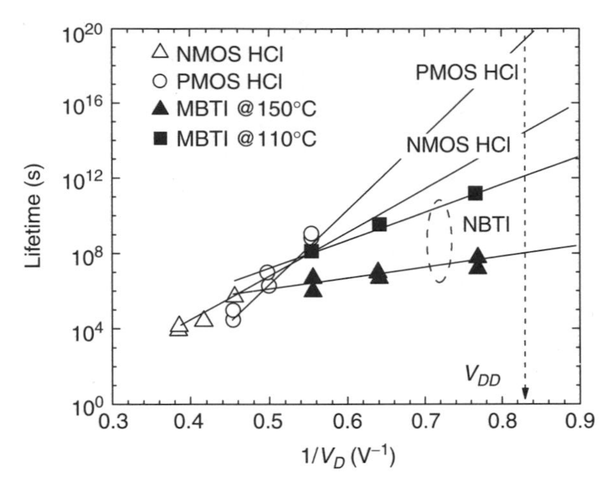

Figure 2.22 Hot-carrier stress lifetime and PMOS NBTI stress lifetime of a 0.13-μm CMOS technology. The projected HCI stress lifetimes of both N- and PMOS under normal V_{dd} values are much longer than that of the PMOS under NBTI stress. (From Ref. 82.)

ring oscillators increases as operation frequency decreases. Also, NBTI stress was found to result in SRAM static noise margin degradation. NBTI is expected to have an increasingly important impact on circuits with technology scaling.

Another recent study showed that with ultrathin nitrided gate oxide, the p-MOSFET lifetime as defined by a certain amount of V_{th} shift could be limited by NBTI. In advanced CMOS technologies, nitrided gate oxide is used widely to suppress boron penetration. Nitridation, however, has been observed to reduce the NBTI lifetime of p-MOSFETs. In a study by Jeon et al. [82], MOSFETs were fabricated using a 0.13-μm CMOS technology with 13-Å nitrided gate oxide. Figure 2.22 shows the lifetimes of p-MOS under NBTI stress, hot-carrier injection (HCI) stress, and NMOS under HCI stress, respectively. It can be seen that the lifetime of p-MOS under NBTI stress is lower than that under HCI stress by several orders of magnitude, thereby becoming the limiting factor in circuit lifetime. Raising the stress temperature from 110°C to 150°C leads to an even lower NBTI lifetime. Very limited data are available for some candidate high-κ gate dielectrics, and comprehensive studies on circuit level are lacking. It can be seen that NBTI reliability will be a very important concern for device technology and circuit design in current and future CMOS technologies.

2.3 BACK-END-OF-LINE TECHNOLOGY

Prior to development of the 350-nm node, advances in interconnect technology have played a key role in the continuous improvement of integrated-circuit density, performance, and cost per function. Driven by performance considerations, the continuous evolution of interconnect scaling has led to taller and narrower wires, decreased line spacing, lower metal resistance per unit length, and faster circuit operating frequency. Multiple-layer copper technology and low-κ dielectrics offer the advantages of faster circuit speed, reduced signal interference, and better reliability over previous aluminum wiring processes. However, back-end-of-line (BEOL) concerns remain and are manifested in the critical issues of electromigration, thermal properties, process control, multilayer integration, and most important, high-speed global signaling in future system-on-chip (SoC) design. In the nanometer regime, wiring delay accounts for most of the overall delay. As benchmarked in Figure 2.23(*a*), wiring delay can exceed gate delay at the 130-nm node and below, even with a copper backend process, and it will account for approximately 75% of the overall delay at the 90-nm node [1]. Moreover, since die size increases with technology due to increasing chip functionality, it becomes more difficult to deliver signals across the chip in one cycle, as shown in Figure 2.23(*b*) [83]. To scale the frequency, global signals need to be pipelined, resulting in higher latency and power consumption when routing signals across function blocks. Because of these scaling effects, the emphasis of overall circuit performance optimization needs to be shifted from gate-level logic to wire-centric design.

2.3.1 Interconnect Scaling

In contrast to FEOL scaling, which inherently enhances circuit performance, reduction in the interconnect size yields larger delays, due to rapid shrinking of the cross-sectional area of the wire. To combat the rising *RC* delay, various advances have been developed from both geometric structure and materials perspective. As shown in Figure 2.24, characteristics of a typical multiple-level structure include:

- *High aspect ratio (defined as the ratio of line thickness to line width).* Since the metal-to-ground capacitance (C_g) is approximately proportional to the

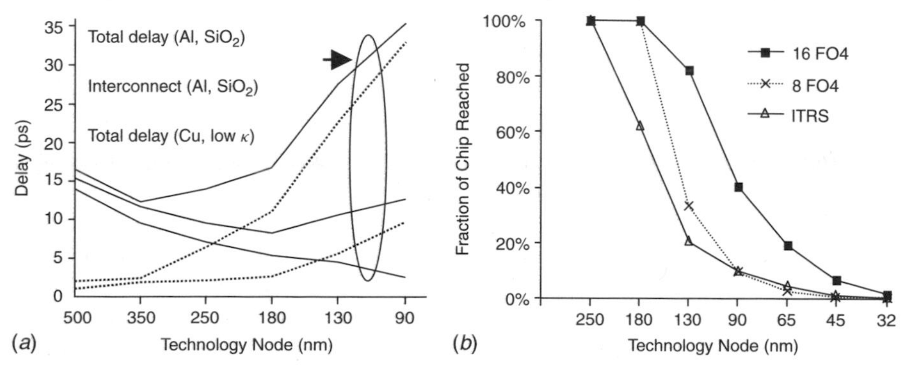

Figure 2.23 Impact of technology scaling on chip performance: wire delay dominated. (*a*) Wire and gate delay in aluminum and copper. (*b*) Fraction of reachable area in one clock cycle.

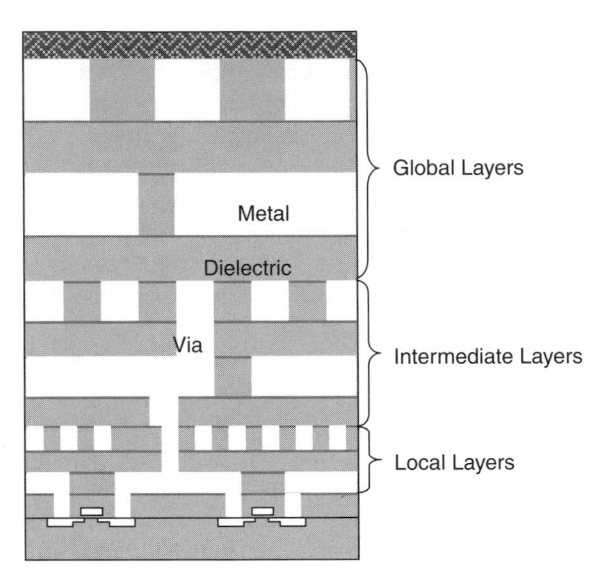

Figure 2.24 Representative cross section of the multiple-level interconnect structure at the 90-nm node.

line width and the metal resistance (R) scales with the cross-sectional area (the product of line width and line thickness), a larger aspect ratio leads to a smaller RC delay when the line pitch decreases. This trend is especially favorable in the case of global interconnects, while a relatively lower aspect ratio is acceptable for local interconnects, whose signal path delay is less sensitive to the increase of RC, due to their short wire length. It should be noted that there are two problems related to high aspect ratio. First, it is difficult to fill a deep and narrow trench completely with metal. As a result, the manufacturing of lines and vias with aspect ratio larger than 4 becomes unreliable, especially for a dual damascene process. Second, with a higher aspect ratio, the increase in line thickness results in a larger coupling capacitance (C_c) to neighboring lines, which increases both the RC delay component and the signal coupling noise. These two undesirable effects limit the practicality of this scaling technique for future technology generations.

- *Hierarchical wiring.* By shrinking local wire pitch and adding successive wire layers with increasing thickness and width, multiple-layer metallization satisfies the need for higher circuit density, reduced RC delay, and smaller resistive loss [83]. The available number of metal wiring layers has tripled over the last 10 years [84] such that current technologies (about 90 to 180 nm) have six to 10 levels, as shown in Figure 2.24. Local and intermediate wires are generally used to connect gates within a functional unit, while global wires serve to connect separate functional units. Besides signal transmission (e.g., databus), the large global wires are also used for clock and power supply distribution, or on-chip implementation of spiral inductors. Metal lines on the same layer are usually routed in parallel, whereas those orthogonal to the lines are routed on the neighboring layers. The recent introduction of diagonal interconnects for global wires can achieve a 19.8% reduction in path delay and a 29.3% reduction in wire length compared to conventional orthogonal interconnect architectures [85]. During the past, however, this technology has posed major challenges to the size of database, optical proximity correction, and mask writing time. Solutions to these problems are keys to enabling future technologies; electronic design automation EDA companies and mask-making houses are currently working toward realizing these solutions.

- *Different approaches to the scaling of local and global interconnect.* In the case of local wires, the pitches and length scales are reduced at a much faster rate than are the vertical dimensions, in order to match the MOSFET density on the substrate and maintain the *RC* delay. For global wires the scaling is determined by the length of the chip edge, which increases as gate dimension scales. As a result, signal delay on global wires increases continuously from one technology generation to the next, limiting overall chip performance enhancement (Figure 2.23). To support frequency scaling for improved performance of high-performance microprocessors, compensation techniques such as adding repeaters and pipelining global signals are required.
- *Copper wiring and low-κ dielectrics.* Since interconnect scaling acts as a significant bottleneck to circuit speed advancement, substantial efforts have been put forth to integrate new BEOL materials into the silicon process. Two major advances include the transition from aluminum to copper as the metal material and the introduction of low-κ dielectrics to replace silicon oxide (SiO_2). These approaches not only effectively reduce interconnect *RC* parasitics per unit length but also benefit BEOL reliability in electromigration. On the other hand, the most difficult challenges for interconnect include process integration, line dimension control, and electrical reliability. As the state of the art enters the nanometer scale, even low-κ copper will be insufficient to enable faster global signaling in the near future [85]. New BEOL technology, including both materials and structures, are necessary to continue the success.

2.3.2 Copper Wire Technology

The fabrication of copper interconnect using a dual-damascene process was introduced by IBM in 1997 [86]. The primary performance advantage of this technology is that the resistivity of copper ($1.8\ \mu\Omega \cdot \text{cm}$, or effectively, $2.2\ \mu\Omega \cdot \text{cm}$ postprocess) is approximately 40% lower than that of aluminum ($3.3\ \mu\Omega \cdot \text{cm}$), so that copper wires exhibit approximately 30 to 40% lower *RC* delay than aluminum wires of the same cross section. Additional benefits include improved electromigration reliability due to the heavier copper atomic mass, and cost reduction due to the simplification of some process steps: Since it is ineffective to pattern copper with the dry etching techniques used in the aluminum process, a chemical–mechanical planarization (CMP)–based damascene process has been developed as the alternative. Overall, with the use of low-κ dielectrics (κ < 3), the transition from Al/SiO_2 to copper/low-κ enables higher wire density, a lowered number of metal layers, faster signal propagation, and larger allowed current density, satisfying the high-performance and high-density requirements for contemporary very large scale integrated systems.

In addition to the advantages achieved by copper technology, there are still serious challenges to be solved from both material and process perspective for successful nanometer-scale integration. From the standpoint of material, there are two major issues: metal resistivity and electromigration. As the cross-sectional dimension of a metal wire scales down to sub-100 nm, electron transportation in copper faces several fundamental physical limitations. Figure 2.25 illustrates the theoretical predictions of copper line resistivity as a function of line width [87]. When line width approaches copper grain size, morphological imperfections (i.e., surface roughness scattering) and grain boundary scattering significantly increase copper resistivity by more than 100% in the sub-100-nm regime (note that a temperature increase of 10°C raises the resistivity by only about 3.6%). These effects can be mitigated by tightly controlling copper growth or incorporating single-crystal copper. However, in the sub-100-nm regime the copper is encased within a narrow groove which limits the grain size and hence the resistivity even after recrystallization. Furthermore, as line dimensions reach the mean free path for electron scattering in copper (39.3 nm) at the 45-nm technology node, metal surface scattering eventually hinders electronic conduction. The dramatic increase

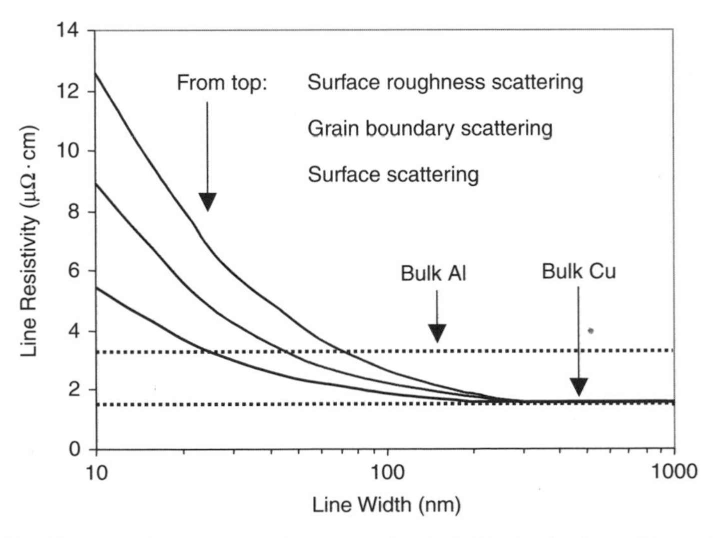

Figure 2.25 Increase in copper resistivity under individual physical effects. The bulk resistivity of copper and aluminum are provided as references [88].

in copper resistivity under these circumstances significantly increases both *RC* signal delay and *IR* drop on power supply.

The other important material property to consider for on-chip metal wire is its immunity to electromigration. When a large current density is driven though the line, metal ions can be carried down the wire by the electrical current, thus accumulating a large concentration downstream or leaving unwanted voids in the line. Due to the heavier atomic mass of copper (63) over aluminum (27), copper line is inherently more robust to these undesirable electromigration effects. Depending on test conditions and metal levels, critical current density to failure, J_{crit}, for copper wire is twofold greater than that for aluminum wire [8]. Experimental data show further that copper has an electromigration lifetime 100-fold longer than aluminum at the same current density [86]. In addition to a higher J_{crit} value, the higher melting temperature of copper (1034°C compared to 660°C for aluminum) also benefits the reliability of electrostatic discharge (ESD). On the other hand, with the continuous evolution of MOSFET density and circuit performance, on-chip interconnect must deliver an increasing level of switching current through a narrower wire. As a result, wire immunity to electromigration is crucial to robust chip operation, and the current copper process still needs improvement, especially in the control of metal surface, grain size, and impurities [89][90].

Copper wires are usually integrated into a multilevel back-end structure using the damascene process, which involves line patterning, trench metal filling, and metal planarization using CMP. Whereas advanced technology demands precise lithography for smaller feature definitions at both the front and back ends of the line (FEOL and BEOL), there are unique requirements for trench filling and CMP control for copper interconnects. For example, during the damascene process, the trench is first etched in the insulator, then filled uniformly with a metallic conductor through electroplating. A similar process is also applied to form vias. A significant problem is the quality of the copper deposition, which should avoid causing pinch-off and creating voids, two unwanted phenomena that result in larger resistance and degraded reliability. Although IBM has demonstrated void-free filling of a 0.2-μm, 4:1 aspect ratio line, the rapid scaling of line width and increase in aspect ratio require a more robust plating process [86,91].

After the copper deposition step is completed, the surface of metal and dielectrics is polished using a chemical–mechanical technique. Although the CMP process has a high copper removal rate, such polishing is the origin of metal erosion and dishing due to the difference in material properties of metal and dielectrics. Figure 2.26 shows the cross-sectional SE micrograph of a 2-μm copper line after the CMP step; physical models have been proposed to describe the effects of metal erosion and dishing [92,93]. Since copper is softer than dielectrics, it is more sensitive to the chemical slurry, and hence its polishing rate is faster. As a result, the metal thickness is lower than expected (erosion effect) and its surface exhibits a hemispherical shape after polishing (dishing effect). A region with higher metal density suffers a greater loss due to erosion, but its effect can be effectively reduced by filling with dummy metal to control the

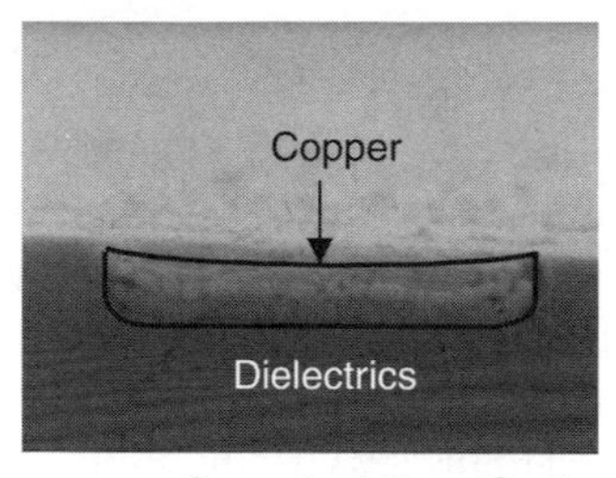

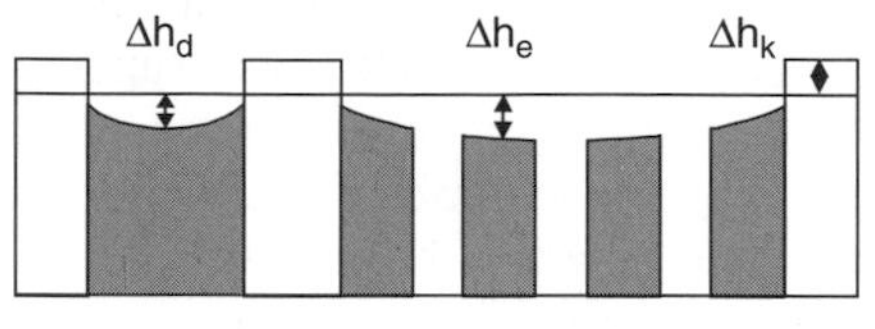

Figure 2.26 SEM cross-sectional picture and model of metal dishing and erosion effects after the CMP process.

uniformity of the metal density. However, grounded metal fill increases the total parasitic capacitance, whereas floating metal fill increases capacitive coupling between lines [92]. The dishing effect is dependent on the width of the line; dishing is more severe for wider lines. To reduce the dishing effect, wide lines are either slotted or perforated with holes (cheesing) to increase its effective rigidity. Usually, these holes are aligned with the direction of current flow to minimize current crowding, which leads to better resistivity when slotting or "cheesing" is applied. This is especially favorable for wide clock lines and power distribution networks. Alternatively, the metal dishing effect can be mitigated by splitting a wide line into several narrower ones. However, this technique consumes more area for the same effective resistance or electromigration limit. Furthermore, the slotting and cheesing steps require postprocessing of the database, and the design rules that guide the process must account for the resulting degradation of electromigration limits and resistivity of the line. Considering the overall area cost and signal delay, studies have shown that the optimal range of wire splitting is two to four times for a typical global interconnect (i.e., the optimal line width after splitting is about 3 to 5 μm) [93]. Note that these approaches are not used solely for process control but have design implications as well, especially for high-frequency signals where the skin depth is less than the wire width.

2.3.3 Low-κ Dielectric Challenges

Stimulated by the demand to reduce wire parasitic capacitance, low-dielectric-constant (κ) insulators were introduced into the multilevel interconnect structure at the 130-nm technology node, so that both *RC* delay and signal coupling for a given metal separation could be decreased. Although SiO_2 as an insulator has many natural advantages in silicon technology, its high relative dielectric constant of 3.9 results in higher power consumption and *RC* delay of global lines. Furthermore, this higher *RC* delay requires more repeaters for global lines in order to meet timing constraints, further leading to larger area cost and power consumption. Advanced low-κ materials are needed to mitigate these problems and therefore support the design of multigigahertz microprocessors. Substantial efforts have been devoted to the search for low-κ dielectrics, such as FSG ($\kappa = 3.5$),

SiOC ($\kappa = 2.9$), SiLK ($\kappa = 2.7$), parylene ($\kappa = 2.3$), Dendriglass ($\kappa = 1.5$), and xerogel ($\kappa = 1.3$). The ultimate goal of low-κ is to find a material whose dialectric constant approaches that of air ($\kappa = 1.0$). Until recently, only minimum-spaced wires were demonstrated to have air gaps, which are really intentional voids in the dielectric. These voids were a nuisance in the past process but have since been harnessed as a means to reduce the effective dielectric constant.

The introduction of low-κ dielectric poses the most difficult challenge to process integration; trade-offs have to be made between dielectric constant and reliability. Since low values of dielectric constant are found primarily among porous insulators (either polymers or silicon compounds), low-κ dielectrics usually have different mechanical and thermal properties than metal lines. Due to the nature of their porosity, it is difficult to control the etch rate of low-κ materials, and that leads to poor sidewall coverage by the liner [94]. In addition, low thermal conductivity of porous low-κ dielectrics causes significant Joule heating effects, introduces thermal and mechanical mismatch stress at the interface between metal copper and insulators [95], and reduces the effective electromigration value of the wire. These problems dramatically degrade the quality of interfacial adhesion and increase the susceptibility of copper penetration into the dielectrics. Consequently, the average lifetime of copper/low-κ was found to be much shorter than that of copper/oxide at test temperatures [96]. In addition to the issue of mechanical reliability, electrical reliability, especially time-dependent dielectric breakdown (TDDB), recently emerged as a significant issue when the separation between metals approaches 100 nm. Although TDDB performance degrades with the reduction in dielectric constant, the failure mechanism is independent of the level of porosity and thermal stress; instead, it is related primarily to the drift of copper ion under an electrical field [97,98].

In summary, although implementation of low-κ dielectric improves circuit performance, the roadblock of thermal and electrical reliability needs to be solved to successfully integrate by 2007 materials whose dielectric constant $\kappa = 2.1$ [85].

2.3.4 Future Global Interconnect Technology

The essential purpose of an interconnect is to provide "communication between distant points with minimal delay" [99]. While transistor scaling continues to achieve smaller feature size and faster operation in the nanometer regime, global interconnect emerges as the bottleneck of signal speed and bandwidth. To solve this problem, various techniques have been incorporated into the multilevel architecture, including copper/low-κ materials, reverse scaling, and on-chip interconnect networks. The intrinsic limitations of material, process, physics, and cost have demonstrated the inadequateness of current approaches beyond 100 nm [99,100].

To enable future high-performance system-on-chip (SoC) integration, active research is under way to develop novel global interconnect technology that is faster, more reliable, and more cost-effective. Particular examples include system-in-package (SiP), radio-frequency (RF) and optical interconnect, and three-dimensional integration. SiP joins global interconnect design with package technology and overcomes yield degradation in SoC integration [101]. Since it links chips and blocks with a similar multilevel structure in a package chip as that of on-chip interconnect, it is relative easy to develop the design principles for SiP. However, the fundamental limitation on signal propagation, such as time of flight, still exists. Radio-frequency (wireless) or optical interconnect improves this limit and brings it closer to the speed of light by implementing wireless networks or optical devices on-chip, respectively. Considering process compatibility, design cost, and performance requirement, they will be favorable when global wires are longer than 32 mm at the 45-nm node [102]. By stacking multiple chips together, three-dimensional integration shortens the longest global wires and hence, reduces signal delay. However, heat removal and input–output interconnections of three-dimensional systems remain quite challenging [99]. As the evolution of interconnect technology continues in the nanometer scale, advanced nanotechnology, such as nanocarbon tube and molecular crystal wire, may provide superior interconnection for future computation systems.

REFERENCES

[1] A. Allan et al., International Technology Roadmap for Semiconductors, *http://public.itrs.net.*

[2] R. Bez et al., Introduction to flash memory, *Proc. IEEE*, Vol. 91, No. 4, 2003.

[3] S. Thompson et al., A 90 nm logic technology featuring 50 nm strained silicon channel transistors, 7 layers of Cu interconnects, low κ ILD, and 1 μm^2 SRAM cell, *IEEE International Electron Devices Meeting*, 2002.

[4] P. A. Kraus et al., Scaling plasma nitrided gate dielectrics to the 65 nm node, *Semiconductor Fabtech*, 19th ed., FT 19-13/1 2003.

[5] M. L. Green et al., Ultrathin (<4 nm) SiO_2 and Si–O–N gate dielectric layers for silicon microelectronics: understanding the processing, structure, and physical and electrical limits, *J. Appl. Phys.*, Vol. 90, No. 5, 2001.

[6] G. D. Wilk, R. M. Wallace, and J. M. Anthony, High-k gate dielectrics: current status and materials properties considerations, *J. Appl. Phys.*, Vol. 89, No. 10, 2001.

[7] S. M. Sze, *Physics of Semiconductor Devices*, Wiley-Interscience, New York, 1981.

[8] D. K. Schroder, *Semiconductor Material and Device Characterization*, Wiley, New York, 1990.

[9] K. F. Schuegraf, C. C. King, and C. Hu, Ultra-thin silicon dioxide leakage current and scaling limit, *Digest of Technical Papers, Symposium on VLSI Technology*, 1992.

[10] T. Hori, *Gate Dielectrics and MOS ULSIs*, Springer-Verlag, New York, 1997.

[11] W. C. Lin and C. Hu, Modeling CMOS tunneling current through ultra thin gate oxide due to conduction- and valence-band electron and hole tunneling, *IEEE Trans. Electron Devices*, Vol. 48, No. 7, 2001.

[12] E. Vogel, Measurement of equivalent oxide thickness, *ITRS Document*, 2003.

[13] K. Ahmed et al., Impact of tunnel currents and channel resistance on the characterization of channel inversion layer charge and polysilicon-gate depletion of sub-20A gate oxide MOSFET's, *IEEE Trans. Electron Devices*, Vol. 46, No. 8, 1999.

[14] S. H. Lo et al., Modeling and characterization of n+- and p+-polysilicon-gated ultra thin oxides (21-26A), *Digest of Technical Papers, Symposium on VLSI Technology*, 1997.

[15] S. H. Lo et al., Quantum-mechanical modeling of electron tunneling current from the inversion layer of ultra-thin-oxide nMOSFET's, *IEEE Electron Dev. Lett.*, Vol. 18, No. 5, 1997.

[16] C. A. Richter et al., A comparison of quantum-mechanical capacitance voltage simulators, *IEEE Electron Device Lett.*, Vol. 22, No. 1, 2001.

[17] J. R. Hauser and K. Ahmed, Characterization of ultra-thin oxides using electrical $C-V$ and $I-V$ measurements, *Characterization and Metrology for ULSI Technology International Conference*, American Institute of Physics, 1998.

[18] G. Lucovsky et al., Separate and independent reductions in direct tunneling in oxide/nitride stacks with monolayer interface nitridation associated with (i) interface nitridation and (ii) increased physical thickness, *J. Vac. Sci. Technol. A*, Vol. 18, No. 4, 2000.

[19] Xin Guo and T. P. Ma, Tunneling leakage current in oxynitride: dependence on oxygen/nitrogen content, *IEEE Electron Device Lett.*, Vol. 19, No. 6, 1998.

[20] P. A. Kraus et al., Fundamental limits of MOSFET scaling with Si–O–N gate dielectrics, *Applied Materials ET Conference*, 2003.

[21] Y. C. Yeo, T. J. King, and C. Hu, MOSFET gate leakage modeling and selection guide for alternative gate dielectrics based on leakage considerations, *IEEE Trans. Electron Devices*, Vol. 50, No. 4, 2003.

[22] P. A. Kraus, K. Ahmed, T. C. Chua, M. Ershov, H. Karbasi, C. S. Olsen, F. Nouri, J. Holland, R. Zhao, G. Miner, and A. Lepert, Low-energy Nitrogen Plasmas for 65-nm node oxynitride gate dielectrics: a correlation of plasma characteristics and device parameters, *Digest of Technical Papers, Symposium on VLSI Technology*, 2003.

[23] E. M. Vogel, W. L. Hill, V. Misra, P. K. McLarty, J. J. Wortman, J. R. Hauser, P. Morfouli, G. Ghibaudo, and T. Ouisse, Mobility behavior of n-channel and p-channel MOSFET's with oxynitride gate Dielectric formed by low-pressure rapid thermal chemical vapor deposition, *IEEE Trans. Electron Devices*, Vol. 43, No. 5, 1996.

[24] A. Hegedus, C. Olsen, N. Kuan, and J. Madok, Clustering of plasma nitridation and post anneal steps to improve threshold voltage repeatability, *IEEE Trans. Semicond. Manuf.*, Vol. 16, No. 2, 2003.

[25] S. Inumiya et al., *Digest of Technical Papers, Symposium on VLSI Technology*, 2003.

[26] C. Hobbs et al., Fermi level pinning at the polySi/metal oxide interface, *Digest of Technical Papers, Symposium on VLSI Technology*, 2003.

[27] E. C. Carr, Role of interfacial nitrogen in improving thin silicon oxides grown in N_2O, *Appl. Phys. Lett.*, Vol. 63, No. 5, 1993.

[28] H. Niimi et al., Ultra thin oxide gate dielectrics prepared by low temperature remote plasma-assisted oxidation, *Surf. Coat. Technol.*, Vol. 98, 1998.

[29] S. V. Hattangady et al., Integrated processing of silicon oxynitride films by combined plasma and rapid-thermal processing, *J. Vac. Sci. Technol. A*, Vol. 14, No. 6, 1996.

[30] H. Stohr and W. Klemm, Uber Zweistoffsystem mit Germanium, I. Germanium/Aluminium, Germanium/Zinn und Germanium/Silicon, *Z. Anorg. U. Allgem. Chem.*, Vol. 241, No. 305, 1939.

[31] C. C. Wang and B. H. Alexander, Energy gap of germanium-Silicon alloys, *American Institute of Mining and Metallurgy Engineering Symposium*, 1954.

[32] E. R. Johnson and S. M. Christian, Some properties of germanium–silicon alloys, *Phys. Rev.*, Vol. 95, No. 560, 1954.

[33] R. Braunstein, R. Moore, and F. Herman, Intrinsic optical absorption in germanium-silicon alloys, *Phys. Rev.*, Vol. 109, No. 695, 1958.

[34] R. Braunstein, Lattice vibration spectra of germanium-silicon alloys, *Phys. Rev.*, Vol. 130, 1963.

[35] R. Braunstein and E. O. Kane, The valence band structure of the III-V compounds, *J. Phys. Chem. Solids*, Vol. 23, 1962.

[36] R. Braunstein, Valence band structure of germanium-silicon alloys, *Phys. Rev.*, Vol. 130, 1954.

[37] F. Herman, Speculation on the energy band structure of Ge–Si alloys, *Phys. Rev.*, Vol. 95, 1954.

[38] R. People, Indirect band gap of coherently strained Ge_xSi_{1-x} bulk alloys on <001> silicon substrates, *Phys. Rev. B*, Vol. 32, No. 2, 1985.

[39] R. People, Physics and applications of Ge_xSi_{1-x} strained-layer heterostructures, *IEEE J. Quantum Electron.*, Vol. 22, No. 9, 1986.

[40] C. G. Van de Walle, Theoretical study of Si/Ge interfaces, *J. Vac. Sci. Technol. B*, Vol. 3, No. 4, 1985.

[41] C. G. Van de Walle, Theoretical calculations of heterojunction discontinuities in the Si/Ge system, *Phys. Rev. B*, Vol. 34, No. 8, 1986.

[42] A. G. O'Neill and D. A. Antoniadis, Deep submicron CMOS based on silicon germanium technology, *IEEE Trans. Electron Devices*, Vol. 43, 1996.

[43] K. L. Wang, S. G. Thomas, and M. O. Tanner, SiGe band engineering for MOS, CMOS and quantum effect devices, *J. Mater. Sci.*, Vol. 6, 1995.

[44] M. C. Ozturk, N. Pesovic, I. Kang, J. Liu, H. Mo, and S. Gannavaram, Ultrashallow source/drain junctions for nanoscale CMOS using selective silicon–germanium technology, *Jpn. Soc. Appl. Phys. IWJT*, 2001.

[45] M. C. Ozturk, J. Liu, H. Mo, and N. Pesovic, Advanced $Si_{1-x}Ge_x$ source/drain and contact technologies for sub-70 nm CMOS, *IEEE*, 2002.

[46] P. Ranade, H. Takeuchi, V. Subramanian, and T.-J. King, A novel elevated source/drain PMOSFET formed by Ge–B/Si intermixing, *Electron Device Lett.*, Vol. 23, No. 4, 2002.

[47] P. Ranade et al., Application of silicon–germanium in the fabrication of ultrashallow extension junctions for sub-100 nm PMOSFETs, *IEEE Trans. Electron Devices*, Vol. 49, No. 8, 2002.

[48] J. M. Harmann et al., Effect of HCl on the SiGe growth kinetics and reduced pressure–chemical vapor deposition, *J. Cryst. Growth*, Vol. 241, 2002.

[49] R. B. Fair, *Rapid Thermal Processing: Science and Technology*, 1993.

[50] B. Ramachandran, H. Forstner, and E. Chiao, Beyond the 100 nm node: single-wafer RTP, *Solid State Technol.*, May 2003.

[51] B. Peuse, G. Miner, and M. Yam, Method and apparatus for measuring substrate temperatures, U.S. patent 5, 660, 472, 1997.

[52] C. Schietinger, B. Adams, and C. Yarling, Ripple technique: a novel non-contact wafer emissivity and temperature method for RTP, *Mater. Res. Symp. Proc.*, Vol. 224, No. 23, 1991.

[53] T. Ito et al., 14 nm-depth low resistance boron doped extension by optimized flash lamp annealing, *International Symposium on Semiconductor Manufacturing*, 2002.

[54] S. Talwar, Y. Wang, and M. Thompson, Laser annealing for junction fabrication in CMOS devices, *Electrochemical Society Spring Meeting*, 2003.

[55] J. Kittl, *VLSI Technology Symposium*, 1996.

[56] J. P. Lu et al., A novel nickel SALICIDE process technology for CMOS devices with sub-40 nm physical gate length, *International Electron Device Meeting*, 2002.

[57] J. Kittl et al., Silicides for 65 nm CMOS and beyond, *Electrochemical Society Meeting*, 2003.

[58] *http://public.itrs.net/.*

[59] S. Kuppurao, H. Joo, and G. Miner, In situ steam generation: a new rapid thermal oxidation technique, *Solid State Technol.*, 2000.

[60] K. Reid et al., Dilute steam rapid thermal oxidation for 30 Å gate oxides, *Electrochemical Society Meeting*, 1999.

[61] P. Meissner, A. Hegedus, J. Madok, R. Thakur, and G. Miner, Thermal technologies for sub-100 nm CMOS scaling: development strategies, *Electrochemical Society Meeting*, 2002.

[62] G. Timp et al., Low leakage, ultra-thin gate oxides for extremely high performance sub-100 nm nMOSFETs, *Technical Digest, IEEE International Electron Devices Meeting*, Vol. 930, 1997.

[63] K. Ahmed et al., Applied Materials internal document.

[64] B. Yu, H. Wang, A. Joshi, Q. Xiang, E. Ibok, and M.-R. Lin, 15 nm gate length planar CMOS transistor, *Technical Digest, IEEE International Electron Devices Meeting*, pp. 937–939, Dec. 2001.

[65] Y.-C. King, H. Fujioka, S. Kamohara, W.-C. Lee, and C. Hu, Ac charge centroid model for quantization of inversion layer in n-MOSFET, *Proceedings of the International Symposium on VLSI Technology, Systems and Applications*, pp. 245–249, June 1997.

[66] International Technology Roadmap for Semiconductors, *http://public.itrs.net.*

[67] Q. Lu, R. Lin, P. Ranade, Y.-C. Yeo, X. Meng, H. Takeuchi, T.-J. King, C. Hu, H. Luan, S. Lee, W. Bai, C.-H. Lee, D.-L. Kwong, X. Guo, X. Wang, and T.-P. Ma, Molybdenum metal gate MOS technology for post-SiO_2 gate dielectrics, *Technical Digest, IEEE International Electron Devices Meeting*, pp. 641–644, 2000.

[68] Y.-C. Yeo, P. Ranade, Q. Lu, R. Lin, T.-J. King, and C. Hu, Effects of high-κ dielectrics on the work-functions of metal and silicon gates, *Proceedings of the Symposium on VLSI Technology*, pp. 49–50, Kyoto, Japan, June 2001.

[69] L. Chang, S. Tang, T.-J. King, J. Bokor, and C. Hu, Gate length scaling and threshold voltage control of double-gate MOSFETs, *Technical Digest, IEEE International Electron Devices Meeting*, pp. 719–722, Dec. 2000.

[70] S. H. Lo, D. A. Buchanan, Y. Taur, and W. Wang, Quantum-mechanical modeling of electron tunneling current for the inversion layer of ultra-thin-oxide nMOSFETs, *IEEE Electron Device Lett.*, pp. 209–211, May 1997.

[71] W.-C. Lee and C. Hu, Modeling gate and substrate currents due to conduction- and valence band electron and hole tunneling, *Proceedings of the Symposium on VLSI Technology*, pp. 198–199, 2000.

[72] Y.-C. Yeo, Q. Lu, W.-C. Lee, T. King, C. Hu, X. Wang, X. Guo, and T. P. Ma, Direct tunneling gate leakage current in transistors with ultra-thin silicon nitride gate dielectric, *IEEE Electron Device Lett.*, Vol. 21, No. 11, pp. 540–542, Nov. 2000.

[73] C. Hu, A compact model for rapidly shrinking MOSFETs, *Technical Digest, IEEE International Electron Devices Meeting*, pp. 285–288, Dec. 2001.

[74] M. V. Dunga, X. Xi, J. He, I. Polishchuk, Q. Lu, M. Chan, A. M. Niknejad, and C. Hu, Modeling of direct tunneling current in multi-layer gate stacks, Workshop on Compact Modeling, *6th International Conference on Modeling and Simulation of Microsystems*, San Francisco, Feb. 2003.

[75] R. Shrivastava and K. Fitzpatrick, A simple model for the overlap capacitance of a VLSI MOS device, *IEEE Trans. Electron Devices*, Vol. 29, No. 12, Dec. 1982.

[76] S. Thompson, P. Packan, T. Chani, M. Stettler, M. Alavi, I. Post, S. Tyagi, S. Ahmed, S. Yang, and M. Bohr, Source/drain extension scaling for 0.1 μm and below channel length MOSFETs, *Proceedings of the Symposium on VLSI Technology*, pp. 132–133, Honolulu, HI, June 1998.

[77] B. Cheng, M. Cao, R. Rao, A. Inani, P. V. Voorde, W. M. Greene, J. M. C. Stork, Z. Yu, P. M. Zeitzoff, and J. C. S. Woo, The impact of high-κ gate dielectrics and metal gate electrodes on sub-100 nm MOSFETs, *IEEE Trans. Electron Devices*, Vol. 46, No. 7, pp. 1537–1542, July 1999.

[78] B. E. Weir, P. J. Silverman, M. A. Alam, F. Baumann, D. Monroe, A. Ghetti, J. D. Bude, G. L. Timp, A. Hamad, T. M. Oberdick, N. X. Zhao, Y. Ma, M. M. Brown, D. Hwang, T. W. Sorsch, and J. Madic, Gate oxides in 50 nm devices: thickness uniformity improves projected reliability, *IEEE International Electron Devices Meeting*, pp. 437–440, 1999.

[79] Q. Lu, H. Takeuchi, X. Meng, T.-J. King, C. Hu, K. Onishi, H.-J. Cho, and J. C. Lee, Improved performance of ultra-thin HfO_2 CMOSFETs using poly-SiGe gate, *Symposium on VLSI Technology*, pp. 86–87, June 2002.

[80] J. Chung, M.-C. Jeng, J. Moon, P. K. Ko, and C. Hu, Low voltage hot-electron currents and degradation in deep submicrometer MOSFETs, *IEEE Trans. Electron Devices*, Vol. 37, No. 7, pp. 1651–1657, July 1990.

[81] V. Reddy, A. T. Krishnan, A. Marshall, J. Rodriguez, S. Natarajan, T. Rost, and S. Krishnan, Impact of negative bias temperature instability on digital circuit reliability, *International Reliability Physics Symposium Proceedings*, pp. 248–254, Apr. 2002.

[82] C.-H. Jeon, S.-Y. Kim, H.-S. Kim, and C.-B. Rim, The impact of NBTI and HCI on deep sub-micron PMOSFETs' lifetime, *IEEE Integrated Reliability Workshop Final Report*, pp. 130–132, 2002.

[83] V. Agarwal, M. S. Hrishikesh, S. W. Keckler, and D. Burger, Clock rate versus IPC: the end of the road for conventional microarchitectures, *Proceedings of the International Symposium on Computer Architecture*, pp. 248–259, 2000.

[84] C. W. Kaanta et al., Submicron wiring technology with tungsten and planarization, *Proceedings of the IEEE International Electron Devices Meeting*, pp. 209–212, Dec. 1987.

[85] M. Igarashi et al., A diagonal-interconnect architecture and its application to RISC core design, *Digest of the IEEE International Solid-State Circuits Conference*, pp. 166–167, Feb. 2002.

[86] D. Edelstein et al., Full copper wiring in a sub-0.25 μm CMOS ULSI technology, *Proceedings of the IEEE International Electron Devices Meeting*, pp. 773–776, 1997.

[87] S. M. Rossnagel and H. Kim, From PVD to CVD to ALD for interconnects and related applications, *Proceedings of the 2001 International Interconnect Technology Conferences*, pp. 3–5, 2001.

[88] A. E. Kaloyeros, E. T. Eisenbraun, J. Welch, and R. E. Geer, Exploiting nanotechnology for terahertz interconnect, *Semicond. Int.*, pp. 56–59, Jan. 2003.

[89] S. Voldman, R. Gauthier, D. Reinhart, and K. Morrisseau, High-current transmission line pulse characterization of aluminum and copper interconnects for advanced CMOS semiconductor technologies, *IEEE 36th Annual International Reliability Physics Symposium*, pp. 293–301, 1998.

[90] C.-K. Hu and J. M. E. Harper, Copper interconnect: fabrication and reliability, *International Symposium on VLSI Technology, Systems, and Applications*, pp. 18–22, 1997.

[91] D. Chung, J. Korejwa, and E. Walton, Introduction of copper electroplating into a manufacturing fabricator, *IEEE/SEMI Advanced Semiconductor Manufacturing Conference*, pp. 282–289, 1999.

[92] B. E. Stine et al., The physical and electrical effects of metal-fill patterning practices for oxide chemical–mechanical polishing processes, *IEEE Trans. Electron Devices*, Vol. 45, No. 3, pp. 664–678, Mar. 1998.

[93] R. Chang, Y. Cao, and C. Spanos, Modeling metal dishing effect for interconnect process-design co-optimization, *IEEE International Electron Devices Meeting*, 2003.

[94] S. Purushothaman et al., Opportunities and challenges in ultra low κ dielectrics for interconnect applications, *IEEE International Electron Devices Meeting*, pp. 529–532, Dec. 2001.

[95] H. Wu, J. Cargo, C. Peridier, and J. Serpiello, Reliability issues and advanced failure analysis deprocessing techniques for copper/low-κ technology, *IEEE 41st Annual International Reliability Physics Symposium*, pp. 536–544, 2003.

[96] K.-D. Lee et al., Electromigration study of Cu/low κ dual-damascene interconnects, *IEEE 40th Annual International Reliability Physics Symposium*, pp. 322–326, 2002.

[97] J. Noguchi et al., Impact of low-κ dielectrics and barrier metals on TDDB lifetime of Cu interconnects, *IEEE 39th Annual International Reliability Physics Symposium*, pp. 355–359, 2001.

[98] E. T. Ogawa et al., Leakage, breakdown, and TDDB characteristics of porous low-κ silica-based interconnect dielectrics, *IEEE 41st Annual International Reliability Physics Symposium*, pp. 166–172, 2003.

[99] J. D. Meindl et al., Interconnecting device opportunities for gigascale integration (GSI), *IEEE International Electron Devices Meeting*, pp. 525–528, 2001.

[100] T. Ohba, Multilevel interconnect technologies in SoC and SiP for 100-nm node and beyond, *6th International Conference on Solid-State and Integrated-Circuit Technology*, pp. 46–51, Oct. 2001.

[101] K. L. Tai, System-in-package (SIP): challenges and opportunities, *IEEE Asia and South Pacific Design Automation Conference*, pp. 191–196, Jan. 2000.

[102] A. V. Mule et al., Towards a comparison between chip-level optical interconnections and board-level exterconnection, *IEEE International Interconnect Technology Conference*, pp. 92–94, 2002.

CHAPTER 3

THEORY AND PRACTICALITIES OF SUBWAVELENGTH OPTICAL LITHOGRAPHY

3.1 INTRODUCTION AND SIMPLE IMAGING THEORY

In the competitive world of chipmaking, the push toward ever smaller and faster devices has been necessary to keep up with the idea of device density doubling every two years. In the following chapters there is an in-depth discussion of photolithography, the key process enabler to density doubling. We focus on both the theory and practical implementation of current resolution enhancement techniques, such as illumination optimization, phase-shift reticles, optical proximity corrections, and subresolution assist features. Special emphasis is placed throughout the book on the interaction between design and resolution enhancement. Understanding the effects of design interaction with optical enhancements and the need for design optimization for resolution enhancement will be keys to successful and manufacturable imaging at 65-nm technologies and beyond. In the 100- and 65-nm nodes, pattern formation is based on optical lithography. Optical systems cannot image arbitrarily small feature sizes, a consequence of the fact that the light used for imaging has a finite wavelength that is comparable to the size of the features being imaged. This is commonly described in the following equation, which relates the wavelength λ of the light used, the numerical aperture

Nano-CMOS Circuit and Physical Design, by Ban P. Wong, Anurag Mittal, Yu Cao, and Greg Starr
ISBN 0-471-46610-7

NA of the projection optics, and the half-pitch d of the feature size used:

$$d = \frac{\kappa\lambda}{\mathrm{NA}}$$

The factor κ that appears in this equation is usually referred to as the κ-factor. The smallest possible value that can be achieved in an optical system is $\kappa = 0.25$. The underlying physics of this equation is depicted schematically in Figure 3.1. A mask pattern of fixed pitch $p = 2d$, when illuminated with coherent light, creates a series of diffracted beams. Their diffraction angles α are given by $n\lambda = p \sin\alpha$, where n can be any integer number. The diffraction angles therefore increase with decreasing pitch. The lens with finite numerical aperture NA refocuses the diverging beams to form an image as a result of interference of these diffraction orders. If the numerical aperture of the lens is too small to capture the diffracted beams corresponding to $n = \pm 1$, the resulting image carries no spatial modulation. In the case depicted, with normal incidence light illuminating the reticle, imaging is achieved with at least three beams: $n = 0, \pm 1$. It is worth mentioning that even if the first diffraction order is captured by the lens, the reconstructed image still suffers from significant image degradation: The original perfectly square transmission characteristics of the mask have turned into a sinusoidal distribution of the image intensity I that retains the spatial distances but has lost all of its edges.

As we progress to patterning of smaller and smaller features, masks are no longer simply illuminated with light at normal incidence. Controlling the angular

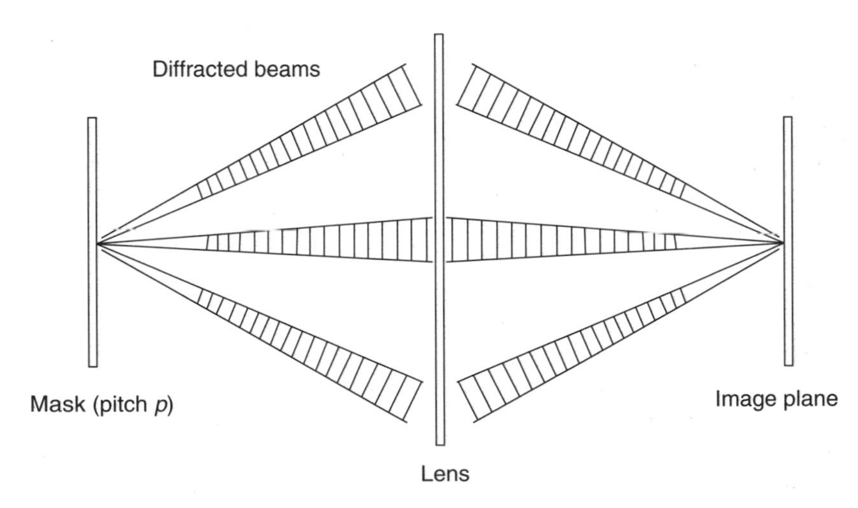

Figure 3.1 Wave optical interpretation of imaging for a fixed-pitch mask pattern. The fixed pattern creates diffracted beams with diffraction angles given by the Bragg condition. The diffracted beams have no internal structure. The lens focuses the diffracted beam. Focusing is achieved through a phase shift that depends on the distance from the center. Interference of diffracted beams re-creates image modulation. If diffracted beams do not pass through the lens, no image is formed.

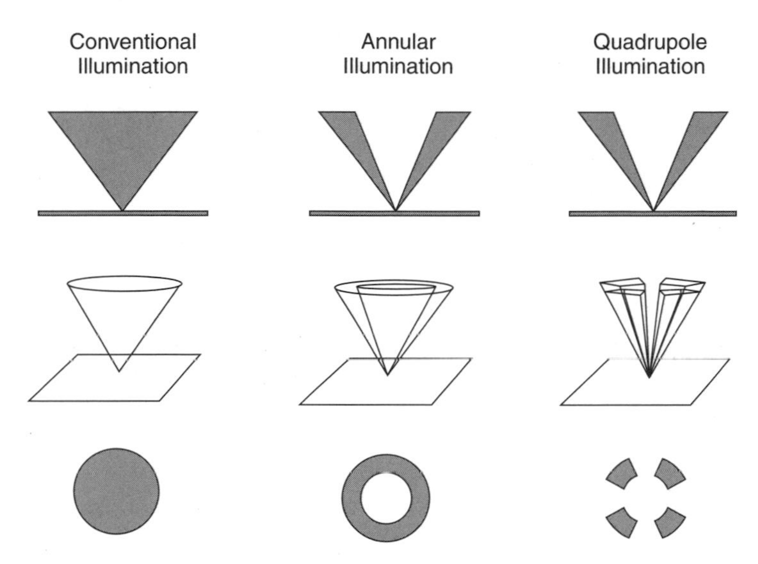

Figure 3.2 Illumination. Schematic representation of various illuminations used in optical lithography. Shown are examples of conventional, annular, and quadrupole illumination. The schematic illustrates the angular distribution of light impinging on the reticle.

distribution of light impinging on the reticle is one of the key process parameters that allow a lithographer to control imaging performance. Figure 3.2 illustrates the concept of conventional illumination. Light up to a certain angle of incidence is illuminating any point on the reticle. A convenient way to illustrate different illumination patterns is to visualize them in a plane, with distance from a center point representing the angle of incidence. The center point corresponds to normal incidence. In this representation, conventional illumination looks like a circle, the outer diameter corresponding to the maximum angle of the cone. There are several other commonly used illuminator patterns—annular, quadrupole, and dipole illuminations—that have recently found more and more widespread use. They are shown in the figure and discussed in more detail in subsequent chapters.

The mechanism by which the illumination pattern affects the imaging process is depicted in Figures 3.3 and 3.4. The diffraction pattern resulting from a mask illuminated with light from various angles is constructed as the superposition of the contributions for each individual angle. In other words, for a illumination pattern such as the one shown in Figure 3.3(*a*) and a mask that has a normal incidence diffraction pattern, such as Figure 3.3(*b*), the resulting overall diffraction pattern is that of Figure 3.3(*c*). As in the simpler cases discussed above, those portions of the diffraction pattern that fall outside the circle defined by the numerical aperture of the system do not contribute to imaging. As the pitch of the mask pattern is decreased the diffraction orders occur at higher and higher angles. As a result larger and larger portions of the first order diffracted light cannot pass through the lens. The resulting degradation in imaging performance is shown in Figure 3.4.

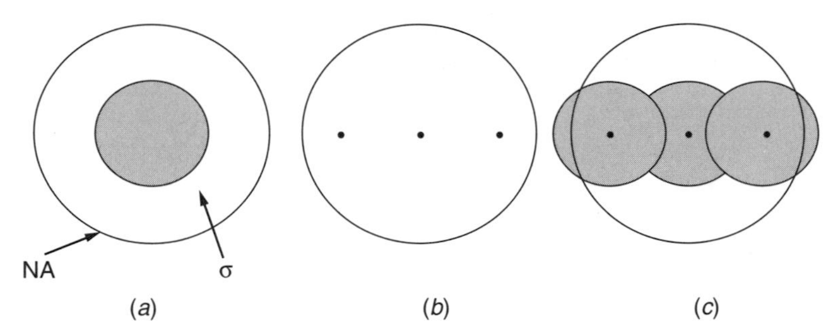

Figure 3.3 Schematic representation of the image formation process in the case of partially coherent light. (*a*) lens and illuminator setting; (*b*) mask diffraction orders; (*c*) diffraction pattern.

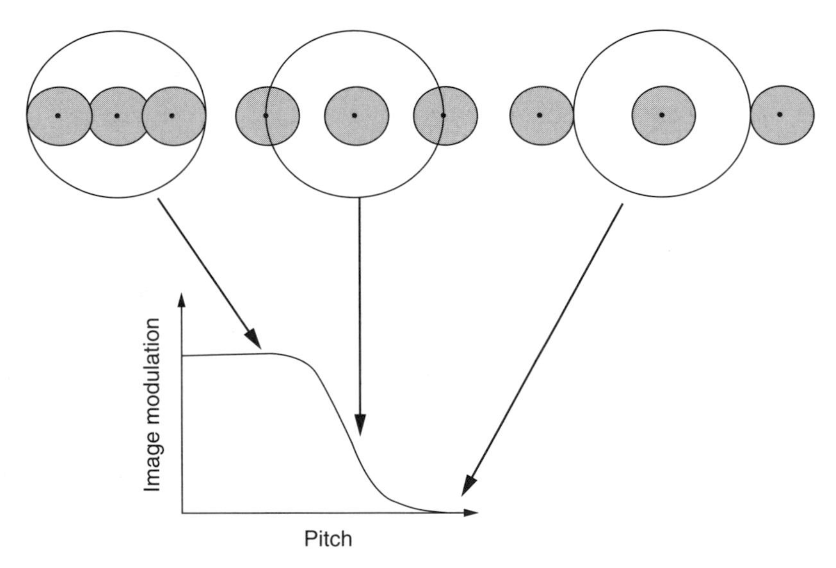

Figure 3.4 Impact of coherence on imaging performance. Image modulation gradually degrades as the pitch of the pattern decreases.

3.2 CHALLENGES FOR THE 100-nm NODE

The challenges facing lithography for the 100-nm node are those inherent to low-κ imaging. Lithography at κ-factors approaching 0.25 is characterized by a significantly reduced ability to maintain desired imaging performance in the face of a wide variety of process variations. The challenges for the 100-nm node are therefore best appreciated through an assessment of the expected κ-factors for this generation, an overview of the type of process variations present in a manufacturing environment, and a discussion of how lower κ-values translate into increased sensitivities to process variations.

3.2.1 κ-Factor for the 100-nm Node

Figure 3.5 shows the growing dilemma that lithography is facing as the various technology nodes progress to ever-smaller feature sizes. The transition to smaller exposure wavelengths has been outpaced by the more rapidly decreasing feature sizes. In the time frame between 1980 and today, the exposure wavelengths used in manufacturing or pilot production have developed from 436 nm through 365, 248, to 193 nm, a drop by approximately a factor of 2.3. In the same time frame, feature sizes have dropped from 1.5 μm to 65 nm, or by a factor of 23, almost an order of magnitude larger. The rapid shrink in feature size dictated by Moore's law has largely been accomplished by decreasing the ratio between image size and exposure wavelength. In about 1995, lithography had entered a regime where the feature sizes were smaller than the wavelength of light used to image them, a regime that is sometimes referred to as the *subwavelength gap* [1]. As we have seen, wavelength alone is not the best means to assess lithographic capability; the κ-*factor*, defined as the ratio between half-pitch scaled with numerical aperture divided by the wavelength, provides a better gauge. The corresponding trend for κ-factors over the same time frame is shown in Figure 3.6. The graph indicates that fortunately, lithographic tool capabilities have improved significantly faster than the wavelength chart would suggest, due essentially to improvements in the numerical aperture of lenses, starting from early 0.28NA tools to 0.85NA tools. Even so, demands on the lithography process have increased dramatically and κ-factors are rapidly approaching the magic barrier of $\kappa = 0.25$. For the 90-nm technology generation, the expected κ-factor is approximately 0.34, under the assumption that critical levels for this technology will be using 193-nm exposure tools at an NA of 0.75. The projections for the 65-nm generation are significantly less clear. Early production will probably rely on ultrahigh-NA ($\geq$0.85) 193-nm tools, equivalent to $\kappa \approx 0.29$. The next-generation exposure tools, at a wavelength of 157 nm, may become the tool of choice for mainstream manufacturing.

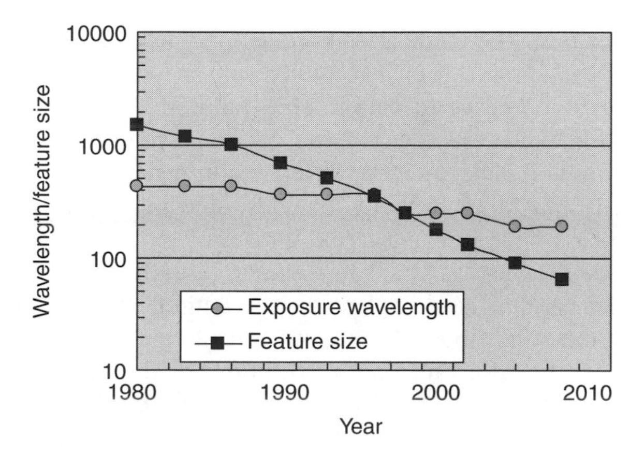

Figure 3.5 Evolution of minimum feature sizes in semiconductor manufacturing relative to the exposure wavelength.

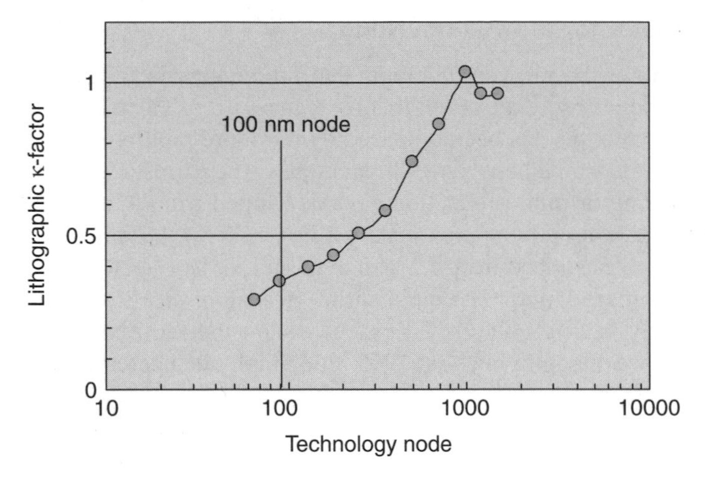

Figure 3.6 κ-Factor versus technology generation.

3.2.2 Significant Process Variations

Before turning to the implications of low-κ imaging on the ability to deal with process variations, we discuss briefly some process variations encountered during lithographic processing. Process variations may be categorized based on the key components of the lithographic process: incoming wafers, mask, illumination system, projection lens, focusing system, and resist, to name the most important. What follows is simply intended to provide a flavor for the types of variations encountered and is by no means an exhaustive account of all possible variations.

Wafers are one of the sources for variations in lithography through a variety of mechanisms. For example, prior patterning steps result in localized variations in reflectivity. Commonly encountered scenarios are resist lines passing over prior-level metal patterns, which will experience increased reflectivity in the vicinity of the underlying pattern. The modulation in reflectivity leads to localized variations in effective light intensities, which in turn result in localized variations in line width and in extreme cases to catastrophic failure [2]. A similar scenario is fairly common for implant resists covering over an active area of silicon as well as over shallow trench dielectric. The difference in materials used in these areas affects reflectivity and thus contributes to undesirable line-width variations. While the primary process option to reduce the impact of reflectivity variations is the use of antireflective layers (ARCs) [3], these are usually not 100% effective. In some cases, such as that with implant resists, antireflective coatings present a problem in that the ARCs must be removed prior to the implant. The additional process to remove the ARCs may not be desirable or too costly. From an aerial image perspective, the effect of reflectivity variations is minimized with a large slope, dI/dx, of the aerial image intensity I. In this case, reflectivity-induced variations in intensity (dI) have a smaller effect on the shift in image critical dimension (dx).

Another fairly common process variation introduced by the wafer are variations in topography both locally and on a wafer scale. In older technologies local topography was simply created by depositing oxide layers over etched metal layers, resulting in hills over the metal areas and valleys over clear areas. These step heights are no longer acceptable, and today's processes use chemomechanical polishing (CMP) to even out the topography present. Even though these processes constitute a significant step forward, their planarization capabilities are not perfect. For example, polishing rates are pattern density–dependent, resulting in film thickness variations on a wafer scale as well as a local scale. Therefore, at chip boundaries and in areas with large deviations from the average pattern density, the polish may leave residual materials. Local wafer topography affects imaging through two mechanisms. First, it leads to local reflectivity variations caused not only by variations in the thickness of the transparent materials but also by variations in the thicknesses of antireflective and resist coatings. Second, it requires that the exposure tool adjust the distance between the wafer stage where the wafer is mounted and the lens to maintain the image at the top surface of the wafer in best focus. Since the autofocus system can adjust only the distance or tilt of the wafer stage, localized wafer height modulations that extend over less than a typical chip size (20 mm $\times$ 20 mm area) are difficult to compensate for. Similarly, wafer curvature as it occurs quite commonly on the edges of a wafer is uncorrectable. The impact of focus variations is a change in critical dimension and in extreme cases, catastrophic pattern failure. Image critical dimension exhibits a nonlinear, typically quadratic dependence on focus, and the magnitude of this variation is a function of the lithographic process and the dimension of the pattern. Typically, the lithographic process has sufficient tolerance to focus induced image changes over a small range of focus variations, but the quadratic dependence quickly leads to unacceptable variations in critical dimensions (CDs) if the focus variations become too large.

Other sources of variations are, for example, nonideal illumination systems. Illuminators, that part of the exposure tool that projects light onto the reticle, need to maintain uniform light intensity across the imaging field, quite a challenge given that a typical slit is about 25 mm wide and uniformities need to be maintained at or below 1% ranges. Nonuniform illuminators cause modifications in local light intensities in different parts of the field for patterns that would otherwise be equivalent. Therefore, they are similar in effect to reflectivity variations and require sufficiently large aerial image slopes to minimize their impact. More difficult to detect and more subtle in their impact are cases where the uniformities occur in the angular distribution rather than the integrated illumination intensity [4]. For example, if the light intensity delivered from one side of the illuminator is stronger than on the other, pattern distortions or shifts are observed most notably in conjunction with focus variation.

The projection optics used to image the mask is also a source of process variations. The imperfections of projection lenses are generally referred to as *aberrations*, an area that has attracted a significant amount of attention [5–9]. They may be caused by imperfect mounting of the various lens elements or deviations from their desired shapes due to the tolerances in lens manufacturing. These deviations may be represented as phase errors with characteristic distributions across the pupil. Aberrations have been categorized based on their rotational symmetry and their radial variation. Figure 3.7 gives a simple example of an aberration. In this case the phase error increases as a function of radius

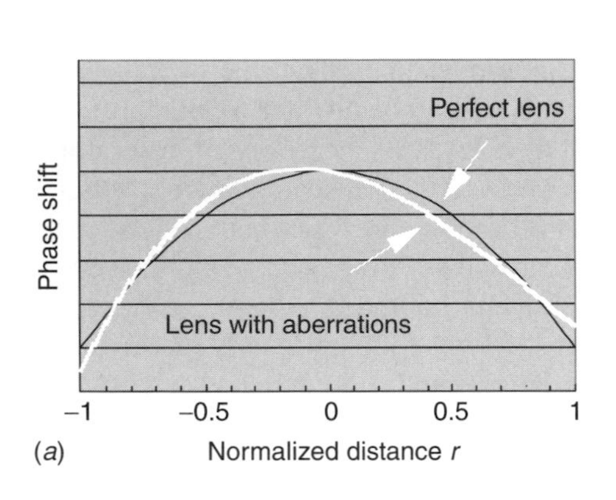

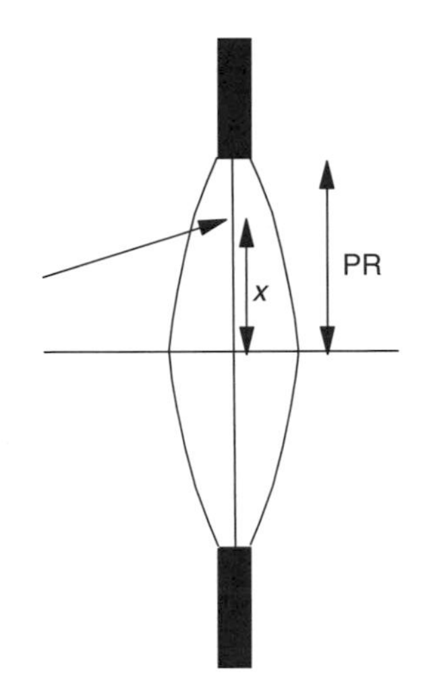

	Polynomial form	Aberration name
1	1	Piston
2	$r \cos \theta$	x-tilt
3	$r \sin \theta$	y-tilt
4	$2r^2 - 1$	Focus
5	$r^2 \cos 2\theta$	Third-order astigmatism x
6	$r^2 \sin 2\theta$	Third-order astigmatism y
7	$(3r^3 - r) \cos \theta$	Third-order coma x
8	$(3r^3 - r) \sin \theta$	Third-order coma y

(*b*)

Figure 3.7 Lens aberrations, signature, types, and interrelationship with diffraction patterns. (*a*) Generic concept for describing aberrations. As demonstrated in Section 3.2, lenses may be treated as phase objects. Perfect performance is achieved only if the lens follows an ideal phase shift versus diffraction angle relationship. Any deviations from the perfect behavior is termed an *aberration*. The normalized distance is the ratio x/PR, where PR is the pupil radius. (*b*) The possible phase deviations may be described as a superposition of a set of basis functions. These basis functions are referred to as *Zernike polynomials*, which are functions of the normalized distance r with values from 0 to 1 and the angle θ in the pupil plane (0 to 360°). The table in part (*b*) provides a flavor for the type of aberrations. (*c*) Generic impact of aberration on dense line-space pattern for two-beam interference: (left) equal phase shift for both orders; (right) different phase shift for both orders. (*d*) Interrelationship between diffraction pattern and aberration (example of astigmatism). Shown is the phase error for a third order astigmatism x. The phase error plotted in the pupil plane has a saddle like shape. The image highlights the interrelationship between symmetry in the aberration and symmetry of the pattern.

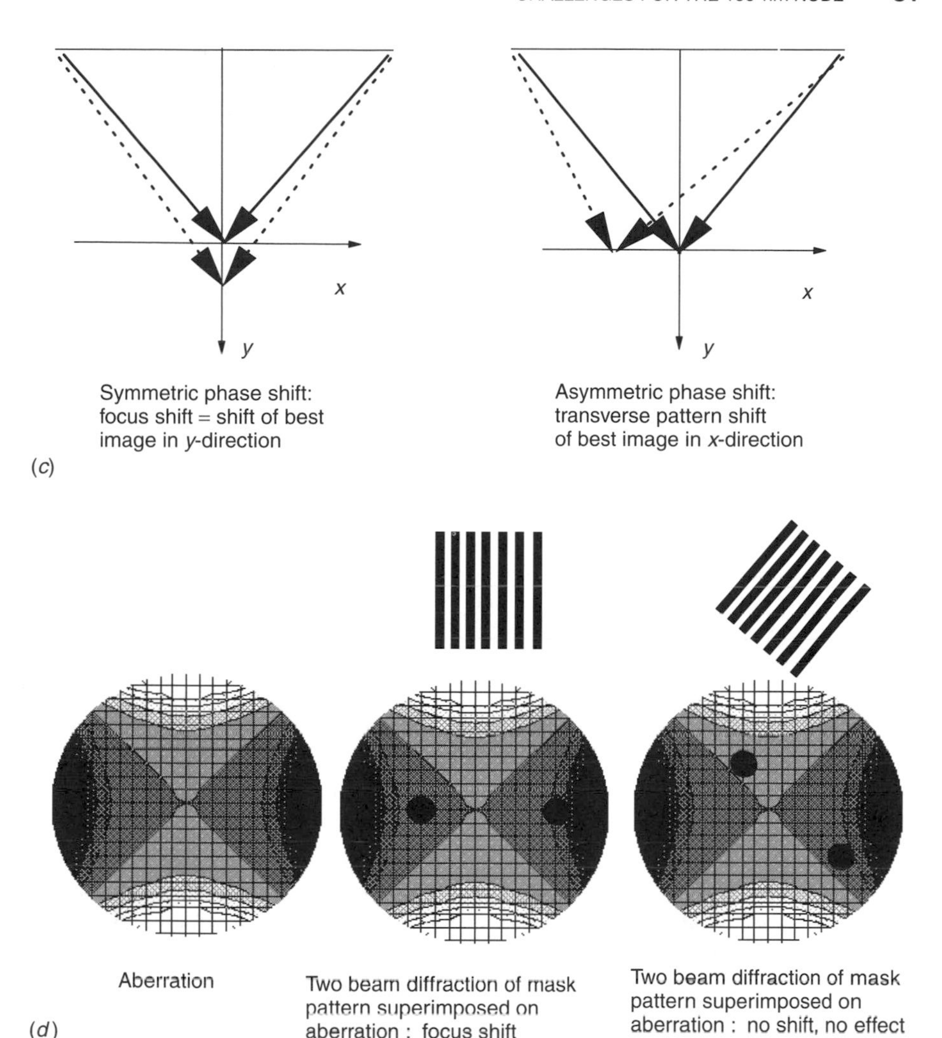

Figure 3.7 *(continued)*

and has twofold symmetry, with negative phase errors introduced on the left side and positive phase error on the right side. Figure 3.7(*d*) is an example of the interaction between the diffraction from a simple mask pattern (line-space pattern in one orientation) and the phase deviation of this aberration. When the mask pattern is oriented vertically, both diffracted beams experience phase shifts of equal amounts, equivalent to a focus shift, as shown in Figure 3.7(*c*). A mask pattern of the same pitch but rotated 90° creates phase shifts that have the same magnitude but opposite sign compared to the vertical case. As a result, for this type of aberration, patterns of different orientation have different best focus condition, an aberration known as *astigmatism*. The effect is minimized if the mask pattern is rotated 45°, in which case the aberration has no impact on imaging since the diffraction orders and the zero order do not experience phase shifts. This is an example for a case where the impact of the aberration depends on the type of pattern to be imaged, the illumination pattern, and the pattern orientation. In general, aberrations result in spatial displacement of patterns either in the direction of the light propagation (or in other words, shifts in best focus) or in a shift perpendicular to the optical axis. Due to significant improvements in lens measurement capabilities, lens manufacturing, and progress in lens mounting techniques, great strides have been made in reducing the overall amount of aberration in lenses used in semiconductor manufacture.

The move from step-and-repeat systems, where the reticle and wafer stage remain stationary during exposure, to step-and-scan systems, where both reticle and wafer are moving simultaneously, introduced a new source of image degradations. Scanning systems have the challenging task of maintaining perfect synchronization of movements between the reticle and the wafer stages during the imaging process. Lack of scan synchronization results in a blurred image. For the common case where the scan synchronization is better in one direction than it is in another, differences in line width will occur for lines parallel and perpendicular to the scan direction [9].

The reticles themselves are a significant source of variation. Masks are created by lithographic processes. A resist layer coated over a chrome-covered quartz plate is exposed by an e-beam or laser scanning tool, developed, and the chrome layer removed in the exposed areas. Common variations found on reticles are differences in feature sizes for features parallel or perpendicular to the scan direction of the mask write tool, variations in line width across the reticle, or a shift in the average dimension of features. Cross-plate variations may be due to inhomogenities of the metal etch process or the develop process.

The process factors mentioned in this section are common sources of process variations that result in deviations that affect imaging during manufacturing processes. Some of these are systematic in that they change little over time, whereas others vary over time, due to factors not always under the control of the process engineering teams. All of them have in common that they cause deviations from the best achievable imaging performance. As a result, electrical parameters of a given chip vary across the chip or over time. If these variations exceed the allowable tolerances, the chip may not be usable. From a chip manufacturing perspective the main impact of low-κ imaging is that expensive control mechanisms have to be put in place to reduce the range of process variations.

3.2.3 Impact of Low-κ Imaging on Process Sensitivities

Having established the expected κ-factor for the 90-nm node and having discussed sources of variations, we turn next to how low-κ imaging affects the tolerance to such process variations. We discuss tolerance to dose and focus variations as well as the impact of reticle CD variations. We also discuss the impact on image fidelity as κ-factors decrease. One of the examples chosen is a change in image

CD for lines of equal size spaced at different pitches in simple one-dimensional structures. The second example is a two-dimensional problem highlighting the impact of low-κ imaging on the capability to print perfect corners.

3.2.4 Low-κ Imaging and Impact on Depth of Focus

Depth of focus (DOF) is the acceptable range of focus variations that still allows maintaining critical dimensions within acceptable CD tolerances. A simple estimate for expected depth of focus (DOF) is given by

$$\mathrm{DOF} \propto \frac{\lambda}{\mathrm{NA}^2}$$

which can be rewritten as

$$\mathrm{DOF} \propto \frac{\mathrm{pitch}^2}{\lambda}$$

This equation indicates that DOF is a lithographic parameter that scales with the square of the minimum dimension in the node. Fortunately, reducing wavelength alleviates the situation somewhat. The decrease in depth of focus is still dramatic, however, as seen in Figure 3.8. For the 90-nm node, the approximate depth of focus for this technology is 0.35 μm. This should be compared to estimates of total expected focus variations arising from all error sources, some of which have been discussed above. A detailed discussion can be found, for example, in Ref. 10. The total range with all error sources accounted for is about 400 nm, comparable with the approximate capability based on the equations above. The step into the 65-nm generation reduces the expected capability to approximately 250 nm. Under the assumption that this generation has to enter early production without being able to rely on the availability of the next-possible exposure

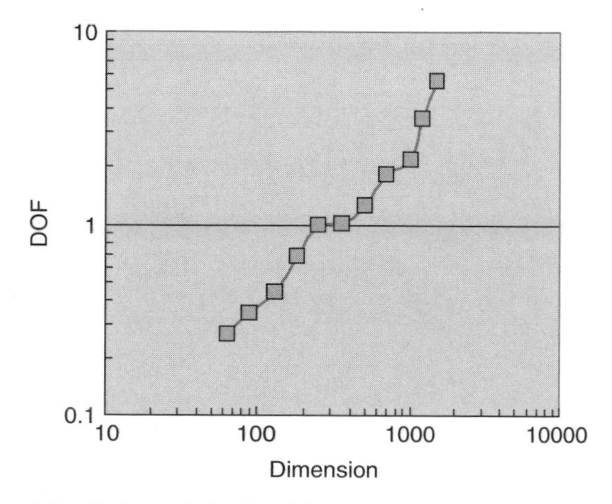

Figure 3.8 Estimated depth of focus as a function of technology.

wavelength at 157 nm, reduction of focus variations will be a challenge. Most significantly, the ability to reduce wafer-induced variations despite the general increase in wafer size will become a particular challenge.

3.2.5 Low-κ Imaging and Exposure Tolerance

As we have seen above, low-κ imaging leads to aerial image degradation through two major mechanisms: On the one hand, the resulting image has a finite slope, a consequence of the fact that diffraction orders higher than 1 are not contributing to the image. On the other hand, the modulation amplitude is reduced since smaller and smaller portions of the diffracted orders contribute to image formation. This implies that any dose changes either driven by changes in the actual dose delivered or due to reflectivity changes of the underlying substrate result in a change in the critical dimensions. For the simplest case of a line-space pattern imaged with incoherent illumination, estimates of the exposure latitude can be made based on the slope of the image. A typical lithographic process will be optimized for the required pitches, and the illumination will be tuned such as to optimize the achievable contrast. Resulting curves for exposure latitude as a function of diminishing κ look like Figure 3.9, indicating the drop in exposure latitude for a dense line-space pattern as a function of κ. Again, these estimates need to be compared to the required exposure latitudes for a 193-nm process estimated to be on the order of 10% [10].

3.2.6 Low-κ Imaging and Impact on Mask Error Enhancement Factor

The mask error enhancement factor (MEEF) describes the relationship between an error in critical dimension on the reticle and the resulting error in the wafer

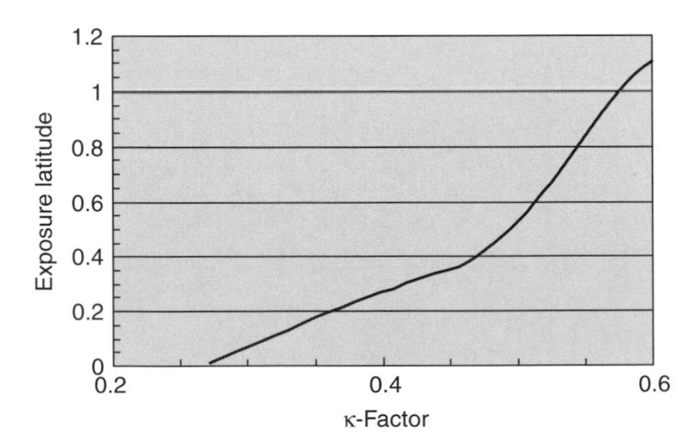

Figure 3.9 Exposure latitude as a function of κ for an equal-line-space pattern. The values are calculated based on a 0.75NA 193-nm exposure tool using conventional illumination, allowing σ to be adjusted for optimum exposure latitude. The exposure latitude calculated is based on an aerial image only.

critical dimension. The MEEF is the ratio of the wafer CD error and the mask CD error. Deviations from the ideal ratio of 1 are due to nonlinearities in the imaging process, which become more pronounced as the κ-factor decreases. Like many other effects, MEEF was not a significant concern for high-κ imaging. MEEF values larger than 1 can qualitatively be explained with graphs like that shown in Figure 3.10, which shows the aerial image as the superposition of a uniform background and the modulation for the case of an equal-line-space pattern. We focus on changes in the aerial image at the equal-line-space point. We assume that the error on the reticle increases the chrome width. In this case there is an overall baseline shift due to a shift in the average intensity transmitted as well as a reduction in the modulation (the strength of the first diffraction order is reduced; equal-size chrome and chrome open result in the best transmission). The slope of the aerial image determines the sensitivity to baseline shifts at the crossover point. For small modulations caused by low-κ imaging, even a small baseline shift results in a fairly large shift of the critical wafer CDs. A large modulation, on the other hand, has larger slopes and therefore exhibits smaller on-wafer CD variations. This graph also qualitatively demonstrates that for a fixed pitch larger MEEFs are to be expected as desired image CDs deviate from the equal-line-space situation. A quantitative estimate for the MEEF of an equal-line-space pattern is given in Figure 3.11. It demonstrates the increase in MEEF for decreasing κ-factor for incoherent imaging ($\sigma = 0.9$), on a 0.75NA 193-nm exposure tool with only the contributions from the aerial image taken into account. As the κ-factor reaches the regime below 0.4, drastic increases in the MEEF are visible. This is particularly a concern for gate-level processes where the tightest CD tolerances are required and mask CD errors become a significant contributor to across-chip line-width variations.

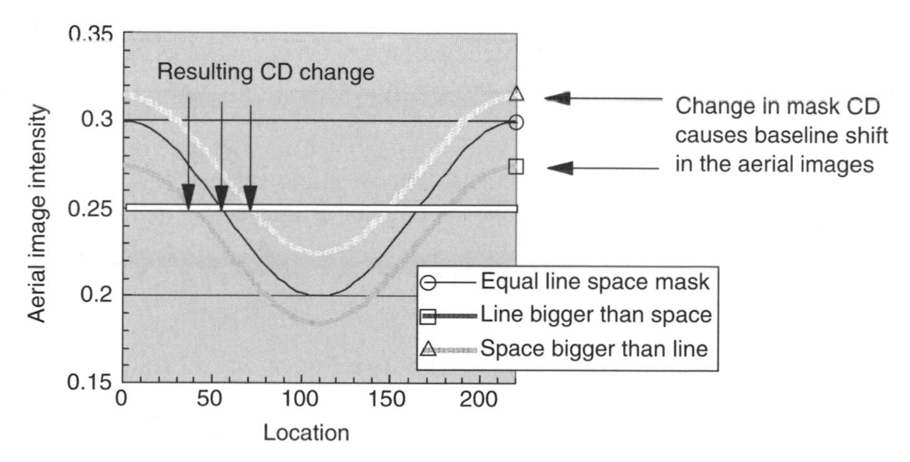

Figure 3.10 Aerial images and MEEF. Changes in mask CD cause a change in baseline, which translates into a CD change. The CD change becomes bigger as the aerial image slope becomes smaller.

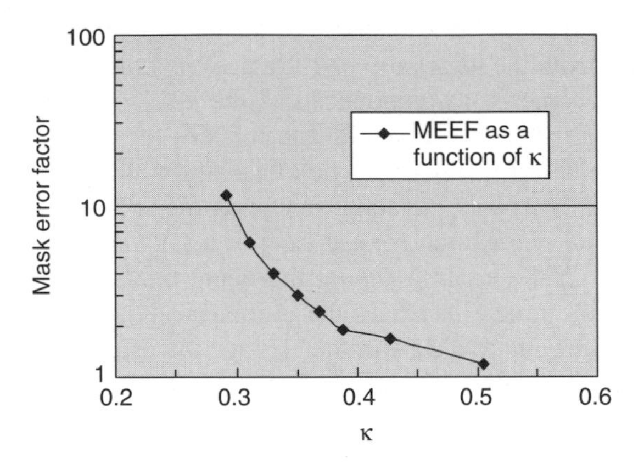

Figure 3.11 MEEF as a function of shrinking dimensions.

3.2.7 Low-κ Imaging and Sensitivity to Aberrations

So far we have treated lenses as perfect, not introducing any errors in the imaging process. As anybody who has used a cheap magnifying glass has found, imaging with real lenses may be anything but perfect. Although lithographic lenses are orders of magnitude better than single-lens magnifying glasses (and also orders of magnitude more expensive), they still exhibit residual errors in imaging performance. These errors, called *aberrations*, can be treated as phase deviations that add to the diffracted order depending on the diffraction angle as well as the orientation of the diffraction order in the plane. They are measured in fractions of a wavelength, and modern lenses achieve wave errors on the order of 20 milliwaves or less. For a 193-nm exposure wavelength, this translates into a tool of approximately 4 nm, indicative of how close to the ideal shape lenses have to be manufactured (and mounted) to avoid aberrations. In general, the wavefront error increases as the diffraction angle becomes larger, which indicates qualitatively that aberration sensitivity increases as κ is decreased. More specific details on the increase in aberration sensitivities are shown in Figure 3.12. These curves have been created through Monte Carlo simulations using the type of aberration and its amplitude as a random variable. The resulting critical dimensions are plotted and then an approximate envelope function is drawn. The progression of Figure 3.12(*a*) to (*d*) demonstrates that the aberration sensitivity increases, as is evident from the increasing spread of CDs indicated by the shaded regions.

3.2.8 Low-κ Imaging and CD Variation as a Function of Pitch

Up to now, only those types of variations have been discussed that are due to process effects. In this section, physical layout is introduced as a potential source of variations. The example is a simple, one-dimensional one, a regular array of lines and spaces. The variation in this case is simply the spacing between

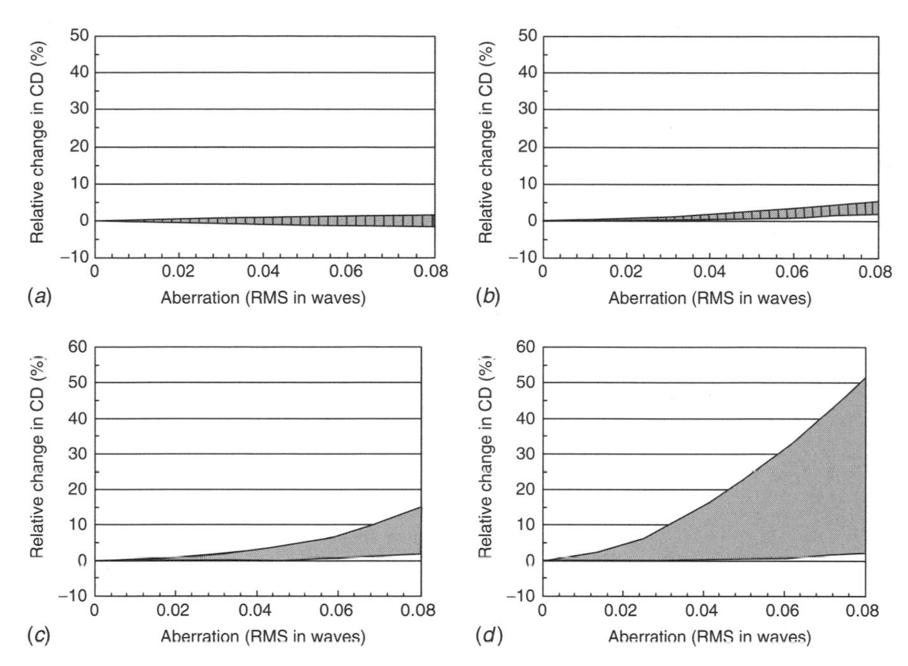

Figure 3.12 Aberration sensitivities for low-κ imaging. Aberrations are measured in fractions of the wavelength (waves). The total wavefront distortion displayed on the x-axis may be composed of a variety of individual aberrations. Depending on the types of aberrations that contribute to the total wavefront distortion, a range of CD variation can be observed on dense line-space patterns. This range widens dramatically as the κ-factor decreases. Shown are four examples, for κ-factors of 0.6 (*a*), 0.5 (*b*), 0.4 (*c*), and 0.35 (*d*).

the lines while the line width remains unchanged. The key effect is highlighted in Figure 3.13: Starting at small pitches, the printed line widths decrease with increasing space between them a result of the nonlinear imaging process in the low κ-regime. This effect becomes more and more pronounced as the κ-factor decreases. Figure 3.13 is based on simulation results for an exposure wavelength of 193 nm, a numerical aperture of 0.75, and conventional illumination with an outer σ value of 0.9. The smallest pitch for each curve corresponds to an equal-line-space pattern and the half-pitch for this layout defines the κ-factor. The overall shape of the through-pitch curve depends on the numerical aperture and the type and coherence of the illumination used. Example of different CD through-pitch curves for different settings of the illuminator are shown in Figure 3.14. The sharp drop in CDs for lines in close proximity is relatively independent of illuminator setting; it is due to negative interference between lines in close proximity to each other. The medium (approximately 500-nm pitch)- and long-range effects are a strong function of the illuminator setting, an effect that is sometimes used to adjust nested to isolated line offsets. Despite the differences induced by different illuminator settings, all of them result in variations

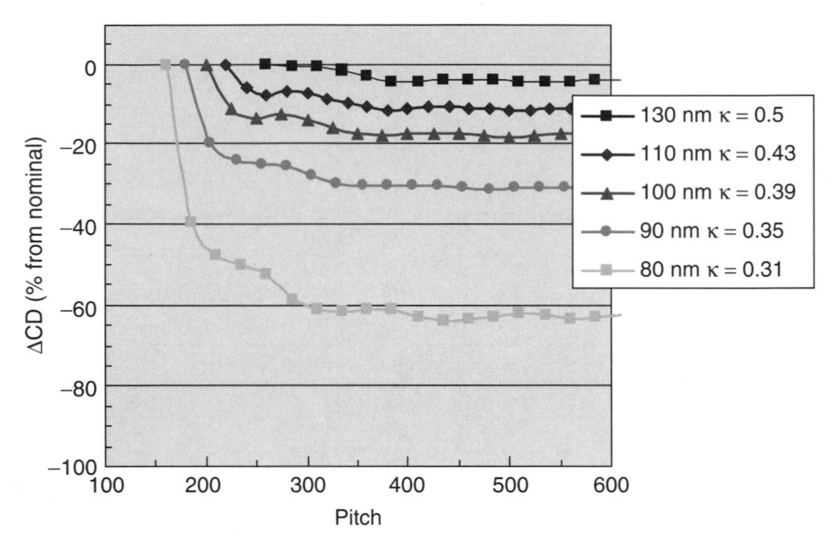

Figure 3.13 CD variation as a function of pitch for a simple one-dimensional array. While the line width remains constant, the pitch is changed. The magnitude of this change is shown for different values of κ.

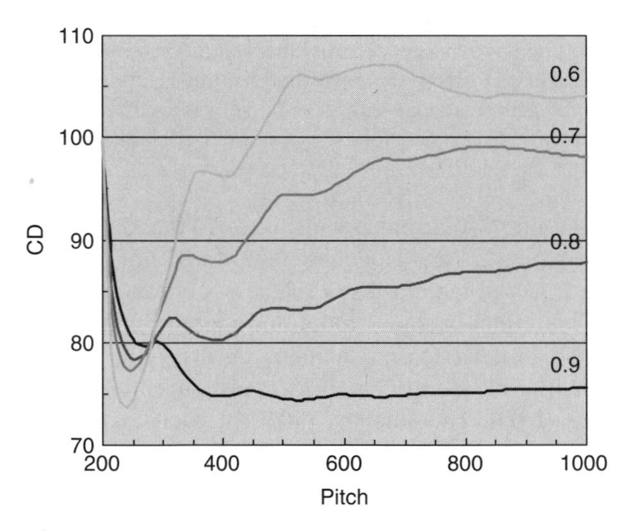

Figure 3.14 Change of CD through pitch curve as a function of illuminator σ.

in the printed CDs, depending on the distance to the nearest neighbor line. This effect also becomes more and more pronounced as the minimum pitch that one attempts to print is reduced. Figure 3.15 shows the maximum delta as a function of κ. For the low-κ regime the expected CD variation approaches the width of the line itself. To maintain a constant line width across different pitches, CD bias

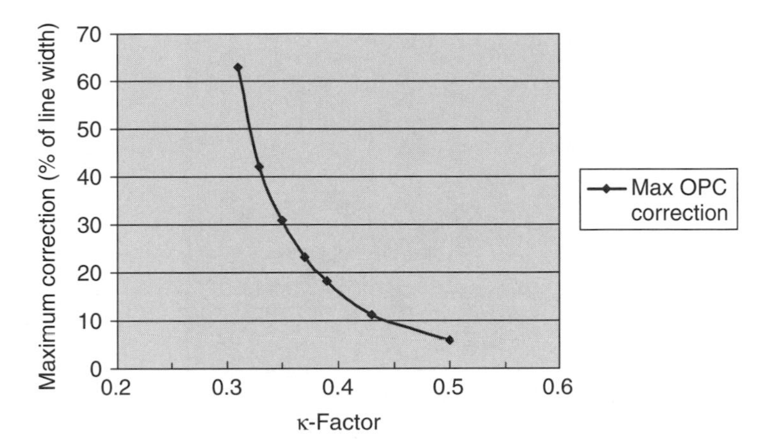

Figure 3.15 Maximum deviation in printed CD as a function of the κ-factor.

is applied to the mask to counteract these variations. Large corrections however are prone to errors. A more detailed description of optical proximity correction, a technique used to correct for such imaging variations, is given in Section 3.3.

3.2.9 Low-κ Imaging and Corner Rounding Radius

Effects similar to those that limit the linear resolution of an optical system also limit its capability to image two-dimensional structures with high fidelity. The fact that the imaging process at its core is described mathematically as a Fourier transform of the mask image explains why aerial images show many of the characteristics of signal processing circuits with finite response time and some dampening. Originally, square input signals are rounded off and output ringing occurs after sharp input signal changes. A sharp corner or turn in a layout is equivalent to supplying a square input signal to an electronic circuit. An example of the resulting aerial image is given in Figure 3.16 for the case of a narrow line emerging from a big block. The contours of the aerial image cannot follow the sharp right angles of the design. Rounding occurs in the vicinity of the corner. Some distance away from the corner the line tends to pinch off, then even farther away becomes wider than nominal, and finally, settles to a fixed value a significant distance from the corner. Quantitatively, this behavior is as shown in Figure 3.17, where the CD of the main line is shown as a function of x. At position $x = -750$, the narrow line emerges out of the chrome block. The aerial image remains significantly larger than the target CD of about 100 nm over quite a significant distance. About 120 nm away from the edge, the image CD reaches the target value of 100 nm but then immediately drops below target over a distance of another 200 nm, followed by a small increase in line width. Similar phenomena may be observed in the case of yet other common design feature: for example, a contact pad placed between two gates. These image deviations

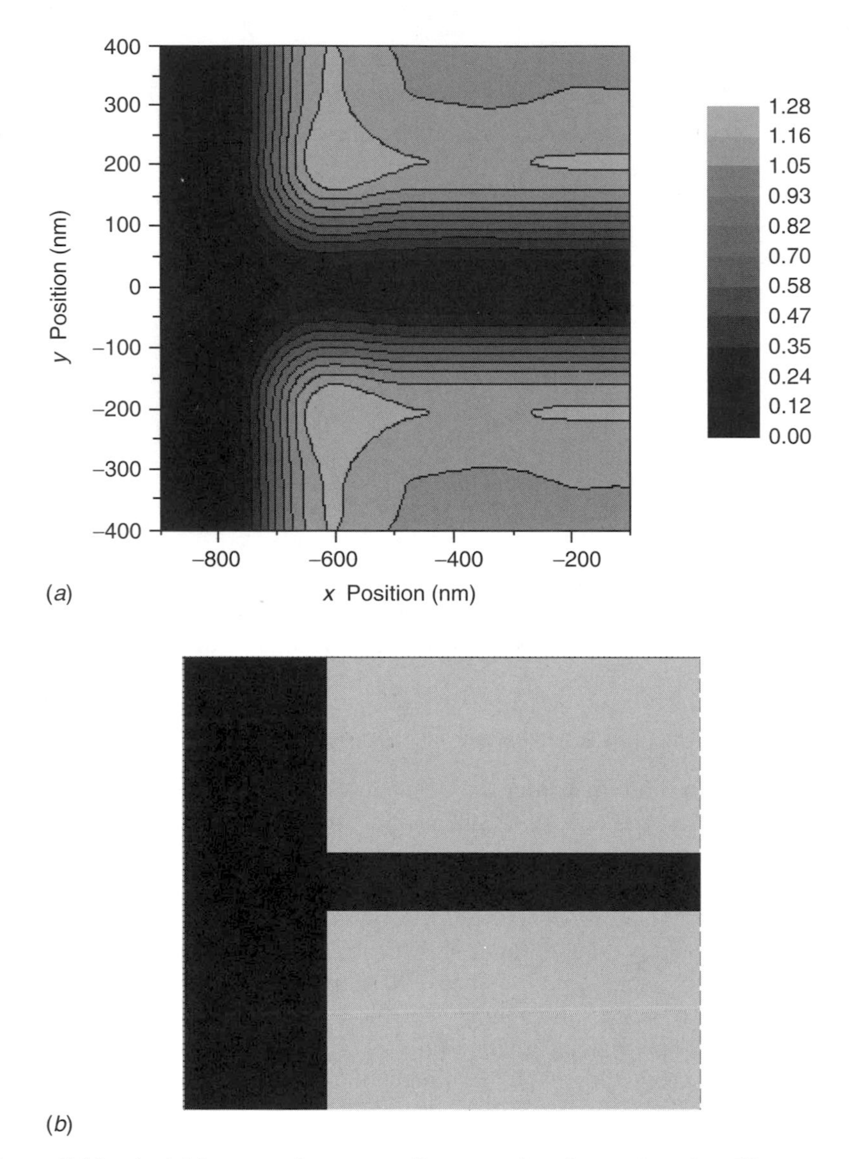

Figure 3.16 Aerial images of a narrow line emerging form a broad padlike structure (*a*). The corresponding layout is shown in (*b*).

have a significant impact on devices. Gates wider than their nominal value have decreased drive currents; on the other hand, increased leakage occurs within the pinch-off area.

The impact of the κ-factor (or resolution) on two-dimensional imaging capabilities is demonstrated in Figure 3.18, where a comparison is made between aerial images for a corner imaged with 0.75NA and 0.5NA 193-nm exposure tools

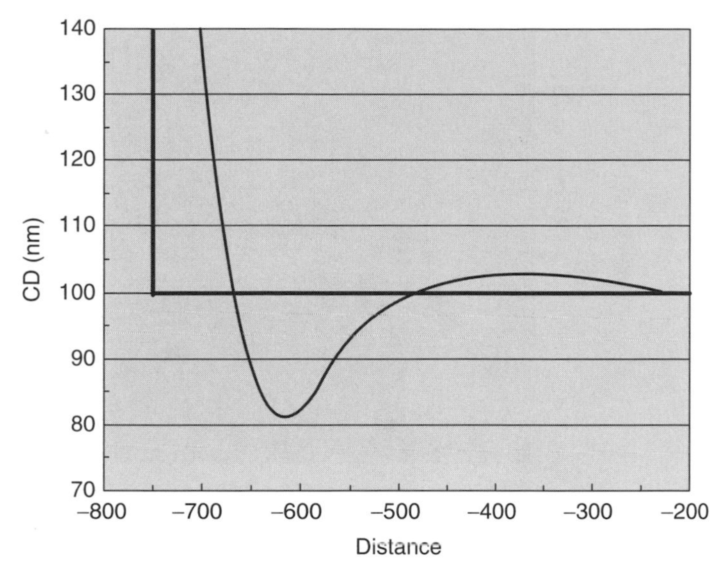

Figure 3.17 CD as a function of distance from the chrome block (layout on Figure 3.16).

using a high-σ-value (σ = 0.9) illumination. The high σ value of the illumination significantly reduces any ringing effects at the corner. The higher-resolution capabilities of the 0.75NA system leads to a drastically reduced corner radius. Low-κ imaging has an additional, somewhat more subtle impact on corner rounding. If, for example, one attempts to image a 240-nm dense line-space pattern together with this structure, the dense structure determines the imaging threshold. The 0.5NA imaging results in a lower threshold than the 0.75NA imaging. However, lower threshold values (contours closer to the blue region) have even larger corner rounding values than those of higher threshold values, thus further deteriorating the performance. A 240-nm pitch structure corresponds to κ = 0.31 imaging for 0.5NA and 0.47 for 0.75NA imaging; corner rounding radii are approximately 210 nm versus 130 nm and thus scale roughly with κ.

3.3 RESOLUTION ENHANCEMENT TECHNIQUES: PHYSICS

In Section 3.2 we described a variety of effects detrimental to lithographic process stability. In the low-κ regime of imaging, the effects are even more significant: reduced ranges for depth of focus and exposure latitude driven by low contrast in the aerial images, increased aberration sensitivities, and reduced pattern fidelity in one- as well as two-dimensional design environments. A variety of techniques have been developed to cope with these effects [11]: (1) specialized illumination patterns, (2) subresolution assist features, (3) phase-shift reticles, (4) optical proximity correction, and (5) optical ground-rule checking.

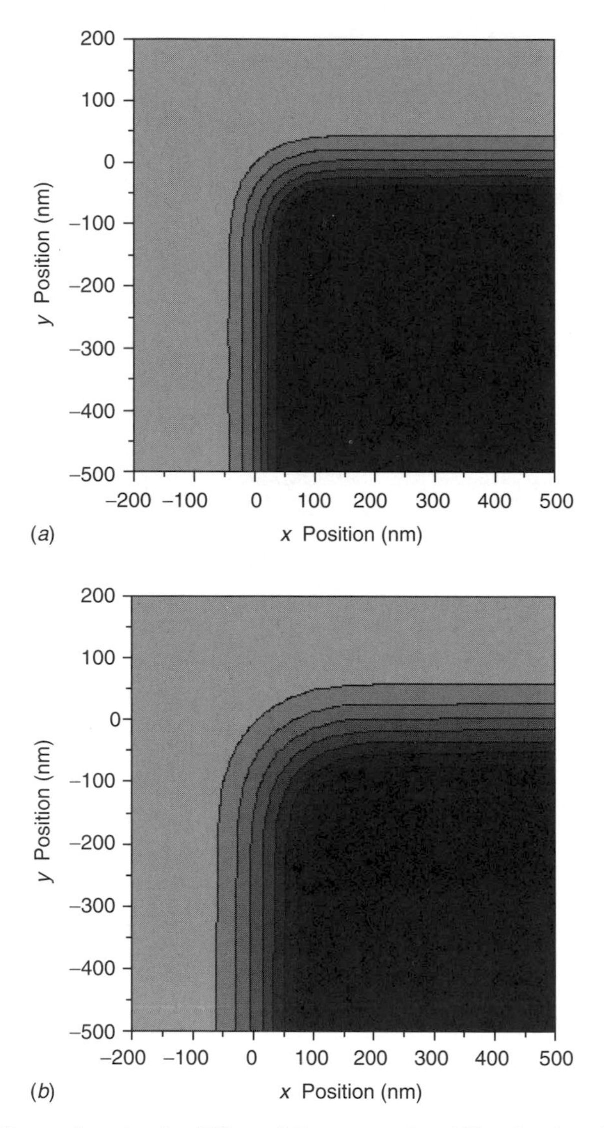

Figure 3.18 Corner imaging for 193-nm 0.9 σ conventional illumination: (*a*) NA = 0.75; (*b*) NA = 0.5.

3.3.1 Specialized Illumination Patterns

In Section 3.1 we have shown that the use of incoherent illumination extends the (theoretical) resolution limit from $\kappa = 0.5$ for coherent illumination to $\kappa = 0.25$ for completely incoherent illumination. The image modulations diminish as we approach the extreme low-κ regime, and there are a variety of negative effects associated with the small image modulations. Any methods that improve image modulation are highly desirable and are, as shown below, the main motivation for using specialized illuminator settings. The mechanism on how to improve aerial image modulation is best demonstrated using the result from convoluting the diffraction pattern with the illuminator. Diffracted light that passes through the lens determines the modulation that forms the image, and the light from the illuminator that does not contribute to diffraction only creates a uniform background and thus also degrades contrast. Thus, inspection of Figure 3.19 indicates that the most serious drawback of large-σ conventional illuminations for imaging tight pitches is that a significant amount of light simply adds a uniform background. Only the crosshatched areas in the figure contribute to modulation; the remainder does not carry spatial information. Following this reasoning, removing the center circular portion of the illuminator is advantageous since it removes a large fraction of the zero-order light, which only contributes to a uniform background intensity while more or less maintaining those portions that contribute to the spatial modulation. This illumination setting, called *annular illumination*, has become a standard offering on most exposure tools [12,13]. Like conventional illumination, annular illumination retains fully rotational symmetry, a feature that is required if the layout allows all orientations of features. Further gains can be made if additional portions of the illuminator are removed. Common illumination patterns are quadrupole and dipole illuminations. They correspond to retaining either four (quadrupole) or two (dipole) segments of the entire annulus. The exact shape of the segments as well as the names used for these illuminator options differs depending on the tool vendor. Simple symmetry arguments suggest that in the case of quadrupole illumination, which has fourfold symmetry, only Manhattan-type geometries will print well. Features drawn at 45° will exhibit the biggest deviation in imaging properties relative to the 0°- and 90°-oriented patterns. Thus, to take advantage of the improved imaging capabilities of quadrupole illumination, the layout has to be limited to Manhattan-type geometries at least for minimum ground-rule features. Printing of 45° lines may be acceptable if they are designed at significantly larger dimensions.

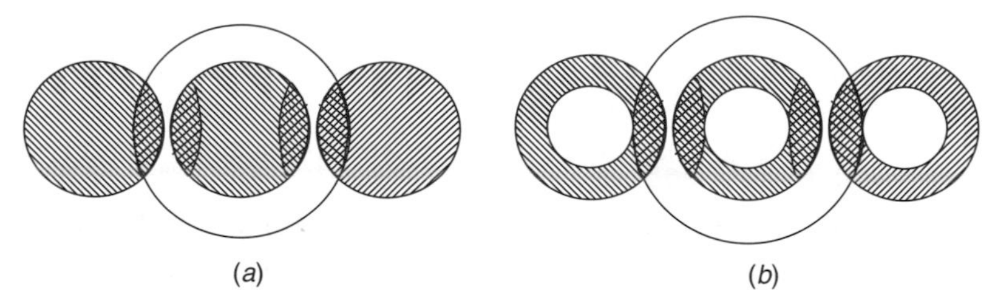

Figure 3.19 (*a*) Diffraction pattern of a dense line-space pattern under conventional illumination. Only the cross hatched areas in the pupil plane contribute to image formation. The hatched areas within the pupil plane only contribute to a uniform background. Modifying the initial illumination pattern to annular reduces the amount of zero-order light that does not contribute to image formation. (*b*) Reduced zero-order contributions using annular illumination.

Dipole illumination has the lowest symmetry of all illumination patterns discussed, and it comes as no surprise that imaging is essentially possible only for lines in one orientation. Even though dipole illumination is the ultimate illumination approach for imaging dense line-space patterns at the resolution limit, the fact that it works only for patterns in one orientation severely limits its applicability. Attempts have been made to circumvent these limitations by using two masks, one for horizontal and one for vertical lines. Unfortunately, double exposures have dramatic cost impacts on the throughput of lithographic tools and therefore on fabrication costs.

Customized illumination patterns provide other advantages that will be illustrated in the dipole case. One important feature is that only one of the two first-order diffracted beams passes through the lens, quite unlike the typical situation for conventional illumination, where both beams contribute. The difference between those scenarios is referred to as two-beam versus three-beam interference. Two-beam interference is sufficient for achieving spatial modulation. More important, two-beam interference is the only approach where a very large depth of focus may be achieved. This large depth of focus occurs in the unique situation where the first diffracted order coincides with one of the poles of the illuminator and the zero-order beam coincides with the other pole. In this case, both first- and zero-order beams have the same angle of incidence. Unfortunately, as the pitch of the patterns increases, zero order and the first diffraction order no longer have the same angle of incidence, and thus the focus insensitivity diminishes as the pitch deviates from the "magic" pitch.

As we have seen, specialized illumination patterns are tuned to optimize imaging performance for very specific layout situations at the price of degraded performance for generic patterns. The layout limitations may be related to orientation as well as pitch. Therefore, the layout chosen becomes a key enabler for this resolution enhancement technique. For the price of a restricted layout space, the lithography process gains exposure as well as focus tolerance.

3.3.2 Optical Proximity Corrections

As seen in Section 3.2, low-κ imaging results in serious deviations of the printed image from the design shape in one- as well as two-dimensional layouts (see Figure 3.20). Changes in line width of one-dimensional patterns due to varying pitches and line widths, corner rounding, and line-end foreshortening in two-dimensional layouts are the main effects encountered. If uncorrected, the impact of these variations is well known. For example, line-width variations at gate level are detrimental to across chip line width variation (ACLV). The same variations in a metal layer may result in increased resistance. Line-end foreshortening will reduce the overlap between contact and the underlying metal line end and thus increase contact resistance. Corner rounding at the inner vertices of an active area may result in increased leakages or device characteristics that shift as a function of overlay.

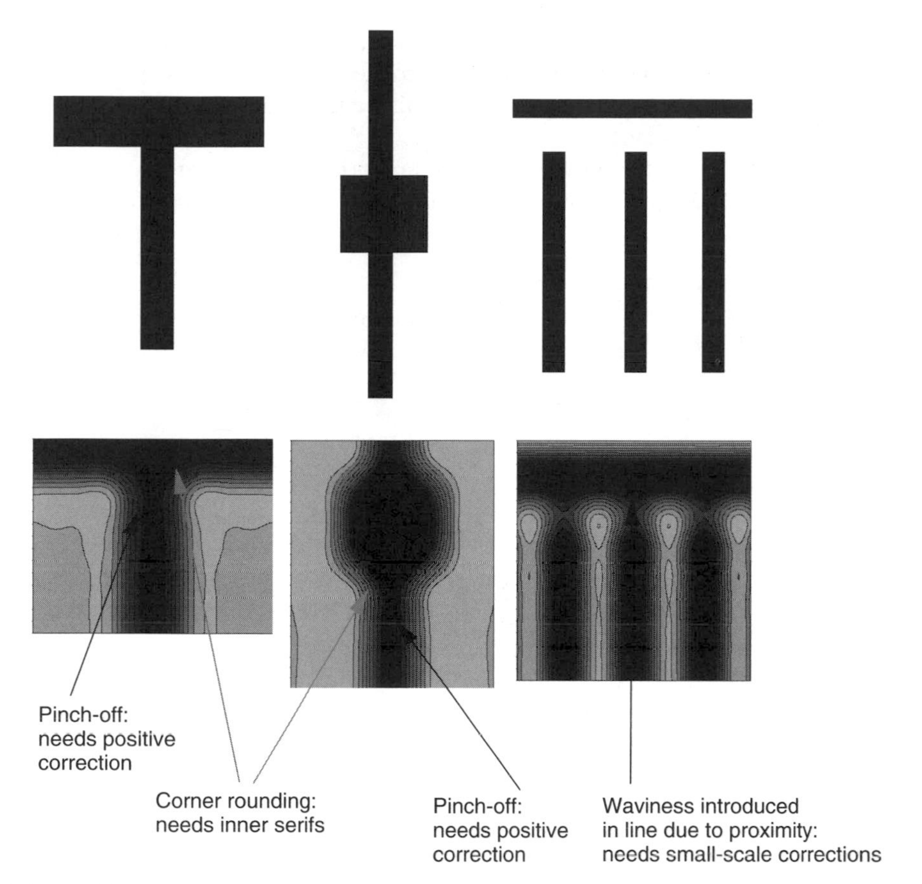

Figure 3.20 Examples of two-dimensional layouts and their OPC impact: T-junction, contact pad, and comb structure. The figure shows the original layout on the top and the resulting aerial image on the bottom. Several variations are visible that need small-scale OPC corrections.

To compensate for these effects, a technique called *optical proximity correction* (OPC) was introduced [14–16]. The name best describes the early implementations, where corrections were made based only on the distance between two lines. Corrections were also made for corner rounding. Nowadays the name is misleading, since effects other than exclusively optical ones are being corrected.

Conceptually, OPC is a computer algorithm that modifies the shapes placed on the mask from those in the original design. The modifications compensate for expected imaging distortions such that the resulting wafer image is as close as possible to the design shape. Figure 3.21 gives an overview of the type of corrections applied. Line-width variations as a function of pitch or line size are corrected by adjusting the width of the feature placed on the reticle. Corner rounding is compensated by adding additional shapes (serifs) to outer corners or removing shapes from inner corners. Line-end foreshortening can be corrected for simply by extending the shape by the appropriate amount, an approach essentially equivalent to the way that line widths are corrected. From a process perspective, it is advantageous to use a combination of line-end extension and serif shapes for correcting corner rounding.

Deviations between the on-wafer pattern and original design are due not only to optical effects; all processing steps involved in the pattern transfer process,

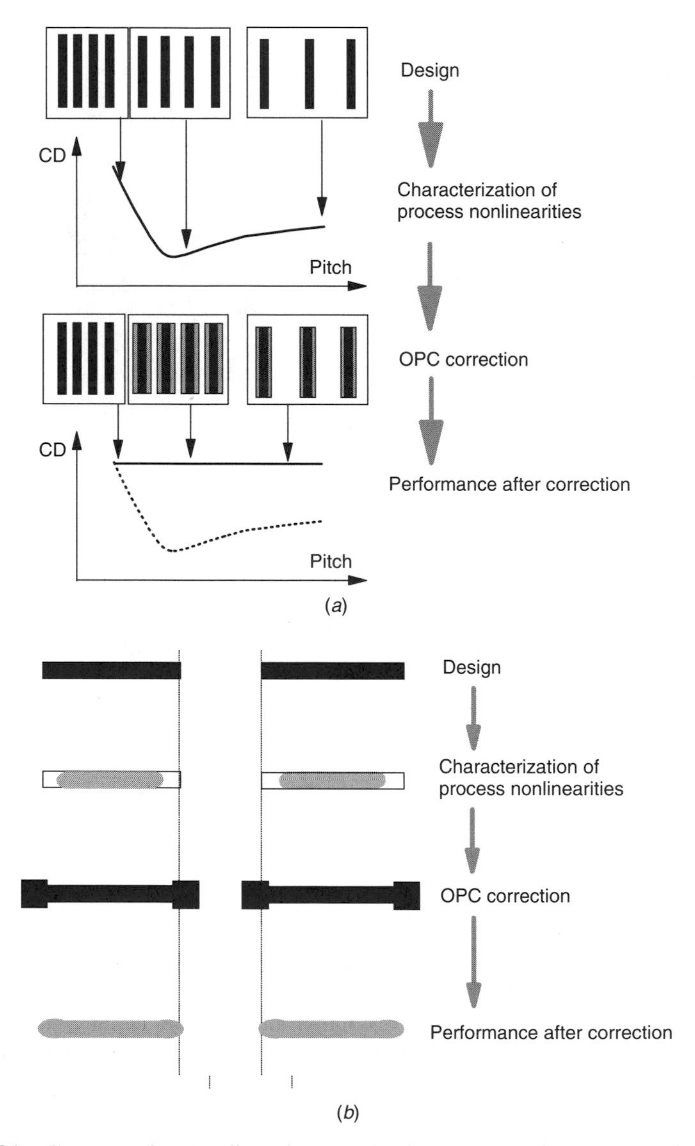

Figure 3.21 Common image distortion mechanisms and their compensation through OPC: (*a*) image CD variations through pitch; (*b*) hammer heads and line-end foreshortening; (*c*) inner and outer corner rounding.

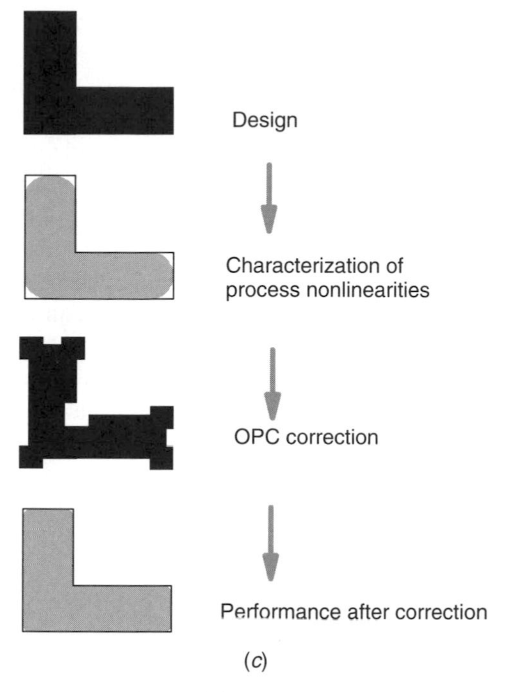

Figure 3.21 *(continued)*

starting with the mask build, wafer etch, and clean processes, contribute to pattern distortions [17]. For a gate-level process, the parameters affecting the electrical gate length go beyond the gate etch process into thermal processing, implants, and spacer processing. A few of these processes will be discussed.

The mask build process is similar to the wafer patterning process in the sense that it also involves exposure and etch steps. For example, perfect corners cannot be transferred onto a mask, due to the finite resolution of the exposure tool that writes the patterns on the reticle. Thus, mask patterns exhibit corner rounding that contributes to line-end foreshortening. Electron beam mask writers exhibit proximity effects similar to optical processes, caused by scattering processes that occur when high-energy electrons hit the resist and the underlying chrome layer. This results in variations of critical dimension for lines at different pitches on the reticle and thus adds to the through-pitch effects caused by the optical tools. Resists are also a key contributor to proximity effects, line-end foreshortening, and corner rounding. The complicated chemical reactions occurring in chemically amplified resists are extremely difficult to model. Image "blur" due to the diffusion processes is only one of the effects that play a role. Wafer etch processes exhibit pattern-dependent effects. For example, etch rates may vary locally, depending on pattern densities, local charging of dielectric materials, and the amount of polymer formation during the etch process. To be correctable through an algorithm that modifies patterns placed on a reticle, the effects discussed have to be systematic, reproducible, and stable. Random variations are not correctable through OPC.

In general, any change in one of the key process components, such as a switch in mask write tool, wafer etch processes, resist changes, and even shifts in the desired critical dimensions, will alter the systematic components and may necessitate a new reticle build. For example, the degradation in through-pitch performance at gate level and thus the degradation in ACLV when increasing the exposure dose in the lithographic process to achieve smaller polysilicon gate dimensions may very well offset the expected chip speed gains.

The optical imaging process is well understood and can be modeled in a fast and accurate manner for any given layout. Nonoptical effects are significantly more complex and more difficult to describe. They are typically handled in a phenomenological manner. A computationally manageable set of parameters is derived from the layout. Based on these parameters, the corrections to the aerial image are computed. Experimental data are generally used to establish the relationship between the parameter and the correction required. Selecting the right set of parameters that allow the most accurate corrections with the least number of parameters is a key ingredient for success, in particular for layers where stringent demands are placed on tolerable error levels. If a large number of parameters are used, it becomes difficult and time consuming to collect the necessary experimental data. Having too few parameters carries the risk that important effects are not captured or cannot be modeled properly.

The general flow of OPC is shown in Figure 3.22. Starting from a specific layout, a certain set of parameters is extracted from this layout. The OPC algorithm then calculates the corrections that need to be applied. Finally a new, modified layout is placed on the reticle. The corrections required are based on experimental data that have been collected and analyzed beforehand. The evolution of layout and the resulting changes to the wafer shapes are also included in Figure 3.21. An OPC algorithm consist of a few key functional units. One such unit divides a hierarchical layout into a number of templates. The corrections for each one of these templates is computed separately allowing for the use of multiplc processors during the computation. Next for each polygon in the layout its segments are further divided into smaller portions, a process called fragmentation. The fragmentation process is required to correct for two-dimensional effects such as the ones seen in Figure 3.16.

At the core of the OPC program is an algorithm that determines how far each segment needs to be moved. This movement may be based on a comparison between the predicted wafer image and it's desired location in which case the procedure is iterated until the desired accuracy has been achieved.

Two different approaches, commonly referred to as model- and rules-based OPC are used, sometimes in combination. In general, *model-based OPC* (MBOPC) derives the corrections based on optical image calculations as well as some physical insight into the non optical mechanisms that cause the pattern deformations. *Rules-based OPC* (RBOPC) tends to be purely empirical. For example, in a rules-based approach, experimental data for the through-pitch curve

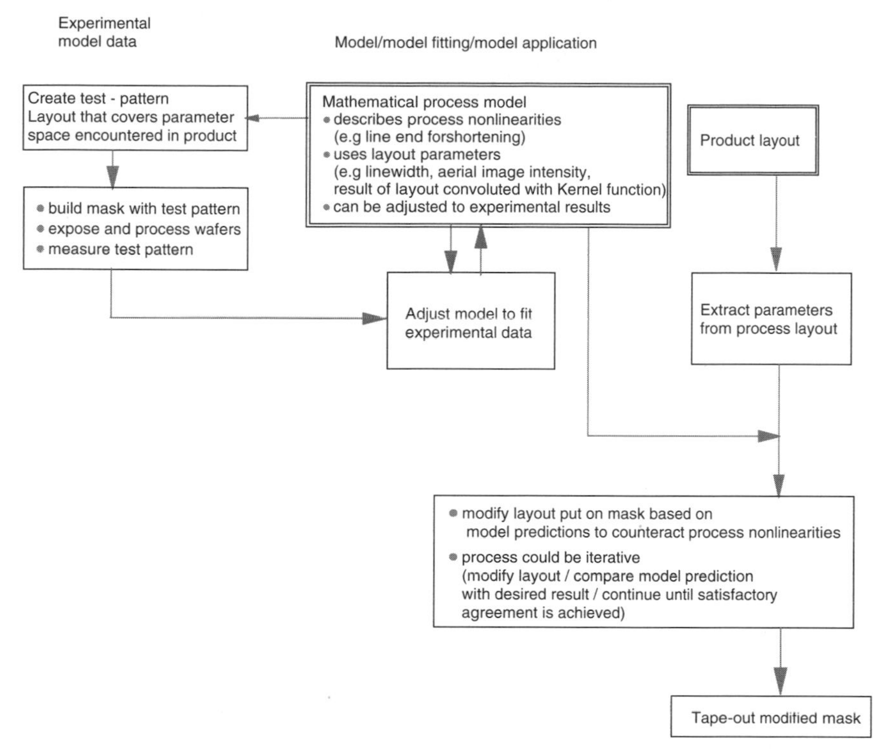

Figure 3.22 Schematic flow diagram of the OPC correction approach.

shown in Figure 3.21 are collected. Then for each combination of line width and space, one determines how much correction is needed to achieve the desired width on wafer. This value is entered into a rules table, one for each combination of line width and space. The algorithm analyzes each layout, determines the line width and the width of the adjacent space, and applies a correction based on the table. Unfortunately, this means that even the simplest one-dimensional cases result in fairly large and complex tables. Line ends and two-dimensional situations create highly complex rule sets. One should keep in mind, however, that rules-based OPC approaches have the highest flexibility of any OPC approach for correcting line-width variations in a simple, one-dimensional layout as well as fairly well-defined and straightforward procedures that allow iterative fine tuning of the corrections. The main disadvantages of RBOPC are the large complexity of the rule sets and the fact that the parameters used for the rules are based on design dimensions and spacings that are not a very efficient set of parameters for defining corrections. Due to these drawbacks, rules-based OPC is unlikely to be used for the 100-nm node and beyond.

The main advantage of model-based OPC is due to its selecting significantly more efficient parameters than rules-based approaches typically do. Physical understanding of the processes involved is used to create numerical models for calculating corrections. Thus, not only are more efficient parameters used, but there are also well-defined formulas that cover a continuum of parameter values without having to list each value separately. Experimental data are collected for selected sampling points in the parameter space and during "model build" are used to adjust a small set of variables in the model.

At the heart of most model-based OPC approaches is a convolution technique that uses the physical layout information and convolutes it with a set of predefined functions called *kernels* [18]. The nth kernel value K_n is given by

$$K_n = \int k_n(x, y) \text{ layout}(x, y)\, dx\, dy$$

where k_n is the nth kernel function. The kernel approach, given a suitable set of kernel functions, is most suitable for calculating aerial images and thus is more adapted to modeling the more difficult layouts, beyond the case of simple lines and spaces. Since the aerial image is a key contributor to the distortion, these type of parameters are a good choice for situations where imaging dominates the distortions. Therefore, model-based corrections do well in two-dimensional situations such as the one shown in Figure 3.16, where image contours oscillate away from the edge, a scenario that is particularly difficult for rules-based corrections. One can argue that the capability to tackle such layouts has been one of the key improvements over older OPC approaches and is of great importance for highly two-dimensional layouts such as SRAMs. Even though the models are not a priori designed to cope with nonoptical effects, a large variety of these effects can be reasonably well modeled using either the kernels themselves or parameters derived from the aerial image. Adequate modeling of nonoptical effects is, however, a significant challenge for model-based OPC [19].

Since the computing effort required to perform optical proximity corrections is quite extensive, the layout is not modified on any point within the design; rather, the program will select sampling points located at the edges of existing layout features. Associated with this point is a line segment of some run length. The program will shift the position of this entire segment based on the calculations performed at one point. Points are added for example based on local variations in line spacing. In addition the examples on CD variations at sharp corners have shown that variations in the local proximity extend over significant distances beyond where the actual change has occurred. To sample these changes adequately, a large number of points has to be added away from any step change or change in local environment. Placement of segmentation points at fixed distances along any layout feature is the simplest and most straightforward strategy. Obviously, the segmentation length has to be chosen small enough so that it accommodates the most rapid possible changes in local environment. The same sampling density is, however, not necessary for long stretches where lines essentially run parallel to each other. As a result, this approach to the placement in sampling points leads to an excessive number of calculation points and thus long run times.

More intelligent placement of sampling points that recognizes changes in local environments and places points accordingly is a more promising alternative.

However, it requires more sophisticated search algorithms that check not only within-shape design variations but also changes in local environment, as pointed out above. The most challenging problem is successful placement of segmentation points on line ends, as this has a significant impact on how the line end is corrected and what final shape can be achieved. A good strategy on how to place the sampling points on the design is as important to successful shape correction as is the ability to model process effects properly.

Model-based OPC is the primary method used to address the one- and two-dimensional image distortions detailed in Section 3.2. Rather than simply transferring the physical layout onto the mask, controlled and deliberate modifications are made to the shapes placed on the reticle. These modifications are intended to counteract imaging distortions. Therefore, model-based OPC has become a key enabling technology for entering the low-κ imaging regime. Its ability to incorporate modeling approaches to process effects gives this technique a key advantage over the older rules-based approaches.

3.3.3 Subresolution Assist Features

Subresolution assist features (SRAFs) are one of the common avenues for enhancing process windows in the lithographic process [20–22]. Unlike optical proximity corrections, where only existing shapes are modified, subresolution assist features are shapes added to a layout that are not present in the design. Essentially, they may be viewed as dummy features added to the mask, based on detailed knowledge of the lithographic process, to enhance its performance. The key mechanism that enables this process window enhancement can be understood by returning to the process window impact of annular illumination. Annular illumination provides a large depth-of-focus enhancement for a specific pitch. This pitch can be defined in the pupil plane as the one where the first-order diffracted beam and the zero-order beam have the same distance from the center. This enhancement effect extends over a limited pitch range around the optimum pitch even though the it tends to drop off rapidly. To extend this benefit to more isolated pitches, one may place a dummy feature between the original lines of the layout such that the resulting pitch is closer to the optimum pitch for the annular illumination condition chosen (see Figure 3.23). Assist features placed at the proper spacing provide a more nested environment for isolated and semi-isolated lines enhancing those spatial frequencies which closely fulfil the two beam condition mentioned above. The concept can be extended by adding one, two, or more dummy features per space. Use of up to four assist features per space has been reported in the literature. It is also evident from the arguments provided above that assist features cannot be used at arbitrarily small pitches. The addition of assist features is useful only if the first-order diffraction pattern created by adding a new feature can pass through the lens. It should be noted that the assist feature generally does not create a new diffraction order but enhances existing high-frequency diffraction orders. This shift provides an advantage only if the gain is significant enough to overcome the loss of modulation at the extreme

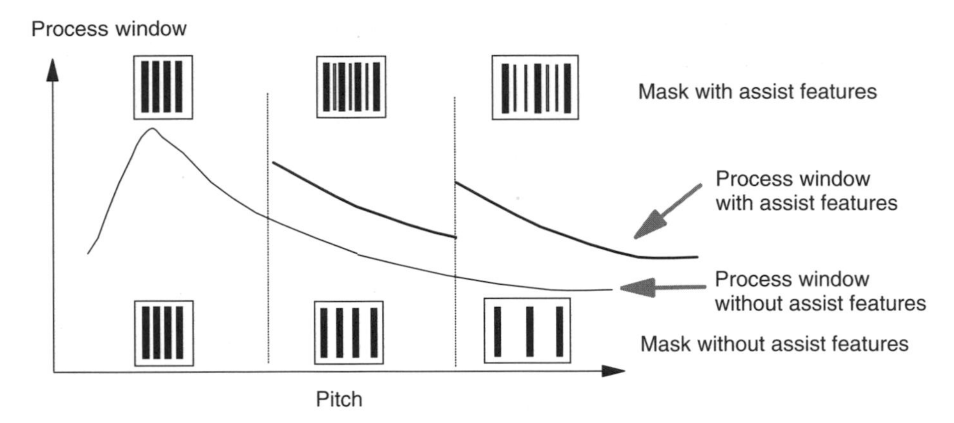

Figure 3.23 Process window as a function of pitch for annular illumination and process window improvement through subresolution assist features.

resolution limits and thus limits the minimum pitch at which assist features may be used.

Assist features are most efficient if their width is large. However, since assist features are not desirable in the final product, they cannot appear in the final resist image. There are two possible strategies to overcome this issue. One of them is to remove the undesired assist features using a second exposure. Unfortunately, such a double exposure drastically reduces exposure tool throughput, increasing fabrication cost, and therefore is generally avoided. The alternative approach is to decrease the size of dummy features to the point where they no longer print on the wafer. This is the most commonly used implementation of assist features and explains the term *subresolution*. Unfortunately, in this implementation one cannot take full advantage of their potential, as their maximum size is limited by printability considerations. Due to their finite size, their efficiency in transferring diffraction intensity from one diffraction order to a higher order is limited. The maximum size of an assist feature that does not print on a wafer depends on a multitude of factors, primarily the NA and type of illumination chosen, the resist used, and the minimum allowed distance between main line and assist feature. The requirement to avoid assist printing also limits how many assist features can intersect at a given point on the reticle. Restrictions to avoid three- and four-way intersections of assist features may have to be implemented. Further restrictions may have to be added based on mask manufacturability and inspection requirements. For example, limits might be placed on the minimum length of an assist feature. Such restrictions are enforced by partially removing assist features after placement, which results in features or part of a layout not being "protected" by assist features. As a result, certain parts of a design may exhibit significantly larger process-induced variations than those of the remainder. Thus, attempts to minimize such layout scenarios are crucial for successful implementation of this technique.

A significant issue for the implementation of assist features is their interaction with model-based OPC programs. For example, the assist rules create discontinuities in through-pitch CD variation. Generally, a very large step is observed in the line-width correction required at the transition between pitches that are not assisted to pitches where an additional assist feature is placed. This step is caused by a topological change in the aerial image. In the case of no assist features, the aerial image intensity increases monotonically as one approaches the midpoint between two lines. The addition of an assist feature, centered between the main features introduces an additional local minimum in the aerial image. Model based OPC approaches that use aerial image parameters to account for non-optical effects may have problems with such drastic changes in the aerial image distributions. Therefore the implementation of assist features requires very carefully calibrated models to avoid large systematic line-width variations caused by OPC errors.

3.3.4 Alternating Phase-Shift Masks

Alternating phase-shift masks, sometimes also referred to as *Levenson phase-shift masks* or *hard phase-shift masks*, are probably the most powerful resolution enhancement technique [23,24]. Not surprisingly, they are also the most difficult resolution enhancement technique to implement, not only from a mask/process technology point of view, but also due to their severe impact on physical layout. The name *phase-shift mask* (PSM) indicates that in these techniques the phase properties of light waves are being utilized. Consider the situation depicted in Figure 3.24. Let us assume that the transparent part of the mask on the right side of the chrome introduces a phase shift of 180° relative to the light passing through the left side. If we now look at the image midway between the two phases, we note that rays from the left and right sides travel the same distance but are 180° phase-shifted relative to each other. The destructive interference of those two beams creates a perfectly dark line centered between the two transparent regions. No other imaging approach achieves a perfectly dark region centered between two closely spaced lines. The occurrence of a line between the two phase regions is an interference phenomenon and the line appears even if the chrome line between the two phase regions is completely removed, a situation referred to as *phase edge imaging*. In this case the line width is determined exclusively by the numerical aperture, dose, and illumination pattern chosen. A look at the diffraction pattern created from an alternating phase-shift mask provides further insight into the mechanism that enhances process windows. We assume, for reasons that will become apparent later, that the illumination pattern used is a low-σ conventional illumination. In the case of a simple chrome-on-glass reticle, the resulting diffraction orders are spaced at integer multiples of 1/pitch. However, for the case of an alternating PSM, the even-numbered diffraction orders and most important, the zero-order diffraction disappear. The argument is simple for the zero-order beam. The intensities from the shifters of equal width on the left and right of the center feature cancel each other, as they have opposite

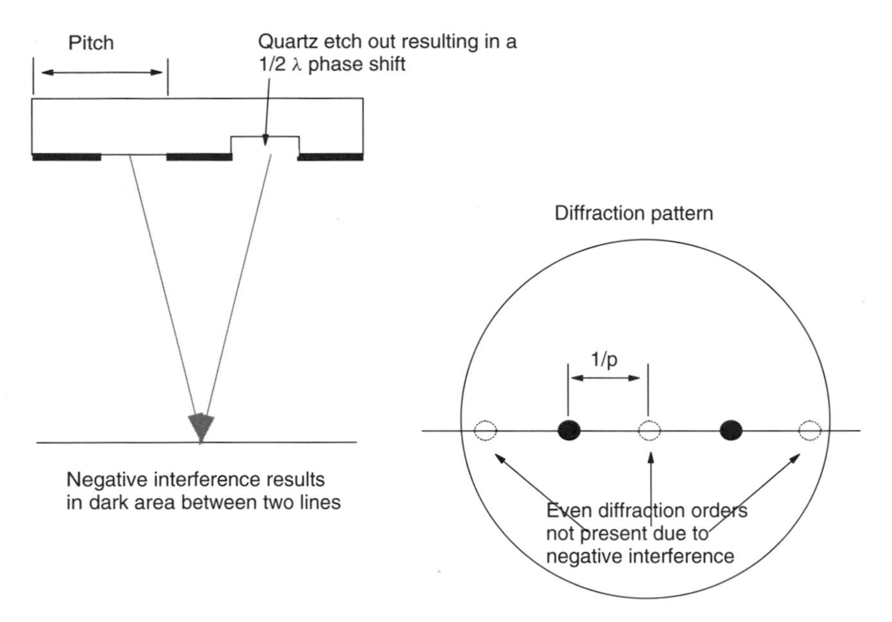

Figure 3.24 Operating principle of an alternating phase-shift mask. Between chrome features, alternating phase-shift masks have clear patterns where the quartz has been recessed and areas without that recess. The recess is achieved by reactive ion etching of the quartz, typically followed by a wet etch. The etch depth into the quartz is adjusted such that light passing through these recessed areas has a $\frac{1}{2}\lambda$ phase shift relative to those areas without the recess. As a result, destructive interference occurs between the shifted and unshifted openings, resulting in zero light intensity between the two openings for the symmetric case shown below. It should be noted that for a dark region to occur, a chrome-covered region between the phase regions is not required; phase edges will result in resist lines printed on the wafer. The diffraction pattern corresponding to a design pitch p is best explained as a diffraction pattern with spacing $1/p$, where the even diffraction orders are eliminated, due to negative interference. Most notably, zero-order diffraction is absent, and thus alternating masks achieve a two-beam interference scenario without the use of annular illumination.

phase. As a result, in the resolution limiting case, where only the first diffraction orders are transmitted through the lens, we have two-beam interference with the two diffraction orders having identical distance from the optical axis. It was shown that in this situation, a large depth of focus can be achieved. Alternating phase shift masks have other advantages over two-beam interference achieved with annular illumination. The depth of focus (DOF) enhancing symmetry of the diffraction orders remains independent of pitch, and the maximum modulation that can be achieved through interference of the $+1$ and -1 diffraction orders is significantly larger than interference between zero and, for example, the $+1$ order, as would be the case for annular illumination.

Alternating phase-shift approaches fall roughly into two categories, a bright-field phase approach and a dark-field phase approach, both depicted in

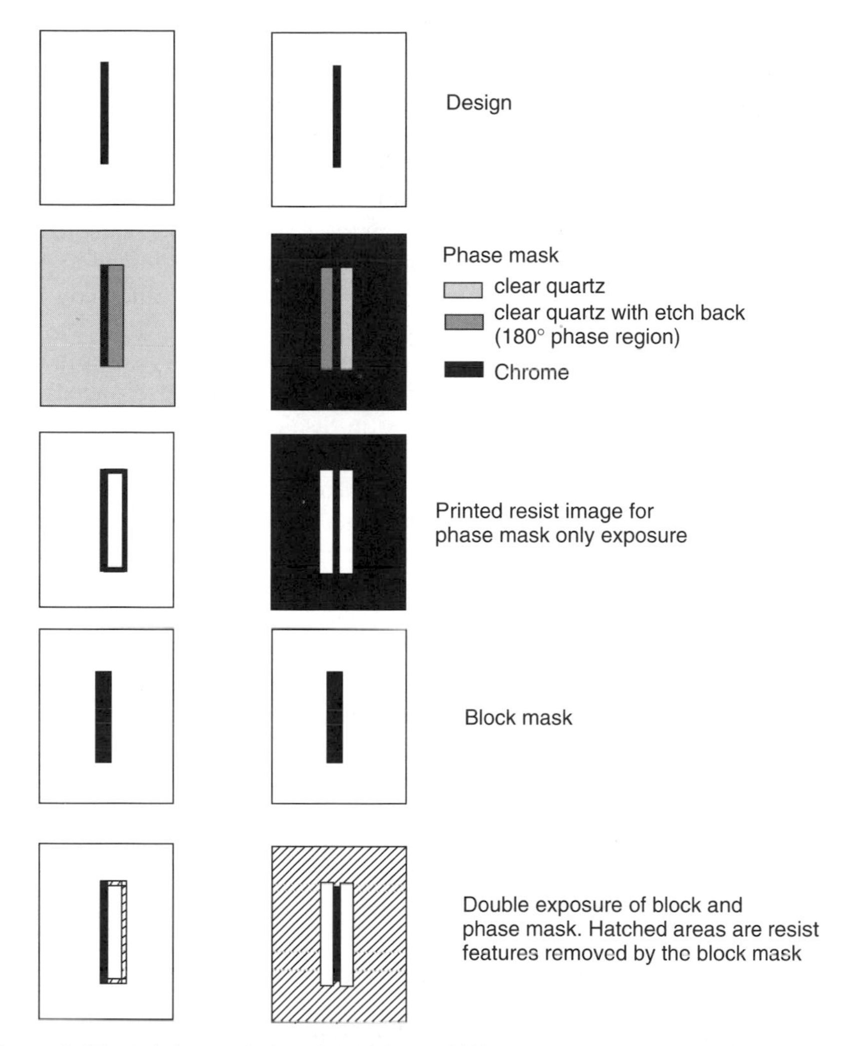

Figure 3.25 Bright- and dark-field phase-shifting approach. Starting from a simple design, the figure shows the patterns on phase mask and the resulting image of the phase-masked imaged in a positive tone resist. In the bright-field mask approach, it should be noted that phase edges result in resist edges that have to be removed. Similarly, for the dark-field phase mask approach, large resist areas have to be removed. This is accomplished through a second exposure using a trim (block) mask. This mask removes undesired resist features while retaining the desired features.

Figure 3.25. In the dark-field approach, two shifters of finite fixed width are placed on both sides of the design line. In the bright-field approach, a 180° phase shifter is placed in a background of 0° openings. In both cases the image created by the phase mask has features not present in the original design. For the bright-field approach a residual line will appear, marking the phase edge at the 0-to-180° transition. For the dark-field approach, a large area of resist is present away from the line that needs to be removed. In both cases, an additional exposure will be required to obtain the final desired pattern. This mask is referred to as a *trim* or *block mask*, since it trims away undesired features, yet protects the desired features.

Implementation of alternating phase-shift masks requires modifications to almost all components of the lithography process. New mask manufacturing, inspection, and repair capabilities are required. Modifications to the exposure tools have to be made to accommodate double exposures efficiently. Data preparation code that can synthesize the required phase and block shapes from an existing layout and the ability of the optical proximity software to account properly for double exposures are required. Finally, the layout itself needs to be *phase enabled*, a point that we will return to in Section 3.4.

With alternating phase-shift masks, exposure tool throughput drops significantly since two exposure passes are required. To minimize delays, the wafer is first exposed with the phase mask and after the first exposure is done remains on the stage. The phase reticle is removed and replaced by the trim reticle, the illuminator setting switched, the wafer exposed with the second reticle and then unloaded from the exposure tool. To reduce the number of reticle exchanges, the next wafer loaded is exposed with the last reticle used on the preceding wafer. Even so, modifications to the exposure tool are required to allow the reticle swap and to minimize delays. The mechanical parts that perform the reticle exchange have to be able to handle a large number of exchanges without excessive wear, reticle alignment and swap have to be performed reasonably fast, and the illuminator changes have to occur at an equally fast rate.

Even more profound than the exposure tool modifications are changes to the data preparation code needed to allow phase-shift mask implementation. Code for creating phase shapes is required. As we will see, except for the simplest possible layouts, this is anything but a trivial task. In situations where the required phase cannot be uniquely determined, phase conflicts are the most common issue. The software has to be able to synthesize the required trim mask shapes. Trim masks typically contain the noncritical design features as well as the trim shapes associated with critical features. Finally, the OPC algorithm needs the ability to simulate double exposures, essentially the incoherent addition of light intensities from the two exposures. Since two masks are available and modifications could be made on either one or both of them, more sophisticated algorithms are necessary than are commonly available.

Finally, phase-shift masks require modified mask manufacturing techniques. The phase-shift characteristic of these masks is created by etching out quartz from the mask substrate. Up to this particular process step, mask manufacturing for a chrome on glass (COG) and an alternating PSM is essentially identical: exposure of resist-coated mask blanks, removal of resist in regions where chrome needs to be etched, chrome etch, resist strip, and mask clean. The alternating PSM masks are then again coated with resist. The subsequent exposure and develop step now opens up those regions on the mask where an additional quartz etch is required to create the 180° phase regions. While it appears that achieving 180° phase shift simply means being able to control the exact etch depth, in reality achieving this requirement is not as simple. The complexity of the optics of phase-shift masks is best appreciated by considering that the light impinging on the quartz opening of the phase shifter creates diffracted light from this opening. The diffracted light enters the cavity below and is reflected off the sidewalls of the quartz or may be blocked on the chrome layer. This simple argument indicates that achieving a 180° phase shift not only for zero-order diffraction but also for the rays diffracted at higher angles is a fairly complex problem. The precise shape of the quartz sidewalls plays a crucial rule. In fact, the only simulations that treat the problem properly take into account the full three-dimensional structure of the reticle and are the subject of extensive research effort [25].

3.4 PHYSICAL DESIGN STYLE IMPACT ON RET AND OPC COMPLEXITY

The necessity to push the lithographic process to smaller and smaller κ-values drives the need for a variety of resolution-enhancement techniques (RETs) such as specialized illumination conditions, model-based OPC, subresolution assist features, and alternating phase-shift reticles. The ability to use these techniques is increasingly intertwined with the particular layout style chosen. There are two approaches for optimizing layout compatibility with RETs. One of them uses the fact that the typical layout is not unique for representing a given circuitry on the chip. Rather, various solutions are possible, and the choice of a particular style is quite often determined by layout convenience. It is this variety of layout styles that is used to optimize a particular design for use with RETs. Although the different layouts may not have pronounced advantages for the designer, sometimes seemingly minor details have a decisive impact on the ability to use RETs. Such rules are most commonly described as recommended rules, design for manufacturing, rules or by use of similar terms and are usually formulated as guidelines rather than as rules. The boundaries are usually fuzzy in the sense that not following these guidelines does not result in catastrophic failures but simply makes the particular layout somewhat less stable to process variations.

There is a second category of layout/resolution enhancement interrelationships. Certain layouts may not be compatible at all with the RETs planned by the process team or result in such severe performance degradations that these layout geometries are disallowed. These design restrictions are typically an issue of controversy between design and process teams. This constitutes a paradigm shift in the data preparation flow and examples will be given in the following sections. In previous technology nodes, the choice of RET has been largely hidden from the design team. The layout team provides a design that then disappears behind a curtain. On the other side of the curtain, the process team implements its RET with essentially little feedback to the design teams. In the low-κ imaging regime this barrier can no longer be maintained and the choice of RET has a profound impact on the layout [26]. In Section 3.5 we provide an overview of the impact of various RETs on layout. For the most part, the philosophy of optimized layouts may be understood by a simple principle: The feature that lithographers can print best is an array of equal lines and spaces. Although the large arrays in DRAMs largely follow this approach, the practice appears to be much more difficult to follow for random logic.

3.4.1 Specialized Illumination Conditions

One of the key enabling techniques for low-κ imaging is the choice of special illumination conditions. As we have seen, the most efficient ones reduce the rotational symmetry of the illumination pattern. The reduction in symmetry is a direct consequence of the fact that light that does not contribute to the pattern formation should be eliminated from the illumination, with the goal of allowing the smallest possible pitch with the best possible process window. Thus, these illumination conditions have one major drawback; they are most efficient only for specific layout geometries. For example, when using quadrupole illumination, Manhattan-type geometries print in a distinctly different manner from patterns rotated 45° relative to them. Based on the orientation of quadrupoles, one of the pattern orientations will have lower contrast and thus significantly degraded CD tolerance and process window performance. Assuming that the quadrupole was optimized for Manhattan geometries, 45° features will typically have a distinctly worse process window and CD tolerance. Dipole illumination has even lower symmetry, and thus only lines in a single orientation can be imaged adequately a quite drastic layout restriction. However the resulting single orientation layouts offer additional advantages, even if they are not used in conjunction with dipole elimination. They eliminate one of the contributions to gate line width variation by disallowing one gate orientation. The necessity for single orientation layouts can be avoided if double exposure is an acceptable option, the resulting technique is sometimes referred to as *double dipole illumination.* Unfortunately double exposures adds significantly to processing cost as it drastically reduces exposure tool throughput. In addition there is added complexity to the data preparation flow. A software algorithm separates features to be imaged with one orientation of the dipole on one of the masks and places the others on the second mask. In its simplest implementation, it may simply separate features based on their orientation. More advanced options have been proposed where the separation is based on the contrast calculated for a pair of design edges [27]. For each design feature the optical contrast is calculated for both orientations of the dipole and placed on the mask that provides the best contrast.

The use of specialized illumination conditions also has implications for what pitches are most easily printable. Figure 3.26 shows representative process windows of fairly simple structure, an array of lines and spaces with fixed line width but varying pitch. The process window is plotted as a function of pitch under the assumption that annular illumination is used. Optimum process windows are achieved in a relatively narrow region around the equal-line-space condition and

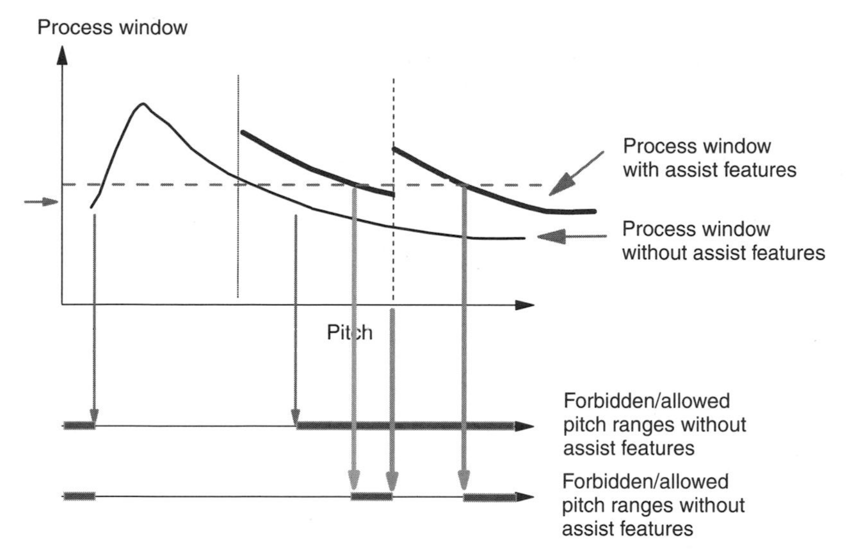

Figure 3.26 Process window for a line-space pattern with fixed line width and varying pitch. A line representing minimum process window requirements has been added to the chart. Based on this requirement, the processing team may establish forbidden zones (i.e., pitches that are to be avoided in the layout for optimum performance).

degrade fairly rapidly away from that condition. Small process windows typically translate into poor CD control or worse, catastrophic failure under normal operating conditions. In addition, these regions typically coincide with regions of large OPC corrections. As a result, layouts using these pitches have increased sensitivity to errors in the OPC model build procedure as well as poor line width control. A possible avenue to circumvent such issues is to introduce *forbidden zones*, which arise from a comparison of achievable process windows with the minimum required process window. Minimum required process windows are derived from a detailed analysis of process variations by the process team. Based on the minimum required process window and figures such as Figure 3.26, those pitches that do not satisfy the requirements can be determined. The details of the forbidden pitches, their ranges and position depend on the type of lithographic approach used and may vary from one fab to another one. A detailed discussion for 0.13- and 0.1-μm generation may be found in [29]. Another consequence from charts such as Figure 3.26 is the generic observation that any minimum isolated features, either line or space, next to large features of the opposite type typically has a severe process window disadvantage. A wide power bus separated by a minimum space from an equally wide ground line are notorious problem areas for lithography and chemomechanical polishing. The lithography team usually has no choice but to back off on their choice of RET in order to maintain yields. A design solution would be much more desirable, and most of the time, even modest increases (10 to 20%) of the minimum line width lead to a dramatic improvement.

Subresolution assist features are used frequently in conjunction with annular illumination. These improve process window performance significantly and help to maintain process windows above the minimum requirements across all pitches. Even so, for very low-κ imaging, process window issues may still remain. The forbidden pitches for an assist feature process tend to be repetitive on the pitch axis. This is a consequence of the fact that the number of assist features is increased at fairly regular intervals on the pitch axis. Process windows tend to be smallest for pitches just below those where an additional assist feature can be introduced. Even if the process window is large enough across the entire pitch range, the discontinuities in the aerial image introduced by the addition of one more assist feature are the most challenging points for any OPC modeling when trying to maintain small CD variation. Therefore, for critical layers such as gate level there is a tendency to try to eliminate these transition regions from the pitches allowed. However, pitch restrictions, are not a single-layer issue. The complete set of ground rules has to be consistent to provide a useful layout framework. An example of how multiple ground rules have to be tuned in unison is shown in Figure 3.27. In this common layout example, two gates are placed next to a corner in the active area, with contacts placed between those gates. In this arrangement a number of ground rules, such as the minimum gate width, minimum contact size, and relative placement of feature have to be consistent with the allowed gate pitches, to avoid having layouts such as the one in Figure 3.27 result in excessive increases in required layout area. So far, our considerations have been for essentially one-dimensional layouts, and issues

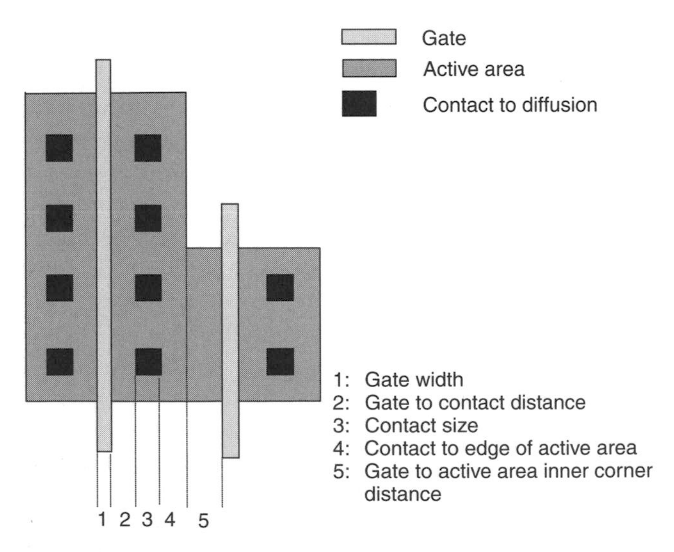

Figure 3.27 Front-end layout example to highlight the interrelationship between gate pitch, contact size, minimum contact to gate, and minimum contact-to-active area edge rules. These ground rules have to be matched to the PC forbidden pitch rules to avoid excessive layout area impact.

associated with process windows and OPC capabilities have been addressed. The most significant improvements in this area that can be made from a design perspective are to maintain a single orientation of the design features as much as possible and to reduce the number of pitches.

3.4.2 Two-Dimensional Layouts

Two-dimensional layouts pose an entire host of additional problems in conjunction with low-κ imaging. For the most part, they strain the capabilities of optical proximity corrections and generally reduce the efficiency of almost any RET. As we have seen, several effects that occur in conjunction with low-κ imaging play a role: Line-end foreshortening, corner rounding, and the fact that any step change in line width causes nonlocal ringing of the image pose challenges for achieving short but straight lines. These effects are less of a concern for long, straight lines. Model-based OPC has improved the capabilities in this area tremendously, although correcting localized line-width variation remains a challenge. Figure 3.20 shows a variety of layouts where the two-dimensional aspects of the layout are significant for OPC. An example is the T-junction, in particular a T-junction with one very short leg, a layout found quite frequently in SRAMs. Also shown is an example of a contact pad placed between two narrow devices. In the last example a series of horizontal gates butted against a vertical gate. The issues with these designs may be highlighted simply by looking at the aerial images for an uncorrected structure. The T-junction exhibits the waviness of the aerial image discussed earlier. Therefore, there is a tendency for pinch off some distance away from the junction. The OPC algorithm attempts to correct for this line-width variation by widening the line in this region. Residual errors in the OPC and imperfect fragmentation lead to reduced line-width control in that portion of the line. If this line is part of an active gate, overcorrecting the pinch-off results in reduced drive currents, and undercorrection leads to increased leakage. The situation is aggravated if one of the segments is short. In this case the OPC correction not only has to compensate for the swing in the aerial image, it also has to compensate for the line-end foreshortening on the short segment. Hammer heads placed on the short segment add further to the swing in the aerial image. To correct these effects with sufficient accuracy the overall result is a highly fractured mask image that not only increases data volumes but is also inherently prone to errors. Since the device is fairly narrow, the performance in terms of leakage or drive current is highly variable. A similar case is that of a contact pad between two narrow devices. The change in width at the center portion of the device leads to a wiggle that occurs in the active gate region. If the gate is fairly short, the hammer heads introduce yet another disturbance to the aerial image on top of the one caused by the contact pad. Improvements in line-width control could be made if the line end and contact pad are kept outside the area where the wiggles in aerial image occurred.

A few examples will be used to highlight the reduced efficiency of RETs when using short-run-length lines. Figure 3.28 shows the layout being considered. The

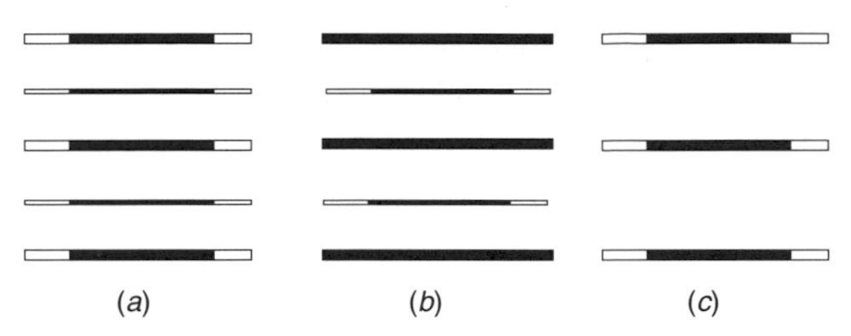

Figure 3.28 Layout example for demonstrating the decreased efficiency of subresolution assist features due to short-run-length lines. These layouts are used for the calculations in Figure 3.29. (*a*) Layout with assist features; the assist feature and the main line are of the same length. (*b*) Layout with assist features; only the assist feature length is shrunk. (*c*) Layout without assist features.

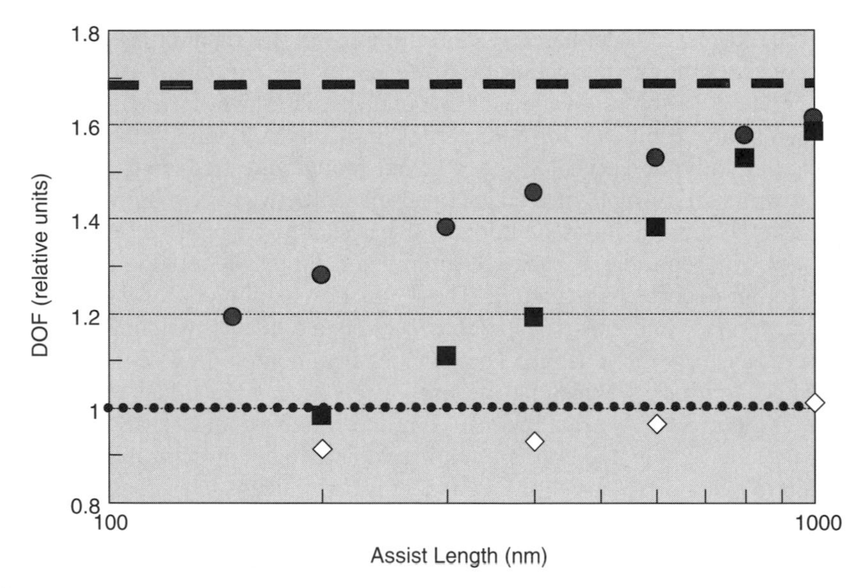

Figure 3.29 Decrease in process window as a function of run length for the layouts in Figure 3.28. This graph demonstrates the diminishing benefit achieved by using short-run-length lines with assist features. The main effect is the diminishing benefit from short-run-length assist features. Full squares, Figure 3.28(*a*); circles, Figure 3.28(*b*); open circles, Figure 3.28(*c*).

design consists of an array of lines spaced at 660 nm, which are being imaged using a 193-nm exposure with a 0.7NA annular illumination. Subresolution assist features are used to enhance the process window for these semi-isolated lines. Two assist features are placed in the space between infinitely long lines with a target CD of 90 nm for each line. Figure 3.29 provides an overview of the process window for the layouts shown in Figure 3.28. For infinitely long lines, the depth of focus gain achieved through the use of subresolution assist features is close to 70%, marked with a dashed line. We now look at a case where both the assist feature and the main line are made shorter and shorter (solid squares in Figure 3.29). The process window advantage starts to drop for feature lengths as large as 1 μm and has vanished completely for lines of run length 200 nm. Even at 400 nm, run lengths of less than one-third of the total process window gain are retained. Closer study of the problem shows that the DOF loss is a combination of two effects, one of them being that the efficiency of the assist features diminishes as they become shorter. This case is represented by solid circles. For this scenario the length of the centerline remains infinite; only the length of the assist feature is shortened, demonstrating purely the effect of shorter assists. The second factor is the length of the main feature, represented by the open diamonds in Figure 3.29. In this case no assist features are used and only the lines are made shorter. Process window loss is moderate but still present. This example is intended to highlight the fact that short-run-length features are detrimental to RETs such as subresolution assist features or specialized illumination conditions.

It should be noted that the process window gain achieved with the assist features and the details of how severe and how fast process window degradation occurs with decreasing runlength are functions of a variety of parameters, such as pitch, illumination condition, assist placement, and size of assist features and varies greatly. However, the generic message that short-run-length features do not lend themselves to efficient RETs remains unchanged.

The effect is linked fundamentally to the fact that well-defined diffraction orders for an array of lines are becoming increasingly less localized as the line length is reduced. RETs, on the other hand, optimize process window performance of patterns with diffraction orders in fairly well localized regions of the pupil plane. As a larger and larger portion of the pattern lies outside the optimum performance region of the resolution enhancement technique, it becomes less and less efficient. The physics of most RETs is based on arrays of equal lines and spaces, long lines on a fixed pitch resulting in highly localized diffraction orders, and this is when they work best. The more that a layout resembles this pattern, the more likely it is to lend itself to RETs and the smaller the strain on the OPC correction. Designs with short-run-length features with many bends and twists are a sure sign of a layout that will put enormous strains on the OPC program and have little to benefit from RETs.

Short-run-length features occur in a multitude of situations, some more obvious than others. For example, the layout scenario shown in Figure 3.30 results in a short-run-length assist scenario. While the array of horizontal lines can be fairly well protected using assist features, assist protection for the vertical line is poor at best. Assist feature codes will attempt to place the assist in the fashion shown in the figure [30]. Horizontal assists have to be cut back to avoid touching the main features. Further complications arise from the fact that three-way intersections are typically not allowed for assist placement as the intersection may no longer be 'subresolution' leave resist residuals on the wafer. If the horizontal line in the center has no assist protection it may turn into a severe process issue. If use of the horizontal line in Figure 5.30 cannot be avoided, increasing the distance between

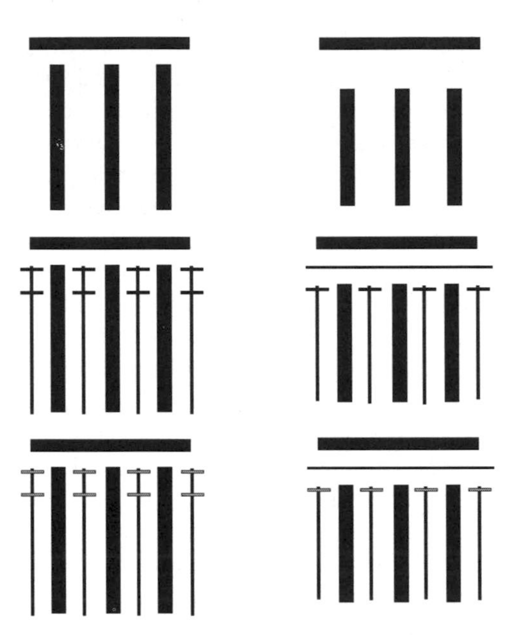

Figure 3.30 Comparison of assist feature placement at two comblike structures. The spacing between the horizontal and vertical lines has been increased in the second example. The top row shows the original layout; the middle row shows the initial assist feature placement; the bottom row shows the remaining assist features after the cleanup code. In the left example the cleanup code has to remove all vertical assist features since they create undesired assist intersections. In the right column the spacing is big enough such that one continuous assist feature can be placed at the horizontal line. The left column results in severe process difficulties. The horizontal line not only has no assist features and thus extremely poor process windows, it also has complicated OPC treatment, due to the strongly varying neighborhood.

the vertical line ends and the horizontal lines to the point where at least one assist feature can be placed is certainly one option. However, this results in a significant area increase. The most area-efficient solution in such cases is typically to change the line width of the center horizontal line to a noncritical line width. Assists are then no longer required to maintain acceptable process windows.

3.4.3 Alternating Phase-Shift Masks

Alternating phase-shift masks have the most radical impact on physical layout. The process window improvement for features where alternating PSM may be applied versus those where they may not is the most pronounced of all the RETs. Layouts that cannot be phase shifted are generally going to be the yield-busting features if kept at minimum dimensions and have to be drawn at dramatically larger dimensions. While non-phase-shifted layouts are highly problematic, and this problem is aggravated by the fact that the implementation of alternating phase-shift masks introduces a range of new issues that increase the likelihood of layouts that cannot be phase shifted. To understand these complications, we start with the fundamental rules associated with alternating phase-shift masks [31].

Each critical line is bordered by two phase regions, a zero phase region on one side, an 180° phase region on the other. A variety of rules usually govern minimum and maximum width of each of the phase regions and minimum distances between phase shapes of different phase. Overlap is allowed only for regions of the same phase, and the densest packing is possible only for this type of tiling. The simplest rules account for the fact that the mask-making process requires a minimum distance between phases of opposite sign. Process considerations may dictate the maximum allowed phase width and the minimum line-width that does not require phase shapes. The basic design implications of these rules are illustrated in Figure 3.31. In addition to these, there is a set of rules intended to prevent design geometries that result in phase conflicts. Phase assignments are made following linear paths through the design; a phase on one side of a critical line implies the opposite phase across the line. Phase conflicts are situations where the phase assignment for a particular space is contradictory, depending on which path is chosen to arrive at the phase region. Common examples of layouts that result in phase conflicts are shown in Figure 3.32: critical T-junctions and embedded line-end scenarios, sometimes also referred to as *belt buckle layouts*. These layouts are only simple examples of a much more generic class of design errors called *cyclic errors* that arise in conjunction with phase-shift masks. A

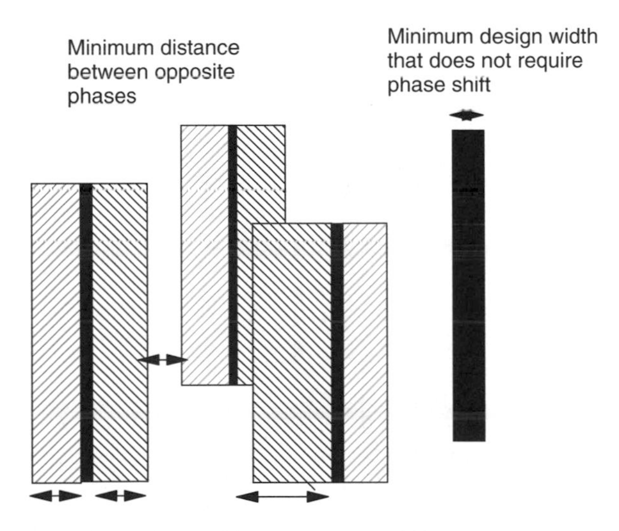

Figure 3.31 Basic layout scenarios for alternating phase-shift masks. Shown are some of the fundamental rules for phase-shift masks: minimum and maximum phase width, minimum distance between opposite phases, and minimum design width that no longer requires phase shift.

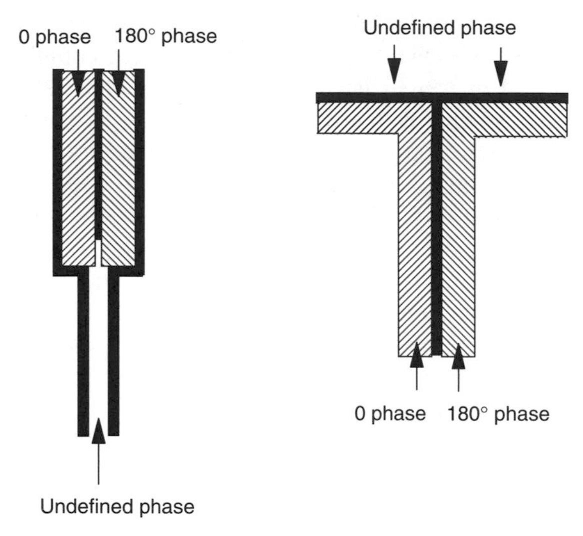

Figure 3.32 Layout situations that result in phase conflicts: T-junction and odd–even path.

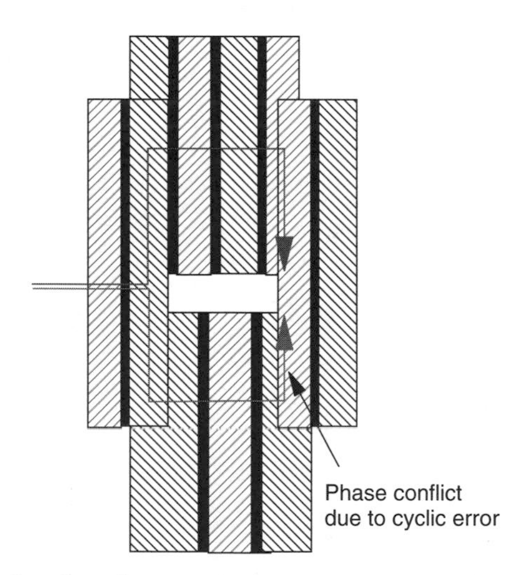

Figure 3.33 Example of cyclic errors. The error occurs as the green diverge and join back together. However, the red path crosses three design lines, whereas the green line only crosses two. These errors pose significant challenges to design rule checking, as they are nonlocal.

case of a cyclic error is depicted in Figure 3.33. Phase assignments are done in sequential fashion. In the layout example of Figure 3.33, these paths split and then recombine in a different location on the layout. Between the point where the path separated and the point where they recombine, each path may have crossed an even or odd number of lines. If one path had an even number of phase transitions and the other one an odd number, a unique phase assignment at the point where the paths recombine is no longer possible. This set of failures highlights the significant challenges facing a ground-rule checking routine. Errors are no longer local in the sense that they can be checked between adjacent features [32].

The difficulties associated with resolving phase conflicts have led to a paradigm shift in the interaction between design and lithography process [26]. Prior technologies have had a clear separation between design space and process space, and the RETs used for the most part had no impact on the design. Limitations and process concerns resulting from the use of RETs were communicated in a set of design rule restrictions similar to those discussed in earlier sections. The complexity of the phase conflict problem forces a change in this methodology. One solution is to move the phase-shift tool into the hands of the design team, providing them with the ability to review phase assignment and potentially, even the expected trim mask shapes. Responsibility for ensuring alternating phase-shift mask compliance is then moved into the hands of the design team. Resolution of errors within the sometimes fairly complex framework of alternating phase-shift rules, combined with the standard ground-rule set, may become a frustrating experience. It is worth mentioning that the alternating phase-shift rules typically include not only rules for phase assignments. There are, for example, additional rules describing minimum distances between phase shapes of opposite sign, rules on overlap of phase shapes, and minimum trim mask sizes, to name a few. As these parameters are likely to be fairly specific, they will not only impede migration from manufacturer to manufacturer but also migration to more advanced lithography approaches. For an ASIC design approach, phase compliance is required at two levels. For one, the layout within individual library elements has to be phase compliant. Once this is accomplished, further complications arise when the individual library elements are assembled. To avoid errors in this phase, individual cells have to be phase neutral to the outside by ensuring that large enough buffer zones are provided around the edge of the cell. Some relief from this can be obtained by attempting to find relatively simple cell boundary colorings so that the rowlike arrangements of cells found in ASIC designs can be put together using relatively simple tiling procedures. This approach is discussed in Ref. 33.

An alternative approach, termed *radical design restrictions*, keeps the complications of phase assignments and trim mask layout away from the design team entirely, yet is no less invasive to traditional design practices [26]. In this approach, sometimes also referred to as the *coarse grid approach*, designs are restricted to one orientation and integer multiples of the contacted pitch. Complicated two-dimensional layout scenarios are avoided in this framework. This approach has several highly attractive advantages. It provides enablement for more than one RET, addresses manufacturability concerns that arise from tight two-dimensional situations and is expected to reduce redesign efforts when migrating to a new technology node. The major challenges for this approach are to keep area impacts to a minimum and to minimize increases in layout complexity on other design levels—for example, the local wiring levels—driven by simplification of the layout at gate level.

As we have seen above, the move to aggressive κ-factors in lithographic processing changes the relationship between design and data processing. Previous technology nodes for the most part communicated only minimum sizes for design features, be it line, space, or area, plus a set of rules that govern the relation between different layers so that overlay tolerances do not lead to chip failure. With the 100- and 70-nm node technologies relying increasingly heavily on RETs, new sets of design rules, driven by the need to maximize the benefit of the new techniques, have been introduced. Restrictions on pattern orientation, forbidden zones eliminating certain space ranges from the design, and restrictions on the minimum run length of features will be required increasingly. Use of alternating PSMs will have an even more fundamental impact on physical layout. In particular, the gate level, with its stringent CD tolerance requirements, will lead the way in this area, with other levels likely to follow, particularly in the 70-nm node. Due to the competitive pressure in the foundry business, companies will be extremely reluctant to pass on these restraints to their customers. The design teams should be aware of the fact that the physics of optical imaging, the tools for exposure, as well as the implementation of RETs are fairly standardized across the industry. An educated customer (i.e., a design team aware of the implications of low-κ imaging and how to optimize layouts for full efficiency of lithographic RETs) will benefit by having designs more tolerant to process variations and therefore more likely to yield. The benefits for the design teams are obvious, a significant improvement in lithographic performance, which is likely to result in improved time to market.

3.4.4 Mask Costs

In the early days of lithography, reticle costs were an insignificant contributor to the total cost of chip manufacturing. This situation has changed dramatically; the cost of a reticle set for a 90-nm node product is on the order of $1 million [34]. Mask cost is the fastest-growing component of the lithography process [35], outpacing the percentage cost increases for wafer processing facility as well as exposure tools. This is of particular concern for ASIC designs, where only a small number of wafers need to be run to satisfy customer needs. A survey of member companies performed by Sematech [36] indicates that the average number of wafers exposed with one mask set is approximately 500 wafers for logic ASICs and about 8000 wafers for DRAM manufacturers. Extrapolation of recently published estimates for wafer processing costs for large-field 90-nm-generation ASIC designs indicates that reticle costs amount to 60% of the total cost of lithography [37].

The push toward low-κ imaging and RETs that enable it have contributed significantly to this cost increase. Mask costs are driven by two principal factors: mask write time and mask yields. Mask yield in the context of this discussion

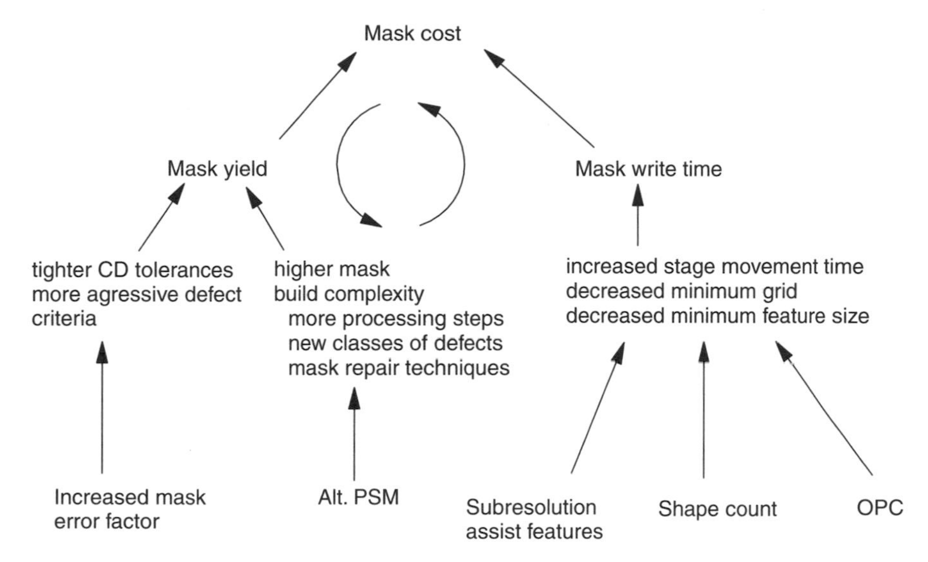

Figure 3.34 Interrelationship between RETs and mask cost. Various RETs (alternating PSMs, assist features, etc.) affect mask yield and mask write times, which in turn affect mask cost.

indicates mask yield to all specifications that include minimum defect criteria and CD requirements. Figure 3.34 summarizes factors and mechanisms that increase mask cost due to the use of low-κ imaging and RETs. Starting from the bottom, a variety of factors arising from the move toward low-κ imaging regime are listed. Some of these are generic to any technology shrink; the increase in the number of shapes per reticle field and tighter image placement requirements are generic to any reduction in ground rules and therefore are mentioned only for the sake of completeness. Both affect mask write times. Tighter placement accuracy increases the amount of stage movements and thus extends write times. Increased shape counts increase the overall data volume that needs to be written. Since advanced mask write tools are among the most expensive pieces of equipment in the semiconductor industry, prolonged mask write times increase mask costs. As we have seen in Section 3.2, low-κ imaging results in increased mask error enhancement factors (MEEFs). Since CD tolerances scale with the technology node, the contribution from the mask to the overall CD budget can only be maintained by tightening mask CD control. More stringent mask CD requirements, in turn, reduce mask yields.

From a defect perspective the mask manufacturer faces two challenges. In the low-κ imaging regime, sensitivities to mask defects are increased. In addition, masks such as alternating PSMs increase the number of processing steps required, with each step adding a small but nonnegligible increase in defectivity. Furthermore, alternating PSMs result in new classes of defects (phase defects) unknown to simpler mask processes and in quite a few cases difficult to repair.

Both the use of optical proximity correction and the addition of subresolution assist features have decreased minimum feature sizes, contributing dramatically to the increase in data volume passed on to the mask write tool, which in turn has drastically increased mask write times. In addition, smaller feature sizes have complicated mask inspections. All these factors have either increased mask write times or decreased mask yields, thus causing mask costs to rise. It is worth noting that extended mask write times in themselves contribute to reduced mask yields, since control of stage movement and stability requirements for the e-beam column become more stringent.

Faced with the dilemma of skyrocketing mask cost there is growing concern in the industry as to the ability to develop new designs. Several concepts and approaches have been developed to address or at least mitigate the impact of rising mask costs. One of the more obvious approaches to reducing effective mask costs is the *multiproject wafer* or *shuttle wafer concept* [38]. Simply put, the chip manufacturer reduces the mask cost per design by incorporating multiple designs on a single reticle. Effectively, the total number of wafers for a mask set has been increased and now is essentially identical to the sum of all wafers required for all the designs. Individual chiplets may be processed through any necessary data preparation and OPC operations separately and then merged before final data preparation. This eases the timing issues when receiving and processing a multitude of designs. Although multiproject wafers appear to be an ideal means of reducing mask costs, a few concerns are worth mentioning. Efficient use of this technology is possible only if wafer processing for the various designs is exactly identical, including all metal levels. Otherwise, a large portion of the wafers are useless, eliminating or reversing the cost advantage. Efficient use of this approach is facilitated greatly if the individual chiplets conform to a standard size, which minimizes the number of good chips that are lost during dicing.

Other avenues for reducing the impact of rising mask costs are to shift away from custom chips toward field-programmable gate arrays (FPGAs), programmable ASICs, or software implementation of otherwise hardwired logic functions on fast microprocessors [39,40]. From a chip manufacturer's perspective, field-programmable gate arrays offer the advantage of completely standardized wafer processing using one mask set. The end users' required functionality is programmed into the individual chip. Unfortunately, their area requirement is still one to two orders of magnitude larger than comparable ASIC circuitry and consumes significantly more power.

Between ASICs, on the one hand, and FPGAs, on the other, there are *structured ASICs*. For structured ASICs, wafers are processed at the chip manufacturer through the entire front end, active area, gate, first contact, and even early metal layers using a standardized mask set. The wafers are then stocked at the wafer manufacturing line for further processing. It is in the top two or three metal layers that the individual logic elements distributed over the chip are connected and the final functionality of the chip is implemented. Each design now essentially has only to carry the mask cost for the last three layers and only a fraction of the mask cost for the other layers.

3.5 THE ROAD AHEAD: FUTURE LITHOGRAPHIC TECHNOLOGIES

Looking at the history of optical lithography, it is obvious that in the past, smaller dimensions have been achieved following a more-or-less evolutionary path:

- Increasing the numerical aperture
- Imaging at smaller κ-factors with increased use of RETs and tighter control over process variations
- Evolutionary reduction of the exposure wavelength

Even though wavelength changes tend to introduce interruptions in the steady flow, these transitions have been made fairly successfully in the past, including the concomitant research and innovation needed in resist materials and optical materials for lenses and reticles to bring the new technology to fruition. Since these approaches have proven so successful, there is a strong tendency to remain on this path as long as possible, and the simple momentum of a multibillion-dollar industry provides enough leverage to overcome many obstacles. Any attempt to switch to a different exposure technology appears very much like an attempt to direct a freight train onto a new track at full speed. Alternative exposure technologies such as x-ray or electron-beam lithography have a long history. Unfortunately, so far the freight train has outpaced all of them and thundered past quite a few of the projected switchover points.

3.5.1 The Evolutionary Path: 157-nm Lithography

Laser light sources with wavelengths below 193 nm include an F_2 laser with an exposure wavelength at 157 nm as well as a potential emission at 126 nm and 157-nm exposure tools are in development at the major exposure tool suppliers. Besides the normal issues with on-time availability of resist materials and laser source reliability, there are some unique problems related to 157-nm lithography: issues associated with optical materials, contamination, and availability of polymer materials used to protect reticles from foreign material [41,42].

At 157-nm wavelength, only two optical materials are sufficiently well developed and have low enough absorption to be usable: calcium fluoride (CaF_2) and fused silica (an amorphous, high-purity form of quartz). Today's 248- and even 193-nm systems use fused silica for most of their optical elements. Fused silica processing technology is highly advanced; the material is completely isotropic in its optical properties and has extremely low thermal expansion coefficients. The most commonly discussed alternatives to fused silica are fluorides. CaF_2 is found in existing 193-nm exposure tools—in particular, in those optical elements that are subject to the highest ultraviolet exposure densities. Its superior resistance to irradiation damage compared with fused silica makes it the material of choice in these circumstances. The availability of two different optical materials also allows correcting for wavelength-dependent focus shifts. Due to its introduction in 193-nm systems, crystal growth of large ingots and polishing techniques for

CaF_2 have made strong advances. Unfortunately, the material was found not to be fully isotropic in its optical properties (intrinsic birefringence), and the effect is drastically greater at 157 nm than it is at 193 nm [43]. This poses a significant challenge for lens design that needs to optimize the orientation in which the materials are used to minimize birefringence effects [44]. Lasers are available, but their bandwidth is too large. Narrow bandwidth is particularly critical since few optical materials are available to correct for chromatic aberrations. To minimize the impact of wider laser line widths, alternative lens designs are under investigation that make use of reflective surfaces [45]. One such design, for example, is a catadioptric system.

Further challenges of this technology are related to contamination issues [46]. Under 157-nm irradiation, residual hydrocarbons in the gases used to purge the lens break down and form deposits on the lens surfaces. These deposits increase absorption of lens elements. It has been found that carbon deposit may be removed using ultraviolet light from deuterium lamps; the carbon is oxidized and removed as gaseous CO_2. However, deposits formed by decomposition of silicones are significantly more troublesome and cannot be removed. Therefore, extreme care must be taken to select materials that do not outgas, and gas purge rates must be designed to keep residual contaminants in the parts per trillion level.

Another major technology challenge to the introduction of 157 nm is related to masks: more precisely, the way in which masks are protected from foreign material (FM). FM on a mask may result in defects that repeat on every exposure and in the worst case render the circuit inoperable. To protect masks from contamination they are covered with a thin organic membrane a few millimeters away from the chrome pattern. This distance keeps defects far enough out of focus so that small particles do not cause printing defects; most larger defects can be removed mechanically. The material used for all technologies, including 193 nm, is a thin polymer membrane resistant to damage from ultraviolet light and sufficiently transparent [47]. Unfortunately, despite a large research effort, an adequate material is not yet known for 157 nm. The alternatives are protective covers made from thin films of fused silica. They are significantly thicker and bend under the influence of gravity, which affects image placement.

The most likely technology to use 157-nm exposure tools are 65- and 45-nm generation. Ultrahigh-NA 193-nm tools, in conjunction with aggressive use of RETs, are a strong contender for the primary lithographic approach at this node. Certainly, the window of opportunity for these tools is narrow, and pilot production of 65 nm is expected to occur on 193-nm tools. Therefore, a few major semiconductor manufacturers have recently removed 157-nm tools from their road maps.

3.5.2 Still Evolutionary: Immersion Lithography

Sometimes, major breakthroughs can be achieved simply by returning to basic assumptions. The proposal for immersion lithography is an example of the value of questioning these assumptions in opening up a new path. Reconsidering the assumptions behind the simple equation describing the resolution capabilities of a system, we realize that it is accurate only in air (i.e., in a material of refractive index 1). More appropriately, the refractive index n should be incorporated, in which case the achievable resolution R is [48]

$$R = \kappa \frac{\lambda}{\mathrm{NA} \cdot n}$$

where NA is the numerical aperture, λ the wavelength of the light source, and κ the κ-factor. This idea opens up completely new avenues for resolution enhancements with significant improvement potential, as the refractive indexes of liquids may differ substantially from 1. For example, the refractive index of water is approximately 1.44 at 193 nm. Thus, 193-nm lithography used for imaging in water is equivalent to 134-nm lithography in air, a dramatic improvement over the capabilities of 193 nm and even over the next available wavelength at 157 nm. This type of lithographic process, proposed by B. Lin, is called *immersion lithography*. A schematic view on how this type of lithography could be implemented is shown in Figure 3.35. A key characteristic of this approach is the presence of

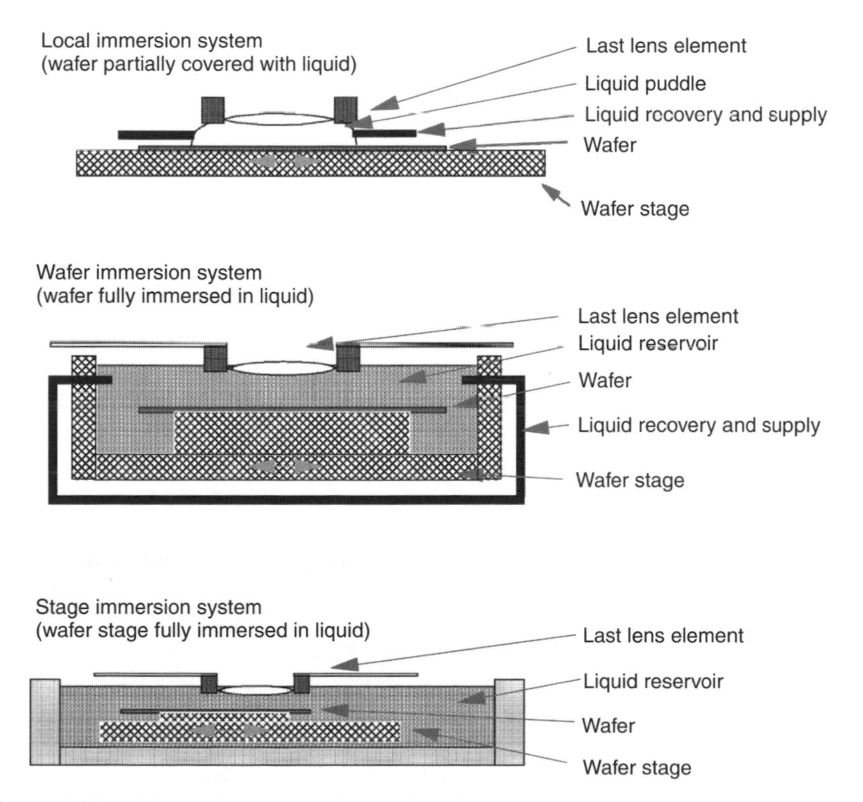

Figure 3.35 Schematic view of immersion lithography. Three different versions are depicted with increasing amounts of immersion.

a liquid between the wafer stage and the lens, and the main issue is to accommodate the normal movements of the stage relative to the lens in a medium of not only higher refractive index but also significantly higher viscosity [49–51]. Several variations have been proposed that differ in how much of the wafer stage should be immersed in the liquid [52]. The proposals range from full immersion of the wafer stage in the liquid to studies where only a liquid bubble is maintained immediately below the lens. The first approach is quite appropriately called the *submarine stage*; the entire wafer stage moves through a liquid. Additional pumps may be required to achieve the liquid flow rates necessary to keep up with the required stage speed. Another alternative is to keep the wafer stage outside the liquid by placing the wafer inside a liquid tank mounted on the wafer stage. Finally, there is a proposal where a liquid reservoir is maintained only below the lens and the liquid puddle moves with the lens across the wafer. Most of the technical problems in this approach arise from the fact that the lens and wafer move relative to each other at fairly high speeds, and the liquid has to accommodate this movement. More specifically, the formation of bubbles during such high-speed movements is a major issue. Other problems are thermal heating of the liquid under exposure and the associated changes in refractive index, as well as availability of resists that can be imaged in water are other potential obstacles to implementing this technology.

This approach has only recently gained a significant amount of momentum and is being investigated seriously by exposure tool suppliers. The proponents of this technology envision its introduction for 65-nm manufacturing, a major challenge given that development for this exposure option is in its infancy.

3.5.3 Quantum Leap: EUV Lithography

Pushing the exposure wavelength, not just by little increments but by more than a full order of magnitude, is the approach used in extreme ultraviolet (EUV) imaging. To gauge the efforts needed to achieve this task, one should keep in mind that since the beginning days of lithography, the exposure wavelength has decreased by a mere factor of 3 from 456 nm to 157 nm. Using this as a measure of difficulty, EUV is a gargantuan task. Everything about this technology is not simply an evolutionary improvement, but a quantum leap: Light sources, exposure tools, masks, mask inspection, lens metrology, and resists are unlike anything that has been used before [53,54].

Wavelengths considered for EUV are in the range 11 to 14 nm, a choice that is determined by the availability of suitable mirrors [55,56]. This wavelength range lies in the transition region between what one might consider ultraviolet and x-rays, so sometimes this is also referred to as *soft x-ray lithography*. Due to the large energy of photons at this wavelength, all materials, including gases, are strong absorbers. Transparent optical elements are therefore not available, resists have to be extremely thin, and the light sources cannot use solid enclosures. Refractive indexes are fairly small, which means that reflectivities from single surfaces are small. Thus, even building mirrors is nontrivial. At the wavelength range 11 to 14 nm, the most promising light sources are high-temperature plasmas of materials such as xenon. These plasmas may be created by focusing high-powered pulsed lasers (Nd:YAG lasers) on liquid or solid targets [57]. Under high-intensity illumination, the material vaporizes and for a short period of time forms an extremely high temperature (200,000 K), fairly high density (10^{17} to 10^{22} electrons/cm^2), small-volume plasma that emits soft x-rays at the desired wavelength. Beside the desired radiation, high intensities of light at other wavelengths is also emitted and needs to be filtered out. Since the starting materials are evaporated, continuous replenishing and replacement of the target is required. One of the most serious drawbacks of these materials is the formation of debris from solid targets. Since transparent materials are not available, protecting the collector mirrors from deposition of debris is difficult. Therefore, frozen noble gases such as xenon are the most promising target materials [58] even though solid materials such as tin are more efficient from a light-emission perspective. Alternative approaches to plasma formation are highly localized, pulsed electric discharges [59]. Source efficiencies for any source type, however, are still about an order of magnitude away from the target values required to achieve the throughputs necessary.

All imaging has to be done using mirrors. Due to the low refractive indexes, single-layer reflectivities are small, and therefore mirrors are based on multilayer systems, where a large number of interfaces add their contributions to the reflection. In this configuration, even if the amount of light reflected on an individual layer is small, the addition of a large number of these layers leads to sizable reflectivities. The thicknesses of the individual layers have to be maintained at a thickness where constructive interference occurs between the beams reflected from each interface. Layer thicknesses are of the order of a few nanometers, and thickness control needs to be within a fraction of a nanometer. Since these mirrors achieve their reflectivity through constructive interference, their reflectivity is highly dependent on the angle of incidence, and thicknesses have to be adjusted. The multilayer mirrors consist of alternating layers of Si and Mo which, due to the differences in atomic weight, have different refractive indices [60].

Like other optical elements in this technology, reflective rather than transparent masks are required [61]. EUV masks are built using a patterned EUV absorber placed on top of a multilayer reflector. The most serious mask issues are related to defects in the multilayer blanks and the fact that there are no known repair techniques for these multilayer blanks.

With only fairly low reflectivity mirrors available, the primary objective of the optical layout of an EUV tool is to minimize the number of optical elements. A schematic of such a system [53] using six reflective surfaces is shown in Figure 3.36. The layout complications inherent in a reflective system capable of imaging relatively large areas limit the numerical aperture to about 0.3 [62], not a significant issue given the large difference in wavelength relative to existing systems.

An EUV engineering test stand has been developed that allows large-field (26×32 mm) exposures at a numerical aperture of 0.1 [63]. Early tools are in

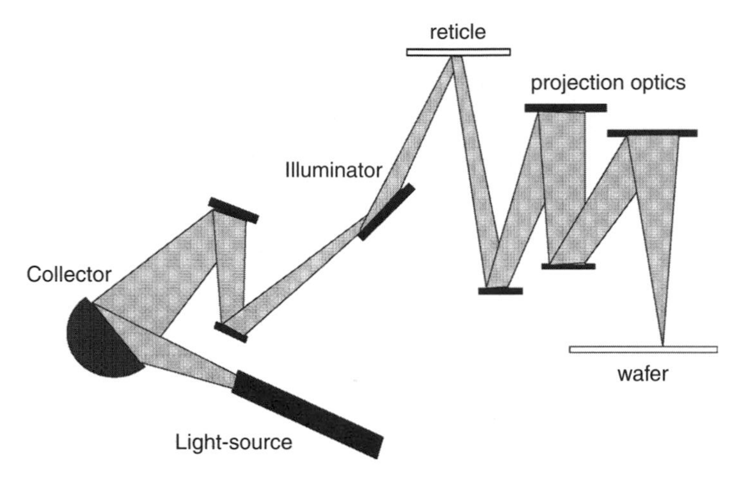

Figure 3.36 Schematic layout of an EUV optical system.

development at ASML [64] and a Japanese consortium that includes Nikon and Canon. The estimated price of one of these systems is about $50 million.

3.5.4 Particle Beam Lithography

Lithography using beams of charged particles (electrons or ions) is an alternative candidate for next-generation lithographic techniques. Broadly speaking, these systems can be differentiated based on the writing strategy (direct write versus projection) and the type of charged particles used (ion versus electron). The resolution of particle beam tools is limited by the fact that a moving particle has a wavelength associated with it given by its mass and velocity. Fortunately, the wavelengths are very small, and resolution on the order of 10 nm could be achieved using electrons.

3.5.5 Direct-Write Electron Beam Tools

Direct-write electron-beam tools have long been used in mask fabrication [65,66]. Figure 3.37 provides a schematic overview of such a system. An electron beam emitted from a source is focused onto a wafer using a set of magnetic lenses. Due to the finite size of the electron beam source there is a trade-off between final achievable spot size and the intensity of the electron beam. Although the final spot size is determined largely by the size of the electron source, high throughputs can only be obtained with high beam currents, requiring large-area electron sources. Originally, electron beam sources were heated tips of lanthanum boride. Electron emission from these tips was achieved through a combination of heat and coverage of the tip by a low-work-function material (i.e., a material with a small energetic barrier for electron emission). These tips did not, however, have the brightness sufficient for higher-resolution systems, and they have been

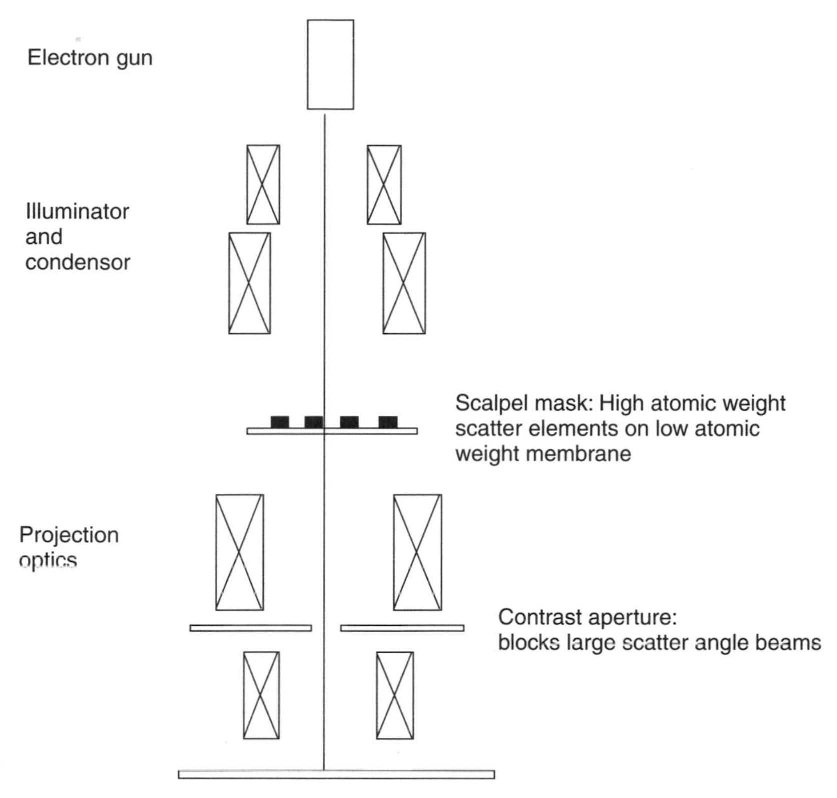

Figure 3.37 Schematic view of an electron-beam projection tool.

replaced by field emission sources [67]. In these sources, electric field crowding at a sharp metal tip is used to extract electrons from a metal tip. Such sources require little or no heating of the tip to achieve electron emission, thus reducing the kinetic energy spread of the electrons emitted. In electron optics, which suffer from high chromatic aberrations, this translates into improved imaging capabilities. Electron optics is based on magnetic fields and the forces they exert on moving charged particles. Electromagnets with appropriately shaped poles can focus diverging electron beams similar to optical lenses [68]. The ferromagnetic materials concentrate the magnetic fields and help achieve the desired field distributions. For applications in electron beam writers, they need to be magnetically soft so that they do not retain significant magnetic fields when the fields' strength needs to be adjusted. Unfortunately, from an aberration perspective the quality of electron optics is fairly poor, even for electrons that travel close to the axis of the system, and it deteriorates rapidly for electron beams traveling at large angles relative to the axis. This restricts the maximum numerical aperture of such lenses and limits the maximum achievable image sizes. Also electron beams exhibit an image blur mechanism that does not have a corresponding phenomenon using light [69]. Due to the mutual repulsion of electrons, an electron beam focused into a small area shows image blur that increases with increasing current. This mechanism limits the imaging capabilities at the high-beam currents that are desirable from a throughput perspective.

Beam deflection [70] is achieved either by electrostatic deflectors or through magnetic deflectors similar to those used in conventional television tubes. Unlike television tubes that do not require nanometer-scale spot sizes, the deflection distance is relatively small (several 10 μm), limited by the rapid increase in aberrations for off-axis electrons. Therefore, only small fields can be written without stage movements. Patterns are written in raster fashion, similar to the way an image is created on a television screen. The beam scans a line with spatial information encoded as beam on–off information. After finishing one line, the next line is written. Areas of approximately 50 μm × 50 μm are scanned at a time, followed by a stage move to the center of the next square.

Among the principal difficulties in applying this technology to lithography is the fact that the data are transferred serially rather than in parallel as is the case for imaging systems. This leads to notoriously low throughput numbers despite significant efforts to increase data transfer rates through parallel electron beam methods. For example, the exposure times for critical reticles are measured in hours. Other issues are related to the fact that the electron beam results in charging of the substrates that may lead to undesirable beam deflections unless the charge deposited can be removed efficiently. High-energy electron beams may also damage the increasingly thinner gate oxides used in advanced technologies. Finally, additional complications arise from the fact that the systems work in vacuum, similar to EUV systems, and thus efficient wafer loading is impeded.

One of the biggest advantages of electron-beam direct-write systems is that they eliminate the need for expensive reticles, and therefore this technology is particularly attractive for ASIC manufacturers faced with extremely high mask costs for very low volume parts. The throughput disadvantage is less of a concern for low pattern density levels, such as the contact layer. A few systems are available on the market. Some of the most exciting developments in this area [71] are attempts to use arrays of electron sources (microcolumns), taking advantage of the manufacturing capabilities developed in the micromechanical area (MEMS). A 64 × 64 array of sources is being discussed with independent deflection control for each source, allowing 250 μm × 250 μm field coverage. It remains to be seen if the challenges of building arrays of electron sources with high enough uniformity and a long enough lifetime can be overcome and reliable tools can be built based on this technology.

The throughput problem can be mediated by adapting the projection approach from optical lithography. Rather than writing each individual pixel separately, small subfields are illuminated and projected onto the wafer [72]. The individual subfields are then stitched together to create the entire exposure field. The precise deflection control of electron beam systems allows the projection of subfields even if they are separated from each other mechanically using struts [73]. The stitching process can be set up to completely eliminate the presence of unpatterned areas between the subfields. This ability is important, as it gives the mask sufficient mechanical stability. The poor mechanical stability of masks thin enough to allow an electron beam to pass through can be circumvented by placing struts between individual subfields.

Mask heating under the high-current electron beams necessary to achieve sufficient exposure tool throughput is one of the limiting effects in this technology. A major breakthrough in this area was made with the Scalpel system, developed at the former Bell Labs [73]. Instead of using absorbing elements on the mask, the electron-beam is diffracted and the diffracted beam absorbed at apertures inside the tool, thus moving the heat-generating absorption to tool components that are not critical for the imaging process. Diffraction is achieved using high-atomic-weight materials.

Although the projection concept improves the throughput of the system significantly, a key remaining limitation is the limited deflection capability, which results in extensive stage movements. This problem is addressed in the PREVAIL (reduction exposure with variable axis immersion lenses) system [74] developed by IBM. In this particular lens design the optical center of the lens that features the smallest aberration necessary for high-resolution imaging can be shifted laterally. This is accomplished through additional deflector fields superimposed onto the focusing lens fields. Through this concept a dramatic improvement in the deflection of the beam can be achieved, up to 10 mm in the reticle field, ± 2.5 mm on the wafer plane.

The technology from the original Scalpel system was transferred to a consortium of ASML and Applied Materials for development of a commercial system. Unfortunately, the joint venture, called eLith, was dissolved at the beginning of 2001 [75]. However, there is still one major exposure tool supplier that intends to provide commercial electron beam projection tools [76].

Ion projection lithography [77] is the last technology to be discussed. Similar to electron beam direct-write tools, ion beams have found applications in mask repair and nondestructive microcharacterization of 300-mm wafers. In these areas the ability of ion beams to deposit as well as remove material is being utilized. The system, developed by an Austrian team, is a projection lithography tool. Direct-write approaches have recently been announced and have found applications in patterning of magnetic storage disks.

Ion projection lithographic systems are comprised of three major sections: source, column, and the chamber in which the wafer resides. Gallium is a common ion used in focused ion beam systems for mask repair, and charged hydrogen or helium atoms are used in lithography tools. In these sources the desired species are introduced in gaseous form and ionized through electron bombardment. Properly laid out electric fields extract the ions from the source and inject it into the electron optical column. The masks used in these systems are stencil masks similar to those used in electron beam projection systems. Unlike the mostly magneto-optical lenses used in electron beam systems, the ion beam that has passed through the reticle enters a multielectrode electrostatic lens system for focusing onto the wafer stage.

REFERENCES

[1] S. Okazaki, Lithography for ULSI, *Proc. SPIE*, Vol. 2440, pp. 18–32, 1995.

[2] H. Levenson, *Principles of Lithography*, SPIE Press, Bellingham, WA, 2001.

[3] R. Dejule, Resist enhancement with antireflective coating, *Semicond. Int.*, July 1996, p. 169.

[4] Y. Borodovsky, Impact of local partial coherence variation on exposure tool performance, *Proc. SPIE*, Vol. 2440, pp. 750–770, 1995.

[5] J. Kirk, Review of photo resist based lens evaluation methods, *Proc. SPIE*, Vol. 4000, pp. 2–8, 2000.

[6] P. Dirksen, C. Juffermans, A. Engelen, P. deBisschops, and H. Muellerke, Impact of high order aberrations on the performance of the aberration monitor, *Proc. SPIE*, Vol. 4000, pp. 9–17, 2000.

[7] N. Farrar, A. Smith, D. Busath, and D. Taitano, Measurement of lens aberration by in-situ interferometer and classification for correct application, *Proc. SPIE*, Vol. 4000, pp. 18–29, 2000.

[8] C. Progler and A. Wong, Zernike coefficients, are they really enough, *Proc. SPIE*, Vol. 4000, pp. 40–52, 2000.

[9] D. Flagello, J. Mulkens, and C. Wagner, Optical lithography into the millennium sensitivity to aberrations, vibrations and polarization, *Proc. SPIE*, Vol. 4000, pp. 172–183, 2000.

[10] M. Maenhoudt, S. Verhaegen, K. Ronse, P. Zandbergen, and E. Murzio, Limits of optical lithography, *Proc. SPIE*, Vol. 4000, pp. 373–387, 2000.

[11] A. K. K. Wit, *Resolution Enhancement Techniques*, SPIE Press, Bellingham, WA, 2001.

[12] W. N. Partlo, P. J. Thompkins, P. G. Dewa, and P. F. Michaloski, Depth of focus and resolution enhancement for i-line and DUV lithography using annular illumination, *Proc. SPIE*, Vol. 1972, pp. 753–764, 1993.

[13] K. Tounai, H. Tanabe, H. Nozue, and K. Kasama, Resolution enhancement with annular illumination, *Proc. SPIE*, Vol. 1674, pp. 753–764, 1992.

[14] Y. Granik, N. Cobb, and T. Do, Universal process modeling with VTRE for OPC, *Proc. SPIE*, Vol. 4691, pp. 377–394, 2002.

[15] K. D. Lucas, J. C. Word, G. N. Vandenberghe, S. Verhaegen, and R. M. Jonckheere, Model-based OPC for first generation 193 nm lithography, *Proc. SPIE*, Vol. 4346, pp. 119–130, 2002.

[16] O. Otto and R. Henderson, Advances in process matching for rules-based optical proximity correction, *Proc. SPIE*, Vol. 2884, pp. 323–332, 1996.

[17] F. X. Zach, D. J. Samuels, A. C. Thomas, and S. A. Butt, Process dependencies of optical proximity corrections, *Proc. SPIE*, Vol. 4346, pp. 113–118, 2001.

[18] J. P. Stirniman and M. L. Rieger, Spatial filter models to describe IC lithographic performance, *Proc. SPIE*, Vol. 3051, pp. 469–478, 1997.

[19] Y. Granik, Correction for etch proximity: new models and applications, *Proc. SPIE*, Vol. 4346, pp. 98–112, 2001.

[20] J. Garofalo, O. Otto, R. Cirelli, R. Kostelak, and S. Vaidya, Mask assisted off axis illumination technique for random logic, *J. Vac. Sci. Technol. B*, Vol. 11, pp. 2651–2658, 1993.

[21] J. Garofalo, O. Otto, R. Cirelli, R. Kostelak, and S. Vaidya, Automated layout of a mask-assisted for realizing 0.5 k1 ASIC lithography, *Proc. SPIE*, Vol. 2440, pp. 302–312, 1995.

[22] A. H. Gabor et al., Subresolution assist feature implementation for high performance logic gate level lithography, *Proc. SPIE*, Vol. 4346, pp. 418–425, 2002.

[23] M. Levenson, N. Viswanathan, and R. Simpson, Improving resolution in photolithography with a phase shift mask, *IEEE Trans. Electron Devices*, Vol. 29, pp. 1812–1846, 1982.

[24] G. Mack, Fundamental issues in phase shifting mask technology, *OCG Microelectronics Conference*, pp. 23–25, 1993.

[25] K. Adam and A. R. Neureuther, Algorithmic implementation of domain decomposition methods for the diffraction simulation of advanced photomasks, *Proc. SPIE*, Vol. 4691, pp. 107–124, 2002.

[26] L. Liebmann, N. A. Northrup, J. Culp, L. Segal, A. Barish, and C. Fonseca, Layout optimization at the pinnacle of optical lithography, *Proc. SPIE*, Vol. 5042, pp. 1–14, 2003.

[27] J. A. Torres, F. Schellenberg, and O. Toublan, Model assisted double dipole decomposition, Vol. 4691, p. 407–417, *Proc. SPIE*, 2002.

[28] S. M. Mansfield, L. Liebmann, A. Molless, and A. Wong, Lithographic comparison of assist feature design strategies, *Proc. SPIE*, Vol. 4000, pp. 63–76, 2000.

[29] R. Socha, M. Dusa, L. Capodieci, J. Finders, F. Chen, D. Flagello, and K. Cummings, Forbidden pitches for 130 nm lithography and below, *Proc. SPIE*, Vol. 4000, p. 1140, 2000.

[30] L. W. Liebmann et al., Optimizing style options for subresolution assist features, *Proc. SPIE*, Vol. 4346, pp. 141–145, 2001.

[31] L. Liebmann, J. Lund, F. L. Hueng, and I. Graur, Enabling alternating phase shift mask designs for a full logic gate level: design rules and design rule checking, *Proc. DAC*, pp. 79–84, 2001.

[32] M. L. Rieger, J. P. Mayhew, and S. Panchapakesan, Layout design methodologies for sub-wavelength manufacturing, *Proc. DAC*, pp. 85–88, 2001.

[33] M. Sanie, M. Cote, P. Hurat, and V. Malhotra, Practical application of full feature alternating phase shift technology for a phase aware standard-cell design flow, *Proc. DAC*, pp. 93–96, 2001.

[34] U. Behringer, Foreword, EMC2003 (European Mask Conference), p. 1, Jan. 13–15, 2003.

[35] A. Balasinski, Mask cost for sub-100 nm technologies: stopping a runaway? *Proc. SPIE*, Vol. 5043, pp. 82–92, 2003.

[36] Lithography CoO analysis, *http://www.sematech.org*.

[37] D. Pramanik, H. Kamberian, C. Progler, M. Sanie, and D. Pinto, Cost effective strategies for ASIC's masks, *Proc. SPIE*, Vol. 5043, pp. 142–148, 2003.

[38] R. D. Morse, *Proc. SPIE*, Vol. 5043, pp. 100–112, 2003.

[39] G. Prophet, Structured ASICs: more gain, less pain? *EDN Europe*, Aug. 7, 2003; *http://www.reed.electronics.com/electronicnews/community/22113/Semiconductors?starting=651*.

[40] B. Dipert, Silicon segmentation, *EDN*, Sept. 18, 2003; *http://www.reed-electronics.com/electronicnews/article/CA321801?stt=000&industryid=22113&industry=Semiconductors.*

[41] S. Dana, Pushing the limits, *OE Mag.*, pp. 20–22, Mar. 2002; *http://www.oemagazine.com/fromTheMagazine/mar02/pushingthelimits.html.*

[42] S. Dana, Progress report: 157 nm lithography prepares to graduate, *OE Mag.*, Feb. 2003; *http://www.oemagazine.com/fromTheMagazine/feb03/157.html.*

[43] J. Burnett et al., *Phys. Rev. B*, Vol. 64, No. 241102, 2001.

[44] W. Ulrich, S. Beiersdorfer, and H. J. Mann, Trends in optical design of projection lenses for UV and EUV lithography, *Proc. SPIE*, Vol. 4146, pp. 13–24, 2000.

[45] J. Burnett, Intrinsic birefringence in calcium fluoride forces optical engineers to use sophisticated design techniques for 157 nm lithography systems, *OE Mag.*, Mar. 2002; *http://www.oemagazine.com/fromTheMagazine/mar02/bire.html.*

[46] R. R. Kunz, V. Liberman, and D. K. Downs, Experimentation and modeling of organic photocontamination on lithographic lenses, *Proc. SPIE*, Vol. 4000, p. 474, 2000.

[47] R. H. French et al., 157-nm pellicles: polymer design for transparency and lifetime, *Proc. SPIE*, Vol. 4691, pp. 576–583, 2002.

[48] B. J. Lin, New λ/NA scaling equations for resolution and depth of focus, *Proc. SPIE*, Vol. 4000, pp. 759–764, 2000.

[49] J. Hoffnagle, W. D. Hinsberg, M. Sanchez, and F. A. Houle, *J. Vac. Sci. Technol. B*, Vol. 17, p. 3306, 1999.

[50] B. Smith, H. Kang, A. Burov, F. Cropranese, and Y. Fen, Water immersion lithography for the 45 nm node, *Proc. SPIE*, Vol. 5040, pp. 679–699, 2003.

[51] Immersion lithography: successor to super high NA 193 nm technology? *http://eedesign.com/pressreleases/bizwire/111879.*

[52] S. Owa and H. Nagasaka, Immersion lithography: its potential performance and issues, *Proc. SPIE*, Vol. 5040, pp. 724–733, 2003.

[53] C. Gwyn and P. Silverman, EUV transition from research to commercialization, *Proc. SPIE*, Vol. 5130, pp. 990–1004, 2003.

[54] N. Harned and S. Roux, Progress report: engineers take the EUV lithography challenge, *OE Mag.*, Feb. 2003; *http://www.oemagazine.com/fromTheMagazine/feb03/euv.html.*

[55] B. Lai and F. Cerrina, Image formation in multilayer optics: the Schwartzschild objective, *Proc. SPIE*, Vol. 563, pp. 174–179, 1985.

[56] T. E. Jewell, J. M. Rodgers, and K. P. Thompson, Reflective systems design for soft x-ray projection lithography, *J. Vac. Sci. Technol. B*, Vol. 8, pp. 1509–1513, 1990.

[57] U. Stamm et al., *Proc. SPIE*, Vol. 4688, pp. 122–133, 2002.

[58] B. A. Hansson, L. Rymell, M. Berglund, O. E. Hemberg, E. Janin, J. Thoresen, S. Mosesson, J. Wallin, and H. M. Hertz, Status of the liquid-xenon-jet laser-plasma source for EUV lithography, *Proc. SPIE*, Vol. 4688, pp. 102–109, 2002.

[59] N. Fornaciari et al., Power scale-up of the extreme-ultraviolet electric capillary discharge source, *Proc. SPIE*, Vol. 4688, pp. 110–121, 2002.

[60] E. Luis et al., Progress in Mo/Si multilayer coating technology for EUV mirrors, *Proc. SPIE*, Vol. 3997, pp. 406–411, 2000.

[61] P. J. Mangat, S. D. Hector, S. Rose, G. F. Cardinale, E. Tejnil, and A. R. Stivers, EUV mask fabrication with Cr absorber, *Proc. SPIE*, Vol. 3997, pp. 76–82, 2000.

[62] T. Oshino et al., Development of illumination optics and projection optics for high NA EUV exposure tool, *Proc. SPIE*, Vol. 5037, pp. 75–82, 2003.

[63] D. J. O'Connel et al., Lithographic characterization of improved projection optics in the EUVL engineering test stand, *Proc. SPIE*, Vol. 5037, pp. 83–94, 2003.

[64] H. Meiling et al., The EUV program at ASML: an update, *Proc. SPIE*, Vol. 5037, pp. 83–94, 2003.

[65] D. R. Herriot, R. J. Collier, D. S. Alles, and J. W. Stafford, EBES: a practical electron lithography system, *IEEE Trans. Electron Devices*, Vol. 22, pp 385–392, 1972.

[66] F. Abboud, J. Poreda, and R. L. Smith, MEBES IV: a new generation raster scan electron beam lithography system, *Proc. SPIE*, Vol. 1672, pp. 111–125, 1992.

[67] M. Gesley, MEBES IV thermal field emission tandem optics for electron beam lithography, *J. Vac. Sci. Technol. B*, Vol. 9, pp. 2949–2951, 1991.

[68] P. W. Hawkes, *Selected Papers on Electron Optics*, SPIE Press, Bellingham, WA, 1994.

[69] L. R. Harriot et al., Space charge effects in projection charged particle lithography systems, *J. Vac. Sci. Technol. B*, Vol. 13, pp. 2402–2408, 1995.

[70] L. H. Lin and H. L. Beauchamp, High speed beam deflection and blanking for electron lithography, *J. Vac. Sci. Technol.*, Vol. 10, pp. 987–990, 1973.

[71] P. Ware, Removing the mask, *OE Mag.*, pp. 26–27, Mar. 2002.

[72] A. E. Novembre et al., Fabrication and commercialization of scalpel masks, *Proc. SPIE*, Vol. 3412, pp. 350–357, 1998.

[73] J. A. Liddle, L. R. Harriot, A. E. Novembre, and W. K. Waskiewicz, SCALPEL: a projection electron beam approach to sub-optical lithography, *http://www1.bell-labs.com/project/SCALPEL/*.

[74] R. S. Dhaliwal et al., PREVAIL: electron projection technology approach for next-generation lithography, *IBM J. Res. Dev.*, Vol. 45, pp. 615–638, 2001.

[75] D. Lammers, EUV gains as venture ends e-beam litho work, *EETimes*, Jan. 2001; *http://www.eetimes.com/story/OEG20010105S0023*.

[76] S. Fukui, H. Shimizu, W. Ren, S. Suzuki, and K. Okamoto, Nikon EB stepper: its system design and preliminary performance, *Proc. SPIE*, Vol. 5037, pp. 504–511, 2003.

[77] H. Loescher, Masked ion beam lithography and direct structuring of curved surfaces, *Proc. SPIE*, Vol. 5037, pp. 156–161, 2003.

CHAPTER 4

MIXED-SIGNAL CIRCUIT DESIGN

4.1 INTRODUCTION

Mixed-signal design on more advanced processes (0.13 μm and below) poses a new host of design issues that must be considered during the initial architectural phase as well as during the implementation cycle. Many new products incorporate analog and mixed-signal components along with larger digital circuits. Microprocessors are a good example of this trend, since they typically include phase-locked loops (PLLs), frequency synthesizers, delayed locked loops, and data converters. Given that the bulk of the die area is comprised of high-speed digital logic, the process is typically developed to address issues associated with digital circuits. This process choice forces analog and mixed-signal designers to develop specialized methodologies to ensure that circuits can meet stringent requirements without the advantage of a specialized process. An overview of the various issues that must be considered to successfully design analog and mixed-signal integrated circuits (ICs) on deep-submicron processes is presented in this chapter, along with recommendations for avoiding various pitfalls.

4.2 DESIGN CONSIDERATIONS

For the technology to continue to progress toward smaller and smaller geometries, several significant advancements have been implemented. These advancements include extremely thin gate oxide [physical thickness or equivalent oxide thickness (EOT) ranges from 1.5 to 2.2 nm for the 130-nm node and 1.1 to 1.7 nm for

Nano-CMOS Circuit and Physical Design, by Ban P. Wong, Anurag Mittal, Yu Cao, and Greg Starr
ISBN 0-471-46610-7

the 90-nm node], shallow source and drain diffusions, shallow trench isolation, multiple oxide thicknesses, multiple threshold voltages for a given oxide thickness, and multiple metal thickness. All of these process advancements have allowed for an explosive growth in the semiconductor marketplace but at a price of increasing complexity in the design methodology and tools required to develop new products successfully. Some of the issues facing mixed-signal designers include:

- Model accuracy
- Leakage current (both source–drain and gate)
- OD stress effects
- Process variation (device to device, die to die, wafer to wafer, lot to lot, and fabrication to fabrication)
- Supply headroom effects
- Device performance
- Noise isolation
- Power

These issues are discussed in this chapter with suggestions on how to design around the problems.

4.3 DEVICE MODELING

Device modeling has become more complex to accurately reflect the physical performance of circuits given the aggressive design targets for next-generation processes. At 90 nm and beyond, enhanced device models are required to design mixed-signal ICs successfully. These models must incorporate several new factors neglected on earlier generations, such as 0.13 μm, in addition to enhancing the modeling of current parameters. The model issues include:

- Shallow trench isolation (STI) stress-induced effect
- Gate leakage current
- Gate-induced drain leakage (GIDL)
- Stress-induced diode leakage
- Polysilicon gate depletion effects
- Shallow source–drain series resistance
- Negative bias temperature instability (NBTI)

Many of these issues have been incorporated in the BSIM3 models, but their effect has become more pronounced with smaller geometries. Figure 4.1 is a simplified diagram of a MOS device cross section with some of the parasitic

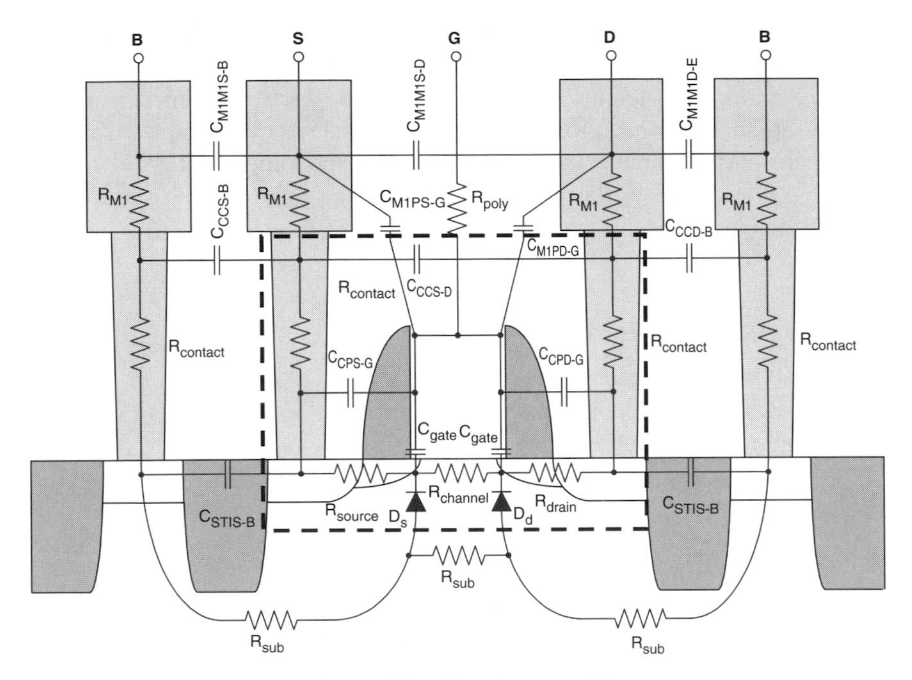

Figure 4.1 Transistor model.

components indicated. On previous-generation processes, many of these components could be neglected since they were small relative to other parasitics (such as the diffusion capacitance), but for the more advanced processes such as 0.13 μm and beyond, they can no longer be neglected. The critical part of the modeling process is to understand what components are included in the base model, what components must be included in a subcircuit for prelayout simulations, and what components must be included as part of the parasitic extraction. The dashed-box region of the device shown in Figure 4.1 indicated the typical region included in the base device model. It is critical to decide what parameters will affect performance the most to minimize the complexity of the model while still yielding reasonable results. Adjusting the aggressive design rules can minimize some of the parasitic components, such as increasing the gate-to-contact spacing. This is equivalent to having two design rules check (DRC) flows for a design, one for the digital portion of the design and a second for the analog portion of the design.

Table 4.1 summarizes the various parasitic components that can be included in the model along with an approximate value based on a typical 90-nm process. The far right column provides the potential error as a reference point for accessing the need to include the parasitic based on an analysis of mixed-signal circuits. The boxes that have been left blank indicate places where it is difficult to provide some boundary on the error since it is very dependent on the specific application.

Figure 4.2 shows the error extracted from a block where the postlayout resistors were set to zero. A total of 72 delays were measured for the block.

TABLE 4.1 Summary of Device Model Parasitics and the Approximate Value for a Typical 90-nm Process[a]

Parameter	Range	Reason for Variation	Neglect	Potential Error
R_{M1}	0.05–0.2 Ω/sq	Metal thickness due to erosion or dishing, and metal CD control	Local routing (<20 μm): typically yes; long routes: no	0–2% for localized routing, 2–40% for longer routes
$R_{contact}$	10–25 Ω/cnt	Contact alignment fill % contact hole etching	Usually part of the base model	
R_{poly}	5–12 Ω/sq	Poly thickness, salicide variation, poly CD control	Small width device: maybe; long width device: no; RF applications: no	
$C_{M1M1S-B}$, $C_{M1M1D-B}$	0.1–0.3 fF/μm	Metal thickness due to erosion, to a lesser extent dishing, and metal CD control	Source–bulk are connect: yes; drain bulk: no, but very layout dependent	1–2%
$C_{M1M1S-D}$	0.05–0.1 fF/μm	Metal thickness, metal CD control	Must include	1–3%
C_{M1PS-G}, C_{M1PD-G}	0.02–0.06 fF/μm	Metal and poly thickness, metal and poly CD control, dielectric (TEOS) thickness variation	Must include	1–3%
C_{CCS-B}, C_{CCD-B}, C_{CCS-D}	0.02–0.05 fF/cnt	Contact hole etching	Typically, neglect	<2%
C_{CPS-G}, C_{CPD-G}	0.02–0.07 fF/cnt	Contact hole etching, poly thickness, poly CD control	Usually, part of the base model	<1%
C_{STIS-B}, C_{STID-B}	0.03–0.07 fF/μm	STI width	Potentially include, depending on layout	<1% unless adjacent device couples noise
R_{sub}	500–1000 Ω/sq	Substrate doping	Potentially, especially for high-speed design and noise simulations	
C_{gate}	~12 fF/μm^2	Oxide thickness, applied voltage	Must include (gate capacitance is lumped together, but in actuality it has several components)	

[a] Values are only approximations based on simulations and hand calculations.

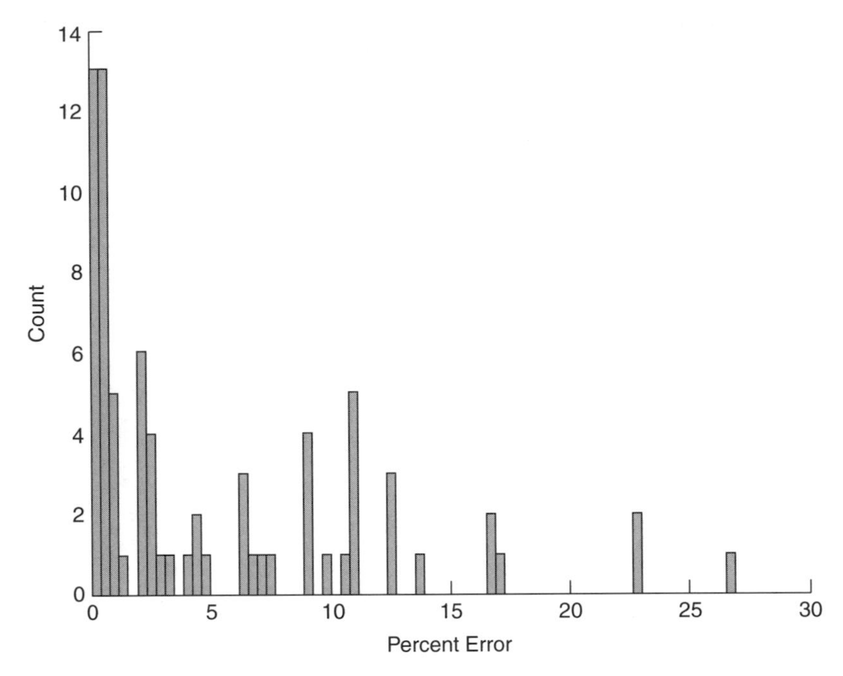

Figure 4.2 Error resulting from neglecting metal and poly resistance on a 90-nm process.

Two-thirds of the paths had an error of less than 5%. This seems quite small, but consider a high-speed clock data recovery chip operating at 10 GHz. The clock period is only 100 ps, so a 5% error would be 5 ps, which could be enough to create timing errors, especially when coupled with other sources of timing errors.

Figure 4.3(*a*) shows the nodal capacitance from a high-speed digital block of a phase-locked-loop circuit on a 90-nm process. The bulk of the nodes have less than 5 fF of capacitance, but several become heavily loaded because of the increased routing parasitics. The $\log_{10}$ of the nodal capacitance for a voltage-controlled oscillator (VCO) cell is shown in Figure 4.3(*b*). The 0 point on the x-axis represents 1 fF; thus, the majority of the capacitance is less than 1 fF. This nodal capacitance results in a slowdown of the VCO cell. Additional effects could include undesired coupling to adjacent nodes that could degrade performance by giving rise to greater phase noise.

The STI stress-induced and gate leakage effects may be accounted for using the newer BSIM4 models. Inclusion of the STI stress effect does not affect the simulation time, but the gate leakage effect can significantly increase simulation time since it is bias dependent and therefore evaluated at each operating point during the simulation run time rather than only at the start of the simulation. The STI stress effect is very layout dependent, requiring knowledge of how a circuit will be laid out to account for it in the prelayout simulations. The stress effect occurs because of the thermal expansion differences between silicon and silicon oxide that results in a compressive force in the silicon [1,37]. This compressive

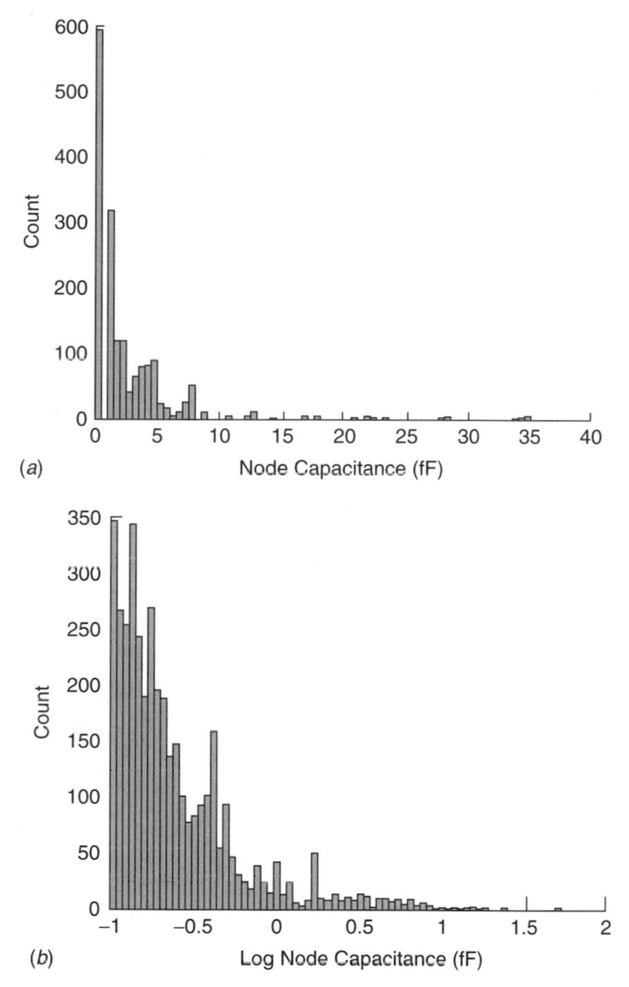

Figure 4.3 Node capacitance for a high-speed digital block (*a*) and a complete voltage-controlled oscillator (*b*) from a phase-locked loop on a 90-nm process.

force causes a shift in the carrier mobility within the channel, which is shown in Figure 4.4. The effect becomes more pronounced for the smaller-geometry devices as the spacing between the two STI regions is decreased. The mobility shift can be modeled using

$$\mu_{\text{eff}} = \mu_{a\,\text{min}} \left(1 + \Delta\mu_{\text{max}} \frac{a - a_{\text{min}}}{a}\right)$$

where $1/a = (1/SA + 1/SB)$, a_{min} is the minimum diffusion spacing supported by the process design rules, $\mu_{a\,\text{min}}$ is the mobility at the minimum gate-to-diffusion spacing, and $\Delta\mu_{\text{max}}$ is the maximum mobility variation relative to $\mu_{a\,\text{min}}$.

For PMOS devices, $\Delta\mu_{max}$ is negative and it is positive for NMOS devices. As the active area decreases, built-in compressive stress increases. This increase in mechanical stress causes a break in crystal symmetry. The affective change in deformation potential leads to a break in the sixfold conduction band and twofold valence band degeneracy. The carrier affective mass as well as the band scattering rates are affected, which directly affects the carrier mobility, resulting in a net increase in the PMOS device mobility and a decrease in the NMOS device mobility. The meaning of SA and SB is indicated in Figure 4.4.

The STI stress effect results in a hole mobility increase and an electron mobility decrease for small outside diameters (ODs). Figure 4.5 shows a plot of the type of I_{DS} change that can occur because of the STI stress effect. The crossover point occurs at the OD size at which the models were generated. The variation between actual current and modeled current can be significant; therefore, the models must include the STI stress effect to ensure good correlation between simulations and silicon characterization. From these data it may appear that the only method to combat this effect is to increase the length of the source and drain regions, but splitting the gate into multiple fingers can also be used to increase the overall OD size, as shown in Figure 4.4.

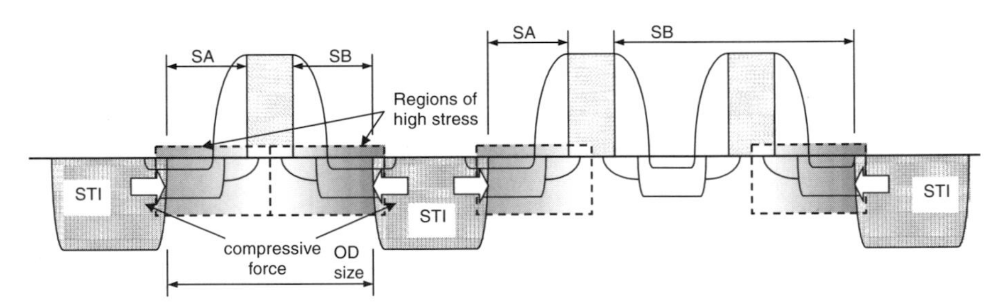

Figure 4.4 NMOS transistor with STI stress shown for a single and multiple gate fingers.

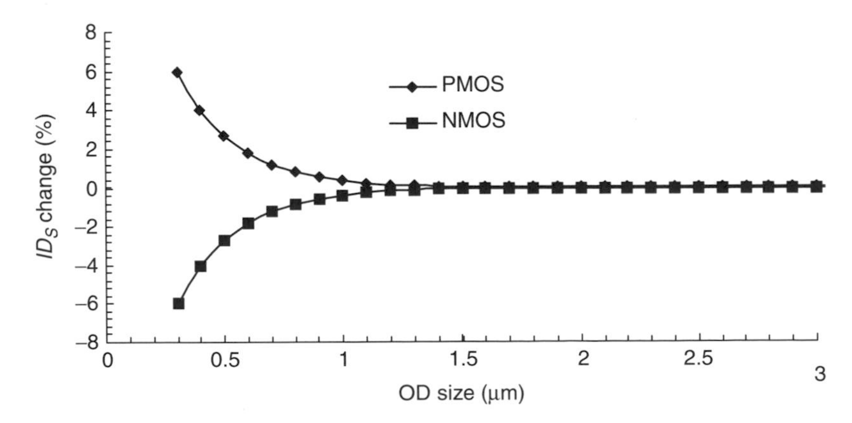

Figure 4.5 $I_{d\ sat}$ variation for an advanced process (0.13 μm and beyond).

Gate leakage has become a problem back as far as 0.15 μm (possibly 0.18 μm). For certain analog components such as loop filter capacitors, the problem has had to be solved by using thick oxide devices, interlayer metal capacitors, or metal comb capacitors. On sub-0.13-μm processes, gate leakage for other circuits will need to be considered as well. On the 90-nm process, gate leakage currents can become as high as 5 nA for a 15 μm × 0.4 μm transistor at 25°C with V_{cc} as the bias voltage. In many applications, such as simple digital logic, this small current does not affect operations seriously, but in low-power designs, where much smaller bias currents are used and devices are operated at subthreshold or very near subthreshold, these leakage currents can become significant and must be considered during the design process. Similar considerations must be given to sample-and-hold circuits as well as to domino logic since the leakage will result in a shift in voltages as the gate leakage removes the stored charge on the hold node. Additionally, the gate leakage current is derived from Fowler–Nordheim tunneling, which can be modeled as shot noise, which is a random event giving rise to white noise [38]. This noise will affect the overall performance of certain circuits, such as low-noise amplifiers and voltage-controlled oscillators. The gate leakage can be broken into three components [2]: two edge direct tunneling (EDT) components (I_{gs0} and I_{gd0}) and gate-to-channel tunneling (I_{gc}) (Figure 4.6). Gate-to-channel tunneling dominants for long-channel devices. In general, edge-direct tunneling is less than gate-to-channel tunneling, so as the length is extended, the leakage current per unit increases.

Pocket implants have been used for some time to reduce threshold voltage roll-off and punch-through. Introduction of pocket implants has a twofold effect on analog circuits. First, it increases the drain-induced threshold voltage shift (DITS), which is more pronounced with the longer-channel-length devices typically used in analog and mixed-signal circuits. This effect can add more error in current mirrors if the source–drain voltage of the mirror devices does not match. A second effect is a reduction in output resistance since the threshold voltage is a function of the drain–source and decreases with increasing drain–source voltage. Since the threshold voltage has a strong dependency on the drain–source voltage,

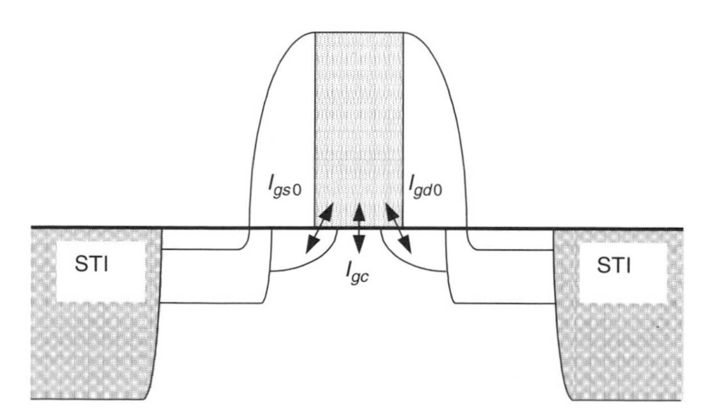

Figure 4.6 Gate leakage components.

it must be considered using the following approximate equation:

$$I_{DS} \simeq \frac{\mu_n C_{\mathrm{ox}}}{2} \frac{W}{L} [V_{GS} - V_t(V_{DS})]^2 (1 + \lambda V_{DS})$$

where μ_n is the electron mobility, C_{ox} the gate oxide capacitance, W the width of the device, L the length of the device, V_{GS} the gate-source voltage applied to the device, V_t the threshold voltage which is a function of the drain–source voltage, and λ the device parameter to account for channel length modulation.

A reduction in the output impedance can have a significant impact on analog circuits since this reduces the effective noise isolation on most analog circuits. At some point it may be necessary to ask for "analog transistors," transistors without pocket implants. However, this will add to the cost and turnaround time in a digital CMOS process, as it may incur additional masking and implantation steps. For the time being it is still possible to design robust circuits based on the methodology we describe in this chapter.

A final effect that should be considered is the negative bias temperature instability (NBTI) influence on long-term threshold voltage shifts[3]. In essence, bias applied to a PMOS device results in the generation of interface traps, which gives rise to a shift of threshold voltage. Unlike the hot-carrier effect case increasing the channel length does not mitigate its effect. The amount of shift depends on the specifics of how the circuit is configured. Switched circuits incur less of a shift since the bias is not supplied continually but still experience the effect. This can still be problematic for high-speed clocking circuits, where small changes in duty cycle can result in erroneous behavior. The potential threshold voltage shift can be in the tens of millivolts, which is significant. One possible way to deal with this problem is to incorporate the threshold voltage shift into process corner models.

4.4 PASSIVE COMPONENTS

Passive components, including resistors, capacitors, varactors, and inductors, are frequently used in analog and mixed-signal design. Resistors have not changed significantly with the shrinking process except for a resistivity shift from the previous generation because of changing doping densities. The primary concern is always with obtaining accurate models and understanding the variability with process, temperature, and applied voltage. In most cases, resistors need to be modeled as multiple resistances to accurately reflect end resistance as well as parasitic capacitances.

Capacitors can offer a host of issues with shrinking technologies, depending on the structure used. Figure 4.7 shows some of the variety of capacitors used in circuit designs. At the 0.18-μm technology node, gate leakage began to be a problem for thin oxide devices used as loop filter capacitors. Gate leakage for thin oxide devices is in the range 5 nA/μm^2 and above at 25°C. A typical number for the capacitance is approximately 12 fF/μm^2. If a PLL filter requires

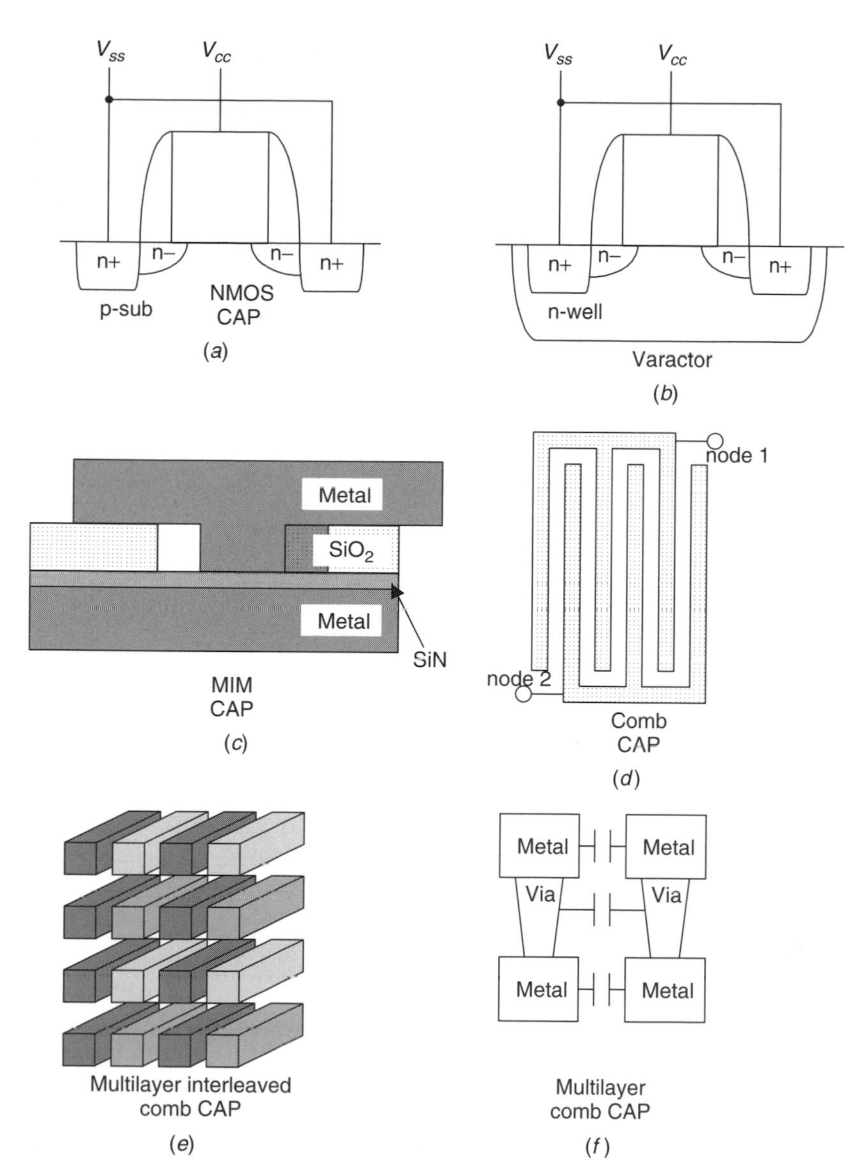

Figure 4.7 Integrated capacitor examples.

200 pF of capacitance, the total leakage current is on the order of 83 μA, which is significant compared to a typical charge pump current. This leakage current will be manifested as excessive deterministic jitter on the PLL output. This has forced the use of thick oxide devices for loop filters, but this creates an additional problem because of the capacitance variation as a function of the applied voltage. This is especially a problem for loop filters since the control voltage can vary over some voltage range. For this reason, varactors or native NMOS devices are

typically used. Varactors are really a special case of a MOS type of capacitor, although specialized structures have been proposed [13].

In some cases, metal–insulator–metal (MIM) capacitors are available if a mixed-signal process is being used. These are typically constructed using a thin nitride insulator between two metal layers and result in a capacitance of approximately 1 to 4 fF/μm^2. One of the major benefits of MIM capacitors is their high Q value (up to 150)[14] since the parasitic resistance can be minimized and their capacitance is constant regardless of the voltage applied.

Another common capacitor structure is the metal comb capacitor, which is really just a variation on metal plate–metal plate capacitors. Several variations are shown in Figure 4.7(*d*)–(*f*). The comb structures can produce a high capacitance per unit area with a high Q value since they are constructed from metal. These capacitors rely on the interconnect capacitance inherent in the close proximity of two adjacent metal lines. A single layer of metal may be used, as shown in Figure 4.7(*d*), if the absolute value of the capacitance is critical, although the capacitance per unit area will not be as high as in multilayer capacitors. The structures illustrated in Figure 4.7(*e*) and (*f*) use additional layers, increasing the capacitance per unit area but also increasing the total capacitance error because of the compounding effect of the metal thickness variation. Accurate prediction of the via-to-via capacitance for the structure shown in Figure 4.7(*f*) can become difficult. Use of a low-κ dielectric will reduce the capacitance per unit area of these structures. The effects of the copper process must also be considered when designing metal layer capacitors since metal density rules must be followed and the metal thickness variation as a function of the metal density must be considered when calculating the true capacitance value. Metal density effects can easily add 2 to 5% error in metal thickness. Given all of the variability of metal capacitors, they are typically used in applications where absolute value is not critical. Even in these cases, it is best to include test wafers in any process characterization, to verify the actual capacitance value. Additional structures may be found in Refs. 15 and 16.

Another common component used in mixed-signal design is the inductor. Inductors are becoming more common on integrated circuits for use in low-noise amplifiers and LC-based VCO cells because of the typical phase noise improvement. Inductors have not faced the same scaling issues as MOS devices since they are formed with metal layers. The major change with inductors has been introduction of the copper process, which has improved the Q value of the inductor by reducing the series resistance. Further improvement may be realized when the low-κ dielectric process becomes more prevalent because of the reduction in parasitic capacitance (assuming that this capacitance is not being used as part of an LC circuit). Inductors are typically formed on a top layer of metal because of the increased thickness (about 1 μm on a standard digital process and up to about 3 μm on a dedicated mixed-signal process). Some consideration must be given to the package since flip-chip packaging is becoming more prevalent. Given the bump sizes (30 to 80 μm in diameter), package routing keep-out zones may be necessary to avoid unwanted coupling between package signals and the inductor.

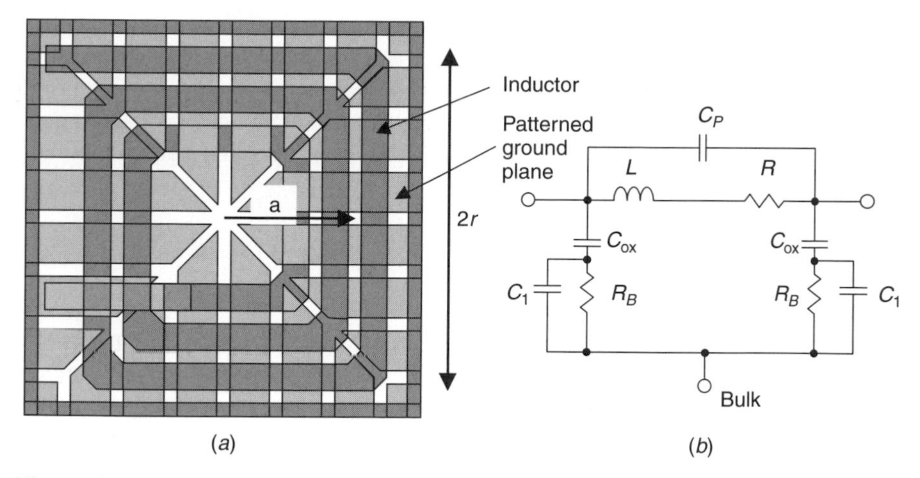

Figure 4.8 (*a*) Spiral inductor with patterned ground plan and (*b*) associated lumped model.

Figure 4.8 is a simplified diagram of a spiral inductor. A patterned ground plane is used to reduce eddy current losses. By slotting the ground plan orthogonal to the inductor winding, the image current induced by the magnetic field of the inductor is cut off, which significantly reduces the negative self-inductance. The various components of the lumped model shown in Figure 4.8 can be approximated as follows [11]:

$$L \simeq \frac{37.5\mu_o n^2 a^2}{22r - 14a} \qquad \text{(spiral inductance)}$$

$$R \simeq \frac{l}{W\sigma\delta(1 - e^{-t/\delta})} \qquad \text{(spiral resistance)}$$

$$\delta = \sqrt{\frac{2}{\omega\mu_o\sigma}} \qquad \text{(skin depth)}$$

$$C_P = \frac{nW^2\varepsilon_{ox}}{t_{ox1}} \qquad \text{(shunt capacitance)}$$

$$C_{ox} = \frac{Wl\varepsilon_{ox}}{t_{ox2}} \qquad \text{(capacitance between spiral and substrate)}$$

$$R_1 \simeq \frac{2}{WlG_{sub}} \qquad \text{(substrate resistance)}$$

$$C_1 \simeq \frac{WlC_{sub}}{2} \qquad \text{(substrate capacitance)}$$

where n is the number of turns, r the spiral radius, a the spiral mean radius, W the wire width, l the wire length, t the wire thickness, σ the conductivity of the wire, δ the skin depth, t_{ox1} the oxide thickness between the main spiral and the cross under, and t_{ox2} the oxide thickness between the spiral and the substrate, μ_o the permeability in vaccum, G_{sub} the substrate conductance per unit area, and C_{sub} the substrate capacitance per unit area. Accurate inductance calculations must be accomplished using three-dimensional field solvers and verified through test structure characterization.

4.5 DESIGN METHODOLOGY

To design mixed-signal circuits successfully for advanced processes requires that careful architectural studies be performed prior to implementation. Several important steps should be taken to ensure that design of specific circuits is possible in these processes. An error in any key decision can result in redesign, which could result in missing a key product window. Evaluation of analog and mixed-signal blocks must begin during the process development phase. Waiting until the process has been defined will result in a significant delay in the overall product development as well as missed opportunities to influence the process itself.

4.5.1 Benchmark Circuits

During the process development phase, it is important to have analog blocks ready to implement on the new process so that they may be added to initial process characterization wafers. Critical parameters should be extracted; for example, simple operational amplifiers can be implemented to study gain, bandwidth, input operating range, supply noise immunity, offset voltage, and so on. Various VCO cells can be added to look at the gain, phase noise, operating range, and supply noise immunity. Current mirrors can be used to estimate output impedances as well as matching characteristics. Bandgap references should also be considered. The key is to keep the cells simple to allow the process to be evaluated, not the specifics of the design itself. This aspect is especially important if test chips are run early during the process development when models are immature and potentially inaccurate. It is also important to look at device-matching issues for a range of device sizes and configurations to evaluate proximity effects. These modules should be implemented using all available oxide thickness options and potential threshold voltages. This allows trade-off studies to be made when deciding which devices will be used for a certain function. Although test chips are the best way to verify a design, having a series of schematics that can be simulated using device models can provide critical insight into potential issues with the process.

4.5.2 Design Using Thin Oxide Devices

One of the key decisions that must be made is whether to design the analog and mixed-signal portions of the overall system using thin oxide devices or one of the thicker oxide devices. Neither approach is without its drawbacks. Consider the option of using thin oxide devices. This will require that all the device issues discussed previously be considered to ensure that the design can be implemented successfully.

Supply headroom can become a major issue. Realistically, the threshold voltage on these more advanced processes is approaching 0.3 V (assuming slow models). If the supply voltage is on the order of 1.0 to 1.2 V, a designer must keep all circuits below three threshold voltages once temperature and supply variation are included. Some supply headroom may be recovered by using lower-threshold thin oxide devices (assuming that this is an option for the process), but this approach results in added complexity in the process corners that must be simulated. If two devices are being used with only the threshold voltage different, the models should be partially correlated. Both devices should experience similar poly critical dimension (CD) control and oxide thickness, but the threshold voltage implant can be different (but is still partially correlated). Understanding how the models correlate can become difficult. For this reason, it is best to avoid using multiple threshold devices in critical circuits if at all possible.

Some of the benefits of using thin oxide devices include a possible reduction in the total area of the mixed-signal block, assuming that a potion is digital, such as the digital filter of a sigma–delta converter. This option also allows easier transfer to the core since no voltage level shifting is required. Level shifting can become especially problematic for phase-locked loops since it can represent a possible location of phase error that is difficult to compensate out correctly.

Some of the architectural trade-offs that must be made include the effects of using thin oxide devices on key circuit blocks such as the VCO. Consider the simple VCO cell shown in Figure 4.9. The two bias voltages, V_{BN} and V_{BP}, must be at least one threshold voltage from the respective supply rail. For a first-pass approximation, the operating range of the control voltage for the VCO is

$$V_{tN} < V_{\text{control}} < V_{CC} - V_{tP}$$

On a 0.13-μm process, V_{CC} may be 1.2 V with V_{tP}, and V_{tN} may be 0.32 V, giving a voltage swing from V_{control} of 0.56 v. On a 90-nm process, V_{CC} may be

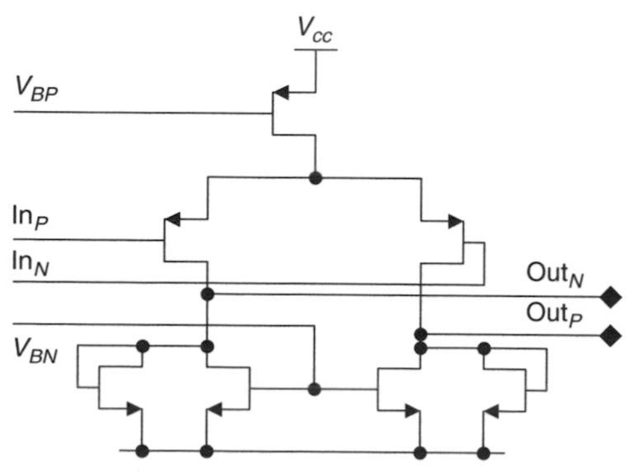

Figure 4.9 Simple voltage-controlled oscillator cell. (From Ref. 12.)

1.0 V with V_{tP} and V_{tN} of 0.3 V, resulting in a voltage swing for V_{control} of 0.4 V. If the VCO must operate over the same range, say 1 GHz to 300 MHz, without any margin included, the VCO gain would be 1250 MHz/V for the 0.13-μm process and 1750 MHz/V for the 90-nm process. The VCO gain is significantly higher for the 90-nm process. If nothing has been done to reduce the noise of the system, we can expect degradation in the overall performance of the phase-locked loop since any noise on the control voltage or supply rails is amplified by the higher VCO gain.

4.5.3 Design Using Thick Oxide Devices

A second option is to use the thicker oxide devices for the analog blocks and possibly some of the digital blocks and then level-shift to the lower core voltage. This approach can increase the headroom for an analog circuit, permitting cascoding of current mirrors and gain stages to increase the noise immunity. Gate leakage problems are reduced significantly, as is the source–drain leakage current. Two problems can occur with taking this approach. First, the analog or mixed-signal block will probably interface with the core at some point. This will require a level shifter to be implemented. For a data converter, these level shifters may not pose a problem, but the entire system must be considered. A data converter may have a reference clock that is derived from another block. This could come from the core power domain, so it must be level shifted. This level shift can represent a point of additional noise, increasing the jitter of the reference clock or degrading the duty cycle, which will reduce the accuracy of the data converter, especially if the duty cycle is important to the design. Similar problems can arise for PLL circuits as well, which has already been discussed. Figure 4.10 shows a possible PLL configuration that uses thick oxide devices for the PLL and then level-shifts (LB block) the output clock to the core voltage level. Figure 4.11 shows a second PLL architecture where only the core analog cells are thick oxide and level shifting is used when transferring between these core cells and the remaining digital portion of the PLL. This architecture still has

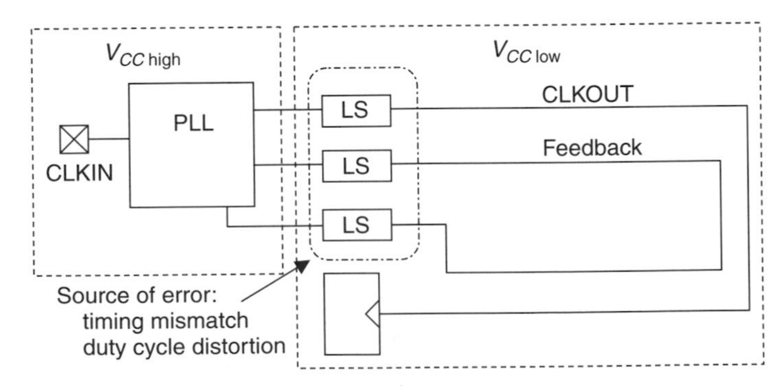

Figure 4.10 Possible PLL architecture where the entire PLL is powered from the higher supply and the core is supplied by the lower supply.

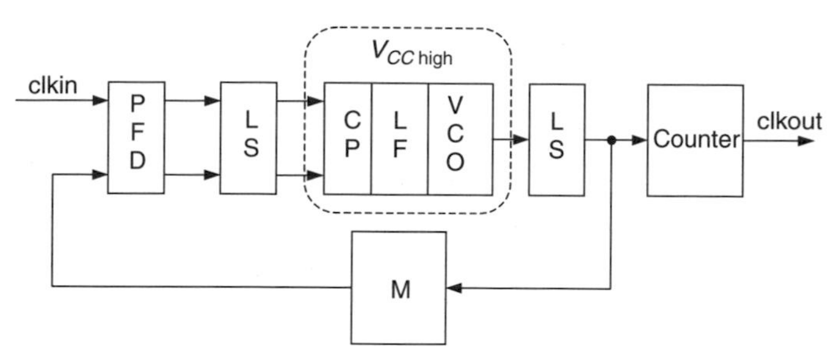

Figure 4.11 Possible PLL architecture where only the PLL core is powered from the higher supply voltage and the remaining logic is supplied from the lower supply.

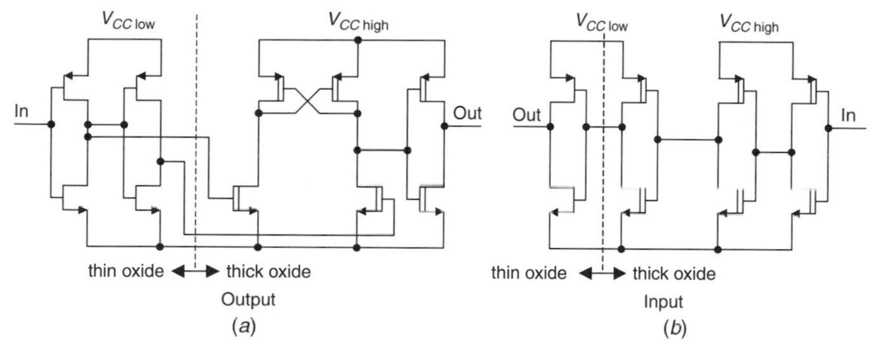

Figure 4.12 Traditional level shifters for (*a*) up conversion and (*b*) down conversion.

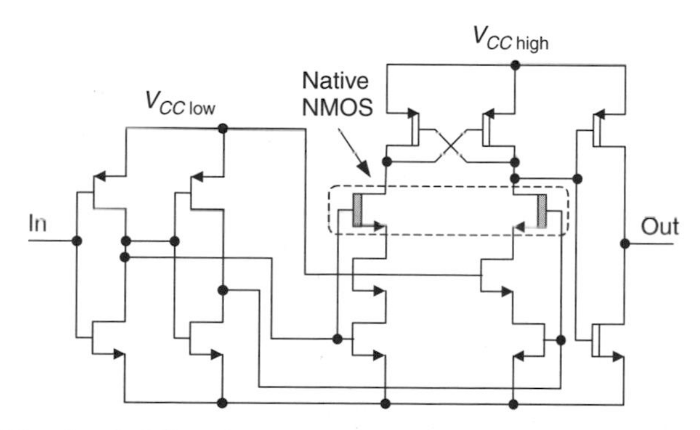

Figure 4.13 New level shifter that provides greater protection to 1.2-V devices. (From Ref. 17.)

problems since duty cycle issues will arise from the level shifters. Figure 4.12 shows the traditional level shifters used in many applications, and Figure 4.13 shows a more recently developed level shifter that takes advantage of thick oxide native NMOS devices to prevent thin oxide devices from "seeing" the higher V_{CC} supply. This thick oxide native NMOS device has a very low threshold voltage, typically close to zero. This allows thin oxide devices to be used for the transfer point between the two supply domains, which helps reduce threshold voltage problems with the thick oxide devices.

From an overall design standpoint, is best to match the core voltage for many circuits, especially phase-locked loops, if possible. This being said, it really depends on the specific application and how the architecture can be modified to alleviate some of the problems from either approach.

4.6 LOW-VOLTAGE TECHNIQUES

Many methods have been developed over the past several years for designing analog and mixed-signal circuits with low supply voltages. The need has arisen because of the thinner oxide thickness, forcing a lower power supply voltage and the portable market, such as cell phones, that operate using low supply voltages. Although the power supply has decreased, the device threshold voltage has not tracked equivalently. On early-generation processes, cascoding was used to generate high-gain circuits because of the increased output impedance. With the reduction in supply voltage, this becomes difficult to achieve, so designers have been forced to cascade stages to obtain the same gain. Although this technique works, each gain stage adds additional poles and zeros, making compensation more difficult[5]. In the following sections we describe some of the various low-voltage building blocks used in analog and mixed-signal design.

4.6.1 Current Mirrors

Current mirrors represent one of the most basic building blocks in any analog or mixed-signal circuit. A good current mirror provides superior replication of the reference current and high output impedance for noise immunity. In the past this was achieved by using cascoding, but this may not be possible given the reduction in headroom. Other current mirrors have been developed to reduce the headroom issue while maintaining high output impedance. The transistor output impedance on these advanced processes has degraded considerably. The circuit shown in Figure 4.14 is an example of a low-voltage, high-output impedance current mirror. By sizing the devices such that MN_2 is one-fourth the size of the other devices, it is possible to reduce the required output voltage to $2V_{d\ \mathrm{sat}}$. There are a few problems with this circuit. First, current matching is not perfect since the V_{DS} values of MN_1 and MN_5 are not the same. This problem is amplified further on the small-geometry process by the threshold voltage sensitivity to source–drain voltage differences. These effects must be considered since they can result in increased offset voltages in operational amplifiers. Also, since the source–drain voltages are different, the various devices will not have the same threshold voltage shift over time, resulting in a further increase in the mismatch.

Figure 4.15 shows two additional low-voltage current mirrors with improved current-matching capability. The current mirror shown in Figure 4.15(a) has been

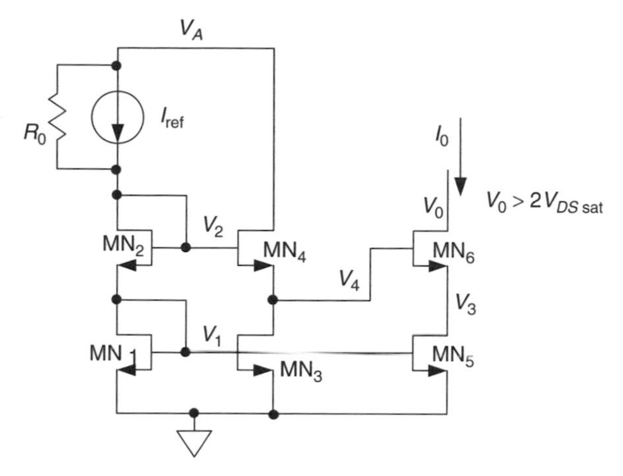

Figure 4.14 Low-voltage high-impedance current mirror. (From Ref. 4.)

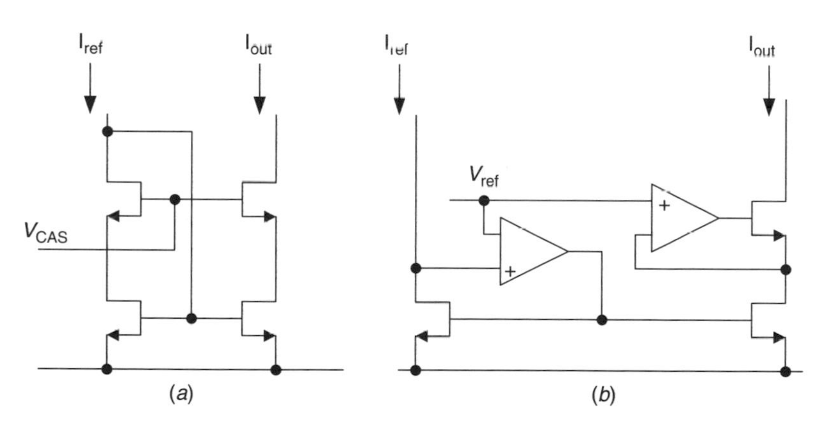

Figure 4.15 Additional low-voltage/current references with improved current matching.

used extensively over the years. By controlling V_{CAS}, it is possible to maximize the output voltage swing while maintaining high output impedance and good current matching. The current mirror shown in Figure 4.15(*b*) offers similar performance but also maintains a fixed voltage on the input node that reduces current errors on the reference side of the circuit[6]. The output node can be made to swing over a wide range by controlling the reference voltage accordingly. Care must be taken when using the current mirror shown in Figure 4.15(*b*) since the active input will require stability compensation.

The current mirrors shown offer the ability to use cascoded structures even with reduced headroom, but gate leakage effects have been neglected up to this point. The gate leakage must be considered when actually implementing the design since it can affect circuit operation, especially for operation in weak inversion, where the reference currents are small but the devices are large.

4.6.2 Input Stages

Several innovative low-voltage rail-to-rail input stages have been developed[8,9,10]. Two general approaches can be used. The first employs using a PMOS and an NMOS differential pair together as shown in Figure 4.16. For low-input voltages close to V_{SS}, only the PMOS differential pair is biased on. For very high input voltages close to V_{CC}, only the NMOS pair is biased on. For voltages in the middle, both the NMOS and the PMOS differential pair are biased on. This can lead to a varying gain as a function of the common-mode range, which is undesirable.

Another newer approach for input stages are bulk-driven MOS transistors. A simple circuit that uses this technique is shown in Figure 4.17. By driving the bulk

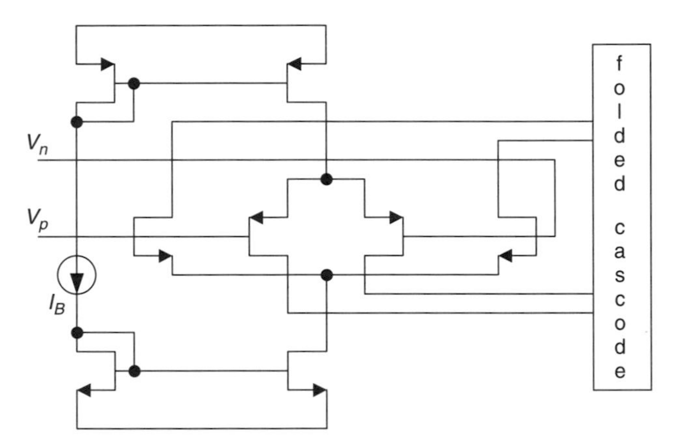

Figure 4.16 Rail-to-rail input stage using PMOS and NMOS pairs.

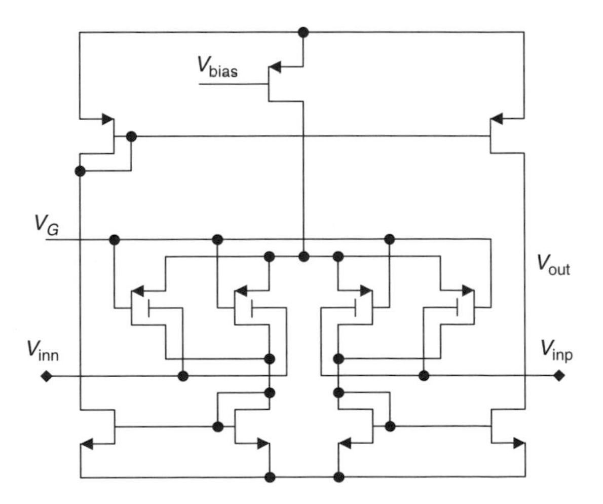

Figure 4.17 Bulk-driven input stage for an operational transconductance amplifier.

of the devices, it is possible to allow negative, zero, or small positive voltages. This can increase the input common-mode range. Some of the drawbacks include a significant decrease in the transconductance compared to traditional differential pairs; a reduction in the maximum operating frequency; potential limitations in the device available for an input pair (i.e., PMOSs are not available unless a deep n-well process is used); the fact that transistors must be laid out in separate n-wells (assuming that PMOS devices are used), which can give rise to increased offset voltage; latch-up problems with turning on the parasitic transistor; and greater equivalent noise.

4.6.3 Output Stages

Output stages must also provide rail-to-rail operation, especially for low-voltage applications. Figure 4.18 shows three different class AB output stages that are

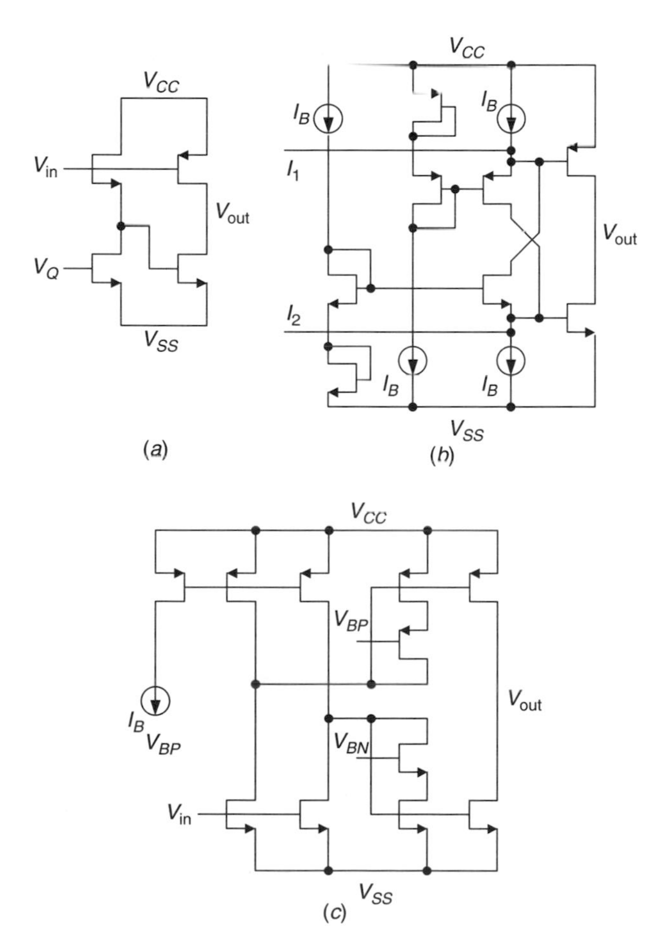

Figure 4.18 Various rail-to-rail output stages for low-voltage applications.

suitable for low-voltage rail-to-rail operation. Figure 4.18(*a*) shows a very simple output stage[10]. The quiescent current is set by the voltage, V_Q. Figure 4.18(*b*) shows Monticelli's output stage[36], which has been widely used in low-voltage applications. This output stage will add additional gain to the amplifier, which can be desirable. The minimum supply voltage is $2V_{\text{th}} + 3V_{DS\ \text{sat}}$, which is the same as the simple stage shown in Figure 4.18(*a*). The final output stage described by You et al. is shown in Figure 4.18(*c*). Although this output stage does not have the gain of the Monticelli output stage, it can operate at a supply voltage as low as $V_{\text{th}} + 2V_{DS\ \text{sat}}$.

4.6.4 Bandgap References

Bandgap references are becoming increasingly difficult to design as supply headroom continues to decrease. One clever circuit used to combat this problem is shown in Figure 4.19. The operational amplifier connects to divided-down versions of the two bias nodes. The voltages at nodes N1 and N2 are equal because of the operational amplifier. If $R_{2\text{A}1}$ and $R_{2\text{A}2}$ are set to be equal to $R_{2\text{B}1}$ and $R_{2\text{B}2}$, nodes N3 and N4 are also equal. This relationship allows the following expression to be derived:

$$V_{\text{OUT}} = \frac{R_3}{R_2}\left(V_{E\text{B}2} + V_T \frac{R_{2\text{A}1} + R_{2\text{A}2}}{R_1} \ln M\right)$$

assuming that a PMOS-based differential pair in the operational amplifier allows the minimum supply voltage to be derived as

$$V_{\text{sup min}} = \frac{R_{2\text{B}2}}{R_{2\text{B}1} + R_{2\text{B}2}} V_{E\text{B}2} + |V_{\text{th}p}| + 2|V_{DS\ \text{sat}}|$$

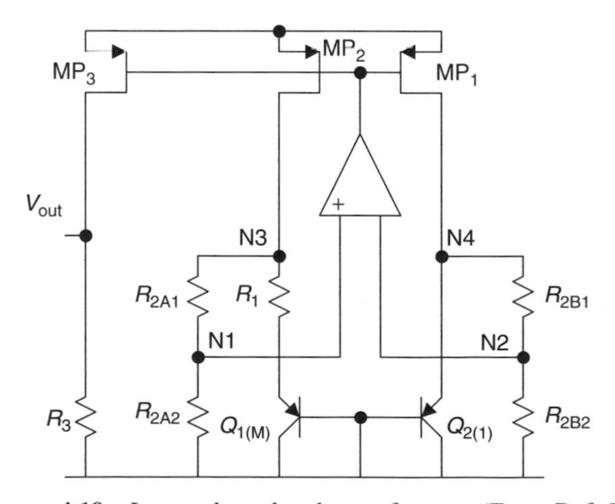

Figure 4.19 Low-voltage bandgap reference. (From Ref. 7.)

if the operational amplifier limits the supply headroom, or

$$V_{\text{sup min}} = V_{EB2} + |V_{DS\ \text{sat}}|$$

if the pnp transistor and the current source limit the supply headroom.

This circuit does have some problems since the drain–source voltages of MP_3 will not match that of MP_1 and MP_2. This mismatch can be reduced by including some type of cascoding, but supply limits may prevent this from being implemented. For extremely advanced processes, the diode path may prove to be the limiting factor, especially at low temperatures.

4.7 DESIGN PROCEDURES

A major problem with the design of these advanced processes is the discrepancy between pre- and postlayout. This issue has arisen as the interconnect parasitics dominant the overall loading on nodes. Initially, the effects of interconnects were neglected completely (0.5 μm and larger). Circuits still functioned adequately using this approach for the larger geometries since the device models included the bulk of the parasitics. As the geometries have continued to shrink, the interconnect has become an even greater portion of the total loading of circuit nodes. Around the 0.25-μm process node, one had to include the parasitics of the interconnect to ensure circuit functionality, but they could typically get away with neglecting the resistance of the interconnect except for very long signal lines. On the 0.15-μm process node, parasitic resistance has become important enough to be included even at a more localized level. For 90 nm and beyond, full netlist extraction must be performed to ensure circuit functionality, but it requires resistor capacitor extraction (RCE) extraction to be accurate. There are three overall techniques for dealing with pre- versus postlayout issues:

1. Attempt to model everything prelayout to minimize the discrepancies between pre- and postlayout.
2. Implement no special modeling; do a preliminary design pass to get the layout completed as quickly as possible, and then optimize postlayout.
3. Use historic data for prelayout estimates to ensure that design targets are met and then get to the layout step as quickly as possible and optimize postlayout.

Approach 1 is the most time-consuming prelayout and the most prone to error since it is sometimes difficult, if not impossible, to know what the layout will look like during the early design phase, and estimating routing parasitics can become difficult without an actual layout or even a floor plan unless the cell is very simple. Approach 2 allows for the quickest way to begin the layout process, but ends up with the greatest error between pre- and postlayout (potentially a 35 to 45% discrepancy for a cell such as a voltage-controlled oscillator). The critical thing here is that a base design is created and a layout can be generated for future simulations, but more iterations of the layout-extracted netlist may be required before a design can be finalized. The difficulty comes from dealing with a layout-generated netlist, which can become difficult to read. Because of that, it is critical to create the schematic hierarchy in a manner that allows small blocks to be extracted to make the netlist more manageable. Approach 3 is really just approach 2, but previous generation cell information is used as a generic guideline to know how much overdesign is required to meet the specification. If on the 0.13-μm process it was observed that a given cell, such as a high-speed counter, degraded by 40% between pre- and postlayout, this can be used as a general guide for what will happen in the next process. This method is quick and no less accurate than spending huge amounts of time trying to model everything accurately prelayout.

There are a few approaches to working with STI stress effects. One approach is to focus on sizing devices to match the device layouts used to create SPICE models. Using this approach will minimize the discrepancy between pre- and postlayout simulations for stress effects. A second (and probably superior) approach is to attempt to include the effects in prelayout simulations by preparing an initial floor plan prior to actual layout so that the OD parameters are known before the actual layout is completed. By floor planning the cell prior to layout, OD effects may actually be used to increase the speed of critical circuits such as a VCO in a phase-locked-loop or clock data recovery circuit. At a minimum, it allows for more accurate modeling prior to layout. Since the layout can greatly affect STI

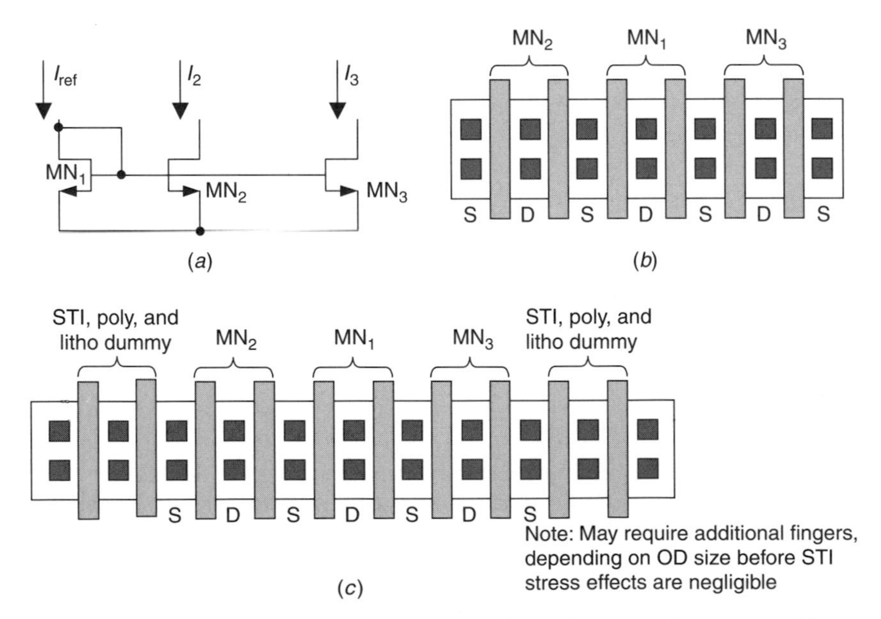

Figure 4.20 (*a*) Simple current mirror, (*b*) small-area layout with poor matching, and (*c*) improved layout with dummy devices.

stress effects, careful consideration must be given to the floor plan prior to the start of layout. Fortunately, many techniques used to generate good analog layouts result automatically in good matching from the STI stress effect. Figure 4.20 shows a simple current mirror and two potential layouts. The layout shown in Figure 4.20(*b*) is compact in size but provides poor matching from a lithographic standpoint as well as STI stress effects. The layout shown in Figure 4.20(*c*) represents an improvement from a matching standpoint but at the cost of a larger area. Dummy devices are used to minimize the STI stress as well as to improve the lithography and etching issues. Neither layout shows what to do on the top and bottom sides of the device; typically, guard rings are added between adjacent cells to maintain a consistent layout on either side of a cell.

4.8 ELECTROSTATIC DISCHARGE PROTECTION

Electrostatic discharge (ESD) protection of critical analog and mixed-signal blocks becomes increasingly difficult on more advanced processes because of the ever-thinning gate oxide. This is coupled with the typically light loading of the analog block relative to the digital blocks. Breakdown voltages are becoming increasingly smaller, quickly approaching less than 6 V for the 130-nm process node. ESD protection must be able to turn on quickly at the desired trigger and hold the voltage at desired levels. Analog and mixed-signal circuits are being forced to integrate with larger digital blocks. This requires the overall topology of the chip to be considered when developing an ESD protection scheme for the analog and mixed-signal portions.

4.8.1 Multiple-Supply Concerns

Figure 4.21 shows an example of a protection scheme if multiple grounds are used in the chip. Back-to-back diodes are used to connect the various grounds together to provide a path for the ESD current to flow since the ESD event can occur between any two pins. Two diodes are used to reduce noise coupling between the various grounds, but this coupling must be considered when determining diode sizes and the number of series diodes. It is imperative to ensure that the diodes are never forward biased during normal operation since this may severely degrade performance of the analog circuit. A few ESD discharge paths are shown as examples of how various components are used during an ESD event.

The I/O signal frequencies are increasing in an attempt to keep up with the clock frequency of the advanced high-performance microprocessors. There is a need to reduce parasitic capacitance on the pad, and ESD protection devices are not exempted. Reduction in the ESD capacitance must not sacrifice the protection level as the gate oxide scales. In fact, the level of protection must be as good or better. This is not new to the radio-frequency (RF) designers who had to battle this issue much earlier on, but we are beginning to experience this difficulty in mixed-signal and microprocessor designs with high-speed external I/O. It is

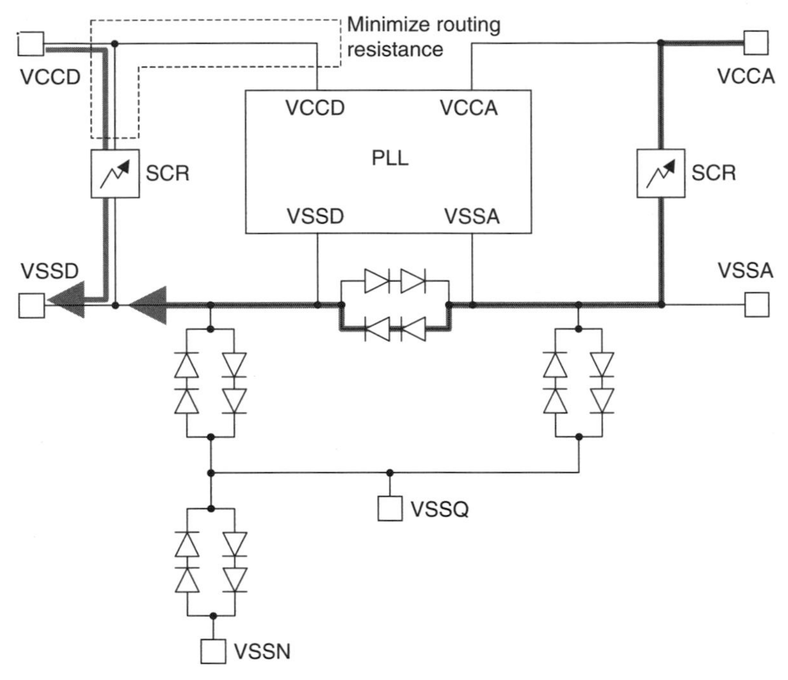

Figure 4.21 ESD protection configuration of a PLL implemented on a large digital IC with separate grounds for the IO (VSSN), core (VSSQ), PLL analog (VSSA), and PLL digital (VSSD).

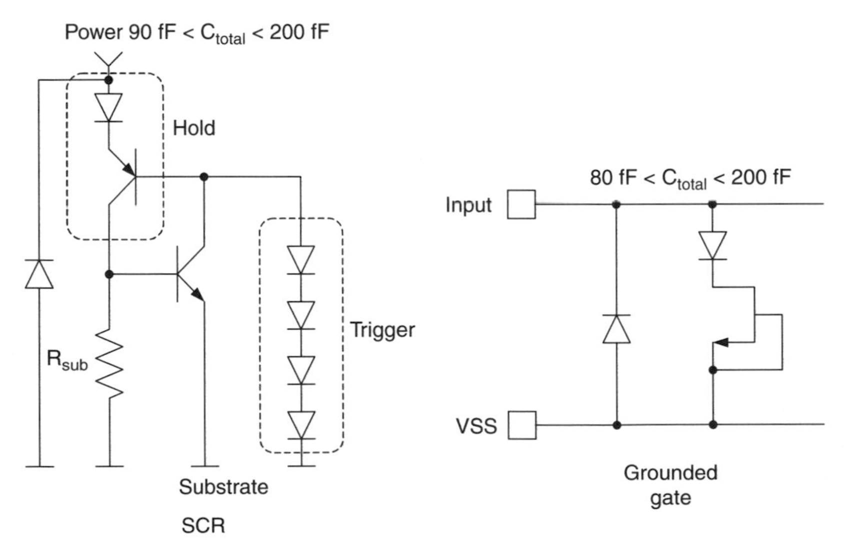

Figure 4.22 Various ESD structures that can be used for protecting high-speed inputs.

critical that any ESD protection added to an input or output pin not affect the overall performance of the primary function. This means that the capacitance of the ESD structure must be constant regardless of the operating voltage. If this is not the case, harmonic distortion may arise that could affect the performance of a high-speed data converter, RF synthesizer, clock data recovery circuit, or phase-locked loops.

Figure 4.22 shows two different structures that can be used for protecting high-speed and analog inputs. Because diodes are connected directly to the high-speed pin, the nodal capacitance can be kept small. Although the device capacitance can be kept small, the overall ESD structure capacitance can become large because of the metal interconnect as well as package parasitics, which can add anywhere from 200 fF to 1 pF to the total nodal capacitance.

4.9 NOISE ISOLATION

Noise isolation for analog and mixed-signal circuits on larger digital ICs has become more difficult given the increasing density of the digital logic and faster signal edge rates, which have led to a substantial increase in the instantaneous current, giving rise to greater noise levels. The decreasing supply voltage that reduces the noise margin magnifies this effect further.

4.9.1 Guard Ring Structures

Guard rings are used extensively for reducing noise within the substrate. Numerous studies have been performed to investigate the benefit of various configurations [18–35]. There is often conflicting information regarding how best to design a guard ring structure to minimize coupling between different portions of a system. Mixed-signal chips have become the norm rather than the exception. Even modern-day microprocessors include some level of analog circuitry in the form of a PLL. Receiver and transmitter chips may contain numerous analog and digital blocks. Coupling between the analog and digital portions of a chip can produce unwanted noise, leading to degraded performance.

As an example, consider a simple analog-to-digital converter circuit. If the usable input range is 2 V and we are using a 12-bit converter, a single bit has a resolution of approximately 488 μV. If noise coupled through the substrate from the digital portion of the circuit is large enough, it can reduce the effective bit resolution of the converter. The coupled noise through the substrate has been shown to be quite high, in the millivolt range for some cases. To design high-performance ICs it is a necessity to consider how to isolate substrate noise during the design phase.

In general, wider guard rings provide better isolation [31]. The reason for this is an increase in resistance between the noise source and the sensitive circuit. The closer the noise source is to the guard ring/substrate tap, the lower the noise [31] because the stray carriers are collected easier at the source before they have an opportunity to begin distributing through the substrate. The distance between the noise source and the receiver makes a significant impact [31]. Reference 23 shows this as being linear with the separation distance, which makes sense from a resistance standpoint because the majority of the current flows near the surface, where it can be collected effectively by guard rings. The ultimate performance and required separation are determined by the substrate doping. The optimum spacing must be determined using the process information. There are several different strategies for isolating circuits. For epitaxial construction, little can be done other some guard rings and close taps for the substrate for noisy circuits. For non-epitaxial construction you can use p^+ guard rings, p^+ guard rings and n-well rings, oxide trenches, silicon on insulator, and deep n-well structures. How well these perform depends on how the power busing scheme is designed and the overall process specifications (doping, etc.). Most studies have shown that the addition of n-well guard rings do little for improving the isolation and may actually degrade the noise performance [19,31].

Having a back-side connection may actually increase the noise for lightly doped substrates [24]. This is probably caused by the increase in resistance to the back-side connection, making carriers more likely to flow to the sensitive circuits rather than flow out of the device. This is very dependent on the location of the sensitive circuit relative to the noise source, wafer thickness, and bonding configuration for back-side connection. For an epitaxial wafer, this may not be true since the substrate has a low resistance.

Connection to the guard rings is critical [28]. Reducing the inductance on guard ring connections can make a significant improvement in the isolation achieved. Also, having a separate guard ring tap on the part can help improve the isolation even more [23]. For low-speed circuits, modeling of the substrate as a resistive network is sufficient, but for very high speed ICs, substrate cannot be modeled as a purely resistive network [18]. Using Pisces and Medici for high-speed simulations may not be feasible because of the long simulation times, but these tools can be used to gain insight into coupling through the substrate.

The premise of a guard ring is to provide a means to collect stray carriers within the substrate such that they do not travel to a sensitive portion of the circuit and introduce undesirable noise. One of the key elements in designing a guard ring structure is the ability to understand what is being designed and how to model it correctly. In many instances it is possible to reach an incorrect conclusion because of improper modeling of the physical design. Figure 4.23 shows a simplified diagram of a potential coupling model. The circuit shown on the left is the noise generator with separate n-well and p^+ guard rings connected to outside supplies. If the resistance in the path of the guard rings is greater than the path through the substrate to the noise-sensitive circuit, the noise will couple to the sensitive circuit. Similarly, if all guard rings are connected, it is possible for current to flow out of one guard ring and into a sensitive circuit through its guard ring. This becomes a greater possibility if the package and die parasitics include the inductance, which can greatly increase the series resistance for high-frequency switching noise.

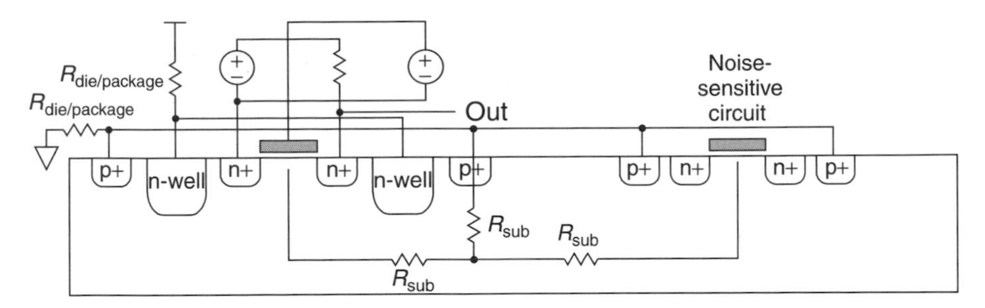

Figure 4.23 Substrate coupling model.

4.9.2 Isolated NMOS Devices

A more recent development has been the inclusion of a deep n-well to provide improved isolation for NMOS devices. This structure is shown in Figure 4.24. The deep n-well region is formed by an ion implant, and then the isolated p-well is created when the n-well is formed. If the doping densities are engineered properly, the NMOS device in the isolated p-well will have characteristics identical to those of a standard NMOS device. This is critical to ensure that devices in this isolated region do not have degraded performance. This structure provides superior isolation for NMOS devices. Typical numbers are 20 to 25 dB for frequencies below 500 MHz and 5 to 10 dB for frequencies above 1 GHz. Some care must be exercised when designing using the deep n-well process to avoid making the deep n-well too large, since this increases the parasitic capacitance to the substrate, which can allow undesirable noise to be coupled into the critical circuitry.

4.9.3 Epitaxial Material versus Bulk Silicon

An epitaxial wafer provides a low-resistance substrate that helps significantly to reduce the possibility of latch-up but effectively shorts all devices together. Because of this, noisy digital blocks will couple into sensitive analog blocks (unless a triple-well process is used). Because of this coupling, it is best to have a single common ground for all blocks on the chip since they are effectively already shorted together through the substrate. Critical block design must begin

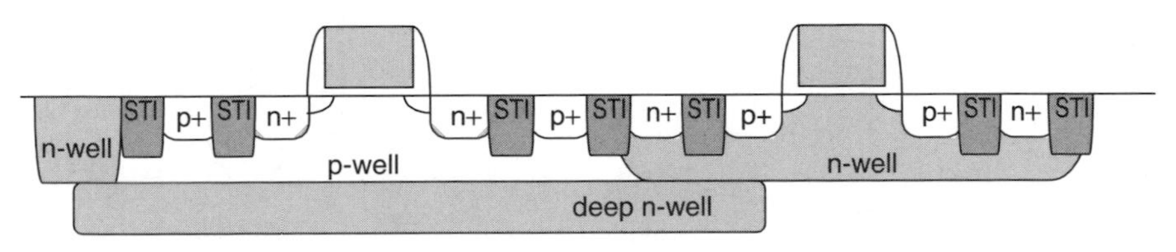

Figure 4.24 Simplified cross section of an NMOS and PMOS device implemented on a process that offers deep n-well implantation.

with knowledge of this heavy coupling on the ground side to ensure that the circuits are designed to minimize sensitivity to ground-side noise. In general, mixed-signal design on an epitaxial process without a triple well is difficult, especially if the digital portion of the chip is large.

4.10 DECOUPLING

Decoupling has become more of an issue for the deep-submicron processes because of the greater device density, faster edge rates, and reduced supply voltage. The instantaneous *IR* drop can be in the volt range for a high-speed output buffer supply and hundreds of millivolts for a low-voltage core supply. This *IR* drop comes from the series resistance and inductance on the supply rails for a given circuit block. Figure 4.25 is a simplified diagram of a board with an IC. Decoupling added at the board level can reduce the supply ripple at the board level but typically has little influence at the IC die level because of the package and die parasitics. Exact values for the various parasitics are difficult to provide, but typical ranges for the inductance can be 5 to 10 nH for a wirebond package and less than 1 nH for a flip-chip package. This figure shows the analog and digital supplies fully isolated.

Two simple equations can be used for investigating the effects of inductance and decoupling on the voltage drop:

$$i = C\frac{dV}{dt}$$

$$V = L\frac{di}{dt}$$

Consider the simple case where the series inductance is 1 nH and the change in current is 1 mA in 30 pS (not an unrealistic number for 0.13 μm and beyond);

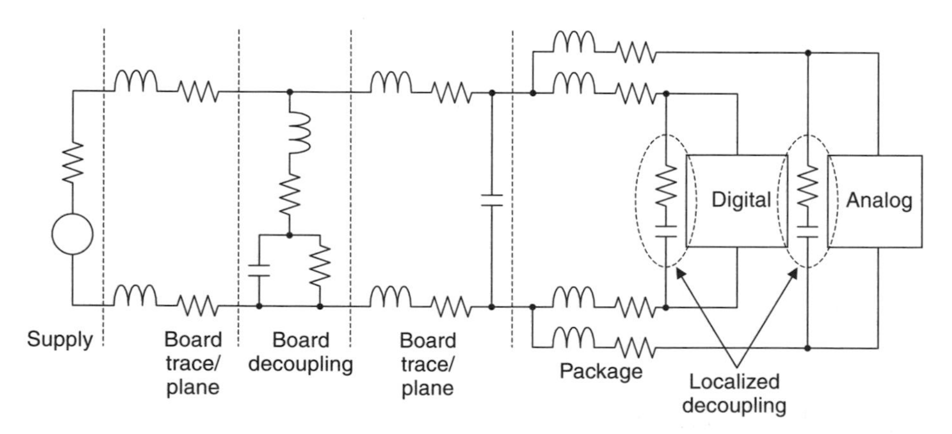

Figure 4.25 Board with an IC.

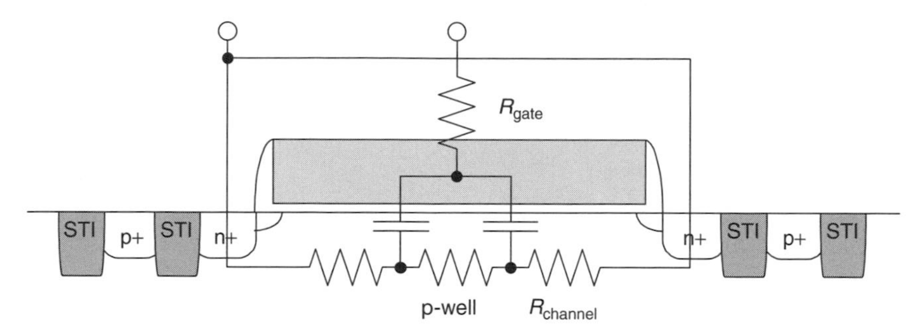

Figure 4.26 N-MOS gate cap.

the *IR* drop across the inductor is approximately 33 mV, which is significant, especially considering how much higher the current can become.

Figure 4.26 is a simplified diagram of an NMOS decoupling capacitor. Gate oxide capacitance is the most common device used as decoupling cells because of the large capacitance per unit area compared to other forms of decoupling. The parasitic components must be considered when designing a decoupling cell as neglecting them will lead to the incorrect assumption that a single large-area gate can be inserted to create a large decoupling cell. This will actually create a poor decoupling cell since the channel resistance will increase across the gate length, giving rise to a large series resistance with the capacitor. Additionally, the poly density rules may prohibit this as well. In effect, only the gate region near the edges will contribute to the capacitance. The only way to minimize this effect is to make the channel length small. This must be coupled with the number of cells being used and the capacitance per cell. The total channel resistance can be extracted from I_{DS} versus V_{DS} curves by looking at the region where V_{DS} is small.

Figure 4.27 is a simplified model of a decoupling cell. Typically, multiple decoupling cells are added in parallel that is equivalent to the circuit shown on the right side of Figure 4.27. By adding multiple decoupling cells in parallel, it is possible to create an effectively small series resistance and larger capacitor. The far-right circuit is used to simulate a few test cases to illustrate the effects of package parasitics and decoupling. The current source is set to drive a current spike of 10 mA with a rising- and falling-edge rate of 30 ps (which is not unreasonable on a 90-nm process). Figure 4.28 shows the effect of inductance on the supply voltage at the current source when the current switches. Figure 4.29 shows how ineffective the decoupling is when it has a high series resistance, which illustrates the importance of minimizing the parasitic resistance of the capacitor cell or making certain that enough are added in parallel. Figure 4.30 shows how minimizing the series resistance allows the capacitor to provide the necessary current to the circuit, reducing the supply droop. Table 4.2 summaries the numerical values for these test cases.

Numerous approaches can be used when adding decoupling to a design. Two of the primary approaches are shown in Figure 4.31. The decoupling fill-in approach is to fill in all white space with decoupling. This is usually an opportunistic

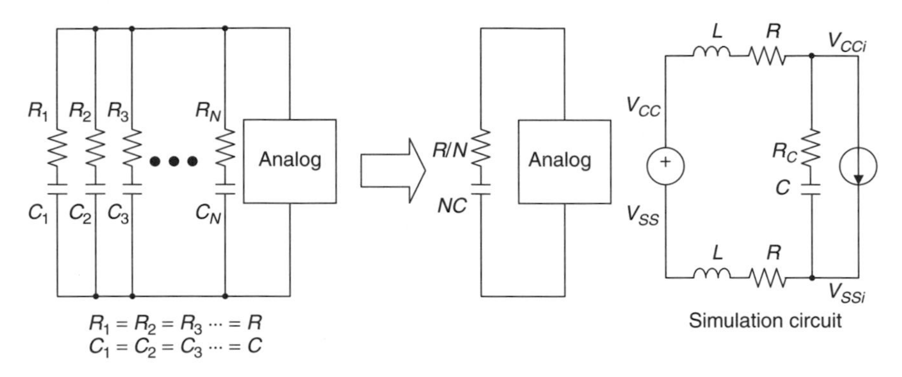

Figure 4.27 Definition of equivalent capacitance.

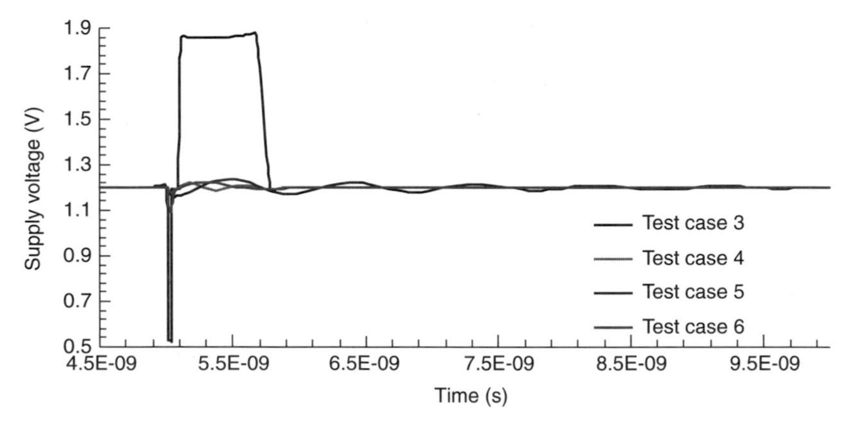

Figure 4.28 Supply voltage variation with no decoupling and high inductance (test case 3), low inductance and some decoupling (test case 4), high inductance and some decoupling (test case 5), and high inductance with some decoupling that has high series resistance (test case 6).

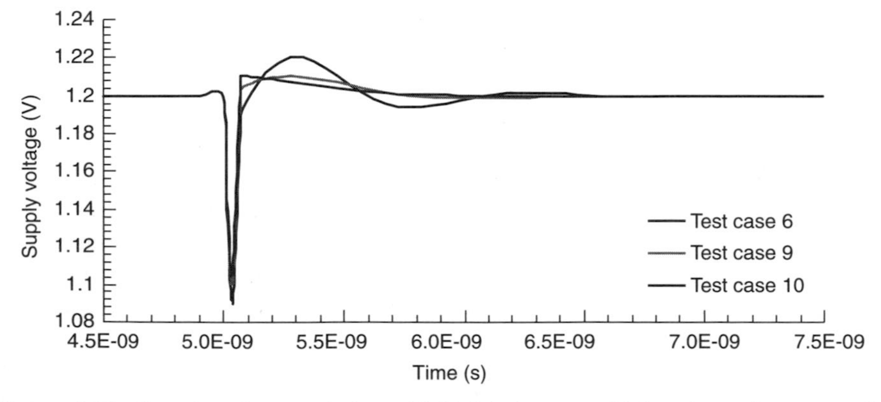

Figure 4.29 Supply voltage variation with high inductance, high series resistance on the decoupling capacitor and 10 pF (test case 6), 20 pF (test case 9), and 40 pF (test case 10).

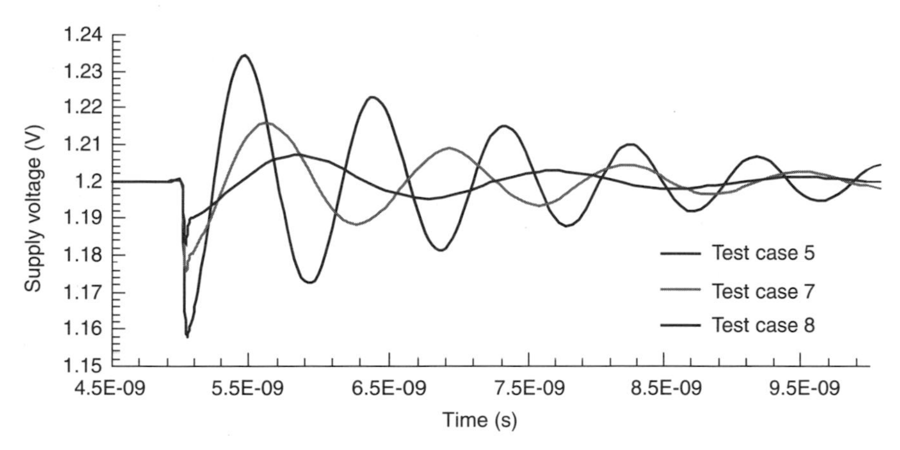

Figure 4.30 Supply voltage variation with high inductance, low series resistance on the decoupling capacitor and 10 pF (test case 5), 20 pF (test case 7), and 40 pF (test case 8).

TABLE 4.2 Summary of Simple Decoupling Simulations

	Test Case									
	1	2	3	4	5	6	7	8	9	10
L (nH)	0	0.1	1	0.1	1	1	1	1	1	1
R (Ω)	0.5	0.5	0.5	0.5	0.5	0.5	0.5	0.5	0.5	0.5
C (pF)	0	0	0	10p	10p	10p	20p	40p	20p	40p
R_C (Ω)	0	0	0	1	1	10	1	1	10	10
Delta	0.01	0.143	1.343	0.050	0.077	0.130	0.040	0.024	0.110	0.106
Max. V_{CC}	1.2	1.267	1.867	1.220	1.234	1.220	1.216	1.207	1.210	1.210
Min. V_{CC}	1.19	1.123	0.523	1.170	1.157	1.090	1.176	1.183	1.101	1.105

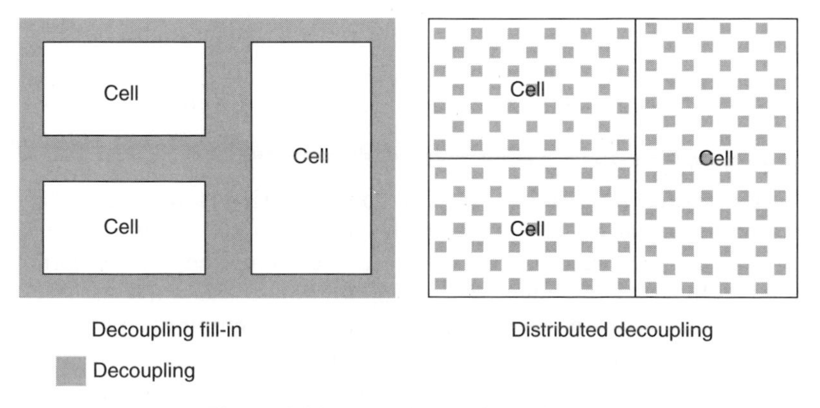

Figure 4.31 Two decoupling schemes.

approach where the decoupling capacitors are added after the layout is completed, but for some circuits, such as a VCO cell, it is common to surround the block with a wide ring of decoupling capacitors. A second approach is to distribute decoupling capacitors throughout the cell. This approach requires much more upfront floor planning and simulations but may result in a more improved performance over the first approach since decoupling is placed next to the devices that need decoupling. The amount of decoupling required is a little more difficult. One method is to extract the average power of the block and then use $C_{\text{decoupling}} = 10P_{\text{avg}}/V_{CC}^2 f$ to estimate the amount of decoupling required. The maximum operating frequency must be used. This provides a good starting point for the decoupling, but in actuality, simulations with the estimated power busing parasitics is the best method to ensure that adequate decoupling is included.

4.11 POWER BUSING

Electromigration can become a problem for analog cells, especially given the sizes of local metal design rules. For most digital logic, electromigration is not an issue on the signal paths, only on the ground paths, since the current is bidirectional and self-heating becomes the major issue. Self-heating failures require a significantly larger current (root mean square) than a dc current, more typical of an analog circuit, and are not typically an issue except for cases where larger buffers are used, such as clock network drivers. It is important to verify what portion of the routing will represent the limiting factor for electromigration. This could be dictated by contacts, thin metal, or vias. Most process technologies guarantee that the metal width is the limiting factor, but this should be confirmed to ensure that via limitations do not force wider metal to be used to meet electromigration rules.

Continuous dc current in analog cells is typically more of an issue than the switched currents of digital circuits, although attention must be paid to the power buses to ensure that they can supply the necessary switched current. Current measurements must be made during the initial design phase to ensure that layout guidelines can be generated. These guidelines must include the minimum number of contacts, minimum bus widths per metal layer, and the minimum number of vias between each layer. The process corner that yields the greatest current should be used to guarantee that adequate signal traces are used. Typically, 20% is added on top of the value calculated, but this should be dictated by how conservative the electromigration rules are.

A simple approach for defining required power busing is to simulate the block at the worst-case current corner (typically, low threshold voltage, high supply voltage, low temperature) to obtain the average current. The average current can be scaled up by 20 to 25%, again depending on how conservative the electromigration rules are for the process. Finally, the resulting current number is used to define the minimum total power bus width per metal layer. This is a conservative approach since it assumes that all current flows from the top layer of metal down to the underlying circuits and that no lateral current flow occurs.

4.12 INTEGRATION PROBLEMS

Integration of a mixed signal block within a larger digital IC can become difficult, and ultimately some degree of trade-off must be made. In this section the various integration problems in a mixed-signal block are discussed along with some basic guidelines for deciding how to determine the potential impact of various options.

4.12.1 Corner Regions

Integration of analog and mixed-signal circuits within a larger digital block can become difficult given all the factors that must be considered. Take, for example, a memory circuit that uses a delay-locked loop (DLL) for critical timing. In most cases there is a strong desire to generate multiple memory sizes using similar floor plans. For this reason, it may be difficult to embed the DLL into the core. A logical placement is the corner of the chip, but the corner represents a region with the highest stress. This stress can give rise to 10 mV or more of threshold voltage shift. A potentially bigger problem is that on the smaller die size, almost no threshold voltage shift may be observed, but on the larger die sizes that are part of the same product family, a varying threshold voltage shift may occur based on the die size, resulting in a variation in performance as a function of die size that is undesirable. Figure 4.32 is a simplified diagram of a corner region of a large die. The most sensitive circuitry is placed as far from the corner as possible, while the less sensitive circuitry is placed closest to the corner. For a PLL, the less sensitive circuitry could be the loop filter (assuming that the capacitance is not affected significantly by the threshold voltage shift).

4.12.2 Neighboring Circuitry

Neighboring circuitry must be considered when floor planning the analog and mixed-signal blocks into a device. If the neighboring circuitry switches extremely large amounts of current, it could significantly affect the analog or mixed-signal

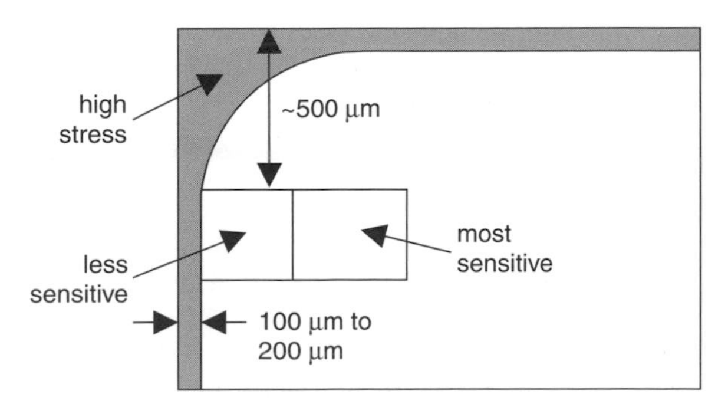

Figure 4.32 Upper right corner of a large die part (1.8 cm × 1.8 cm) with the highest stress region indicated.

block performance by injecting noise into the substrate, disturbing the local supply voltage, and introducing large temperature gradients. Substrate noise has already been discussed, so the other two effects are discussed in the following paragraphs.

A sensitivity factor relative to temperature and supply voltage can be determined for each block within an analog or mixed-signal circuit. For example, a high-speed counter in a PLL may have a sensitivity factor of 0.25 ps/°C and 0.4 ps/mV. By first determining the sensitivity of a circuit to these factors, a more intelligent decision can be made on how to lay out the floor plan for the cells. If a specific cell is insensitive to temperature gradients, placing it by cells that can generate large temperature gradients should have little impact on the cells. The same is true for cells that may disturb the local supply voltage.

4.13 SUMMARY

Designing analog and mixed-signal circuits on deep-submicron processes presents a host of issues that have been discussed. To ensure success requires the designer to understand the problems associated with the process. Knowing and understanding what parameters are included in the models is crucial for ensuring that simulations accurately predict what the silicon performance will be. Becoming involved with process development even if the design is being done through a foundry is becoming more critical to ensure that potential issues are identified before the process is fully defined.

Proper architectural analysis must be completed as early in the project as possible to allow investigation of as many options as possible. Currently, it appears that use of thin oxide devices is possible for many of the mixed-signal circuits necessary for system-on-a-chip ICs, at least at the 90-nm technology node, but it is not clear if this will hold true for the 65-nm technology node. Having the necessary base circuit schematics ready to simulate when device models become available will be crucial for making architectural trade-offs before finalizing a design.

All aspects of a design must be considered during the initial architectural phase. This includes the floor plan, power busing, noise isolation, and decoupling. Potential design pitfalls can be avoided by planning out all aspects of a design carefully before implementation is begun.

REFERENCES

[1] R. A. Bianchi, G Bouche, and O. Roux-dit-Buisson, Accurate modeling of trench isolation induced mechanical stress effects on MOSFET electrical performance, *IEDM '02 Digest, Electron Device Meeting*, pp. 117–120, 2002.

[2] C. H. Choi, K. Y. Nam, Z. Yu, and R. W. Dutton, Impact of gate tunneling current in scaled MOS on circuit performance: a simulation study, *IEEE Trans. Electron Devices*, Vol. 48, No. 12, Dec. 2001.

[3] G. Chen, M. F. Li, C. H. Ang, J. Z. Zheng, and D. L. Kwong, Dynamic NBTI of p-MOS transistors and its impact on MOSFET scaling, *Electron Device Lett.*, Vol. 23, No. 12, pp. 734–736, Dec. 2002.

[4] T. C. Choi, R. T. Kaneshiro, R. W. Brodersen, P. R. Gray, W. B. Jett, and M. Wilcox, High-frequency CMOS switched-capacitor filters for communications application, *IEEE J. Solid-State Circuits*, Vol. 18, pp. 652–664, Dec. 1983.

[5] S. Yan and E. S. Sinencio, Low voltage analog circuit design techniques: a tutorial, *IEICE Trans. Analog Integrat. Circuits Syst.*, Vol. E00-A, No. 2, pp. 1–17, Feb. 2000.

[6] T. Serrano and B. Linares-Barranco, The active-input regulated-cascode current mirror, *IEEE Trans. Circuits Syst. II: Analog Digital Signal Process.*, Vol. 41, pp. 464–467, June 1994.

[7] K. N. Leung and P. K. T. Mok, A sub-1-V 15 ppm/°C CMOS bandgap voltage reference without requiring low threshold voltage device, *IEEE J. Solid-State Circuits*, Vol. 37, No. 4, pp. 526–530, Apr. 2002.

[8] T. Stockstad and H. Yoshizawa, A 0.9-V 0.5-μA rail-to-rail CMOS operational amplifier, *IEEE J. Solid-State Circuits*, Vol. 37, No. 3, pp. 286–292, Mar. 2002.

[9] J. F. Duque-Carrillo, J. L. Ausin, G. Torelli, J. M. Valverde, and M. A. Dominguez, 1-V rail-to-rail operational amplifiers in standard CMOS technology, *IEEE J. Solid-State Circuits*, Vol. 35, No. 1, pp. 33–44, Jan. 2000.

[10] J. M. Carrillo, J. F. Duque-Carrillo, G. Torelli, and J. L. Ausin, Constant-g_m constant-slew-rate high-bandwidth low-voltage rail-to-rail CMOS input stage for VLSI cell libraries, *IEEE J. Solid-State Circuits*, Vol. 38, No. 8, pp. 1364–1372, Aug. 2003.

[11] T. H. Lee, *The Design of CMOS Radio-Frequency Integrated Circuits*, Cambridge University Press, Cambridge, 2000.

[12] J. G. Maneatis, Low-jitter process-independent DLL and PLL based on self-biased techniques, *IEEE J. Solid-State Circuits*, Vol. 31, No. 11, pp. 1723–1732, Nov. 1996.

[13] J. Maget, M. Tiebout, and R. Kraus, Influence of novel MOS varactor on the performance of a fully integrated UMTS VCO in standard 0.25 μm CMOS technology, *IEEE J. Solid-State Circuits*, Vol. 37, No. 7, pp. 1–6, July 2002.

[14] D. Coolbaugh, E. Eshun, R. Groves, D. Haranel, J. Johnson, A. Harnmad, Z. He, V. Ramachandran, K. Stein, S. St. Ongel, S. Subbanna, D. Wang, R. Volant, X. Wang, and K. Watson, Advanced passive devices for enhanced integrated RF circuit performance, *Technical Digest, 2002 IEEE MTT-S International Microwave Symposium*, pp. 187–191, Seattle, WA, June 2002.

[15] R. Aparicio and A. Hajimiri, Capacity limits and matching properties of integrated capacitors, *IEEE J. Solid-State Circuits*, Vol. 37, No. 3, pp. 384–393, Mar. 2002.

[16] H. Samavati, A. Hajimiri, A. R. Shahani, G. N. Nasserbakht, and T. H. Lee, Fractal capacitors, *IEEE J. Solid-State Circuits*, Vol. 33, No. 12, pp. 2035–2041, Dec. 1998.

[17] W.-T. Wang, M.-D. Ker, M.-C. Chiang, and C.-H. Chen, Level shifters for high-speed 1-V to 3.3-V interfaces in a 0.13 μm Cu-interconnection/low-κ CMOS technology, *Proceedings of Technical Papers, VLSI Technology, Systems, and Applications International Symposium*, pp. 307–310, Apr. 18–20, 2001.

[18] M. Pfost and H.-M. Rein, Modeling and measurement of substrate coupling in Si-bipolar ICs up to 40 GHz, *IEEE J. Solid-State Circuits*, Vol. 33, pp. 582–591, Apr. 1998.

[19] H. M. Chen, M. H. Wu, L. Chang, and C. F. Wu, The study of substrate noise and noise-rejection-efficiency of guard-ring monolithic integrated circuits, pp. 123–128, *Electromagnetic Compatibility*, 2000, IEEE International Symposium, Volume 1, 21–25 Aug 2000.

[20] J. P. Z. Lee, F. Wang, A. Phanse, and L. C. Smith, Substrate cross talk noise characterization and prevention in 0.35 μm CMOS technology, *IEEE Custom Integrated Circuits Conference*, pp. 479–482, 1999.

[21] A. Samavedam, A. Sadate, K. Mayaram, and T. S. Fiez, A scalable substrate noise coupling model for design of mixed-signal ICs, *IEEE J. Solid-State Circuits*, Vol. 35, pp. 895–904, June 2000.

[22] X. Aragones and A. Rubio, Experimental comparison of substrate noise coupling using different wafer types, *IEEE J. Solid-State Circuits*, Vol. 34, pp. 1405–1409, 1999.

[23] D. K. Su, M. L. Loinaz, S. Masui, and B. A. Wooley, Experimental results and modeling techniques for substrate noise in mixed-signal integrated circuits, *IEEE J. Solid-State Circuits*, Vol. 28, pp. 420–430, Apr. 1993.

[24] S. Masui, Simulation of substrate coupling in mixed-signal MOS circuits, *Digest of Technical Papers, IEEE Symposium on VLSI Circuits*, pp. 42–43, 1992.

[25] M. J. Chen, C. Y. Huang, P. N. Tseng, N. S. Tsai, and C. Y. Wu, Design model for minority-carrier well-type guard-rings in CMOS circuits, *IEEE Custom Integrated Circuits Conference*, pp. 4.5.1–4.5.4, 1991.

[26] R. Singh and S. Sali, Substrate noise issues in mixed-signal chip designs using SPICE, *10th International Conference on Electromagnetic Compatibility*, pp. 108–112, Sept. 1–3 1997.

[27] M. Ingels and M. S. J. Steyaert, Design strategies and decoupling techniques for reducing the effects of electrical interference in mixed-mode ICs, *IEEE J. Solid-State Circuits*, Vol. 32, pp. 1136–1141, July 1997.

[28] K. M. Fukuda, T. Kikuchi, and M. Hotta, Measurement of digital noise in mixed-signal integrated circuits, *IEEE J. Solid-State Circuits*, Vol. 30, pp. 87–92, Feb. 1995.

[29] R. Gharpurey and R. G. Meyer, Modeling and analysis of substrate coupling in integrated circuits, *IEEE J. Solid-State Circuits*, Vol. 31, pp. 344–353, Mar. 1996.

[30] A. J. Rainal, Eliminating inductive noise of external chip interconnections, *IEEE J. Solid-State Circuits*, Vol. 29, pp. 126–129, Feb. 1994.

[31] K. Joardar, A simple approach to modeling cross-talk in integrated circuits, *IEEE J. Solid-State Circuits*, Vol. 29, pp. 1212–1219, Oct. 1994.

[32] B. R. Stanisic, R. A. Rutenbar, and L. R. Carley, Addressing noise decoupling in mixed-signal ICs: power distribution design and cell customization, *IEEE J. Solid-State Circuits*, Vol. 30, pp. 321–326, Mar. 1995.

[33] B. R. Stanisic, N. K. Verghese, R. A. Rutenbar, L. R. Carley, and D. J. Allstot, Addressing substrate coupling in mixed-mode ICs: simulation and power distribution synthesis, *IEEE J. Solid-State Circuits*, Vol. 29, pp. 226–238, Mar. 1994.

[34] N. K. Verghese, D. J. Allstot, and M. A. Wolfe, Verification techniques for substrate coupling and their application to mixed-signal IC design, *IEEE J. Solid-State Circuits*, Vol. 31, pp. 354–365, Mar. 1996.

[35] M. Pfost, H. M. Rein, and T. Holzwarth, Modeling substrate effects in the design of high-speed Si-bipolar ICs, *IEEE J. Solid-State Circuits*, Vol. 31, pp. 1493–1501, Oct. 1996.

[36] D. M. Monticelli, A quad CMOS single-supply op amp with rail-to-rail output swing, *IEEE J. Solid-State Circuits*, Vol. 21, pp. 1026–1034, Dec. 1986.

[37] P. Smeys, P. B. Griffin, Z. U. Rek, I. D. Wolf, and K. C. Saraswat, Influence of process-induced stress on devices characteristics and its impact on scaled device performance, *IEEE Trans. Electron Devices*, Vol. 46, pp. 1245–1252, June 1999.

[38] P. R. Gray and R. G. Meyer, *Analysis and Design of Analog Integrated Circuits*, Wiley, New York, 1993.

CHAPTER 5

ELECTROSTATIC DISCHARGE PROTECTION DESIGN

5.1 INTRODUCTION

Electrostatic discharge (ESD) protection has been implemented on integrated circuits (ICs) for more than 30 years. ESD design has faced three significant problems related to scaling processes. First, the gate oxide of the core devices has continued to shrink, reducing the oxide breakdown voltage to a level that makes it difficult to protect. This is shown in Figure 5.1, where the breakdown voltage of the thin oxide devices is less than 5 V and the operating supply may be as high as 1.5 V. The ESD protection device must operate between these two voltages, which becomes progressively more difficult to achieve as the two boundaries move closer together. A second effect has been the ever-increasing I/O speed, making large-I/O-buffer ESD devices impractical because of the high capacitive loading. A final effect is the introduction of several supply voltages, which makes the overall ESD protection scheme more complicated. Some supplies may be lightly loaded, making it difficult for the ESD protection to turn on fast enough to prevent damaging some of the devices. In this chapter, various ESD protection schemes are presented that have been implemented for several process generations and are still applicable to the 90-nm technology node and beyond.

Nano-CMOS Circuit and Physical Design, by Ban P. Wong, Anurag Mittal, Yu Cao, and Greg Starr
ISBN 0-471-46610-7

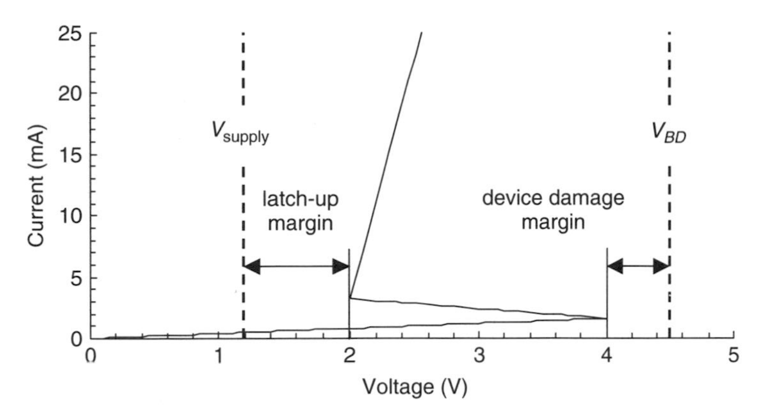

Figure 5.1 ESD protection scaling issue for deep-submicron processes.

5.2 ESD STANDARDS AND MODELS

The ESD phenomenon often happens between two or more objects with different electrostatic potentials. The ESD phenomenon is known to be a serious problem for IC products fabricated by advanced deep-submicron semiconductor process technologies. In scaled-down CMOS processes, MOS devices with shallower junction depth, thinner gate oxide, lightly doped drain (LDD) structure, and silicided diffusion have higher operating speed and lower operating power, but they become weaker to ESD stresses. Devices are usually damaged by ESD due to the rapidly generated heat or the rapidly created high electrical field. To predict the ESD immunity level or to find the ESD-sensitive (weak) point of the ICs, there are several organizations that make ESD standards. They are ESDA (Electrostatic Discharge Association), AEC (Automotive Electronics Council), EIA/JEDEC (Electronic Industries Alliance/Joint Electron Device Engineering Council), and MIL-STD (U.S. Military Standards). The dominant ESD test methods on component-level IC products are known as HBM (human body model) [1], MM (machine model) [2], and CDM (charged device model) [3].

5.3 ESD PROTECTION DESIGN

5.3.1 ESD Protection Scheme

The concept of on-chip ESD protection design, shown in Figure 5.2, is used to avoid damage from the HBM/MM ESD stresses under almost-random pin combinations. For every input or output pin, there are ESD clamp devices placed from the pad to VDD and VSS power lines to discharge the four modes of ESD stresses on the I/O pin. To overcome the ESD stresses in pin-to-pin ESD stresses and VDD-to-VSS ESD stresses, the power-rail ESD clamp circuit must be placed between the VDD and VSS power lines of the IC [4,5]. For most logic

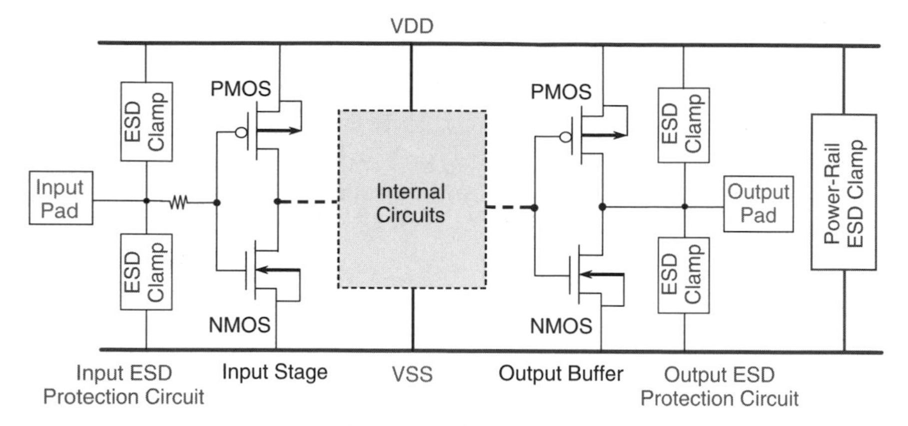

Figure 5.2 Concept of on-chip ESD protection design.

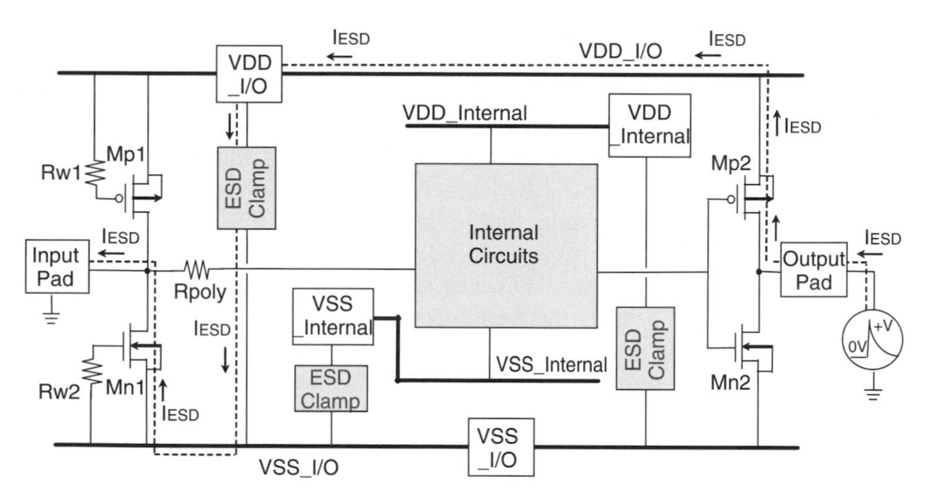

Figure 5.3 Whole-chip ESD protection scheme for an IC with separated power pins for I/O circuits and internal circuits.

ICs, the power pins for I/O circuits are often separated from the power pins of the core circuits to avoid noise coupling issues and to reduce ground bounce. With the separated power pins, the typical whole-chip ESD protection scheme is shown in Figure 5.3. Besides the ESD clamp devices at the input and output pads, the most important design to achieve whole-chip ESD protection for all devices and circuits in an IC against ESD damage (especially against pin-to-pin and VDD-to-VSS ESD stresses) is the arrangement on the power lines and the power-rail ESD clamp circuits between the separated power lines. As shown by the dashed lines in Figure 5.3, the ESD current discharging path or the IC under pin-to-pin ESD zapping can be built up by using I/O ESD devices, power metal lines, and power-rail ESD clamp circuits. Only by having a successful whole-chip

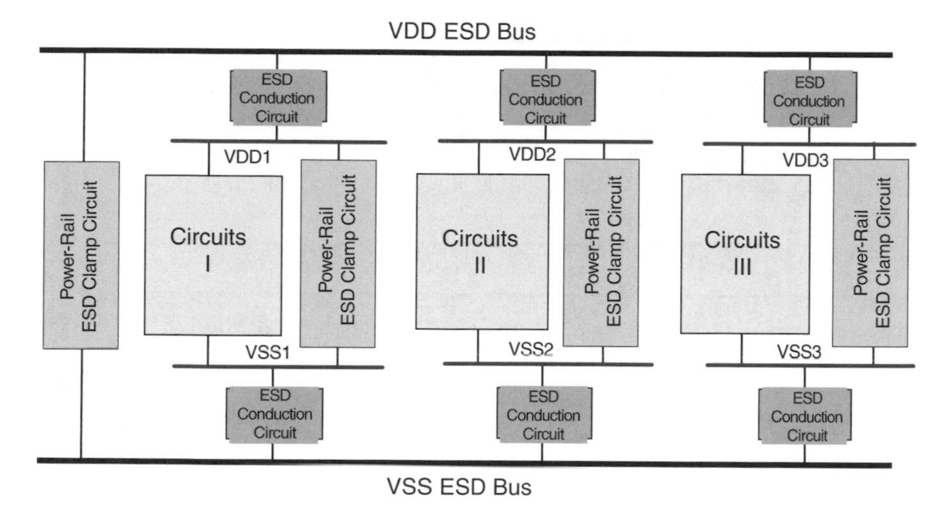

Figure 5.4 Suggested whole-chip ESD protection scheme for an IC with more separated power lines. (From Ref. 6.)

ESD protection scheme can the internal (core) circuits be protected effectively by ESD clamp devices located at the I/O pads and ESD clamp circuits located between the power rails.

Most traditional ESD protection designs were focused on ESD protection circuits or devices for I/O pins, which can provide protection against ESD stresses on an I/O pin under the four modes of pin combination. But ESD protection circuits or devices located at I/O pads cannot provide enough protection for internal circuits against pin-to-pin and VDD-to-VSS ESD stresses, which often cause an ESD failure located in the internal circuits but not on I/O ESD devices. For ICs with more independent power supplies, the whole-chip ESD protection scheme suggested is shown in Figure 5.4, which has been included in the design rules of one famous semiconductor foundry. The ESD conduction circuits between the VDD ESD bus and the separated power supplies (VDD1, VDD2, VDD3) can be realized by using stacked diodes [7] or even bidirectional silicon-controlled rectifier (SCR) devices [8,9]. For an IC [such as a system on a chip (SoC)] with more complex power supplies with different voltage levels, a whole-chip ESD protection scheme with multiple ESD buses is shown in Figure 5.5. With successful arrangement of the whole-chip ESD protection scheme, the internal circuits can still be safely protected by ESD clamp devices located at I/O pads and between the power rails.

5.3.2 Turn-on Uniformity of ESD Protection Devices

The devices that can be realized in general CMOS processes are the resistor, diode, NMOS/PMOS, field oxide device (FOD), vertical/lateral bipolar junction transistor (BJT), SCR device (p-n-p-n structure), capacitor, and inductor. ESD

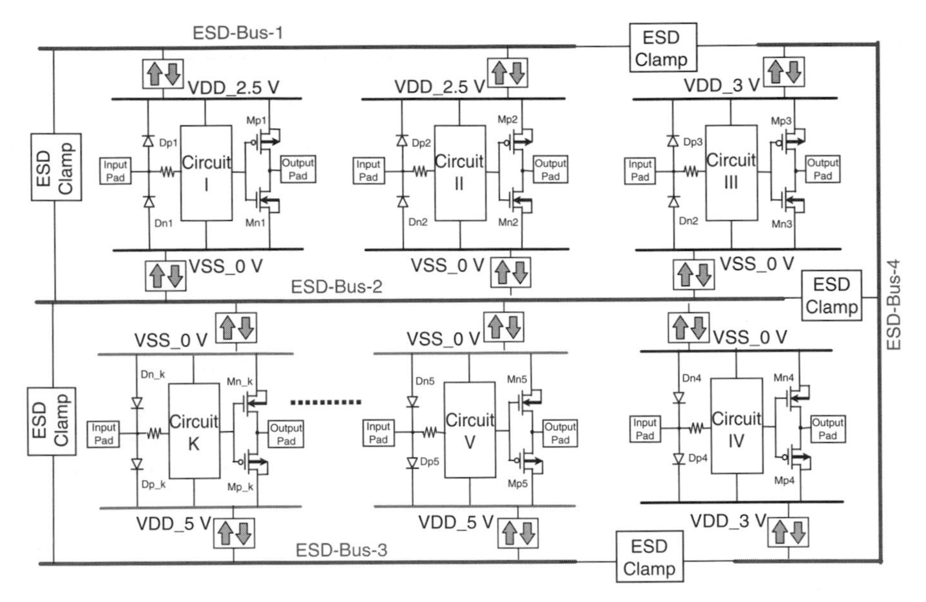

Figure 5.5 Whole-chip ESD protection scheme with multiple ESD buses for an IC with a more complex power supply configuration and different voltage levels.

clamp devices or circuits are therefore built up with those devices to sustain the ESD levels requested (such as an HBM of 2000 V, an MM of 200 V, and a CDM of 1000 V). To sustain a high-enough ESD level, the devices used in ESD clamp circuits should be drawn with a corresponding large-enough device dimension or size to sustain the ESD-induced heat without causing burnout or damage. However, the layout skill needed to draw ESD clamp devices in a reasonable silicon area to sustain a high-enough ESD level must be optimized [10–12]. Some advanced area-efficient designs for the layout of ESD clamp devices to have a high-enough ESD level within a smaller layout area had been reported in [13–16].

PMOS and NMOS devices in the output buffers are also often working as the ESD clamp devices to protect the I/O pin without adding an extra ESD clamp device to the pad. To provide enough driving current to the external load and to withstand high-enough ESD stresses, the NMOS/PMOS of output buffers are often designed with a channel width of several hundreds of micrometers. Such ESD protection devices with larger device dimensions are often realized with multiple fingers to reduce the total layout area [17]. But during ESD stress, the multiple fingers of an ESD protection MOSFET cannot be turned on uniformly. If only some MOSFET fingers are turned on, the device can be damaged by the ESD event [18]. This often causes a low ESD level in an ESD protection circuit, even if the MOSFET has large device dimensions. To improve the turn-on uniformity among multiple fingers, gate-driven design [19–22] and substrate-triggered design [24–28] were reported to increase the ESD robustness of the large-device-dimension NMOS. Recently, the ESD robustness of gate-driven NMOS has been found to decrease dramatically when the gate voltage is somewhat increased [16,22]. The gate-driven design causes large ESD current discharging through the strong-inversion channel of NMOS [29]; therefore, the NMOS device is easily burned out by ESD energy. Gate-driven and substrate-triggered techniques can improve the ESD robustness of large-dimension ESD protection devices. But the higher gate bias can induce larger channel current and a higher electric field across the gate oxide to damage MOSFETs from the explication of energy band diagrams [29]. This effect causes degradation of ESD robustness in gate-driven devices. Compared to gate-driven design, substrate-triggered design can avoid the formation of channel current and enhance the space-charge region to sustain higher ESD current far away from the channel surface. Therefore, substrate-triggered design can be one of the most effective solutions for improving the ESD robustness of CMOS devices in nanoscale-CMOS technologies [30].

5.3.3 ESD Implantation and Silicide Blocking

In addition to the layout or triggering techniques used to improve the ESD robustness of ESD clamp devices within a limited layout area, some process modifications had been developed with extra masking layers to further improve the ESD robustness of I/O. To enhance ESD robustness of these clamp devices, ESD implantations had been reported for inclusion into process flow to modify device structures for ESD protection [31–37]. N-type ESD implantation was used to cover the LDD peak structure and to make a deeper junction in NMOS devices for ESD protection [31,32]. P-type ESD implantation under the drain junction of the NMOS was used to reduce the reverse junction breakdown voltage and to turn on the parasitic lateral bipolar of the NMOS device earlier [33,34]. With higher doping concentration, P-type ESD implantation can also be used to reduce the reverse junction breakdown voltage of diode or field oxide devices and to allow the devices to sustain higher ESD stresses under reverse-biased conditions [35]. Moreover, both N- and P-type ESD implantations were used in NMOS devices to ensure higher ESD robustness [36]. Experimental comparisons among various ESD implantations for ESD protection in the same CMOS process have been made [37]. A modified design of ESD implantation for NMOS to improve, especially, MM ESD robustness has been reported [38,39].

Another process modification for improving ESD robustness of I/O devices is to use extra mask layers to block the formation of silicided diffusion in the drain regions of devices connected to an I/O pad [40–46]. A similar approach has been implemented without the additional mask layers by drawing an N-well into the drain region of NMOS to block the formation of silicided diffusion in the drain regions of I/O devices to improve ESD robustness [47,48]. Recently, a dummy gate has also been used to block formation of silicided diffusion in the drain regions of I/O devices without using an extra masking layer [49,50].

5.3.4 ESD Protection Guidelines

Some guidelines in the design of ESD clamp devices or circuits for I/O pins or power pins follow.

1. Provide the IC with an efficient ESD protection scheme to bypass any ESD stress while the IC is under ESD stress conditions.
2. Pass the normal I/O signals and keep inactive while the IC is operating under normal conditions.
3. Reduce (to the minimum possible) the input capacitance and resistance to permit acceptable I/O signal delay.
4. Have a high ESD robustness within a reasonable (as small as possible) layout area.
5. Maintain a high latch-up immunity in CMOS ICs (all the I/O devices and ESD protection devices have to be surrounded by guard rings). Fabrication of such ESD protection devices should be compatible with the process technology (without additional mask layers or modified process steps, if possible).
6. ESD circuits must not affect I/O circuit functionality during normal operation (such as high-voltage-tolerant I/O applications or power-down operations).

5.4 LOW-*C* ESD PROTECTION DESIGN FOR HIGH-SPEED I/O

5.4.1 ESD Protection for High-Speed I/O or Analog Pins

Conventional ESD protection design with a two-stage structure for a digital input pin is shown in Figure 5.6, where a gate-grounded short-channel NMOS is used as a secondary protection device to clamp the overstress voltage across the gate oxide of the input circuits. To provide a high ESD level, a robust device (such as an SCR, field oxide device, or long-channel NMOS) is used as the main discharge element in the primary protection stage to bypass ESD current on the input pad. Between the primary and secondary stages of the input ESD protection circuit, a resistor is added to limit ESD current flowing through short-channel NMOS in the secondary stage. The resistance value of this resistor depends on the turn-on voltage of the ESD clamp device in the primary stage and the secondary breakdown current of the short-channel NMOS in the secondary stage. The primary ESD clamp device must be triggered on to bypass ESD current before the gate-grounded NMOS (ggNMOS) in the secondary stage is damaged by the overstress ESD current. If the primary ESD clamp device has a high turn-on voltage, the resistance should be large enough: on the order of several hundreds of ohms. Such two-stage ESD protection design can provide a high ESD tolerance level for digital input pins. But the large series resistance and large junction capacitance in ESD clamp devices cause

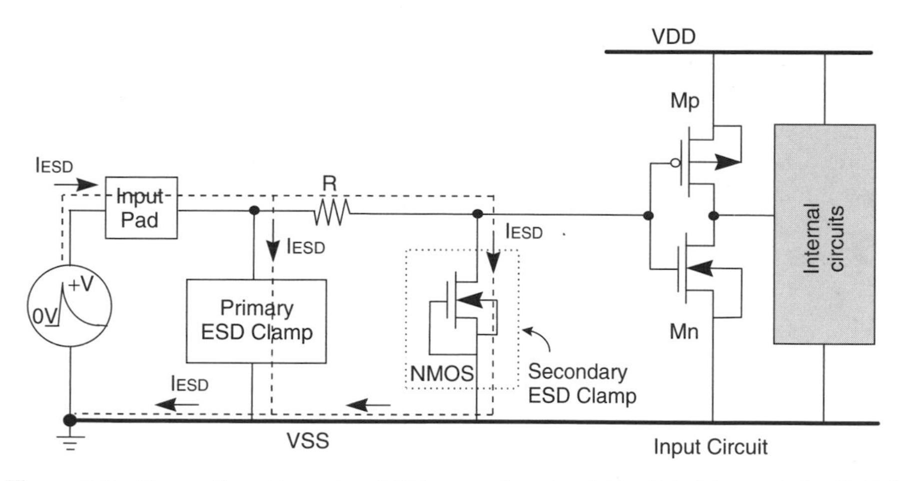

Figure 5.6 Conventional two-stage ESD protection circuit for digital input pin in CMOS ICs.

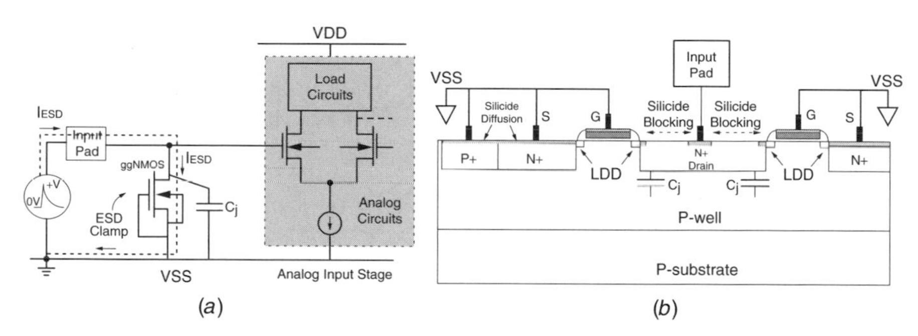

Figure 5.7 (*a*) Single-stage ESD protection circuit for an analog input pin in CMOS ICs; (*b*) cross-sectional view of a ggNMOS with a silicide-blocking drain region.

a long *RC* delay to the input signal. It is not suitable for analog signals or high-speed I/O applications, which are becoming much more prevalent in the more advanced processes.

For analog input signal or high-speed I/O applications, series resistance between the input pad and internal circuits is typically forbidden. Therefore, the two-stage ESD protection design in Figure 5.6 is no longer suitable for analog or high-speed I/O pins. To protect the analog or high-speed I/O pin, an ESD protection circuit with a single-stage ESD protection design is used, as shown in Figure 5.7(*a*), where a ggNMOS is used as the ESD clamp device. Lack of series resistance to limit ESD current toward the ggNMOS, as well as the ESD robustness of the NMOS device, seriously degrade ESD protection on advanced deep-submicron CMOS technologies [51,52]. A ggNMOS is often designed with large device dimensions and wide drain-contact-to-poly-gate layout spacing to sustain an acceptable ESD level [16,52,53]. The additional silicide-blocking masks [40–44] have been included in the deep-submicron CMOS process to increase the robustness of the ESD clamp device. A schematic cross-sectional view of a ggNMOS with a silicide-blocking drain region is illustrated in Figure 5.7(*b*). But a ggNMOS with larger device dimensions and a wider drain region contributes a larger parasitic drain capacitance to the input pad. Such a parasitic junction capacitance is nonlinear and dependent on the input voltage level, making it unacceptable for truly high-speed I/O and analog inputs.

For some high-resolution data converter circuits, the input capacitance of an analog input pin is required to be kept as constant as possible within the input voltage range. A major source of distortion in high-speed analog circuits, especially in single-ended input implementations, is the voltage-dependent nonlinear input capacitance associated with ESD clamp devices at the analog input pad. Typical degradation on circuit performance due to the nonlinear input capacitance of the input ESD clamp devices has been reported in Ref. 65, where the input capacitance varies from 4 to 2 pF due to the input voltage swing from 0 to 2 V. This capacitance variation caused an increase in harmonic distortion in an analog-to-digital converter (ADC) and therefore degraded the precision of the ADC from 14 bits to 10 bits. Thus, it has been an emerging challenge to design an effective ESD protection circuit for high-precision analog applications in scaled-down CMOS technologies. In Section 5.4.2, low-*C* ESD protection design, with the advantages of small input capacitance, no series resistance, and a high ESD level, is described in detail.

5.4.2 Low-*C* ESD Protection Design

The low-*C* ESD protection circuit for high-speed I/O or analog pins is shown in Figure 5.8(*a*). The corresponding layout of this low-*C* ESD protection circuit in a 0.35-μm silicided CMOS cell library is shown in Figure 5.8(*b*) for reference. In Figure 5.8(*a*), the Dp1 (Dn1) is the parasitic junction diode in the drain region of an Mp1 (Mn1) device. To reduce the input capacitance of a high-speed I/O or analog pin, Mn1 and Mp1 are both designed with much smaller device dimensions (W/L): only 50/0.5 (μm/μm). The HBM ESD level of a stand-alone NMOS with a device dimension of 50/0.5 (μm/μm) is less than 500 V in the typical 0.35-μm silicided CMOS process when an NMOS is zapped in PS-mode ESD stress. But such a small NMOS can sustain an HBM ESD level of 8000 V in the same 0.35-μm silicided CMOS process, while the NMOS is zapped in NS-mode ESD stress. In PS-mode (NS-mode) ESD stress, NMOS is operated in its drain-breakdown condition (drain diode forward-bias condition) to bypass ESD current. The power dissipation located on the ESD clamp device is equal to the product of ESD current and the operating voltage of the device during ESD events. Therefore, an NMOS has a significantly different ESD stress level between PS and NS modes. Similarly, a stand-alone PMOS with a small device dimension also has a high ESD stress level in the PD mode but a much lower ESD stress level in ND-mode ESD stress.

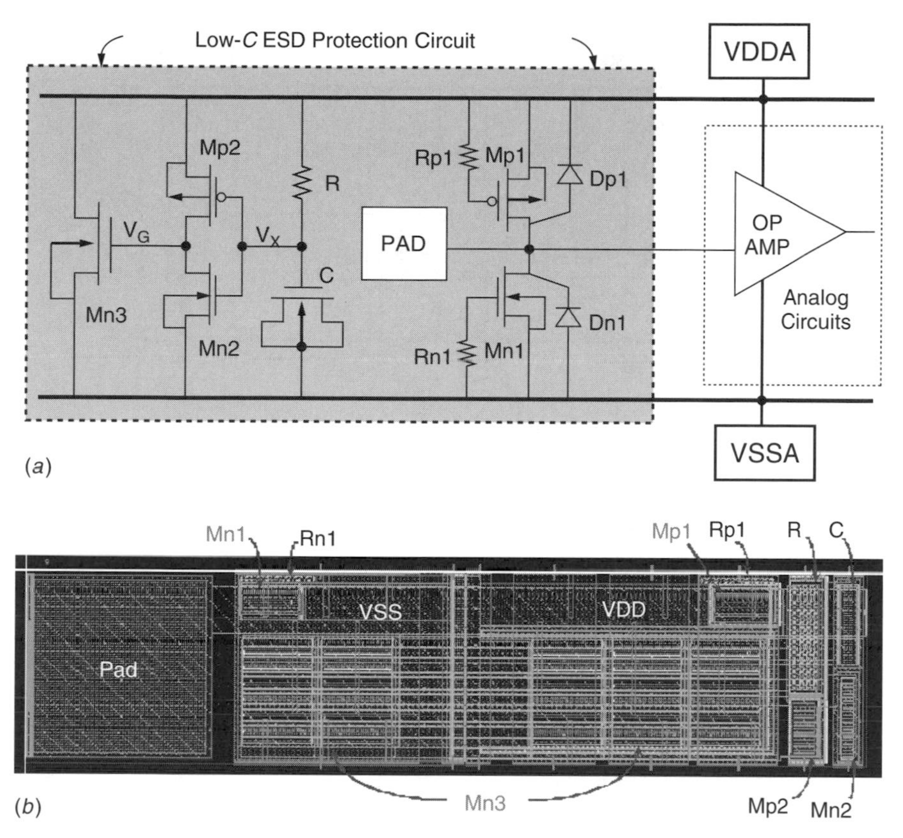

Figure 5.8 (*a*) Low-*C* ESD protection circuit for high-speed I/O or analog pins; (*b*) layout example of a low-*C* ESD protection circuit for high-speed I/O or analog pins in a 0.35-μm silicided CMOS process. [Part (*a*) from Ref. 55.].

To avoid the small-dimension Mn1 and Mp1 going into the drain-breakdown condition during PS- and ND-mode ESD stresses, an efficient ESD clamp circuit between the power rails is co-constructed into an analog ESD protection circuit to increase the overall ESD protection level. In Figure 5.8(*a*), the *RC*-based ESD detection circuit [5] is used to trigger on the Mn3 device when the pad is zapped in PS- or ND-mode ESD stresses. The ESD current paths in this analog ESD protection circuit are illustrated by the dashed lines in Figure 5.9(*a*) and Figure (*b*), respectively, when the analog pin is zapped in the PS and ND modes. Because the Mn1 in PS-mode (Mp1 in ND-mode) ESD stress is not operated in the drain-breakdown condition, the ESD current is bypassed through the forward-biased drain diode Dp1 in Mp1 (Dn1 in Mn1) and the turned-on Mn3. The Mn3 is especially designed with a larger device dimension [$W/L = 1800$ μm/0.5 μm in Figure 5.8(*b*)] to sustain a high ESD level. Although the large-dimension Mn3 has a large junction capacitance, this capacitance does not contribute to the analog

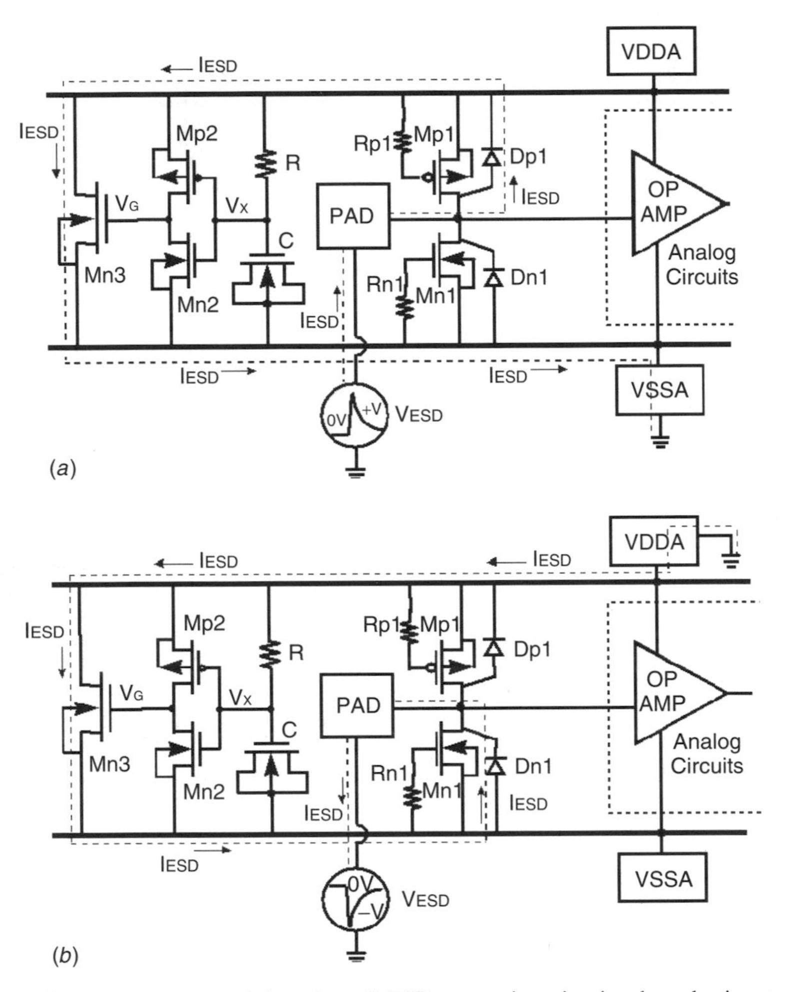

Figure 5.9 ESD current path in a low-*C* ESD protection circuit when the input pin is zapped in (*a*) PS-mode and (*b*) ND-mode ESD stress.

pad. Therefore, the analog pin can sustain a much higher ESD level in the four-mode ESD stresses but only with a very small input capacitance. The parasitic resistance between the input pin protection devices and the power ESD device is critical and must be minimized to ensure that the devices fire fast and the *IR* drop is minimized during the ESD event.

When the input pins are zapped in analog pin-to-pin ESD stress, the ESD current path along this proposed analog ESD protection circuit is as shown in Figure 5.10. During pin-to-pin ESD stress, both the VDDA and VSSA power lines in the IC are floating. The ESD current is first conducted from the zapped pad to the VDDA power line through the junction diode Dp1 in the Mp1 of an input ESD protection circuit. Therefore, the VDDA line is charged by the ESD

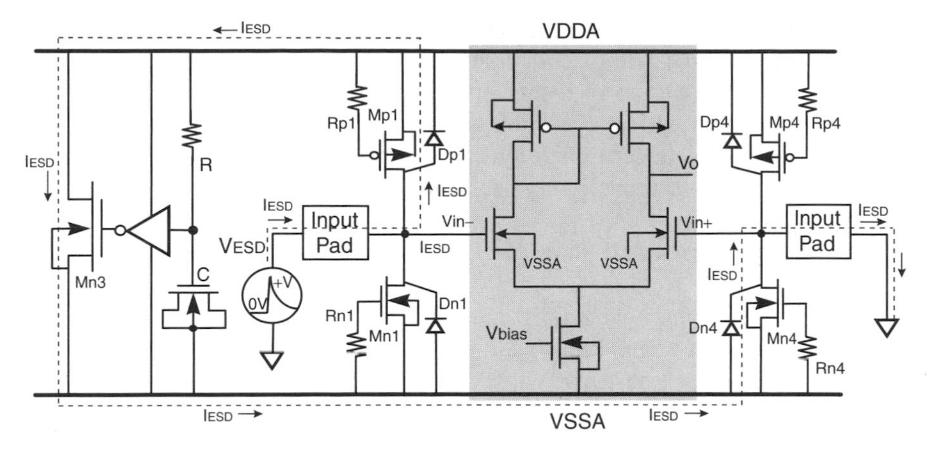

Figure 5.10 ESD current path in the low-C ESD protection circuit when the input pins are zapped in analog pin-to-pin ESD stress.

energy. The VSSA line initially has a voltage level near to ground because the VSSA line is connected to a grounded pad through the diode Dn4 in Mn4 of another input ESD protection circuit. The pin-to-pin ESD stress voltage across the two pins of differential input stage therefore comes across the VDDA and VSSA power lines. The Mn3 device connected between the VDDA and VSSA power lines is turned on by the RC-based ESD detection circuit to bypass the ESD current from VDDA to VSSA. Finally, the ESD current flows out the chip from the VSSA power line to the grounded pad through the forward-biased diode Dn4 in Mn4. With suitable design of the ESD detection circuit to turn on the Mn3 quickly [10], the pin-to-pin ESD stress can quickly be discharged away from the gate oxide of the differential input stage. By using this design, the gate oxide of the analog differential input stage can be fully protected without adding any series resistance between the input pad and analog internal circuits. Therefore, the input signal can have the widest bandwidth from the pad to the internal circuits because this low-C ESD protection circuit protects them.

5.4.3 Input Capacitance Calculations

The input capacitance of this low-C ESD protection circuit for high-speed I/O or analog pins can be calculated as

$$C_{\text{in}} = C_{\text{pad}} + C_n + C_p$$

where the C_{pad} is the parasitic capacitance of the bond pad. C_p (C_n) is the drain junction capacitance and drain-to-gate overlapped capacitance in the Mp1 (Mn1). The drain junction capacitance of a single NMOS or PMOS is strongly bias dependent. The input capacitance of the previous ESD protection design with a single NMOS in Figure 5.7(a) varies extensively when the input signal has a different voltage level. But the input capacitance of this low-*C* ESD protection circuit [Figure 5.8(*a*)] with a complementary PMOS and NMOS structure can be kept almost constant even if the input signal has a voltage swing from 0 V to V_{dd} (3 V). The total input junction capacitance of the low-*C* ESD protection circuit with different device dimensions are accurately calculated in the frequency domain by using pin-capacitance-measurement simulation [56] in the STAR-HSPICE CAD tool.

The simulated results are shown in Figure 5.11, where the channel widths of Mn1 and Mp1 vary from 50 to 400 μm with a fixed channel length of 0.5 μm under different voltage levels on the input pad. The drain-contact-to-poly-gate spacing in both Mn1 and Mp1 is drawn as 3.4 μm, whereas the source-side spacing is drawn as 1.55 μm. With device dimensions of 50/0.5 (μm/μm) in both Mn1 and Mp1, the input capacitance of the low-*C* ESD protection circuit is varying from 0.37 to 0.4 pF for an input voltage swing of 0 to 3 V. But the input capacitance of the traditional ESD protection circuit in Figure 5.7(*a*) with a ggNMOS of $W/L = 400/0.5$ (μm/μm) varies from 1.83 to 1.12 pF for an input voltage swing of 0 to 3 V.

The layout size of the metal bond pad for wire bonding in the 0.35-μm CMOS process is specified as 96 × 96 μm^2, which contributes a parasitic C_{pad} value of 0.67 pF. So the total input capacitance of this low-*C* ESD protection circuit, including the bond pad, is only about 1.04 to 1.07 pF, even if the input signal has a voltage swing from 0 to 3 V. With such a small and almost constant

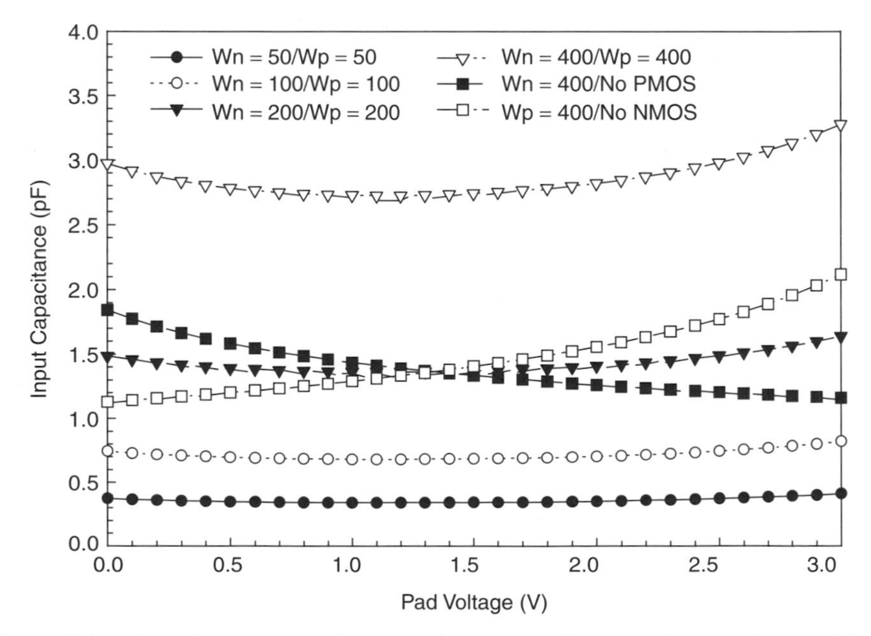

Figure 5.11 Input junction capacitance of the low-*C* ESD protection circuit with different device dimensions in Mn1 and Mp1 during different input voltage levels on the pad.

input capacitance, this low-*C* analog ESD protection circuit is more suitable for high-precision and high-speed I/O applications in both analog and digital pins. To further reduce the parasitic capacitance generated from the bond pad, low-*C* bond pad structures have been developed [57]. An additional reduction in the capacitance can be achieved by adding diodes between the ESD devices (Mn1 and Mp1) and the input pad. Polysilicon diodes with a stacked configuration have been developed in a 0.25-μm CMOS process to protect the RF input pin of a RF low-noise amplifier (LNA) operating at 2.4 GHz [58].

The main nonlinear source in the input capacitance is the bias-dependent junction capacitance at the drain regions of Mn1 and Mp1 of the low-*C* ESD protection circuit shown in Figure 5.5(*a*). When the input signal at the input pad has an increasing voltage, the drain junction capacitance of Mn1 decreases but the drain junction capacitance of Mp1 increases. On the contrary, the drain junction capacitance of Mn1 increases but the drain junction capacitance of Mp1 decreases when the input signal at the input pad has a decreasing voltage. From the complementary structure of this low-*C* ESD protection circuit, the input capacitance can be kept almost constant if suitable layout dimensions and spacings are selected to draw the Mn1 and Mp1 devices. A design model to optimize the layout dimensions and spacing of ESD clamp devices had been developed to keep the input capacitance almost constant for this low-*C* ESD protection circuit [70]. Variation on the total input capacitance of this low-*C* ESD protection circuit can be designed below 1%. The absolute error must be determined by considering the difference in the doping densities of NMOS and PMOS devices. Mismatches between assumed and actual doping densities will result in an increase in capacitance nonlinearity.

5.4.4 ESD Robustness

The low-*C* ESD protection circuit has been practically fabricated in a 0.35-μm silicided CMOS process with an operational amplifier as its input circuit. In this test chip, both the inverting and noninverting input pins are protected by the analog ESD protection circuit proposed. The silicide-blocking mask is also used on the device's Mn1, Mp1, and Mn3 to improve their ESD robustness, but without using the extra ESD implantation process modification.

The fabricated analog ESD protection circuits are zapped by an ESD simulator in both the HBM (human-body model) and MM (machine model) ESD stresses. The ESD test results of the maximum sustaining voltage are summarized in Table 5.1 which includes the analog pin-to-pin ESD stress. The failure criterion is defined as leakage current at the pad that exceeds 1 μA under 5-V voltage bias after any ESD zapping. As shown in Table 5.1, the low-*C* ESD protection circuit can successfully provide the analog pins with an HBM (MM) ESD level above 6000 V (400 V) in all ESD stress conditions but without adding a series resistor between the pad and the internal circuits.

The conventional ESD protection design shown in Figure 5.7(*a*) with a ggNMOS of $W/L = 480/0.5$ (μm/μm) for an analog input pin is also fabricated in

TABLE 5.1 ESD Level of a Low-*C* ESD Protection Circuit in HBM and MM ESD Testing

	Pin Combination in ESD Test				
	PS Mode	NS Mode	PD Mode	ND Mode	Pin-to-Pin
HBM (V)	6000	−8000	7000	−7000	6000
MM (V)	400	−400	400	−400	400

the same test chip as a reference. The HBM PS-mode ESD level of the design in Figure 5.7(*a*) is about 3 kV, but its analog pin-to-pin HBM ESD level is below 500 V. The pin-to-pin ESD damage location is founded on the poly gate of the first input stage in the operational amplifier circuit. So the conventional ESD protection design cannot protect the thinner gate oxide of the differential input stage in deep-submicron CMOS technologies during pin-to-pin ESD stress.

Although this test structure has been fabricated on a 0.35-μm process, it is still applicable to more advanced processes. The primary difference will be the physical device sizes necessary to match the diffusion capacitance of the NMOS and PMOS devices.

5.4.5 Turn-on Verification

To verify the turn-on efficiency of the low-*C* ESD protection circuit during pin-to-pin ESD stress, a square-wave voltage pulse generated from a pulse generator (HP 8118A) is applied to the inverting pin of an operational amplifier, while the noninverting pin of the operational amplifier is grounded and both the VDDA and VSSA pins are floating. The experimental setup to verify the turn-on efficiency in the positive and negative pin-to-pin ESD stress conditions is shown in Figure 5.12(*a*) and (*b*), respectively.

The measured voltage waveforms in positive pin-to-pin ESD stress conditions are shown in Figure 5.13(*a*) and (*b*). The voltage waveform in Figure 5.13(*a*) is the original voltage pulse generated from a HP 8118A pulse generator with a pulse height of 8 V and a pulse width of 200 ns. The 8-V voltage pulse has a rise time of about 10 ns, which is similar to the rise time of an HBM ESD pulse. The drain breakdown voltage of the NMOS Mn1 in the 0.35-μm silicide CMOS process without an extra ESD implantation process modification is about 8.5 V. Therefore, the voltage pulse with a pulse height of 8 V does not cause the drain breakdown in Mn1 in an analog ESD protection circuit. By applying such a voltage pulse to the analog pin, the turn-on efficiency of the proposed analog ESD protection circuit can be practically verified. Although this positive voltage pulse is applied to the input pin as shown in Figure 5.12(*a*), the low-*C* ESD protection circuit clamps it, and the degraded voltage waveform is shown in Figure 5.13(*b*). The voltage waveforms in the negative pin-to-pin ESD stress condition are also measured and shown in Figure 5.14(*a*) and (*b*).

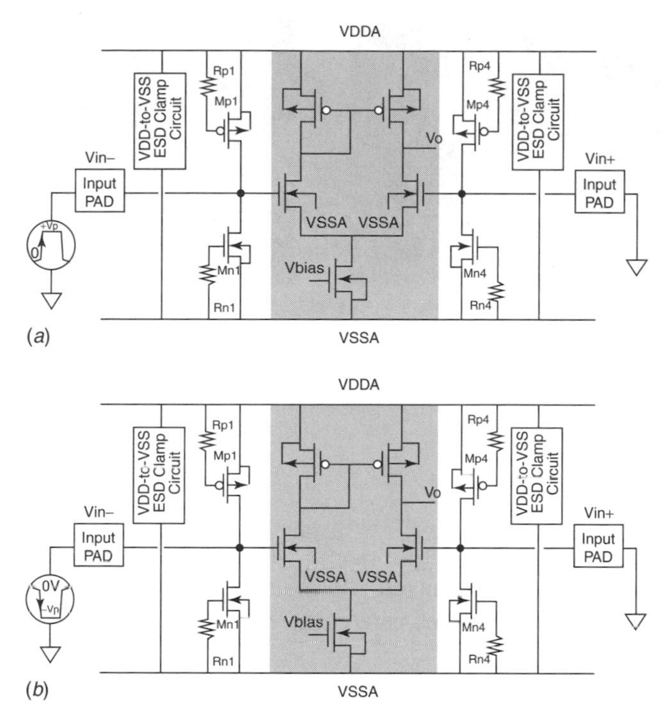

Figure 5.12 Experimental setup to verify the turn-on efficiency of the low-C ESD protection circuit during (a) positive and (b) negative pin-to-pin ESD stress conditions.

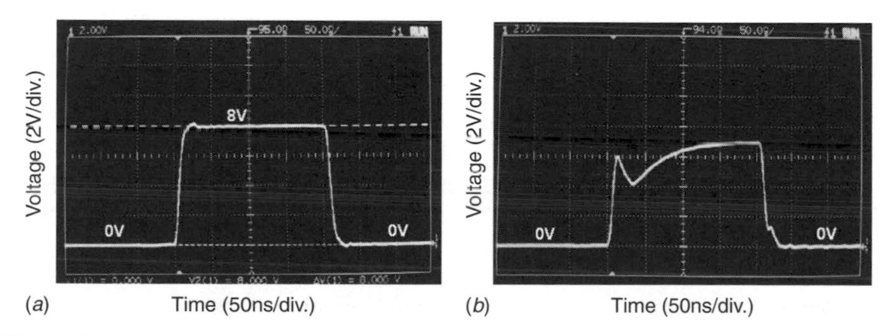

Figure 5.13 (a) Measured voltage waveform of the original 8-V voltage pulse generated from a pulse generator; (b) degraded voltage waveform when the 8-V voltage pulse is applied to the analog inverting input pin in the pin-to-pin stress condition shown in Figure 5.12(a).

The voltage waveform in Figure 5.14(a) is the original negative voltage pulse generated from the pulse generator with a pulse height of −8 V and a pulse width of 200 ns. When this negative voltage pulse is applied to the input pin as shown in Figure 5.12(b), it is clamped by the analog ESD protection circuit to a

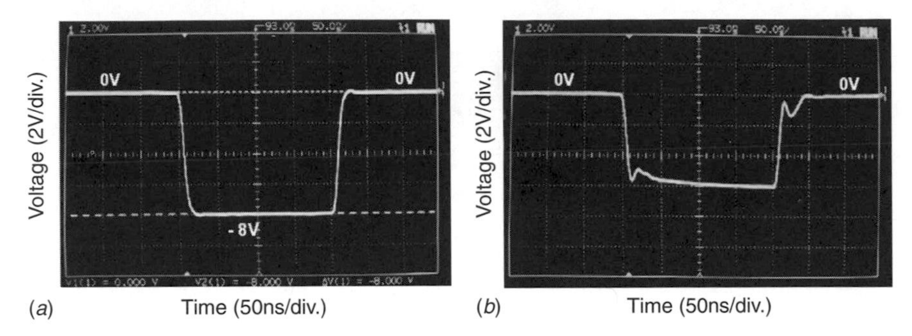

Figure 5.14 (*a*) Measured voltage waveform of the original negative (−8 V) voltage pulse generated from a pulse generator; (*b*) degraded voltage waveform when the voltage pulse is applied to an analog inverting input pin in the pin-to-pin stress condition shown in Figure 5.12(*b*).

voltage level of only −5 V, where the degraded voltage waveform is shown in Figure 5.14(*b*). From Figures 5.13(*b*) and 5.14(*b*), the voltage pulses are actually clamped by the low-*C* ESD protection circuit during the pin-to-pin ESD stress conditions. Therefore, the thinner gate oxide of the input stage can be protected during pin-to-pin ESD stress.

The turn-on efficiency of the low-*C* ESD protection circuit during PS- and ND-mode ESD stresses are also verified. The experimental setups are shown in Figure 5.15(*a*) and (*b*) for the PS and ND modes, respectively. When a positive voltage pulse with a voltage level of 8 V [shown in Figure 5.15(*a*)] is applied to the input pin in the PS-mode condition, the low-*C* ESD protection circuit clamps the positive voltage pulse and the degraded voltage waveform on the pad is as shown in Figure 5.16(*a*). While the negative voltage pulse shown in Figure 5.15(*b*) is applied to the input pin in the ND-mode condition, the negative (−8 V) voltage pulse is clamped by the low-*C* ESD protection circuit and the degraded voltage waveform on the pad is as shown in Figure 5.16(*b*). This has practically verified the turn-on efficiency of the low-*C* ESD protection circuit.

From the experimental verification above, positive or negative voltage pulses are clamped by the low-*C* ESD protection circuit through the VDD-to-VSS ESD clamp device Mn3. Mn1 and Mp1 in the low-*C* ESD protection circuit are operated in the junction diode forward-based condition rather than the drain-breakdown condition; therefore, the proposed analog ESD protection circuit can sustain a much higher ESD level even though the Mn1 and Mp1 devices have much smaller dimensions than those of traditional protection devices. This low-*C* ESD protection circuit without the series resistance between the pad and the internal circuits can safely protect the thinner gate oxide of the input stage. With much smaller device dimensions in Mn1 and Mp1, the total input junction capacitance connected to the pad can be reduced for high-frequency analog I/O or high-speed digital I/O applications.

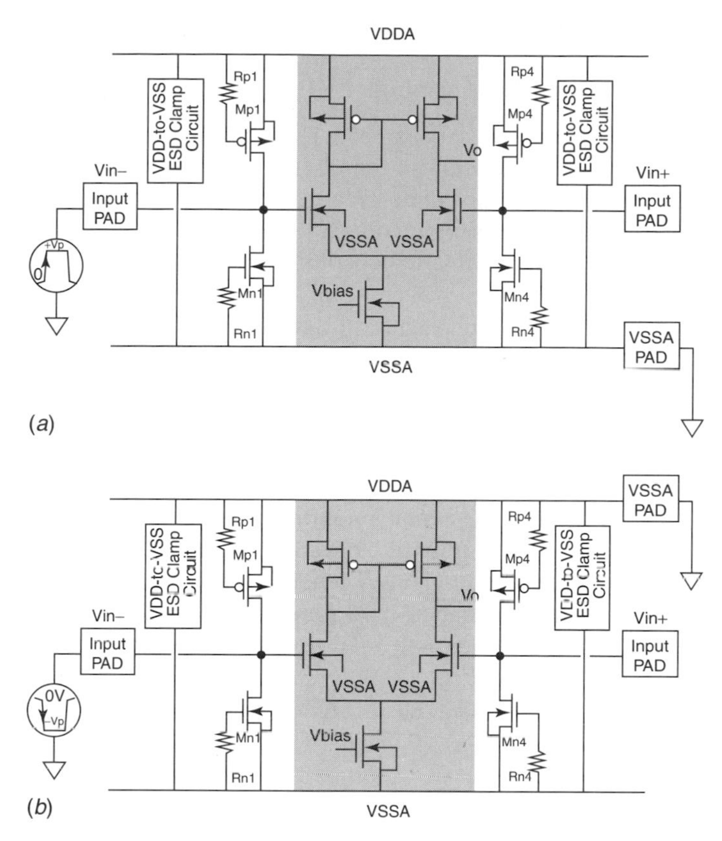

Figure 5.15 Experimental setup to verify the turn-on efficiency of the low-*C* ESD protection circuit during (*a*) PS- and (*b*) ND-mode ESD stress.

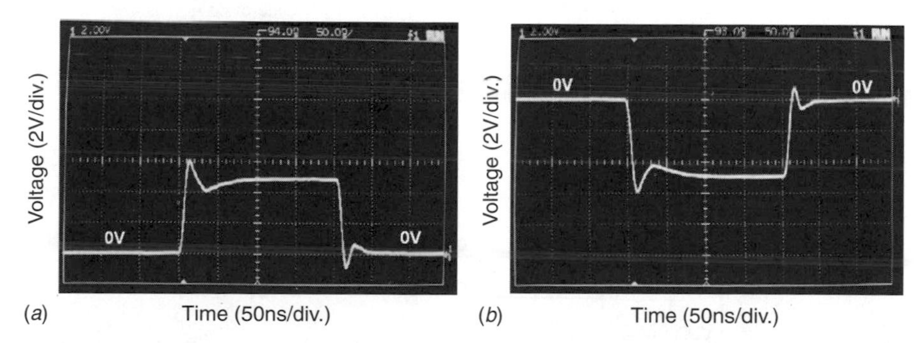

Figure 5.16 (*a*) Degraded voltage waveform when a positive 8-V voltage pulse is applied to the input pin in the PS-mode ESD stress condition; (*b*) degraded voltage waveform when a negative −8-V voltage pulse is applied to the input pin in the ND-mode ESD stress condition.

5.5 ESD PROTECTION DESIGN FOR MIXED-VOLTAGE I/O

5.5.1 Mixed-Voltage I/O Interfaces

The transistor dimensions have been scaled down into the nanometer region, resulting in a significant reduction in the circuit power supply voltage as well. Obviously, the smaller transistor dimension makes the chip area smaller, saving silicon cost. The lower power supply voltage results in lower power consumption (or reduced increase assuming that the operating increases). With the advance of modern CMOS technology, chip design quickly migrates to the lower voltage level, but some peripheral components or other ICs are still operated at higher voltage levels (3.3 or 5 V). In other words, chips with different supply voltages coexist in a system. Because of this mixture of supply voltages, most microelectronic systems require interfacing of semiconductor chips or subsystems with different internal power supply voltages. With the mix of power supply voltages, chip-to-chip interface I/O circuits must be designed to avoid electrical overstress across the gate oxide [60], to avoid hot-carrier degradation [61] on the output devices, and to prevent undesirable leakage current paths between the chips [62,63]. For example, 3.3-V interfacing is generally required for ICs implemented in CMOS processes with a normal internal power supply voltage of 1.0 or 1.5 V. The traditional CMOS I/O buffer with VDD of 3.3 V is shown in Figure 5.17(*a*),

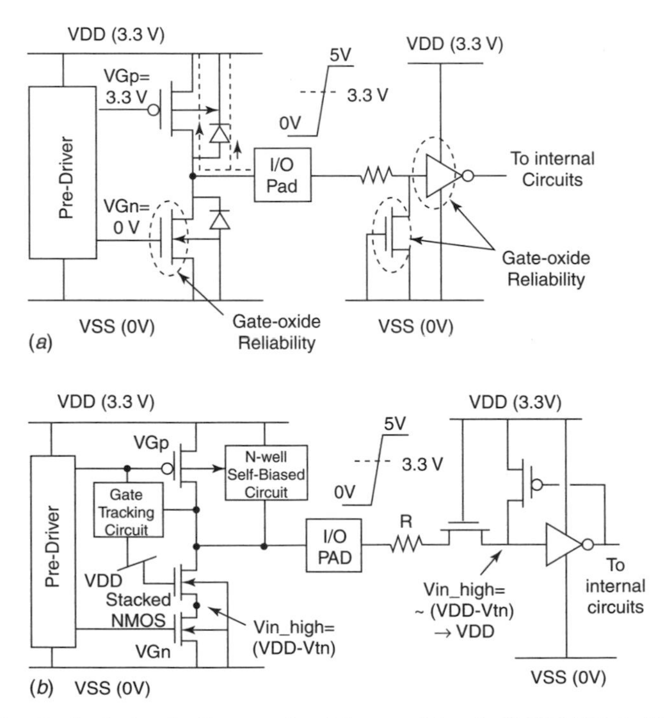

Figure 5.17 Typical circuit diagram for (*a*) a traditional CMOS I/O buffer and (*b*) a mixed-voltage I/O buffer with stacked NMOS and the self-biased-well PMOS.

where there is an output stage and an input stage. When a 5-V signal is applied to the I/O pad, the channel of the output PMOS and the parasitic drain-to-well junction diode in the output PMOS generate leakage current paths from the I/O pad to VDD [the dashed lines shown in Figure 5.17(*a*)]. Moreover, the gate oxides of the output NMOS, gate-grounded NMOS for input ESD protection, and input inverter stage are overstressed by the 5-V input signal. To solve the gate oxide reliability issue without using the additional thick gate oxide process (called dual-gate oxide in some CMOS processes [64,65]), the stacked-MOS configuration had been widely used in mixed-voltage I/O buffers [66–72] or even in power-rail ESD clamp circuits [73]. The basic circuit diagram of a typical 3 V/5 V-tolerant mixed-voltage I/O circuit is shown in Figure 5.17(*b*) [67]. The pull-up PMOS, connected from the I/O pad to VDD power line, has self-biased circuits for tracking its gate and n-well voltages, when the 5-V input signals enter the I/O pad. A detailed diagram of a circuit that implements the gate-tracking function and the n-well self-biased circuit block is shown in [72].

5.5.2 ESD Concerns for Mixed-Voltage I/O Interfaces

ESD stresses on an I/O pad have four pin-combination modes: positive-to-VSS (PS mode), negative-to-VSS (NS mode), positive-to-VDD (PD mode), and negative-to-VDD (ND mode) ESD stress conditions. To have sufficiently high ESD robustness of the CMOS output buffer, the CMOS buffer is generally drawn with larger device dimensions and a wider spacing from the drain contact to the poly gate, which often occupy a larger layout area in the I/O cell. Without increasing device dimensions in the I/O cells, the VDD-to-VSS ESD clamp circuits across the power lines of CMOS ICs had been used successfully to improve ESD robustness of CMOS I/O buffers [55]. The ESD current paths for a traditional CMOS output buffer under the positive-to-VSS ESD stress condition is illustrated by the dashed lines in Figure 5.18(*a*), where most ESD current is

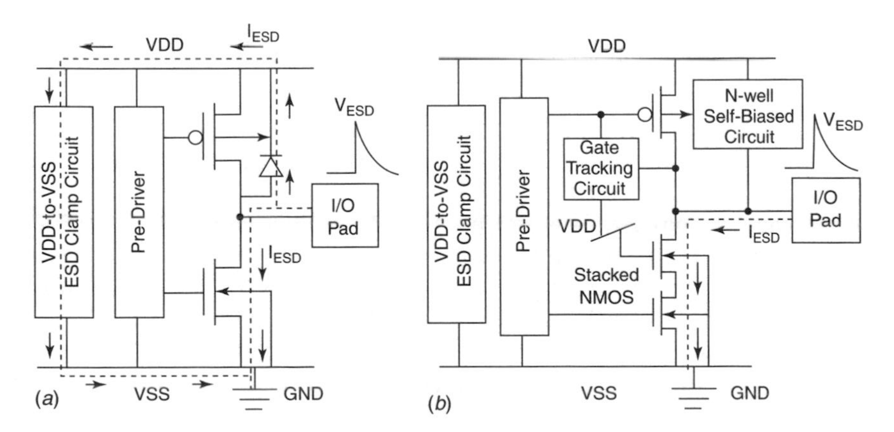

Figure 5.18 ESD current paths along (*a*) a traditional CMOS output buffer and (*b*) a mixed-voltage output buffer, under positive-to-VSS ESD stress conditions.

discharged through the parasitic diode of PMOS and the VDD-to-VSS ESD camp circuit to ground. Therefore, the traditional CMOS output buffer can sustain a high ESD stress. But due to the leakage current issue in the mixed-voltage I/O buffer, there is no any parasitic diode connected from the I/O pad to the VDD power line. Because of the limitation of placing a diode from the pad to VDD in mixed-voltage I/O circuits, the positive-to-VSS ESD voltage zapping on the I/O pad cannot be diverted from the pad to the VDD power line, and cannot be discharged through the additional power-rail (VDD-to-VSS) ESD clamp circuit. Such positive-to-VSS ESD current on the I/O pad is totally discharged through the stacked NMOS in the snapback breakdown condition. The ESD current under this positive-to-VSS ESD stress condition along the mixed-voltage output buffer is shown by the dashed line in Figure 5.18(*b*). The NMOS devices in a stacked configuration have a higher trigger voltage (V_{t1}), a higher snapback holding voltage (V_{sb}), and a lower secondary breakdown current (I_{t2}) than that of the single NMOS [74]. Therefore, such mixed-voltage I/O circuits with stacked NMOS often have a much lower ESD level than that of I/O circuits with a single NMOS [74,75]. So the mixed-voltage I/O circuit often has the lowest ESD level (often <2 kV in the human-body-model ESD test) under a positive-to-VSS ESD stress condition. Without the parasitic diode connected from the I/O pad to the VDD power line, the mixed-voltage I/O circuit also has a lower ESD level under a positive-to-VDD ESD stress condition. Therefore, ESD protection design on the mixed-voltage I/O circuits is mainly focused to improve the ESD level under positive ESD stress conditions.

To increase the ESD level of such mixed-voltage I/O circuits, some designs with extra multiple diodes in stacked configuration had been added from the I/O pad to the VDD power line [63,64]. However, while mixed-voltage I/O circuits are operating in a high-temperature environment with a high-voltage input, the forward-biased leakage current from the pad to V_{DD} through the stacked diodes must be reduced by additional circuit designs [76–79].

5.5.3 ESD Protection Device for a Mixed-Voltage I/O Interface

In this section, a new ESD protection design is presented to improve the ESD robustness of mixed-voltage I/O buffers significantly by using a stacked-NMOS triggered SCR device [80,81]. The new ESD protection circuit for mixed-voltage I/O circuits, which combines a stacked-NMOS structure with the gate-coupling circuit technique into the SCR device, is fully process-compatible to general CMOS processes without causing a gate oxide reliability problem. Without using thick gate oxide, the new ESD protection design for a 3 V/5 V-tolerant mixed-voltage I/O buffer has been verified successfully in a 0.35-μm CMOS process and should be applicable to the smaller geometry processes.

The cross-sectional view and corresponding layout pattern of a stacked-NMOS triggered silicon-controlled rectifier (SNTSCR) device are shown in Figure 5.19(*a*) and (*b*), respectively. This SNTSCR device structure can be realized in general CMOS processes without extra process modification. The

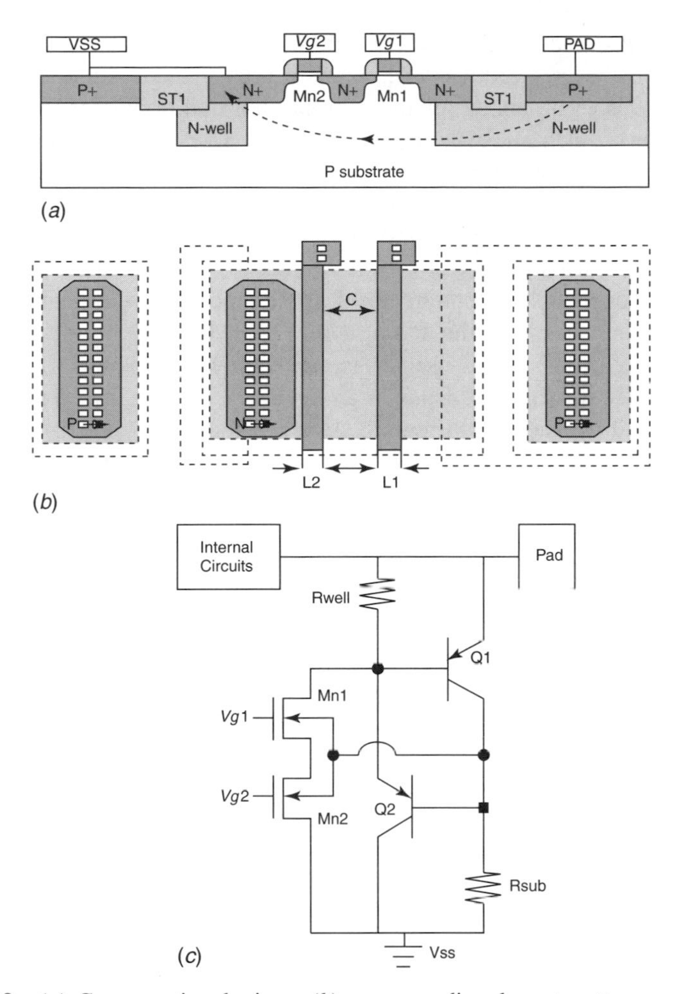

Figure 5.19 (*a*) Cross-sectional view, (*b*) corresponding layout pattern, and (*c*) equivalent circuit of a proposed SNTSCR device in a p-substrate CMOS process.

SNTSCR device is disposed on a bond pad to protect mixed-voltage I/O circuits from ESD damage. The corresponding equivalent circuit of this SNTSCR device is shown in Figure 5.19(*c*).

In the SNTSCR device, two NMOS transistors (Mn1 and Mn2) are stacked in the cascoded configuration, where the drain of Mn1 is across the junction between an N-well region and the p-substrate. The p^+ diffusion, N-well, p-substrate, and n^+ diffusion to form a lateral SCR device between the I/O pad and VSS is indicated by the dashed line in Figure 5.19(*a*). The purpose of Mn1 and Mn2 connected in stacked configuration is to sustain the high voltage level of input signals without causing a gate oxide reliability issue in the SNTSCR device under normal circuit operating conditions. If a single NMOS is inserted in the lateral SCR device, such as the traditional LVTSCR [82], the voltage across gate oxide will be greater than VDD when a high-voltage signal enters the I/O pad. This causes a gate oxide reliability issue in a traditional LVTSCR for long-term operation in such mixed-voltage I/O circuits. During ESD stress conditions, Mn1 and Mn2 are both turned on by a suitable gate-biased design, triggering the lateral SCR on for discharging ESD current. Without using thick gate oxide in the CMOS process, the proposed SNTSCR device has no gate oxide reliability issue when used to protect mixed-voltage I/O circuits. This ESD structure has become increasingly important for deep-submicron processes where thinner oxide devices are used for the I/O structure to reduce the pin capacitance for high-speed applications and still interface with the higher-voltage ICs.

To investigate the characteristics of an SNTSCR device, three layout parameters [C, L1, and L2 shown in Figure 5.19(*b*)] of the layout pattern are adjusted. C is the poly-to-poly spacing across the center common n+ diffusion; L1 and L2 are the channel lengths of Mn1 and Mn2, respectively. Such SNTSCR devices with different layout parameters but with a fixed channel width of 60 μm have been fabricated in a 0.35-μm CMOS process. The measured $I-V$ characteristics of an SNTSCR device with C = 0.5 μm and L1 = L2 = 0.35 μm under different gate biases of V_{g1} and V_{g2} are shown in Figure 5.20. The trigger voltage (V_t) of the SNTSCR device decreases from 10 V to 6 V when the gate bias increases from $V_{g1} = V_{g2} = 0$ V to $V_{g1} = V_{g2} = 0.5$ V. As $V_{g1} = V_{g2} > 0.6$ V, both Mn1 and Mn2 are turned on to trigger SNTSCR on; therefore, V_t decreases to around

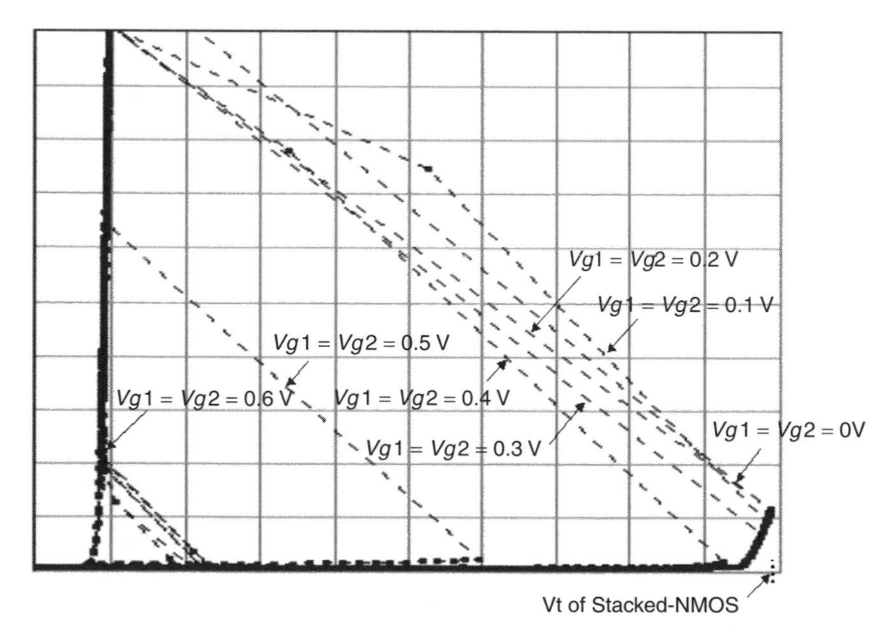

Figure 5.20 Measured $I-V$ curves of fabricated SNTSCR devices with C = 0.5 μm and L1 = L2 = 0.35 μm under different gate biases (x-axis: 1 V/div.; y-axis: 1 mA/div.).

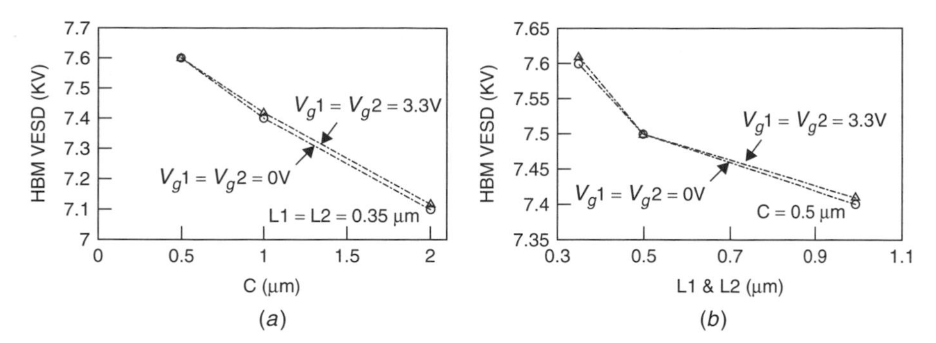

Figure 5.21 (*a*) Dependence of the HBM ESD level on the layout parameter C of SNTSCR devices under different gate biases ($V_{g1} = V_{g2}$). Failure criterion: $I_{\text{leakage}} > 1\ \mu\text{A}$ at $V_{\text{bias}} = 5$ V. (*b*) Dependence of the HBM ESD level on layout parameters L1 and L2 of SNTSCR devices under different gate biases ($V_{g1} = V_{g2}$). Failure criterion: $I_{\text{leakage}} > 1\ \mu\text{A}$ at $V_{\text{bias}} = 5$ V.

1 to 2 V. With suitable gate biases on Mn1 and Mn2, the trigger voltage of an SNTSCR device can be reduced lower than the snapback breakdown voltage of the stacked NMOS (about 10 V) in the mixed-voltage I/O buffer. Therefore, the new proposed ESD protection circuit with a SNTSCR device can effectively protect mixed-voltage I/O buffers.

The impact of layout parameters and gate biases on the positive-to-VSS HBM ESD level of an SNTSCR device are shown in Figure 5.21(*a*) and (*b*). Figure 5.21(*a*) shows the dependence of the HBM ESD level on layout parameter C with fixed channel lengths of L1 = L2 = 0.35 μm in an SNTSCR device. Figure 5.21(*b*) shows the dependence of the HBM ESD level on the channel lengths of L1 and L2 with a fixed layout parameter C = 0.5 μm in an SNTSCR device. The failure criterion is defined at the leakage current greater than 1 μA under the voltage bias of 5 V. The HBM ESD robustness of an SNTSCR device is slightly degraded when the layout parameters C, L1, and L2 are increased. With shorter layout parameters (C, L1, and L2), an SNTSCR device has a narrower anode-to-cathode spacing, which implies a smaller turn-on resistance. Therefore, an SNTSCR device drawn with shorter layout parameters (C, L1, and L2) has a higher ESD level. The gate biases on V_{g1} and V_{g2} do not obviously improve the ESD level of an SNTSCR device (Figure 5.21), but it can trigger on the SNTSCR earlier to discharge ESD current. Thus, an SNTSCR device with a suitable gate-biased design can effectively protect mixed-voltage I/O buffers of CMOS ICs.

5.5.4 ESD Protection Circuit Design for a Mixed-Voltage I/O Interface

Based on experimental investigation on SNTSCR devices, an ESD protection design with an SNTSCR for protecting mixed-voltage I/O buffers is shown in Figure 5.22(*a*). An ESD detection circuit is designed to provide suitable gate

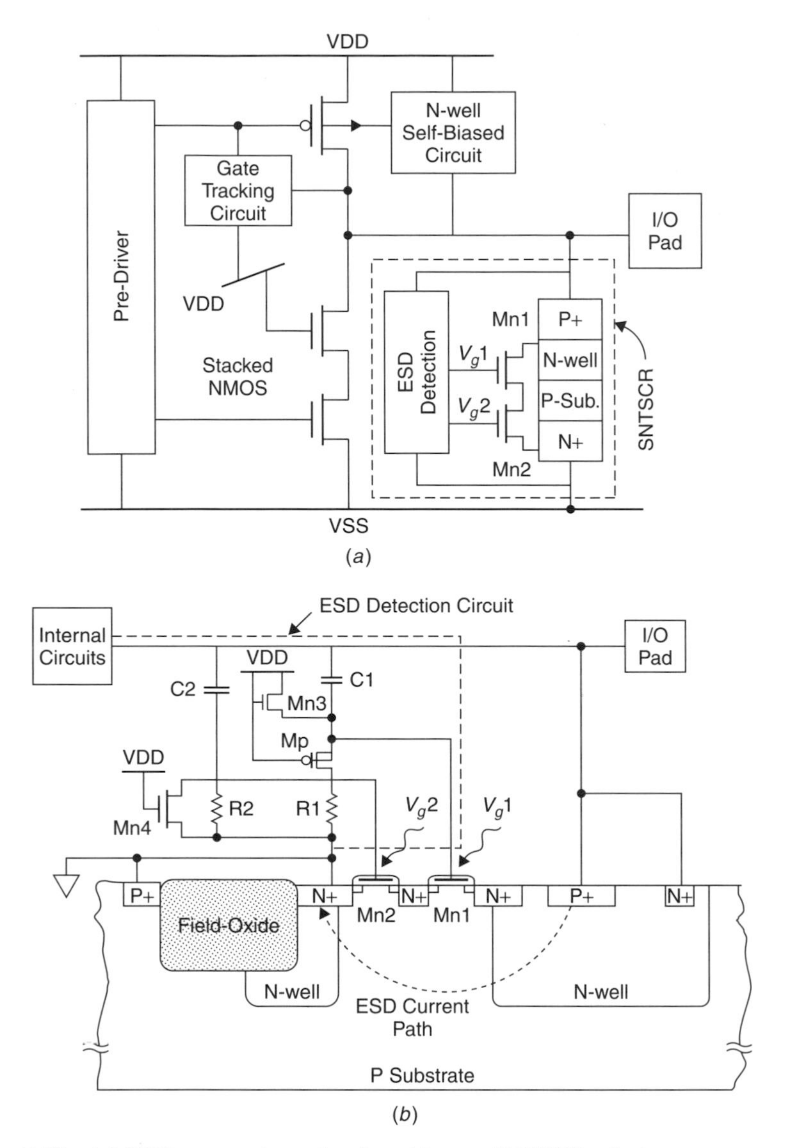

Figure 5.22 (*a*) ESD protection circuit with an SNTSCR device to protect the mixed-voltage I/O buffer; (*b*) design of an ESD detection circuit with an SNTSCR device with a larger design region on the sustaining resistance and coupling capacitance for protecting mixed-voltage I/O buffers.

biases to trigger on the SNTSCR device during ESD stress conditions. On the contrary, this ESD detection circuit must keep the SNTSCR off when the IC is under normal circuit operating conditions. An ESD detection circuit designed using a gate-coupling technique with consideration of gate oxide reliability issues is shown in Figure 5.22(*b*).

In normal circuit operating conditions the SNTSCR is kept off so that it does not interfere with the voltage levels of signals on the I/O pad. In this normal operating state, Mn3 in Figure 5.22(*b*) acts as a resistor to bias the gate voltage (V_{g1}) of Mn1 at VDD. But the gate of Mn2 is grounded through resistor R_2. When the I/O pad is applied with a high input voltage (5 V), the center common n+ region between the Mn1 and Mn2 transistors has a voltage level about VDD − Vthn (Vthn is the threshold voltage of an NMOS). So all the devices in the ESD protection circuit can meet the limited electrical-field constraint of gate oxide reliability during normal circuit operating conditions. The coupled voltage through capacitor C_1 could also increase the gate voltage of Mn1 when the voltage on I/O pad transfers from 0 V to 5 V. The PMOS device (Mp) in Figure 5.22(*b*) is therefore designed to clamp the excessive voltage once the voltage of Mn1's gate increases to VDD + V_{thp} (V_{thp} is the magnitude of the threshold voltage of Mp). Proper design of an ESD detection circuit ensures that the SNTSCR structure can be kept off under normal circuit operating conditions. Moreover, the PMOS (Mp) can further clamp the gate voltage of Mn1 to ensure gate oxide reliability on Mn1 even if the I/O pad has a high input voltage level.

The capacitors (C_1 and C_2), resistors (R_1 and R_2), Mp, and Mn3 compose an ESD detection circuit for providing suitable gate biases to the SNTSCR device. Capacitor C_1 (C_2) is designed to couple the ESD transient voltage to the gate of Mn1 (Mn2) to lower the trigger voltage of the SNTSCR device. Resistor R_1 (R_2) is designed to maintain the coupled voltage longer on the gate of Mn1 (Mn2) for triggering the SNTSCR device into its holding region. To further improve the design region for easily choosing the suitable sustaining resistance and coupling capacitance in general CMOS processes, device Mn4 is added across the sustaining resistor R_2, which is located between the gate of Mn2 and VSS. The gate of Mn4 is biased at VDD, but it is better to connect the gate to the VDD power line through a diffusion resistor for consideration of the antenna rule issue associated with the process. Under the normal circuit operating condition, Mn4 is always turned on to clamp the coupling voltage V_{g2} below the threshold voltage (V_{thn}), and to keep Mn2 off. Therefore, the SNTSCR can be guaranteed off under normal circuit operating conditions.

During positive-to-VSS ESD stress conditions, a positive high ESD voltage is applied to the I/O pad with VSS grounded but VDD floating. In this ESD stress condition, the gate of Mp is grounded since the initial voltage level on the floating VDD power line is zero. So Mp is turned on, but Mn3 is off. The capacitors, C_1 and C_2, are designed to couple the ESD transient voltage from the I/O pad to the gates of Mn1 and Mn2, respectively. The coupled voltage should be designed higher than the threshold voltage to turn on Mn1 and Mn2 for triggering the SNTSCR device on before the devices in the mixed-voltage I/O circuit are damaged by ESD energy. Under ESD stress conditions, Mn4 is off since the initial voltage level on the floating VDD power line is zero. The voltage (V_{g2}) coupled to the gate of Mn2 is determined by the sustaining resistance (R_2) and the coupling capacitance (C_2). When the SNTSCR is triggered on, the ESD current is discharged primarily from the I/O pad to VSS through this SNTSCR device. The characteristics of lower trigger voltage and low holding voltage of the gate-coupling SNTSCR device can safely protect the thin gate oxide of mixed-voltage I/O circuits as well as sustaining a high ESD level within a smaller silicon area.

The purpose of the ESD detection circuit is to provide suitable gate biases for an SNTSCR device under the normal circuit operating conditions and ESD stress conditions. To obtain suitable gate biases, it is important to determine the values of the coupling capacitors (C_1 and C_2) and sustaining resistors (R_1 and R_2). Based on the foregoing operating principles, suitable values of C_1, C_2, R_1, and R_2 to meet the circuit operations desired in various CMOS processes can be adjusted and finely tuned using HSPICE simulation.

5.5.5 ESD Robustness

Positive-to-V_{SS} human-body-model (HBM) ESD levels of mixed-voltage I/O buffers with or without an ESD protection circuit are measured and compared in Figure 5.23(*a*). Mixed-voltage I/O buffers with different stacked-NMOS channel widths are also tested as a reference. The HBM ESD level of mixed-voltage I/O buffers (with a stacked-NMOS channel width of 120 μm) can obviously be improved from the original ~2 kV to become greater than 8 kV using the proposed ESD protection circuit with an SNTSCR device. In Figure 5.23(*a*), all the mixed-voltage I/O buffers protected by the proposed ESD protection circuit have the same SNTSCR device width of 60 μm.

The positive-to-VSS machine-model (MM) ESD levels of mixed-voltage I/O buffers with or without an ESD protection circuit are measured and compared in Figure 5.23(*b*). Mixed-voltage I/O buffers with various stacked-NMOS channel widths are also tested as a reference. From the measured results, the MM

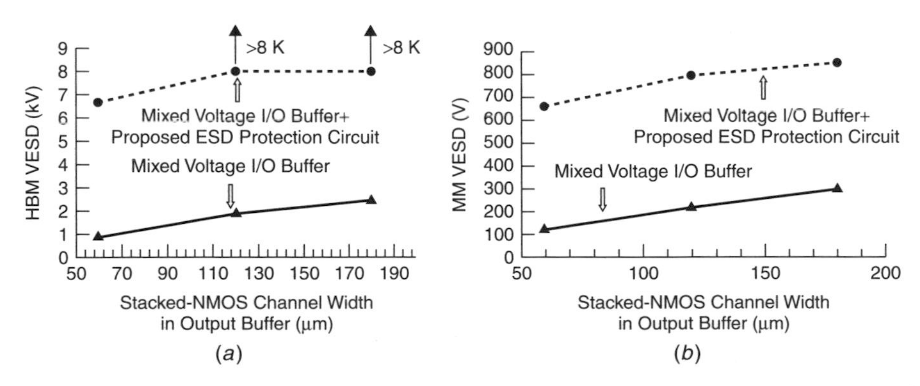

Figure 5.23 (*a*) Comparison of the positive-to-VSS HBM ESD robustness of mixed-voltage I/O buffers with or without an ESD protection circuit under various channel widths of a stacked NMOS in mixed-voltage I/O buffers. (*b*) Comparison of the positive-to-VSS MM ESD robustness of mixed-voltage I/O buffers with or without an ESD protection circuit under various channel widths of a stacked NMOS in mixed-voltage I/O buffers.

ESD level of a mixed-voltage I/O buffer with a channel width of 120 μm in a stacked NMOS can be improved significantly from the original ~200 V to become ~800 V by an ESD protection circuit with an SNTSCR device width of only 60 μm.

5.5.6 Turn-on Verification

To verify the turn-on efficiency of an ESD protection circuit for mixed-voltage I/O, a 0- to 8-V sharply rising voltage pulse with a rise time of 10 ns is applied to the I/O pad when VSS is relatively grounded but VDD is floating (to simulate the positive-to-VSS ESD stress condition). The stacked NMOS in the output buffer has a snapback breakdown voltage of about 10 V. Such a 0- to 8-V pulse applied to the I/O pad does not break down the stacked NMOS of the mixed-voltage output buffer. But the 0- to 8-V pulse can trigger on the ESD protection circuit to cause a degraded voltage waveform such as that shown in Figure 5.24(*a*), where the applied 0- to 8-V voltage pulse is clamped to about

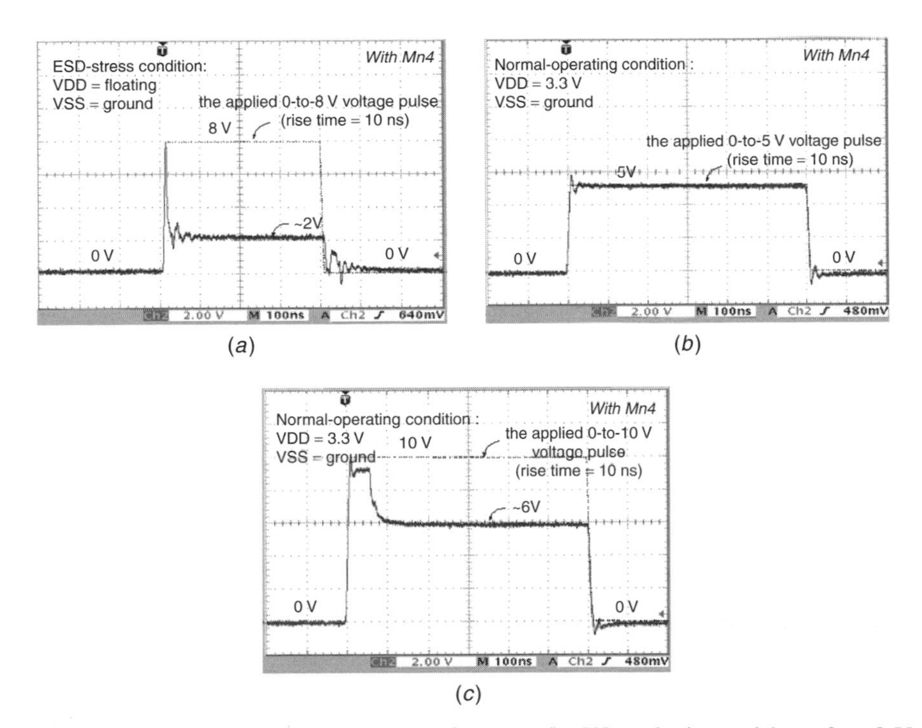

Figure 5.24 (*a*) Measured voltage waveform on the I/O pad triggered by a 0-to-8 V voltage pulse with a rise time of 10 ns, under positive-to-VSS ESD stress conditions (y-axis: 2 V/div.; x-axis: 100 ns/div.); and the measured voltage waveforms on the I/O pad triggered by (*b*) 0- to 5-V and (*c*) 0- to 10-V voltage pulses with a rise time of 10 ns under normal operating conditions with VDD = 3.3 V and VSS = 0 V (y-axis: 2 V/div.; x-axis: 100 ns/div.).

2 V by the SNTSCR device. The transition time of about 10 ns from 8 V to 2 V in Figure 5.24(*a*) is the corresponding turn-on speed of the SNTSCR device realized in this 0.35-μm CMOS process.

In the normal circuit operating condition with VDD (VSS) biased at 3.3 V (0 V), a 0- to 5-V input voltage pulse with a rise time of 10 ns is applied to the I/O pad while a digital oscilloscope monitors voltage on the I/O pad. The applied 0- to 5-V waveform is not degraded, as shown in Figure 5.24(*b*). During normal circuit operating conditions, the gate of Mn4 is biased at VDD (3.3 V). The coupled gate voltage of Mn2 through capacitor C_2 is discharged to ground by the turned-on Mn4, so the SNTSCR device is not triggered on by normal input signals of 5 V. If the applied voltage pulse is increased further, to 10 V, under normal circuit operating conditions, the measured voltage waveform on the I/O pad is as shown in Figure 5.24(*c*). The applied 0- to 10-V pulse is clamped to about ~6 V in Figure 5.24(*c*), but not to the voltage level of about 2 V as that shown in Figure 5.24(*a*). The stacked NMOS has a snapback breakdown of about 10 V and a snapback holding voltage of about 6 V in this 0.35-μm CMOS process. Therefore, the degraded voltage level of about 6 V in Figure 5.24(*c*) is clamped by the stacked NMOS of the mixed-voltage I/O buffer in the snapback region. If the SNTSCR device in the ESD protection circuit is triggered on, the voltage level should be clamped to its holding voltage of about 2 V. This result has further confirmed that the Mn4 device in the ESD detection circuit can safely apply the SNTSCR device to protect the mixed-voltage I/O buffer without being unexpectedly triggering under the normal circuit operating condition.

5.6 SCR DEVICES FOR ESD PROTECTION

Due to the low holding voltage (V_h, about 1.5 V in general CMOS processes) of a silicon-controlled rectifier (SCR) device, the power dissipation (power $\simeq I_{\mathrm{ESD}} \times V_h$) located on the SCR device during ESD stress is significantly less than that located on other ESD protection devices compared to a diode, MOS, BJT, or field oxide device. The SCR device can sustain a much higher ESD level within a smaller layout area in CMOS ICs, so it has been used to protect internal circuits against ESD damage for a long time. But the SCR device still has a higher switching voltage (i.e., trigger voltage, about 22 V) in the sub-$\frac{1}{4}$ − μm CMOS technology, which is generally greater than the gate oxide breakdown voltage of the input stages. Furthermore, gate oxide thickness, time to breakdown (t_{BD}), or charge-to-breakdown (Q_{BD}) will also be decreased with shrinking CMOS technologies. This trend makes it imperative to reduce the switching voltage of SCRs and to enhance their turn-on speed for efficient protection of ultrathin gate oxide from latent damage or rupture, especially against fast charged-device-model (CDM) ESD events, which is becoming the most difficult specification to meet in deep-submicron processes.

An overview of SCR-based devices for on-chip ESD protection is given in this section. In addition, solutions to avoid the transient-induced latch-up issue of SCR-based devices in CMOS IC products with a maximum voltage supply greater than 1.5 V are also discussed. However, such latch-up problems will certainly vanish when the maximum voltage supply of IC products is smaller than the holding voltage of SCR devices. For example, a single SCR with a holding voltage of about 1.6 V can safely be used as ESD protection without latch-up danger in a 0.13-μm CMOS process with a maximum voltage supply of 1.2 V.

5.6.1 Turn-on Mechanism of SCR Devices

The equivalent circuit schematic of a SCR device is shown in Figure 5.25(*a*). The SCR device consists of a lateral NPN and a vertical PNP bipolar transistor and forms a two-terminal, four-layer PNPN (P+/N-well/P-well/N+) structure, which is inherent in CMOS processes. The switching voltage of the SCR device is dominated by the avalanche breakdown voltage of an N-well/P-well junction, which is about 22 V in a 0.25-μm CMOS process or 18 V in a 0.13-μm CMOS process. When a positive voltage applied to an SCR anode is greater than the breakdown voltage and its cathode is grounded, for example, the hole and electron current will be generated through the avalanche breakdown mechanism. The hole current will flow through the P-well to P+ diffusion connected to ground, whereas the electron current will flow through the N-well to N+ diffusion connected to the SCR anode. As long as the voltage drop across the P-well resistor (R_{pwell}) [N-well resistor (R_{nwell})] is greater than 0.7 V, the NPN (PNP) transistor will be turned on to inject the electron (hole) current to further bias the PNP (NPN) transistor and initiate SCR latching action. Finally, the SCR will successfully be triggered

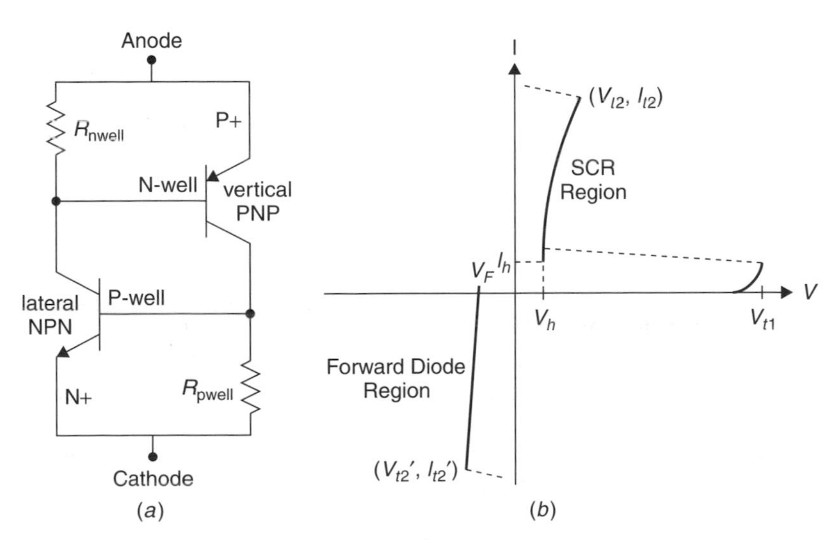

Figure 5.25 (*a*) Equivalent circuit schematic of a SCR device. (*b*) $I-V$ characteristics of an SCR device in a CMOS process under positive and negative voltage biases.

on into its latching state to discharge ESD current through the positive-feedback regenerative mechanism [94,95].

The dc $I-V$ characteristic of SCR device is shown in Figure 5.25(*b*). Once the SCR is triggered on, the required holding current to keep the NPN and PNP transistors on can be generated through the positive-feedback regenerative mechanism of latch-up without involving the avalanche breakdown mechanism. So the holding voltage (V_h) of an SCR can be reduced to a lower voltage level of about 1.5 V, typically. When negative voltage is applied on the anode terminal of SCR, the parasitic diode (N-well/P-well junction) inherent in SCR structure will be forward biased to clamp the negative voltage at a lower voltage level of about 1 V (cut-in voltage of a diode). Whether the ESD energy is positive or negative, the SCR device can clamp ESD overstresses to a lower voltage level, so the SCR device can sustain the highest ESD robustness within a smaller layout area in CMOS ICs.

5.6.2 SCR-Based Devices for CMOS On-Chip ESD Protection

Low-Voltage Triggering SCRs [82,85] To protect input and even output stages more effectively, a low-voltage triggering SCR (LVTSCR) has been developed. The device structure of the LVTSCR is illustrated in Figure 5.26(*a*) and the corresponding $I-V$ characteristic of the LVTSCR in a 0.25-μm CMOS process is shown in Figure 5.26(*b*). An example of using the LVTSCR device as the input ESD protection circuit is shown in Figure 5.26(*c*). In some applications, the N-well of an LVTSCR is connected to the input pad. The switching voltage of the LVTSCR (about 7 V) is equivalent to the drain breakdown or punch-through voltage of a short-channel NMOS device, which is inserted into the LSCR structure rather than the original switching voltage of the LSCR device (about 22 V). With such a low switching voltage, the LVTSCR can provide effective ESD protection for the input or output stages of CMOS ICs without a secondary ESD protection circuit. Therefore, the total layout area of the ESD protection circuits with the LVTSCR can be saved. Furthermore, to protect both PMOSs and NMOSs in the input or output stages of CMOS ICs, a complementary LVTSCR structure [99] has been invented to provide better ESD protection.

Gate-Coupled LVTSCRs [86] To effective protect ultrathin gate oxide in a deep-submicron CMOS process, the gate-coupling technique was applied to further reduce the switching voltage of the LVTSCR without involving avalanche breakdown mechanism. The ESD protection circuit for an input or output pad with complementary gate-coupled LVTSCR devices [NMOS-triggered LSCR (NTLSCR) and PMOS-triggered LSCR (PTLSCR)] is shown in Figure 5.27(*a*). The device structure of a complementary gate-coupled LVTSCR is illustrated in Figure 5.27(*b*), and the $I-V$ characteristics of a gate-coupled LVTSCR in a 0.25-μm CMOS process is shown in Figure 5.27(*c*). The dependence of the switching voltage of an SCR device on the gate bias voltage of an NTLSCR device is shown in Figure 5.27(*d*). The capacitances (C_n and C_p) in Figure 5.27(*a*) must

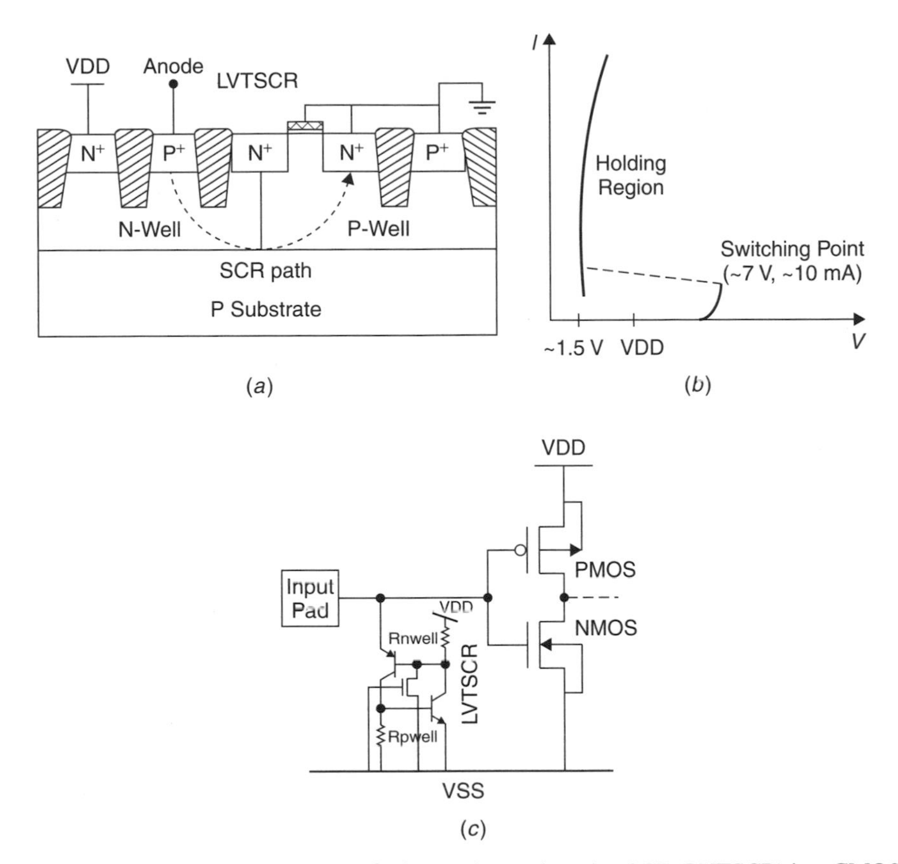

Figure 5.26 (*a*) Device structure of a low-voltage triggering SCR (LVTSCR) in a CMOS process. (*b*) $I-V$ characteristics of an LVTSCR in a 0.25-μm CMOS process. (*c*) Input ESD protection circuit with an LVTSCR device.

be designed at a suitable value where the coupled voltage under normal circuit operating conditions is smaller than the threshold voltage of NMOS/PMOS but greater than the threshold voltage of NMOS/PMOS under ESD zapping conditions [20]. The switching voltage of a gate-coupled NTLSCR (PTLSCR) can be adjusted using the coupled voltage on the gate of the short-channel NMOS (PMOS) in the SCR device structure. The higher coupled voltage on the gate of the short-channel NMOS/PMOS in the LVTSCR leads to lower switching voltage in the LVTSCR. Therefore, gate-coupled LVTSCR devices can quickly discharge ESD current to protect the ultrathin gate oxide of the input or output stages more effectively.

Grounded-Gate NMOS Triggered SCRs [88,89] A grounded-gate NMOS triggered SCR (GGSCR) is another choice for on-chip ESD protection circuit. A NMOS transistor, which resembles a GGNMOS configuration, is used as an external trigger device to trigger on the GGSCR. In contrast to the LVTSCR, the

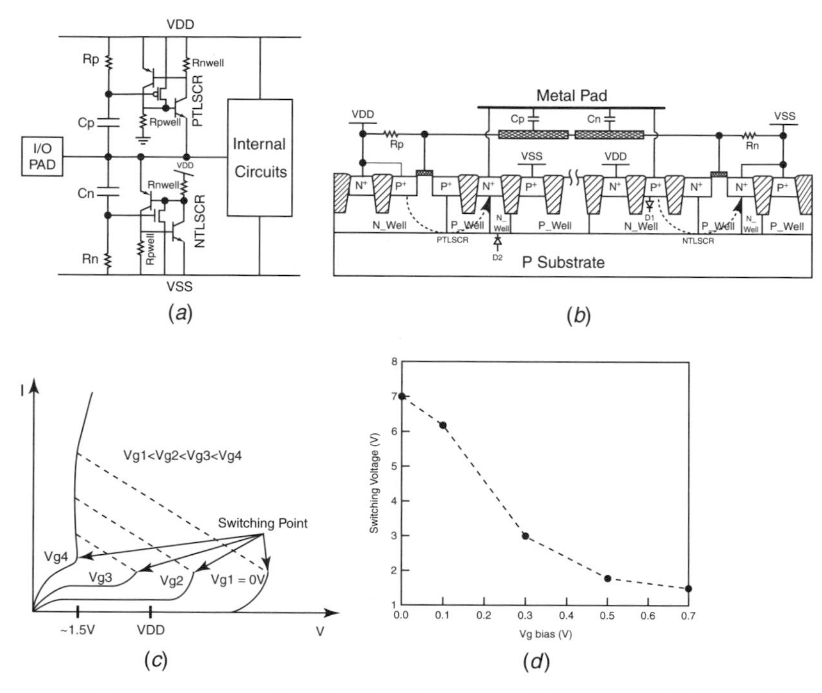

Figure 5.27 (*a*) ESD protection circuit with gate-coupled LVTSCR devices. (*b*) Device structure of a gate-coupled LVTSCR device in a CMOS process. (*c*) I–V characteristics of a gate-coupled LVTSCR device in a 0.25-μm CMOS process. (*d*) Dependence of switching voltage of an SCR device on the gate bias voltage of an NTLSCR device.

drain of the external trigger NMOS in GGSCR is coupled directly to the pad, and its gate and source are coupled into the P-substrate (the base of the NPN). An example of using a GGSCR device as an input ESD protection circuit is shown in Figure 5.28(*a*). The layout top view of the GGSCR is illustrated in Figure 5.28(*b*) [105]. When an ESD event is applied to the I/O pad in Figure 5.28(*a*), the external trigger NMOS will enter avalanche breakdown first to inject the triggering current into the P-substrate and poly resistor. As long as the base voltage of NPN is greater than 0.7 V, the GGSCR will be triggered on. The poly resistor in Figure 5.28(*b*) controls the triggering and holding current and prevents false triggering of the GGSCR. From the experimental results, a GGSCR designed with a shorter anode-to-cathode spacing will have a lower holding voltage, higher I_{t2}, better dV/dt triggering ability, and faster turn-on speed than those of an LVTSCR with longer anode-to-cathode spacing.

Substrate-Triggered SCRs [90,91] The turn-on mechanism of an SCR device is essentially a current triggering event. While a current is applied to the base or substrate of an SCR device, it can be quickly triggered into its latching state. With the substrate-triggered technique, p-type substrate-triggered SCR (P_STSCR) and n-type substrate-triggered SCR (N_STSCR) devices for ESD protection were

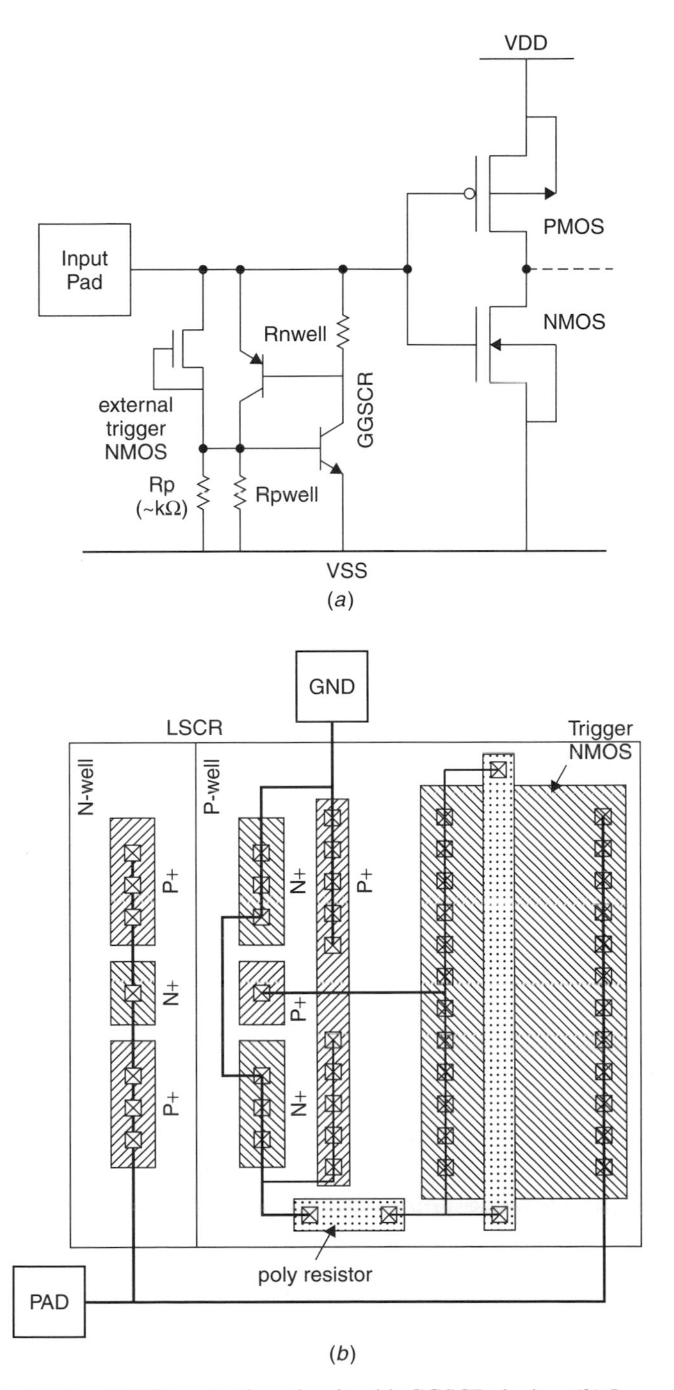

Figure 5.28 (*a*) Input ESD protection circuit with GGSCR device. (*b*) Layout top view of GGSCR in a CMOS process.

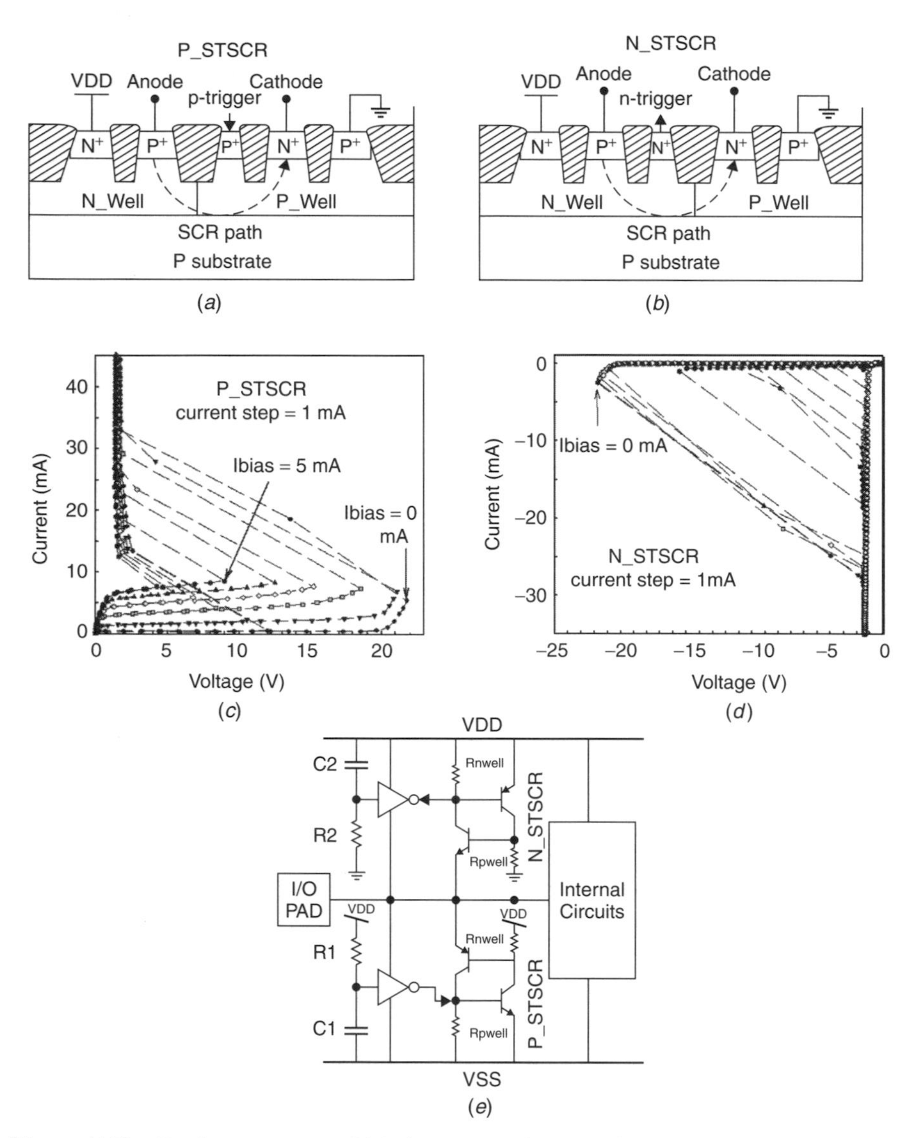

Figure 5.29 Device structures of (*a*) the p-type substrate-triggered SCR (P_STSCR) and (b) the n-type substrate-triggered SCR (N_STSCR) devices. The I-V characteristics of (c) the P_STSCR and (d) the N_STSCR devices in a 0.25-μm CMOS process. (e) The ESD protection circuit with the P_STSCR and N_STSCR devices for I/O pad.

reported. The device structures of the P_STSCR and N_STSCR are illustrated in Figure 5.29(*a*) and (*b*), respectively. Comparing this structure to the traditional lateral SCR device structure, an extra P+ diffusion is inserted into the P-well of the P_STSCR device structure and connected out as the p-trigger node of the P_STSCR device. For the N_STSCR, an extra N+ diffusion is inserted into the

N-well of the N_STSCR device structure and connected out as the n-trigger node of the N_STSCR device. The I–V characteristics of the P_STSCR and N_STSCR are shown in Figure 5.29(*c*) and (*d*), respectively. With the increase in substrate-/well-triggered current, the switching voltage of the P_STSCR/N_STSCR device can be reduced to its holding voltage. The turn-on time of the STSCR can be also reduced to about 10 ns under a 5-V voltage pulse with a 10-ns rise time in a 0.25-μm CMOS process. With the lower switching voltage, the STSCR device can clamp the ESD voltage to a lower voltage level more quickly, to fully protect the ultrathin gate oxide of input stages from ESD overstress. The ESD protection circuit for an input or output pad with P_STSCR and N_STSCR devices is shown in Figure 5.29(*e*). In normal circuit operating conditions with VDD and VSS power supplies, the input of inv_1 (inv_2) is biased at VDD (VSS). Therefore, the output of the inv_1 (inv_2) is biased at VSS (VDD), due to the turn-on of NMOS (PMOS) in the inv_1 (inv_2), whenever the input signal is logic high or logic low. The p-trigger (n-trigger) node of the P_STSCR (N_STSCR) device is kept at VSS (VDD) by the output of the inv_1 (inv_2), so the P_STSCR and N_STSCR devices are guaranteed off under normal circuit operating conditions. Under the positive-to-VSS (PS) ESD-zapping condition (with grounded VSS but floating VDD) shown in Figure 5.29(*e*), the input of the inv_1 is initially floating with a zero voltage level; therefore, the PMOS of the inv_1 will be turned on due to the positive ESD voltage on the pad. So the output of the inv_1 is charged up by the ESD energy to generate the trigger current into the p-trigger node of the P_STSCR device. Therefore, the P_STSCR device is triggered on, and the ESD current is discharged from the I/O pad to the grounded VSS pin through the P_STSCR device. A similar operating principle can be applied to N_STSCR under the negative-to-VDD (ND) ESD-zapping condition (with grounded VDD but floating VSS). Furthermore, an STSCR device with a dummy-gate structure has been invented [108,109] to further reduce the switching voltage and to improve the turn-on speed of STSCR. The bipolar current gain of an STSCR with a dummy-gate structure is larger than that of STSCR with a shallow trench isolation (STI) structure, so the triggering efficiency of STSCR with a dummy-gate structure is better than that of STSCR with STI.

Double-Triggered SCRs [94] Another method to reduce the switching voltage of an LSCR device and to further enhance the turn-on speed of an LSCR device more efficiently is the double-triggered technique. The device structure of a double-triggered SCR (DTSCR) is shown in Figure 5.30(*a*). The extra P+ and N+ diffusions are inserted into the P-well and N-well of a DTSCR device structure and connected out as the p- and n-trigger nodes of the DTSCR device. The dependence of the switching voltage of the DTSCR device on the substrate-triggered current under different N-well triggered currents is summarized in Figure 5.30(*b*). The switching voltage of the DTSCR can be reduced to a lower voltage level more efficiently if the substrate and N-well triggered currents are applied synchronously to the p- and n-trigger nodes, respectively. The I/O ESD protection circuit realized with DTSCR devices is shown in Figure 5.30(*c*).

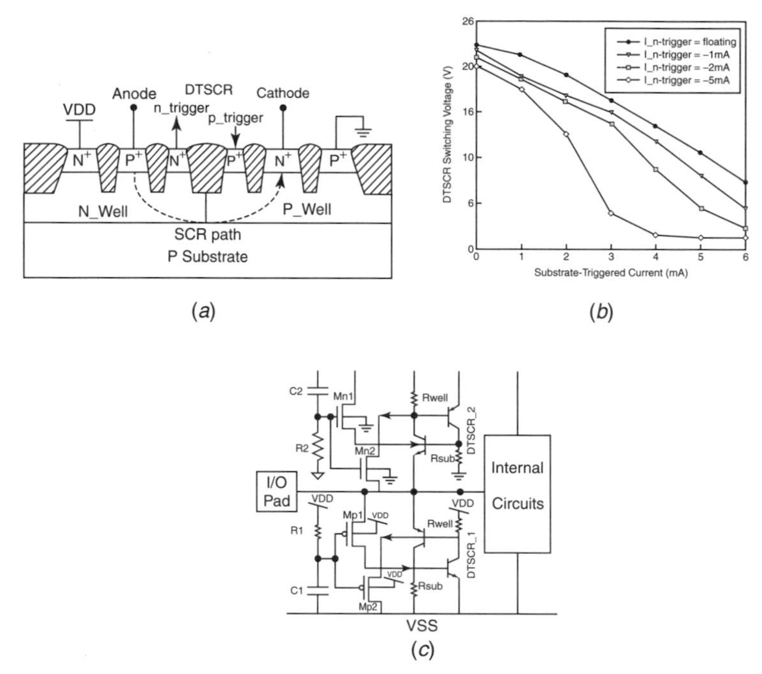

Figure 5.30 (*a*) Device structure of a double-triggered SCR (DTSCR). (*b*) Dependence of the switching voltage of a DTSCR device on substrate-triggered current under different N-well triggered currents in a 0.25-μm CMOS process. (*c*) ESD protection circuit with DTSCR devices for an I/O pad.

Based on the *RC* delay principle, the substrate and N-well triggered currents can be generated by Mp1 and Mp2 (Mn1 and Mn2), respectively, under PS (ND) ESD-zapping conditions. Therefore, the DTSCR with the double-triggered currents shown in Figure 5.30(*c*) can be triggered on more quickly to discharge ESD current. In normal circuit operating conditions with VDD and VSS power supplies, the gates of Mp1 and Mp2 (Mn1 and Mn2) are biased at VDD (VSS) through the resistor R_1 (R_2). Therefore, the Mp1, Mp2, Mn1, and Mn2 are all in the off state whenever the input signal is high or low. The p-trigger (n-trigger) node of the DTSCR device is kept at VSS (VDD) through the parasitic resistor R_{sub} (R_{well}), so such DTSCR is guaranteed to be kept off under normal circuit operating conditions. From the experimental results in a 0.25-μm CMOS process, when a 0- to 5-V voltage pulse is applied to the anode of DTSCR, the turn-on time of the DTSCR is about 37 ns under a positive voltage pulse of 1.5 V at the p-trigger node of the DTSCR. But the turn-on time of the DTSCR can be further reduced, to about 12 ns, while a negative voltage pulse of −5 V is synchronously applied to its n-trigger node. The dummy-gate structure used to block the STI in the SCR device can be applied to the DTSCR structure to further reduce the switching voltage and to enhance the turn-on speed of the DTSCR more efficiently.

Native-NMOS-Triggered SCRs [95] Native NMOS is built directly in a lightly doped p-type substrate in a sub-$\frac{1}{4}$-μm CMOS process, whereas normal NMOS (PMOS) is in a heavily doped P-well (N-well) in P-substrate twin-well CMOS technology. The native NMOS and lateral SCR can be merged to form a new ESD protection device, native-NMOS-triggered SCR (NANSCR), which has the advantages of lower switching voltage and faster turn-on speed. The device structure of an NANSCR is illustrated in Figure 5.31(a). The gate of native NMOS is connected to a negative bias circuit (NBC) [96] to turn off the NANSCR under normal circuit operating conditions. A comparison of dc $I-V$ curves between NANSCR and LVTSCR is shown in Figure 5.31(b). The switching voltage (about 4 V) of an NANSCR with a channel length of 0.3 μm is smaller than that (about 5 V) of an LVTSCR with a channel length of 0.13 μm in a 0.13-μm silicide CMOS process under the same channel width. With the substrate-triggered technique, the switching voltage of an NANSCR can be reduced further with an increase in the W/L ratio of native NMOS. The ESD protection circuit for an input or output pad with NANSCR devices is shown in Figure 5.31(c). In normal circuit operating conditions, the gates of native NMOS in all NANSCR devices are biased by the same NBC to turn off the NANSCR devices. So the NANSCR devices NANSCR_1 and NANSCR_2 will not

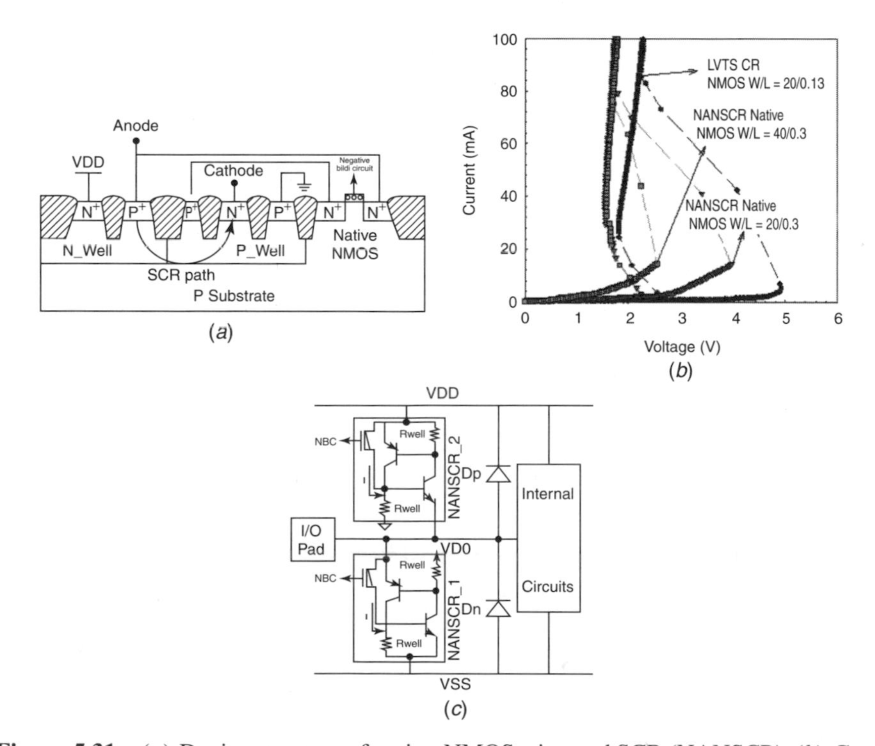

Figure 5.31 (a) Device structure of native-NMOS-triggered SCR (NANSCR). (b) Comparison of dc $I-V$ curves between NANSCR and LVTSCR in a 0.13-μm CMOS process. (c) ESD protection circuit with NANSCR devices.

interfere with the functions of I/O circuits. Under PS ESD-zapping conditions, the gate of native NMOS in NANSCR_1 is floating. NANSCR_1 is triggered on quickly by the substrate-triggering current generated from the already-on native NMOS. So the positive ESD current can be discharged quickly from the I/O pad through NANSCR_1 to the grounded VSS line. From the experimental results in a 0.13-μm CMOS process, the turn-on speed of NANSCR is faster than that of LVTSCR. Moreover, NANSCR can sustain a higher CDM ESD level (5 V/μm^2) than that (2.33 V/μm^2) of an LVTSCR. So it is more suitable for protecting the ultrathin gate oxide in nanoscale CMOS technologies.

5.6.3 SCR Latch-up Engineering

To make SCR-based devices with low enough switching voltage for effective ESD protection, transient-induced latch-up problems [87] must be avoided. There are two solutions to avoid having SCR-based devices with low switching voltage triggered on accidentally by a noise pulse when the CMOS ICs are under normal circuit operating conditions. Figure 5.32(*a*) shows one method for avoiding latch-up by increasing the triggering current of low-voltage-triggered SCR-based devices, but the switching and holding voltages are kept the same. With a higher triggering current, low-voltage-triggered SCR-based devices such as LVTSCRs have enough noise margin against overshooting or undershooting noise pulses on the pads. A high-current NMOS-triggered lateral SCR (HINTSCR) [99] device has been designed successfully by adding a bypass diode to the LVTSCR structure to increase its triggering current up to 218.5 mA in an 0.6-μm CMOS process. Such an HINTSCR has a noise margin greater than VDD + 12 V in 3-V applications. In addition, a high-holding-current SCR (HHI-SCR) device [89] modified from a GGSCR was reported with a holding current of about 70 mA in a 0.1-μm CMOS process by adjusting the external poly resistance from kilohms in a GGSCR to only about 10 Ω in an HHI-SCR.

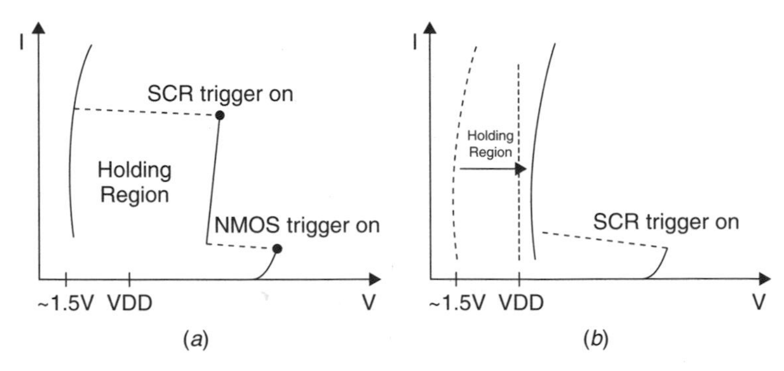

Figure 5.32 Two solutions to overcome latch-up issue in the ESD protection design with SCR-based device: (*a*) increasing the trigger current; (*b*) increasing holding voltage to avoid the SCR-based devices being accidentally triggered on by noise pulse.

Another method immune from latch-up is to increase the holding voltage of SCR-based devices to be greater than the maximum VDD voltage supply, as shown in Figure 5.32(*b*). By using an epitaxial substrate, the holding voltage of an SCR device can be increased to avoid latch-up problems [116]. But the fabrication cost of the CMOS wafer will also be increased. By stacking the voltage drop elements (such as diodes or SCR devices), an SCR-based device can elevate its total holding voltage in a bulk CMOS process. The switching voltage and current can still be kept at a lower voltage level by suitable trigger-assisted circuit design. A cascaded-LVTSCR [117] structure was designed to increase the holding voltage (>VDD) without degrading its ESD robustness in a 0.35-μm silicided CMOS technology. In addition, ESD protection circuits designed with stacked STSCR devices [90] or designed with an STSCR device and a stacked diode string [91] have been reported to have a 7-kV HBM ESD level and are free of latch-up problems in a 0.25-μm silicided CMOS technology. Recently, the holding voltage of a single SCR device has been adjusted dynamically for ESD protection (with a low holding voltage) and for normal circuit operations (with a higher holding voltage) [102]. A dynamic holding voltage SCR (DHVSCR) was reported to be an ESD protection device with high latch-up immunity. The device structure of an DHVSCR is shown in Figure 5.33(*a*). A PMOS and a NMOS are inserted into a DHVSCR device structure rather than an LSCR structure. The $I-V$ characteristics of the DHVSCR under normal circuit operating conditions and ESD-zapping conditions in a 0.25-μm CMOS process are shown in Figure 5.33(*b*). Under normal circuit operating conditions, the gates (V_{g1} and V_{g2}) of the PMOS and NMOS are biased at 2.5 V (VDD), but are biased at 0 V under ESD-zapping conditions. The holding voltage and holding current of the DHVSCR under normal circuit operating conditions are 2.8 V and 172 mA, respectively. Thus, the DHVSCR will not be kept in the latch-up state under normal circuit operating conditions. However, the holding voltage and holding current of the DHVSCR under ESD-zapping conditions drop to 2.2 V and 91 mA,

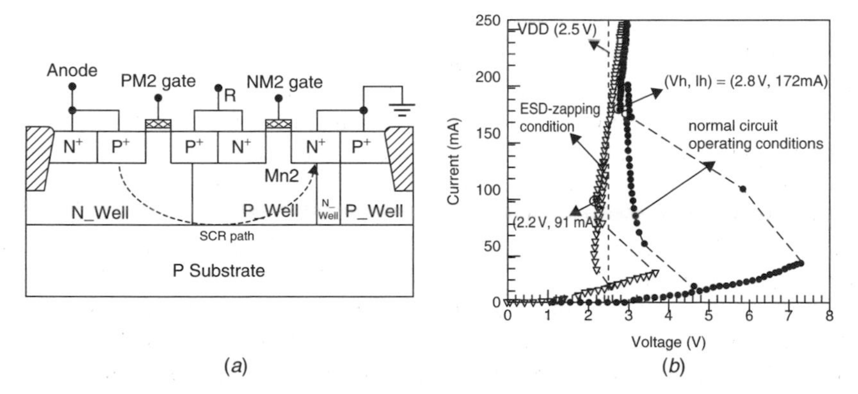

Figure 5.33 (*a*) Device structure of a dynamic holding voltage SCR (DHVSCR) in a CMOS process. (*b*) $I-V$ characteristics of the DHVSCR under normal circuit operating conditions and ESD-zapping conditions in a 0.25-μm CMOS process.

respectively. So the DHVSCR can also clamp the ESD overstress to a lower voltage level to sustain the higher ESD levels. The holding voltage and holding current of the DHVSCR can be adjusted by controlling the gate voltages of the PMOS and NMOS, which are merged with the SCR structure.

However, with scaled-down CMOS technologies, the power supply voltages in CMOS ICs have also been scaled down to follow the constant-field scaling requirement and to reduce power consumption. For CMOS IC products realized in a 0.13-μm silicided CMOS process, the maximum supply voltage for the internal circuit has been reduced to 1.2 V, so the latch-up concern inherent in SCR-based devices will certainly vanish. Therefore, SCR-based devices with lower switching voltages can be a great candidate for on-chip ESD protection, due to its highest ESD robustness, smallest layout area, and lack of latch-up danger compared with other ESD protection devices. However, in such a 0.13-μm CMOS process with ultrathin gate oxide, the turn-on speed of SCR devices should be enhanced to discharge ESD overstress voltage quickly for protecting such thinner gate oxide. Use of the NANSCR [95] and dummy-gate structure [92] to improve the turn-on speed of SCR-based devices will be a better choice to protect such an ultrathin gate oxide in nanoscale CMOS processes.

The switching voltage and turn-on speed of SCR-based devices must be suitably designed to fully and effectively protect the ultrathin gate oxide of input stages, especially against fast CDM ESD events. The switching voltage and turn-on speed of SCR-based devices will be the dominate factors in the overall performance of on-chip ESD protection circuits with SCR-based devices in nanoscale CMOS processes with a maximum voltage supply below 1.2 V. For a nanoscale CMOS process, SCR-based devices with a low-enough switching voltage and fast-enough turn-on speed are required to protect the ultrathin gate oxide of I/O circuits.

5.7 SUMMARY

With continued scaling of CMOS technologies into nanoscale dimensions, more circuits and functions have been integrated into a chip, such as the system on a chip (SoC). With scaled-down CMOS processes, the gate oxide becomes much thinner, which is more easily ruptured by ESD overstress. An SoC integrates more circuits and functions into a chip and often has a larger die size, which provides a larger body capacitance for storing CDM charges in the body of the IC. Additionally, SoCs often have a pin count of several hundreds. The layout area for each I/O cell in a high-pin-count SoC is significantly limited, which also limits the layout area for drawing ESD protection devices in such a high-pin-count SoC. An SoC often has multiple separated power pins, which cause unexpected ESD failures at the interface circuits between circuit blocks. Therefore, the ESD issue will become worst as we scale to deep-submicron processes, especially for CDM ESD events.

To effectively protect SoCs fabricated in nanoscale CMOS processes, the turn-on speed of ESD protection devices must be enhanced to discharge ESD current quickly before the internal circuits are damaged by ESD stresses. If an on-chip ESD clamp device (NMOS) has a negative threshold voltage (or a near-0-V threshold voltage), the ESD clamp device can stay in the already-on condition to discharge ESD current when the IC is being zapped by ESD pulses. But during normal circuit operating conditions, an extra negative bias can be used to turn off the ESD clamp device. With the already-on characteristics, the ultrathin gate oxide of SoCs in nanoscale CMOS processes can still be safely protected.

For an SoC with multiple separated power pins, a whole-chip ESD protection scheme is very important to avoid ESD failure located at internal or interface circuits of the SoC. The layout arrangement of power lines will have a strong impact on the ESD level of an SoC. Some suitable ESD connecting cells should be used to connect the separated power lines in the chip. Among ESD protection devices the SCR has the highest area efficiency to sustain ESD overstress. But an SCR with a four-layer structure often has a slow turn-on speed, which could be too slow to protect an SoC with ultrathin gate oxide. If an SCR device with some modified design or triggering circuits can have a fast enough turn-on speed, it will be the best choice for ESD protection in the SoC with VDD voltage levels below 1.2 V, without causing latch-up danger.

REFERENCES

[1] *ESD Association Standard Test Method for Electrostatic Discharge Sensitivity Testing: Human Body Model-Component Level*, ESD STM 5.1, ESD Association, 2001. Rome, NY

[2] *ESD Association Standard Test Method for Electrostatic Discharge Sensitivity Testing: Machine Model-Component Level*, ESD STM 5.2, ESD Association, 1999. Rome, NY

[3] *ESD Association Standard Test Method for Electrostatic Discharge Sensitivity Testing: Charged Device Model-Component Level*, ESD STM 5.3.1, ESD Association, 1999. Rome, NY

[4] C. Duvvury, R. N. Rountree, and O. Adams, "Internal chip ESD phenomena beyond the protection circuit," *IEEE Trans. Electron Devices*, Vol. 35, pp. 2133–2139, 1988.

[5] M.-D. Ker, Whole-chip ESD protection design with efficient VDD-to-VSS ESD clamp circuit for submicron CMOS VLSI, *IEEE Trans. Electron Devices*, Vol. 46, No. 1, pp. 173–183, Jan. 1999.

[6] M.-D. Ker and H.-H. Chang, ESD bus lines in CMOS ICs for whole-chip ESD protection, U.S. patent 6,144,542, Nov. 2000.

[7] M.-D. Ker, ESD protection circuit for mixed mode integrated circuits with separated power pins, U.S. patent 6,075,686, June 2000.

[8] M.-D. Ker and H.-H. Chang, ESD protection scheme for mixed-voltage CMOS integrated circuits, U.S. patent 6,002,568, Dec. 1999.

[9] M.-D. Ker and H.-H. Chang, Whole-chip ESD protection for CMOS ICs using bi-directional SCRs, U.S. patent 6,011,681, Jan. 2000.

[10] C. Jiang, E. Nowak, and M. Manley, Process and design for ESD robustness in deep submicron CMOS technology, *Proceedings of the IEEE International Reliability Physics Symposium*, pp. 233–236, 1996.

[11] C. Duvvury, R. N. Rountree, and R. A. McPhee, ESD protection: design and layout issues for VLSI circuits, *IEEE Trans. Ind. Appl.*, Vol. 25, pp. 41–47, 1989.

[12] A. Stricker, D. Gloor, and W. Fichtner, Layout optimization of an ESD-protection n-MOSFET by simulation and measurement, *Proceedings of the EOS/ESD Symposium*, pp. 205–211, 1995.

[13] M.-D. Ker, C.-Y. Wu, and T.-S. Wu, Area-efficient layout design for CMOS output transistors, *IEEE Trans. Electron Devices*, Vol. 44, No. 4, pp. 635–645, Apr. 1997.

[14] M.-D. Ker, T.-Y. Chen, and H.-H. Chang, New layout design for submicron CMOS output transistors to improve driving capability and ESD robustness, *J. Microelectron. Reliab.*, Vol. 39, No. 3, pp. 415–424, June 1999.

[15] M. Mergens, K. Verhaege, C. Russ, J. Armer, P. Jozwiak, G. Kolluri, and L. Avery, Multi-finger turn-on circuits and design techniques for enhanced ESD performance and width-scaling, *Proceedings of the EOS/ESD Symposium*, pp. 1–11, 2001.

[16] T.-Y. Chen and M.-D. Ker, Analysis on the dependence of layout parameters on ESD robustness of CMOS devices for manufacturing in deep-submicron CMOS process, *IEEE Trans. Semicond. Manuf.*, Vol. 16, No. 3, pp. 486–500, Aug. 2003.

[17] C. Jiang, E. Nowak, and M. Manley, Process and design for ESD robustness in deep submicron CMOS technology, *Proceedings of the IEEE International Reliability Physics Symposium*, pp. 233–236, 1996.

[18] C. Duvvury, C. Diaz, and T. Haddock, Achieving uniform nMOS device power distribution for sub-micron ESD reliability, *Technical Digest, IEEE International Electron Devices Meeting*, pp. 131–134, 1992.

[19] C. Duvvury and C. Diaz, Dynamic gate coupling of NMOS for efficient output ESD protection, *Proceedings of the IEEE International Reliability Physics Symposium*, pp. 141–150, 1992.

[20] M.-D. Ker, C.-Y. Wu, T. Cheng, and H.-H. Chang, Capacitor-couple ESD protection circuit for deep-submicron low-voltage CMOS ASIC, *IEEE Trans. VLSI Syst.*, Vol. 4, No. 3, pp. 307–321, 1996.

[21] H.-H. Chang and M.-D. Ker, Improved output ESD protection by dynamic gate floating design, *IEEE Trans. Electron Devices*, Vol. 45, No. 9, pp. 2076–2078, Sept. 1998.

[22] J. Chen, A. Amerasekera, and C. Duvvury, Design methodology and optimization of gate-driven NMOS ESD protection circuits in submicron CMOS processes, *IEEE Trans. Electron Devices*, Vol. 45, No. 12, pp. 2448–2456, 1998.

[23] M.-D. Ker, T.-Y. Chen, and C.-Y. Wu, Design of cost-efficient ESD clamp circuits for the power rails of CMOS ASICs with substrate-triggering technique, *Proceedings of the IEEE International ASIC Conference and Exhibit*, pp. 287–290, 1997.

[24] M.-D. Ker, T.-Y. Chen, C.-Y. Wu, H. Tang, K.-C. Su, and S.-W. Sun, Novel input ESD protection circuit with substrate-triggering technique in a 0.25-μm shallow-trench-isolation CMOS technology, *Proceedings of the IEEE International Symposium on Circuits and Systems*, Vol. 2, pp. 212–215, 1998.

[25] C. Duvvury, S. Ramaswamy, V. Gupta, A. Amerasekera, and R. Cline, Substrate pump NMOS for ESD protection applications, *Proceedings of the EOS/ESD Symposium*, pp. 7–17, 2000.

[26] M.-D. Ker, T.-Y. Chen, and C.-Y. Wu, Substrate-triggered ESD clamp devices for using in power-rail ESD clamp circuits, *Solid-State Electron.*, Vol. 46, No. 5, pp. 721–734, Apr. 2002.

[27] M.-D. Ker and T.-Y. Chen, Substrate-triggered ESD protection circuit without extra process modification, *IEEE J. Solid-State Circuits*, Vol. 38, No. 2, pp. 295–302, Feb. 2003.

[28] M.-D. Ker and T.-Y. Chen, Substrate-triggered technique for on-chip ESD protection design in a 0.18-μm salicided CMOS process, *IEEE Trans. Electron Devices*, Vol. 50, No. 4, pp. 1050–1057, Apr. 2003.

[29] T.-Y. Chen and M.-D. Ker, Investigation of the gate-driven effect and substrate-triggered effect on ESD robustness of CMOS devices, *IEEE Trans. Device Mater. Reliab.*, Vol. 1, No. 4, pp. 190–203, Dec. 2001.

[30] M.-D. Ker and H.-C. Jiang, Whole-chip ESD protection strategy for CMOS integrated circuits in nanotechnology, *Proceedings of the IEEE International Conference on Nanotechnology*, pp. 325–330, 2001.

[31] J.-S. Lee, Method for fabricating an electrostatic discharge protection circuit, U.S. patent 5,672,527, Sept. 1997.

[32] T.-Y. Huang, Method for making an integrated circuit structure, U.S. patent 5,529,941, June 1996.

[33] C.-C. Hsue and J. Ko, Method for ESD protection improvement, U.S. patent 5,374,565, Dec. 1994.

[34] R.-Y. Shiue, C.-S. Hou, Y.-H. Wu, and L.-J. Wu, ESD implantation scheme for 0.35 μm 3.3 V 70Λ gate oxide process, U.S. patent 5,953,601, Sept. 1999.

[35] T. Lowrey and R. Chance, Static discharge circuit having low breakdown voltage bipolar clamp, U.S. patent 5,581,104, Dec. 1996.

[36] J.-J. Yang, Electrostatic discharge protection circuit employing MOSFETs having double ESD implantations, U.S. patent 6,040,603, Mar. 2000.

[37] M.-D. Ker and C.-H. Chuang, ESD implantations in 0.18-μm salicided CMOS technology for on-chip ESD protection with layout consideration, *Proceedings of the International Symposium on the Physical and Failure Analysis of Integrated Circuits*, pp. 85–90, 2001.

[38] M.-D. Ker, T.-Y. Chen, and H.-H. Chang, ESD implantation method in deep-submicron CMOS technology for high-voltage-tolerant applications with light-doping concentrations, U.S. patent 6,514,839, Feb. 2003.

[39] M.-D. Ker, H.-C. Hsu, and J.-J. Peng, ESD implantation for sub-quarter-micron CMOS technology to enhance ESD robustness, *IEEE Trans. Electron Devices*, in press, Volume 50, Issue 10, Oct 2003, pg 2126–2134. Oct. 2003.

[40] K. Mistry, N-channel clamp for ESD protection in self-aligned silicided CMOS process, U.S. patent 5,262,344, Nov. 1993.

[41] H. Ooka, Semiconductor integrated circuit device including two types of MOSFETs having source/drain region different in sheet resistance from each other, U.S. patent 5,283,449, Feb. 1994.

[42] T. Randazzo and B. Larsen, Input/output transistors with optimized ESD protection, U.S. patent 5,493,142, Feb. 1996.

[43] J.-J. Wang, P.-C. Shieh, and P.-N. Tseng, Method of forming a resistor for ESD protection in self aligned silicide process, U.S. patent 5,547,881, Aug. 1996.

[44] J.-S. Lee, Method for fabricating an electrostatic discharge protection circuit, U.S. patent 5,672,527, Sept. 1997.

[45] M. Ma, ESD protection using selective siliciding techniques, U.S. patent 5,744,839, Apr. 1998.

[46] T.-Y. Chen, M.-D. Ker, and H.-H. Chang, Method of fabricating ESD protection device by using the same photolithographic mask for both the ESD implantation and the silicide blocking regions, U.S. patent 6,444,404, Sept. 3, 2002.

[47] D. Scott, P. Bosshart, and J. Gallia, Circuit to improve electrostatic discharge protection, U.S. patent 5,019,888, May 1991.

[48] K. Lee, A. Lee, M. Marmet, and K. Ouyang, Electro-static discharge protection circuit with bimodal resistance characteristics, U.S. patent 5,270,565, Dec. 1993.

[49] G.-L. Lin and M.-D. Ker, Fabrication of ESD protection device using a gate as a silicide blocking mask for a drain region, U.S. patent 6,046,087, Apr. 2000.

[50] C.-S. Kim et al., A novel NMOS transistor for high performance ESD protection devices in 0.18 μm CMOS technology utilizing salicide process, *Proceedings of the EOS/ESD Symposium*, pp. 407–412, 2000.

[51] A. Amerasekera and C. Duvvury, The impact of technology scaling on ESD robustness and protection circuit design, *Proceedings of the EOS/ESD Symposium*, pp. 237–245, 1994.

[52] S. Daniel and G. Krieger, Process and design optimization for advanced CMOS I/O ESD protection devices, *Proceedings of the EOS/ESD Symposium*, pp. 206–213, 1990.

[53] S. G. Beebe, Methodology for layout design and optimization of ESD protection transistors, *Proceedings of the EOS/ESD Symposium*, pp. 265–275, 1996.

[54] I. E. Opris, Bootstrapped pad protection structure, *IEEE J. Solid-State Circuits*, pp. 300–301, Feb. 1998.

[55] M.-D. Ker, T.-Y. Chen, C.-Y. Wu, and H.-H. Chang, ESD protection design on analog pin with very low input capacitance for high-frequency or current-mode applications, *IEEE J. Solid-State Circuits*, Vol. 35, No. 8, pp. 1194–1199, Aug. 2000.

[56] *Star–Hspice User's Manual: Applications and Examples*, Avanti, Campbell, CA 1998.

[57] M.-D. Ker, H.-C. Jiang, and C.-Y. Chang, Design on the low-capacitance bond pad for high-frequency I/O circuits in CMOS technology, *IEEE Trans. Electron Devices*, Vol. 48, No. 12, pp. 2953–2956, Dec. 2001.

[58] M.-D. Ker and C.-Y. Chang, ESD protection design for CMOS RF integrated circuits using polysilicon diodes, *J. Microelectron. Reliab.*, Vol. 42, No. 6, pp. 863–872, June 2002.

[59] M.-D. Ker and T.-Y. Chen, Layout design to minimize voltage-dependent variation on input capacitance of an analog ESD protection circuit, *J. Electrostat.*, Vol. 54, No. 1, pp. 73–93, Jan. 2002.

[60] T. Furukawa, D. Turner, S. Mittl, M. Maloney, R. Serafin, W. Clark, J. Bialas, L. Longenbach, and J. Howard, Accelerated gate-oxide breakdown in mixed-voltage I/O circuits, *Proceedings of the IEEE International Reliability Physics Symposium*, pp. 169–173, 1997.

[61] E. Takeda and N. Suzuki, An empirical model for device degradation due to hot-carrier injection, *IEEE Electron Device Lett.*, Vol. 4, pp. 111–113, 1983.

[62] S. H. Voldman, ESD protection in a mixed voltage interface and multi-rail disconnected power grid environment in 0.5- and 0.25-μm channel length CMOS technologies, *Proceedings of the EOS/ESD Symposium*, pp. 125–134, 1994.

[63] S. Dabral and T. J. Maloney, *Basic ESD and I/O Design*, Wiley, New York, 1998.

[64] M. Hargrove et al., High-performance sub-0.08 μm CMOS with dual gate oxide and 9.7 ps inverter delay, *Technical Digest, IEEE International Electron Devices Meeting*, pp. 22.4.1–22.4.4, 1998.

[65] S. Poon et al., A versatile 0.25-micron CMOS technology, *Technical Digest, IEEE International Electron Devices Meeting*, pp. 28.2.1–28.2.4, 1998.

[66] M. Takahash, T. Sakurai, K. Sawada, K. Nogami, M. Ichida, and K. Matsud, 3.3 V–5 V compatible I/O circuit without thick gate oxide, *Proceedings of the IEEE Custom Integrated Circuits Conference*, pp. 23.3.1–23.3.4, 1992.

[67] M. Pelgrom and E. Dijkmans, A 3/5 V compatible I/O buffer, *IEEE J. Solid-State Circuits*, Vol. 30, pp. 823–825, 1995.

[68] J. Conner, D. Evans, G. Braceras, J. Sousa, W. Abadeer, S. Hall, and M. Robillard, Dynamic dielectric protection for I/O circuits fabricated in a 2.5-V CMOS technology interfacing to a 3.3-V LVTTL bus, *Technical Digest, International Symposium on VLSI Circuits*, pp. 119–120, 1997.

[69] G. Singh and R. Salem, High-voltage-tolerant I/O buffers with low-voltage CMOS process, *IEEE J. Solid-State Circuits*, Vol. 34, pp. 1512–1525, 1999.

[70] H. Sanchez, J. Siegel, C. Nicoletta, J. Nissen, and J. Alvarez, A versatile 3.3/2.5/1.8-V CMOS I/O driver built in a 0.2-μm 3.5-nm Tox 1.8-V CMOS technology, *IEEE J. Solid-State Circuits*, Vol. 34, pp. 1501–1511, 1999.

[71] A.-J. Annema, G. Geelen, and P. de Jong, 5.5-V I/O in a 2.5-V 0.25-μm CMOS technology, *IEEE J. Solid-State Circuits*, Vol. 36, pp. 528–538, 2001.

[72] M.-D. Ker and C.-S. Tsai, Design of 2.5 V/5 V mixed-voltage CMOS I/O buffer with only thin oxide device and dynamic *n*-well bias circuit, *Proceedings of the 2003 IEEE International Symposium on Circuits and Systems*, Vol. 5, pp. 97–100, 2003.

[73] T. Maloney and W. Kan, Stacked PMOS clamps for high voltage power supply protection, *Proceedings of the EOS/ESD Symposium*, pp. 70–77, 1999.

[74] W. Anderson and D. Krakauer, ESD protection for mixed-voltage I/O using NMOS transistors stacked in a cascode configuration, *Proceedings of the EOS/ESD Symposium*, 1998, pp. 54–71.

[75] J. Miller, M. Khazhinsky, and J. Weldon, Engineering the cascoded NMOS output buffer for maximum V_{t1}, *Proceedings of the EOS/ESD Symposium*, pp. 308–317, 2000.

[76] T. J. Maloney and S. Dabral, Novel clamp circuits for IC power supply protection, *Proceedings of the EOS/ESD Symposium*, pp. 1–12, 1995.

[77] S. H. Voldman, G. Gerosa, V. Gross, N. Dickson, S. Furkay, and J. Slinkman, Analysis of snubber-clamped diode-string mixed voltage interface ESD protection network for advanced microprocessors, *Proceedings of the EOS/ESD Symposium*, pp. 43–61, 1995.

[78] T. J. Maloney, K. Parat, N. Clark, and A. Darwish, Protection of high voltage power and programming pins, *Proceedings of the EOS/ESD Symposium*, pp. 246–254, 1997.

[79] M.-D. Ker and W.-Y. Lo, Design on the low-leakage diode string for using in the power-rail ESD clamp circuits in a 0.35-μm silicide CMOS process, *IEEE J. Solid-State Circuits*, Vol. 35, pp. 601–611, 2000.

[80] M.-D. Ker and C.-H. Chuang, Stacked-NMOS triggered silicon-controlled rectifier for ESD protection in high/low-voltage-tolerant I/O interface, *IEEE Electron Device Lett.*, Vol. 23, No. 6, pp. 363–365, June 2002.

[81] M.-D. Ker and C.-H. Chuang, ESD protection design for mixed-voltage CMOS I/O buffers, *IEEE J. Solid-State Circuits*, Vol. 37, No. 8, pp. 1046–1055, Aug. 2002.

[82] A. Chatterjee and T. Polgreen, A low-voltage triggering SCR for on-chip ESD protection at output and input pads, *IEEE Electron Device Lett.*, Vol. 12, pp. 21–22, 1991.

[83] M.-D. Ker and C.-Y. Wu, Modeling the positive-feedback regenerative process of CMOS latch-up by a positive transient pole method: I. Theoretical derivation, *IEEE Trans. Electron Devices*, Vol. 42, pp. 1141–1148, 1995.

[84] M.-D. Ker and C.-Y. Wu, Modeling the positive-feedback regenerative process of CMOS latch-up by a positive transient pole method: II. Quantitative evaluation, *IEEE Trans. Electron Devices*, Vol. 42, pp. 1149–1155, 1995.

[85] M.-D. Ker, C.-Y. Wu, and H.-H. Chang, Complementary-LVTSCR ESD protection circuit for submicron CMOS VLSI/ULSI, *IEEE Trans. Electron Devices*, Vol. 43, pp. 588–598, 1996.

[86] M.-D. Ker, H.-H. Chang, and C.-Y. Wu, A gate-coupled PTLSCR/NTLSCR ESD protection circuit for deep-submicron low-voltage CMOS IC's, *IEEE J. Solid-State Circuits*, Vol. 32, pp. 38–51, 1997.

[87] G. Weiss and D. Young, Transient-induced latch-up testing of CMOS integrated circuits, *Proceedings of the EOS/ESD Symposium*, pp. 194–198, 1995.

[88] C. Russ, M. P. J. Mergens, J. Armer, P. Jozwiak, G. Kolluri, L. Avery, and K. Verhaege, GGSCR: GGNMOS triggered silicon controlled rectifiers for ESD protection in deep submicron CMOS processes, *Proceedings of the EOS/ESD Symposium*, pp. 22–31, 2001.

[89] M. P. J. Mergens, C. C. Russ, K. G. Verhaege, J. Armer, P. C. Jozwiak, and R. Mohn, High holding current SCRs (HHI-SCR) for ESD protection and latch-up immune IC operation, *Proceedings of the EOS/ESD Symposium*, pp. 10–17, 2002.

[90] M.-D. Ker and K.-C. Hsu, Substrate-triggered SCR device for on-chip ESD protection in fully silicided subquarter-micrometer CMOS process, *IEEE Trans. Electron Devices*, Vol. 50, pp. 397–405, 2003.

[91] M.-D. Ker and K.-C. Hsu, Latch-up-free ESD protection design with complementary substrate-triggered SCR devices, *IEEE J. Solid-State Circuits*, Vol. 38, No. 8, pp. 1380–1392, Aug. 2003.

[92] K.-C. Hsu and M.-D. Ker, Improvement on turn-on speed of substrate-triggered SCR device by using dummy-gate structure for on-chip ESD protection, *Proceedings of the International Conference on Solid State Devices and Materials*, pp 440–441, Sept. 16–18 2003.

[93] M.-D. Ker and G.-L. Lin, Low-voltage-triggered electrostatic discharge protection device and relevant circuitry, U.S. patent 6,465,848, Oct. 2002.

[94] M.-D. Ker and K.-C. Hsu, SCR device with double-triggered technique for on-chip ESD protection in sub-quarter-micron silicided CMOS process, *IEEE Trans. Device Mater. Reliab.*, in press, Volume 3, Issue 3, Sept. 2003, pg 58–68. Sept. 2003.

[95] M.-D. Ker and K.-C. Hsu, Native-NMOS-triggered SCR (NANSCR) for ESD protection in 0.13-μm CMOS integrated circuits, Proc. of 2004 IEEE International Reliability Physics Symposium (IRPS), Phoenix, Arizona, 2004, pp. 381–386, April 25–29.

[96] M.-D. Ker, C.-Y. Chang, and H.-C. Jiang, Design of negative charge pump circuit with polysilicon diodes in a 0.25-μm CMOS process, *Proceedings of the IEEE AP-ASIC Conference*, pp. 145–148, 2002.

[97] A. Z. H. Wang and C.-H. Tsay, An on-chip ESD protection circuit with low trigger voltage in BiCMOS technology, *IEEE J. Solid-State Circuits*, Vol. 36, pp. 40–45, 2001.

[98] M.-D. Ker and C.-Y. Wu, CMOS on-chip electrostatic discharge protection circuit using four-SCR structures with low ESD-trigger voltage, *Solid-State Electron.*, Vol. 37, pp. 17–26, 1994.

[99] M.-D. Ker, Lateral SCR devices with low-voltage high-current triggering characteristics for output ESD protection in submicron CMOS technology, *IEEE Trans. Electron Devices*, Vol. 45, pp. 849–860, 1998.

[100] G. Notermans, F. Kuper, and J.-M. Luchis, Using an SCR as ESD protection without latch-up danger, *Microelectron. Reliab.*, Vol. 37, pp. 1457–1460, 1997.

[101] M.-D. Ker and H.-H. Chang, How to safely apply the LVTSCR for CMOS whole-chip ESD protection without being accidentally triggered on, *Proceedings of the EOS/ESD Symposium*, pp. 72–85, 1998.

[102] Z.-P. Chen and M.-D. Ker, Dynamic holding voltage SCR (DHVSCR) device for ESD protection with high latch-up immunity, *Proceedings of the International Conference on Solid State Devices and Materials*, in press, pp. 160–161 Sept. 16–18 2003.

CHAPTER 6

INPUT/OUTPUT DESIGN

6.1 INTRODUCTION

Design of input and output buffers has become increasingly difficult for the more advanced processes. Designers face a huge number of variables that must be prioritized when deciding how to design I/O buffers for a given process. Some of the issues that must be considered include:

1. I/O standards
2. Signal transfer between the core logic and the I/O buffers, assuming that the two are at different voltage levels
3. An electrostatic discharge (ESD) protection scheme
4. Performance requirements
5. Pin capacitance (which is tightly coupled with issues 3 and 4)
6. Electromigration
7. I/O switching noise
8. Layout
9. Termination
10. Impedance matching
11. Preemphasis
12. Equalization

Nano-CMOS Circuit and Physical Design, by Ban P. Wong, Anurag Mittal, Yu Cao, and Greg Starr
ISBN 0-471-46610-7

At each process node, the performance of I/O buffers has had to increase. This has become progressively more difficult given how fast the core performance has scaled relative to the I/O buffers. This is driven in large part because the I/O standards are typically at a higher voltage than the core supply. This invariably requires a thicker oxide device (or device cascading), resulting in reduced drive strength for an equivalently sized gate. Larger output devices may be required to meet edge-rate specifications; therefore, the pin capacitance is higher because of the increased diffusion area, which ultimately slows down the edge rates. Eventually, a saturation point is reached where increasing the buffer size does not result in faster edge rates. In the following sections we summarize some of these issues and present current trends among IC designers to improve I/O performance on these more advanced processes. This chapter is on the basis of a standard CMOS process. If a more exotic process is used, such as the silicon–germanium process, a whole host of new design issues arise.

6.2 I/O STANDARDS

Numerous I/O standards have been generated over the past 30 years. Some of the major standards include advanced graphics port (AGP), current mode logic (CML), emitter-coupled logic (ECL), positive referenced emitter-coupled logic (PECL), Gunning transceiver logic (GTL), high-speed transceiver logic (HSTL), stub series terminated logic (SSTL), low-voltage CMOS (LVCMOS), low-voltage differential signal (LVDS), hypertransport (which is a subset of the LVDS standard), low-voltage positive referenced emitter-coupled logic (LVPECL), and low-voltage transistor–transistor logic (LVTTL). Many of these standards are single ended and have severe performance limitations. The major I/O standards being used for truly high-speed interfaces are HSTL, SSTL, LVDS, LVPECL (and other forms of ECL), and CML. CML buffers are used in the highest-speed applications.

Two basic signaling schemes are used for I/O buffers today: single ended and differential. Single-ended standards are limited in operating speeds up to approximately 600 to 800 Mbps, which is driven primarily by memory interfaces although some techniques have been looked at to go to higher speeds [5]. The differential standards can provide data rates of 1 to 1.5 Gbps for the LVDS standard and up to 10 Gbps (and potentially higher) for the CML standard. The differential I/O standards can be broken into two basic categories: parallel buses and serial links. The LVDS standard is more typically a parallel bus, whereas CML is a serial link. Figure 6.1 shows a typical high-speed serial system. Parallel data are sent to the chip on the left side using LVDS as the standard. The data stream is encoded using an 8B/10B standard before transmitting on the serial side. This encoding scheme is used to guarantee $0 \rightarrow 1$ and $1 \rightarrow 0$ transitions for clock recovery and maintain dc balance by having equal numbers of 0's and 1's on average. This encoding scheme is the most popular form of line coding, although is does add some overhead to the bandwidth. For example, if 8B/10B

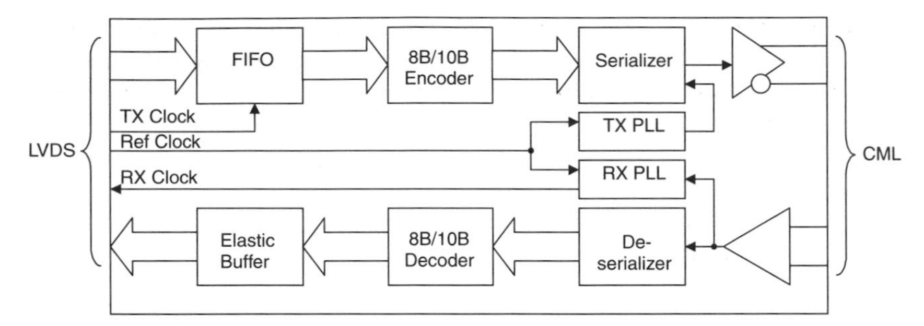

Figure 6.1 Typical serial system diagram with LVDS signaling for the parallel data and CML signaling for the serial data.

encoding is used and the desired data rate is 1 Gbps, the actual transmission rate must be 1.25 Gbps to account for the overhead associated with the encoding.

6.3 SIGNAL TRANSFER

One of the first decisions that must be made is what oxide thickness will be used for the I/Os. Several thickness may be available for a given process node, but minimizing the number of options can greatly reduce the overall cost of the product from both a processing cost and yield standpoint. In most cases, using two oxides allows the best trade-off between core performance and the voltage levels supported by the I/O. In general, using the thinnest oxide possible for the I/Os allows the highest performance because of the greater current drive capability of a smaller device, which can greatly reduce the pin capacitance. If the core operates at a voltage different from that at which the I/Os operate, transfer between these two power domains can become a problem. This issue was discussed at length in Chapter 4. In many cases, the I/Os are operated on a totally separate supply and ground than the core, to prevent the I/O noise from coupling into the core. Figure 6.2 is a simplified diagram of a potential scheme for transferring data from the core to the output buffer. Here, the signal is still

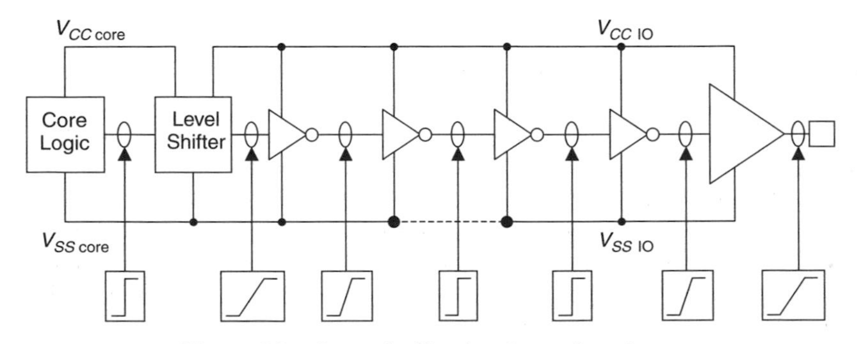

Figure 6.2 Output buffer signal transfer scheme.

referenced relative to the quieter supply of the core until the edge rate can be increased before transferring to the noisier ground of the output buffer. Ground bounce effects can have a huge impact on signal jitter when transferring between the two grounds, but maintaining a fast edge rate can mitigate this effect. A second approach is to create a very localized short between the IO ground ($V_{SS\ \mathrm{IO}}$) and the core ground ($V_{SS\ \mathrm{core}}$). This is shown by the dashed line between these two nodes in Figure 6.2. This localized short provides a path for the current to return to the driving source, which can reduce the ground bounce effects. This approach has some risk, since unwanted current can flow from the I/O ground to the core ground, creating unwanted noise in the core. The critical concept to take away from this section is that the translation between the core voltage and the I/O voltage is not simple and must be given careful consideration. Simulations *must* be run to access the overall performance of this translation stage. A good design can produce a sensitivity factor in the range 20 to 100 ps/V.

6.3.1 Single-Ended Buffers

The basic structure of single-ended buffers has not changed significantly with the scaling process. The primary change has come from the higher data rates that must be supported at each technology node. The performance issue can be reduced if a fixed interface will be used, such as a 1.8-V SSTL. This allows the output buffers to be designed using thinner oxide devices, which affords higher performance and lower capacitance. A more typical case is one requiring the chip to interface with several different I/O standards, requiring I/Os that are compliant with 3.3- and 1.5-V I/Os. Designers must either use an oxide thickness that can support the higher voltage and suffer performance problems for the lower voltage standards or use a thinner oxide device and create a complex structure to avoid the devices from an overvoltage condition for the 3.3-V standards. The higher threshold voltage of the thicker oxide devices and the lower supply voltage required by the interface cause a performance problem. Figure 6.3 is a simplified diagram of a stacked output buffer block diagram from Ref. 4. This architecture was developed for using a 1.8-V transistor to support 1.8-, 2.5-, and 3.3-V I/O standards. Stacked I/O structures offer the possibility of supporting multiple I/O standards with a single thinner oxide device that offers potentially improved I/O performance. By providing the appropriate bias voltage at P_{bias} and N_{bias}, it is possible to generate a controlled impedance for matching the board-level signal trace and load, which will reduce reflection and improve signal integrity. Taking advantage of the stacked structure and using one set of devices to achieve the desired impedance can improve output impedance control. Some of the problems with this approach include hot-socket protection, ESD protection, and tristating the output when the buffer is not being used.

6.3.2 Differential Buffers

Differential buffers are the only option for truly high-speed data transfer. Figure 6.4 is a simplified diagram of an LVDS output buffer with common-mode

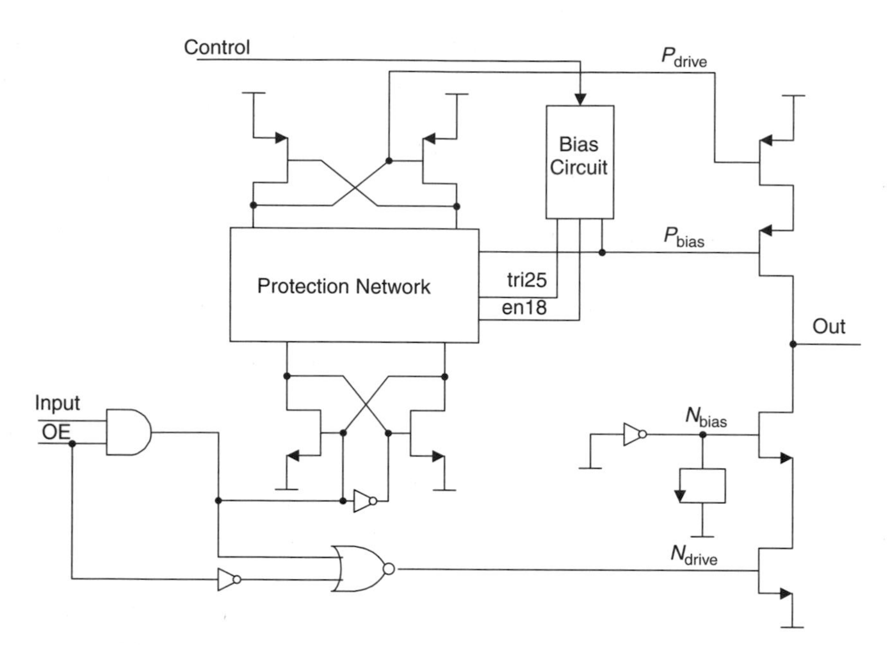

Figure 6.3 Stacked output buffer block diagram. (From Ref. 4.)

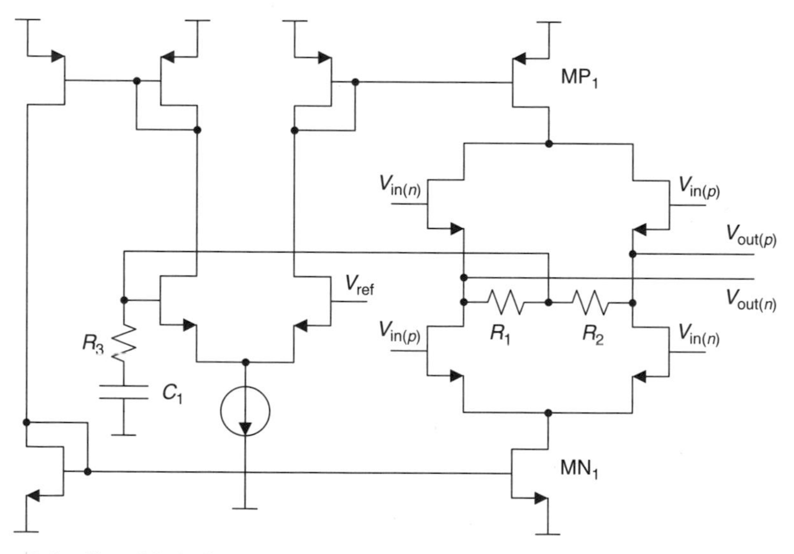

Figure 6.4 Simplified diagram of an LVDS output buffer with active common-mode feedback. (From Ref. 13.)

feedback [13]. The two resistors, R_1 and R_2, are used to provide the common-mode level of the output buffer to the feedback amplifier that adjusts the bias voltage of MN_1 and MP_1 to maintain a constant common-mode voltage. The reference voltage is typically generated by a bandgap reference that can be scaled to provide the desired common-mode voltage. Use of a bandgap reduces the common-mode variation and supply noise sensitivity. This technique is reasonably popular in actual design because the feedback loop provides good common-mode voltage control. The values of R_1 and R_2 must be large to avoid loading the output buffer. If at all possible, adding cascading to the two current sources (MP_1 and MN_1) can further improve the performance of the buffer because of the superior current matching, and increase the output impedance. Ensuring that the two current sources (MP_1 and MN_1) are equal is important for the overall performance of the buffer. Refer to Chapter 10 for more details on minimizing device mismatch. Mismatch between these two current sources leads to a shift in the common-mode voltage of the output signal.

Current Mode Logic Buffer Given the high performance of CML buffers, they have gained wide acceptance for many applications, including most high-speed serial interfaces, because of their ability to reach speeds of 10 Gbps and potentially higher on a standard CMOS process [3]. This trend does not appear to be changing for the 90-nm node and potentially, the 65-nm node. Current-mode logic buffers can be either ac or dc coupled, although ac coupling tends to be more popular. Figure 6.5 is a simplified diagram of a tapered CML output buffer. Tapering is used because the final stage typically drives a large load. To reduce the delay through the buffer, tapering is implemented similar to an inverter chain. The minimum delay occurs when the delay through each stage is equal [11,12]. An inductor is inserted on the intermediate stages to increase the edge rate of the signals by delaying the current through the resistor to allow faster capacitor charging [6]. These inductors are in the range 2 to 10 nH. These inductors can become large and are difficult to characterize, so they are often implemented by using an active load that acts like an inductor. The capacitors that couple the gate of one device to the drain of the opposite device are used to reduce the gate drain overlap capacitance effect, which causes input signal coupling to the output node. The addition of these capacitors helps reduce this effect. Given the speed requirements typical of this buffer type, NMOS devices are used because

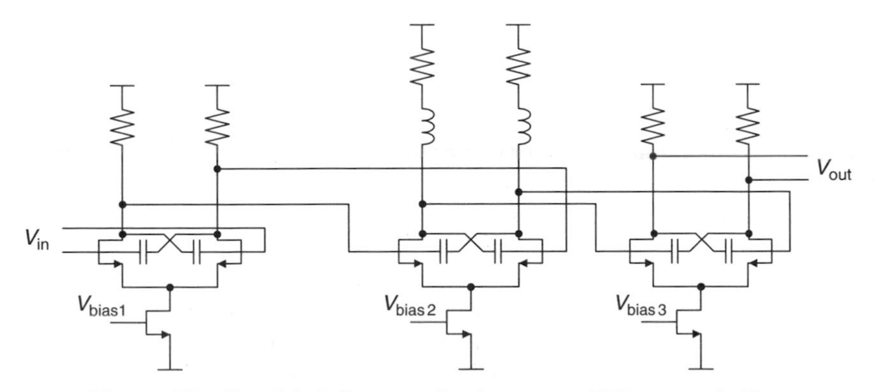

Figure 6.5 Simplified diagram of a three-stage CML output buffer.

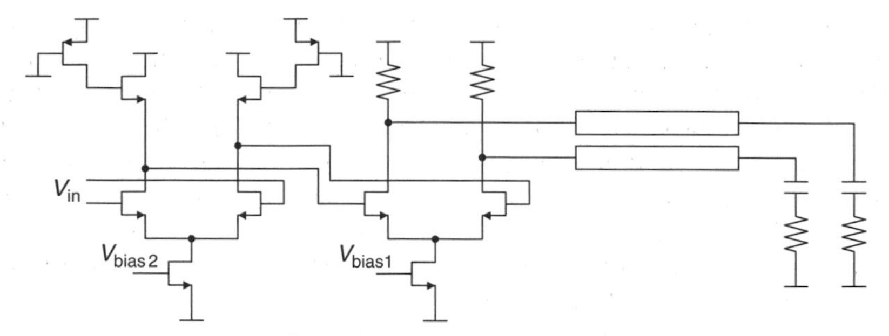

Figure 6.6 CML buffer driving an off-chip load that is ac coupled.

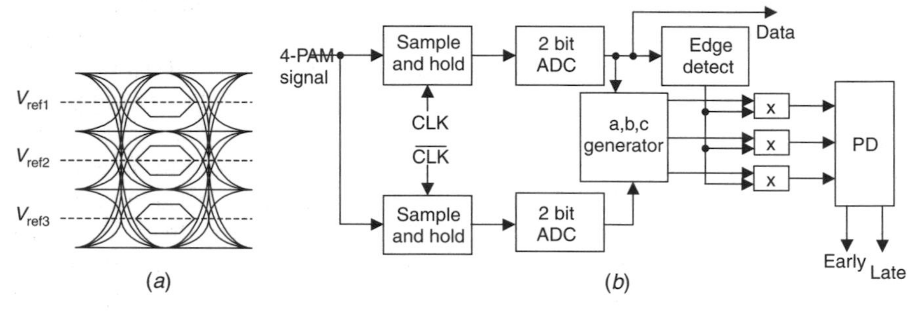

Figure 6.7 Multilevel pulse amplitude modulation signaling waveform and system diagram.

of their increased drive capability. These devices must be operated in saturation at all times to ensure high-speed performance.

Figure 6.6 is a simplified diagram of the final stage of a CML buffer driving a load that is ac coupled. Use of ac coupling allows incompatible voltage levels to be connected together. If ac coupling is used, the data must be encoded using a standard such as 8B/10B to prevent the common-mode voltage from shifting around. If the receiver has any delay sensitivity to the common-mode voltage, any variation in the common-mode voltage will manifest itself as jitter on the signal received.

Figure 6.7(*a*) is a simplified diagram of a multilevel pulse amplitude modulation scheme that is used to increase the effective bandwidth of an output and input buffer by encoding four levels per transmission. This waveform is constructed by overlaying multiple bit transitions on top of one another. This approach allows the data rate to be effectively doubled without having to increase the frequency by a factor of 2. Of course, this scheme complicates the input and output buffer significantly since four discrete voltage levels must be transmitted and received rather than the two-level scheme discussed previously. The clock recovery unit becomes more complicated since these four levels must be decoded to determine the phase of the incoming signal relative to a reference clock. One of the big drawbacks to this signaling scheme is that there is no generally accepted standard, making proliferation of this technique difficult at this time. Figure 6.7(*b*) is a simplified diagram of a 4-pulse amplitude modulation (PAM) receiver that utilizes an Alexander phase detector [10], which is part of a clock data recovery system. A two-bit analog-to-digital converter is used to sample the data to determine the effective level to decode it into two bits of data.

6.4 ESD PROTECTION

ESD is discussed in Chapter 5, so it is mentioned only briefly here. ESD protection on these advanced processes has not changed for the thicker oxide devices. The same protection schemes used on 0.5-, 0.35-, and 0.25-μm processes are still valid. The major changes in the ESD protection scheme have centered on protecting I/O constructed using thinner oxide devices, given their greater susceptibility to ESD. A second goal has been how to reduce the additive capacitance from ESD protection at the input and output buffers to improve the overall system speed. It is also important to make sure that the capacitance does not vary with the applied voltage; otherwise, it will induce additional jitter on the signal since the delay will vary as a function of the voltage.

Figure 6.8 is a simplified diagram of how output buffers can be designed to improve ESD performance [9]. This structure does not use the minimum gate to contact spacing, but rather, relies on spacing that can be considerably greater than the minimum design rule: in most cases, greater than 1 μm. This technique effectively adds a resistor between the contact and the drain and source that acts as a ballast to improve the ESD performance of the device. This figure shows the resistor on both the source and drain sides of the device, but there is some indication that having it on the drain side is sufficient for protecting the device because the series resistance reduces the drain voltage, which ultimately reduces the stress on the gate. This resistor also acts to reduce the drain current during the second breakdown during snapback [15]. In any event, increasing the contact-to-gate spacing increases the capacitance at the output of the device, thereby degrading the performance of the output buffer. A second drawback to this approach is the large area required for the output buffer. A final drawback is the requirement to have the area unsalicided, which adds mask layers.

A second approach for protecting the output buffers is to use the approach discussed in Ref. 14. Here, a silicon-controlled rectifier (SCR) structure is added

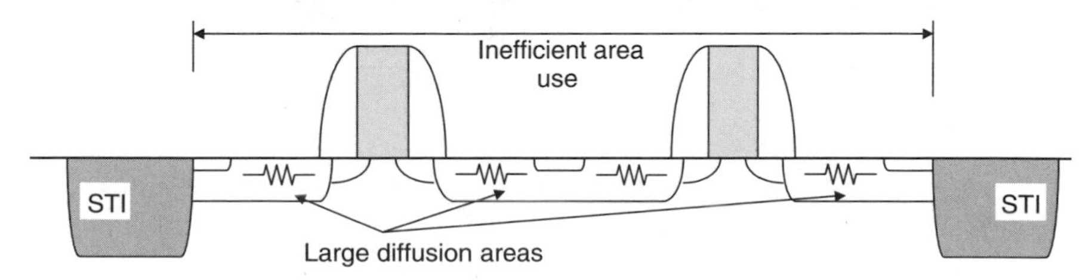

Figure 6.8 Simplified diagram of an output buffer transistor with two fingers.

to the output pad to protect the devices connected to the pad from an ESD event. This scheme works well as long as the SCR structure trigger voltage occurs before the output device breakdown voltage. Using this scheme allows the output devices to be drawn much smaller, reducing the pin capacitance. A low-capacitance SCR structure can be employed to further reduce the pin capacitance.

6.5 I/O SWITCHING NOISE

The switching noise of I/O buffers has become an even bigger issue on these advanced processes because of the higher I/O counts, increased edge rates, and increased operating frequencies. Previous generations of designs were able to rely on decoupling provided at the board level, but these high-speed I/Os now *require* on-die decoupling to meet the stringent requirements of modern advanced systems. This point is illustrated in Figure 6.9, which shows one output of a

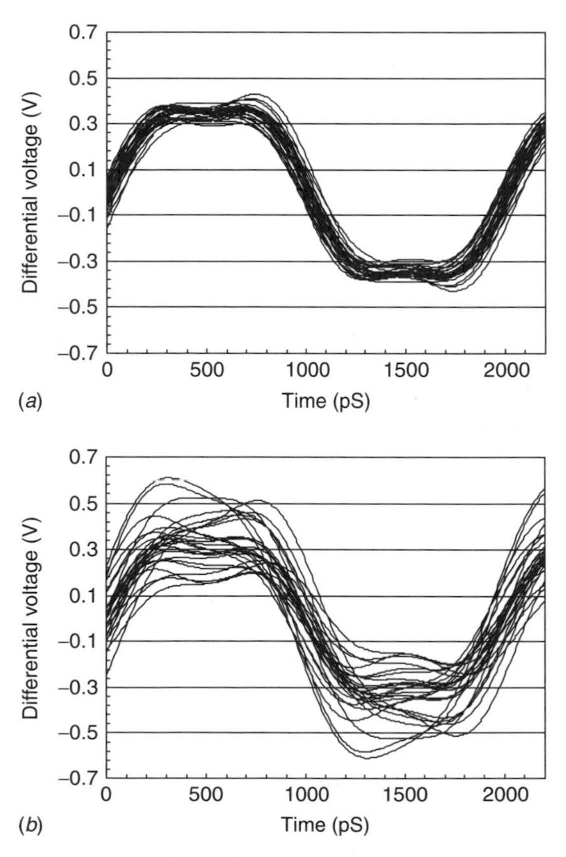

Figure 6.9 Simulation of a 16-bit-wide high-speed parallel interface with a maximum bit rate of 1 Gbps (*a*) with and (*b*) without on-die decoupling.

high-speed interface that is toggling at the maximum data rate (1 Gbps) while the remaining 15 I/Os toggle with a pseudorandom bit pattern. The intersymbol interference is apparent for the case where no on-die decoupling is included. Without this on-die decoupling, supply droop and ground bounce can become significant enough to cause output buffers to fail, especially when traversing from all I/Os driving from a high to a low or a low to a high. Efficient board-level decoupling has become difficult to implement because of the high pin counts, making it difficult to place decoupling capacitors close enough to the device, especially if off-chip termination is required. Additionally, even given the significant improvements with device packages to reduce parasitic resistances and inductances, they are still not sufficient for current I/O needs. Again, it cannot be stressed enough that on-die I/O decoupling is absolutely required on these advanced processes if the I/O must support any of the high-speed interfaces.

Figure 6.10 shows a simplified diagram of a model for an I/O bank with eight I/Os and the associated parasitic components. The package parasitic capacitance has been neglected. This simplified diagram is expanded to enable simulation of a more realistic I/O bank size of 68 I/Os within a flip-chip package. The I/O switching effects on the supply voltage at the I/O are shown in Figures 6.11 and 6.12. This model grouped I/Os into clusters of four to simplify the simulation. Package and die parasitics were estimated based on a physical design of a 90-nm process. Varying amounts of decoupling are allocated per I/O to illustrate the importance of on-die decoupling. The two figures show the results of adding 9 pF per I/O. The plot of the average supply voltage shown in Figure 6.11 reveals that the power bus is not balanced precisely, which is why the *IR* drop is greater at one end than the other, although the supply difference is quite small, so the average supply voltage is reasonably constant. The maximum supply voltage variation shown in Figure 6.12 is calculated by subtracting the V_{CC} and V_{SS} supply levels at each buffer cluster (four buffers in one group to form the cluster) and finding

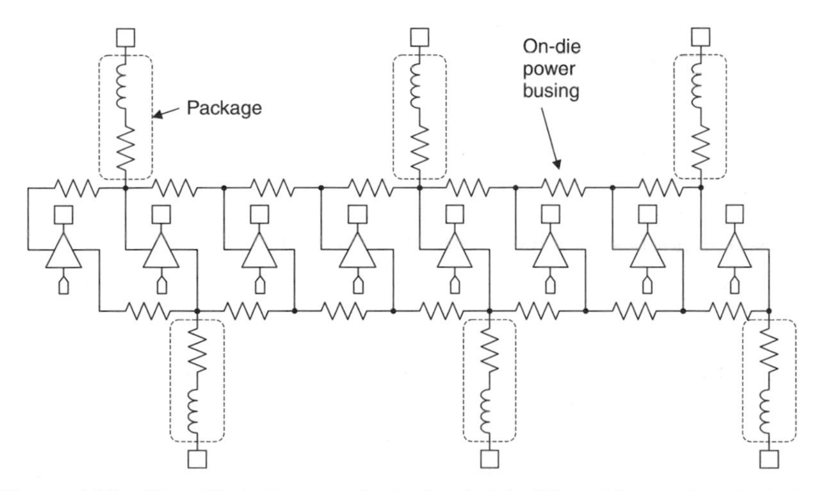

Figure 6.10 Simplified diagram of a bank of eight I/Os with parasitics included.

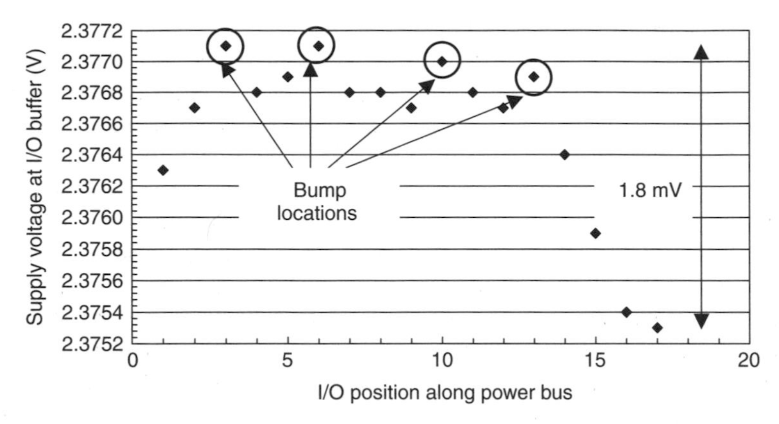

Figure 6.11 Average supply voltage across the I/O when half the I/Os switch in a bank of 68 I/Os at a frequency of 100 MHz with 9 pF per I/O buffer.

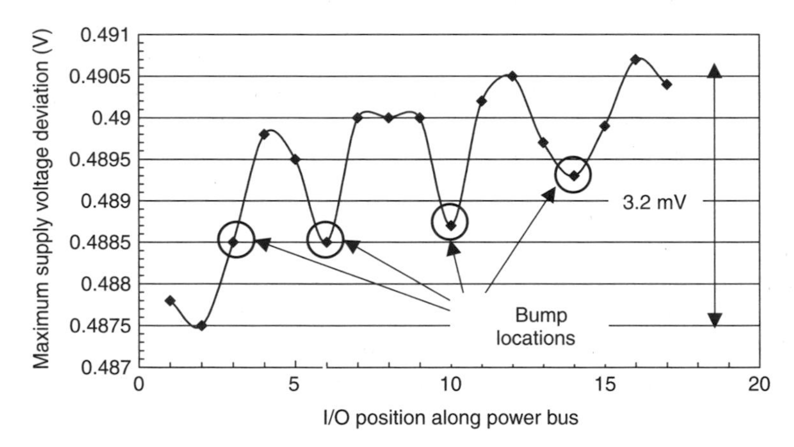

Figure 6.12 Maximum supply voltage variation when half of the I/Os switch in a bank of 68 I/Os with 9 pF per I/O buffer.

the maximum deviation. Table 6.1 summarizes the effects of various amounts of decoupling on the I/O bank supply voltage.

It is apparent that the localized inductive and resistive drop on an I/O bank can become significant based on these simulations. On-die decoupling is the only way to effectively minimize this *IR* drop, even when low-inductive packages are used. To design an I/O system successfully, the following steps are recommended.

1. Estimate the current draw of each I/O. It is important to capture the correct timing of the current draw to permit accurate modeling. This should be done for all process corners.
2. Determine the amount of metalization available for the I/O power bus. This is typically based on the previous-generation power busing. This allows

TABLE 6.1 Summary of the Supply Voltage Variation as a Function of the Amount of Decoupling

Decoupling per I/O (pF)	Average Supply Voltage (V)	Maximum Supply Voltage Deviation (V)
4.5	2.38	1.03
9.0	2.38	0.49
18.0	2.38	0.26

construction of a realistic resistive network for the I/O power bus. This step must be coupled with step 1 to ensure that electromigration rules are not violated. Electromigration must be monitored during this stage of the development to ensure that adequate power busing is available at all levels (on-die, bumps, package, and balls).

3. Estimate the bump routing to the I/O power bus to generate a resistive connection between the bump and the I/O power mesh. This assumes that flip-chip packaging is used, but the same approach applies for wire-bond packages.
4. Develop/acquire models for the bumps. The bumps can have resistances in the range 0.01 to 0.1 Ω and inductances in about the 0.1 nH range, so they cannot be neglected.
5. Estimate the package resistive network. Even if a plane is used in the package, there will be some resistance and inductance.
6. Develop/acquire models for the package balls. The balls can have resistances in the range 0.01 to 0.1 Ω and inductances in the 0.2 nH+ range, so they cannot be neglected.

Once these tasks have been completed, trade-off studies can be performed to look at the effects of increasing or decreasing the I/O power bus, adding bumps, reducing the routing between the bumps and the power bus, increasing the package routing/plane, and adding additional balls. Relative placement of the bumps and balls can be made based on this analysis. Additionally, trade-offs can be made between adding additional on-die decoupling capacitance, power busing, and package bumps and balls. This analysis must be done at the start of the project in order to make the appropriate trade-offs.

A generic model for I/O switching noise effects that accounts for negative feedback from the increase in supply noise is presented in Ref. 16. Here the supply noise can be determined by using

$$V_n = V_k + \frac{T}{L_p}\frac{p}{nK}\left(1 - \sqrt{1 + 2\,V_k\frac{L_p n K}{pT}}\right)$$

where V_k is $V_{cc} - V_{tn}$, T is the rise time for the output buffer to reach its maximum value, L_p the inductance per pin or ball, p the number of pins or balls, n the number of output buffers switching at the same time, and $K = \mu_n C_{ox}(W/L)$. This simple equation does not provide the necessary accuracy for doing detailed design of an I/O power bank, but the general form of the equation is quite useful for defining the supply noise effects of multiple I/O switching at the same time. By modifying the equation so that it has the form

$$V_n = A + \frac{B}{n}(1 - \sqrt{1 + Cn})$$

and then using SPICE simulations to determine the coefficients A, B, and C, it is possible to have a simple expression to determine the effects of I/O switching noise.

6.6 TERMINATION

Termination has become a major issue for large systems on a single chip. Figure 6.13 is a simplified diagram of some single-ended termination schemes. This termination is used to match the impedances between the driver, the transmission line, and the load to minimize reflections that are caused by impedance discontinuities. These reflections will manifest themselves as intersymbol interference. With I/O counts in the hundreds for a reasonably sized ASIC, it has become extremely difficult, if not impossible, to provide adequate

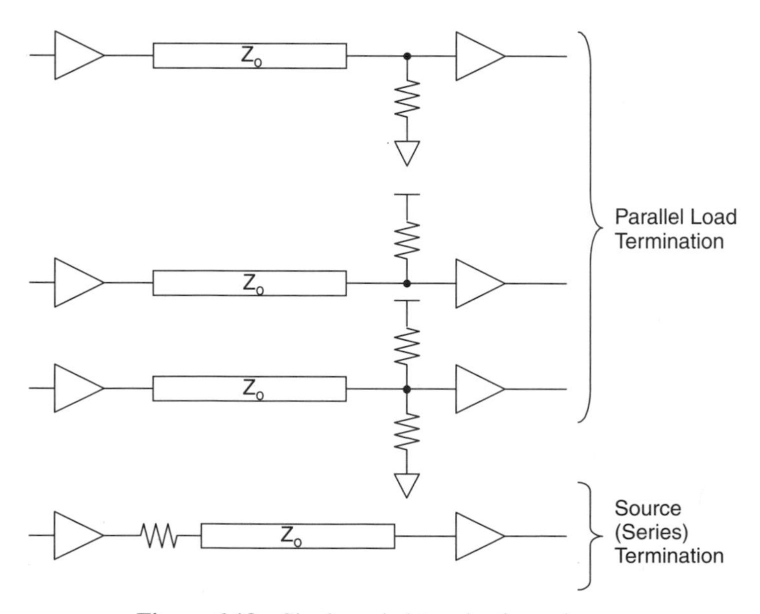

Figure 6.13 Single-ended termination schemes.

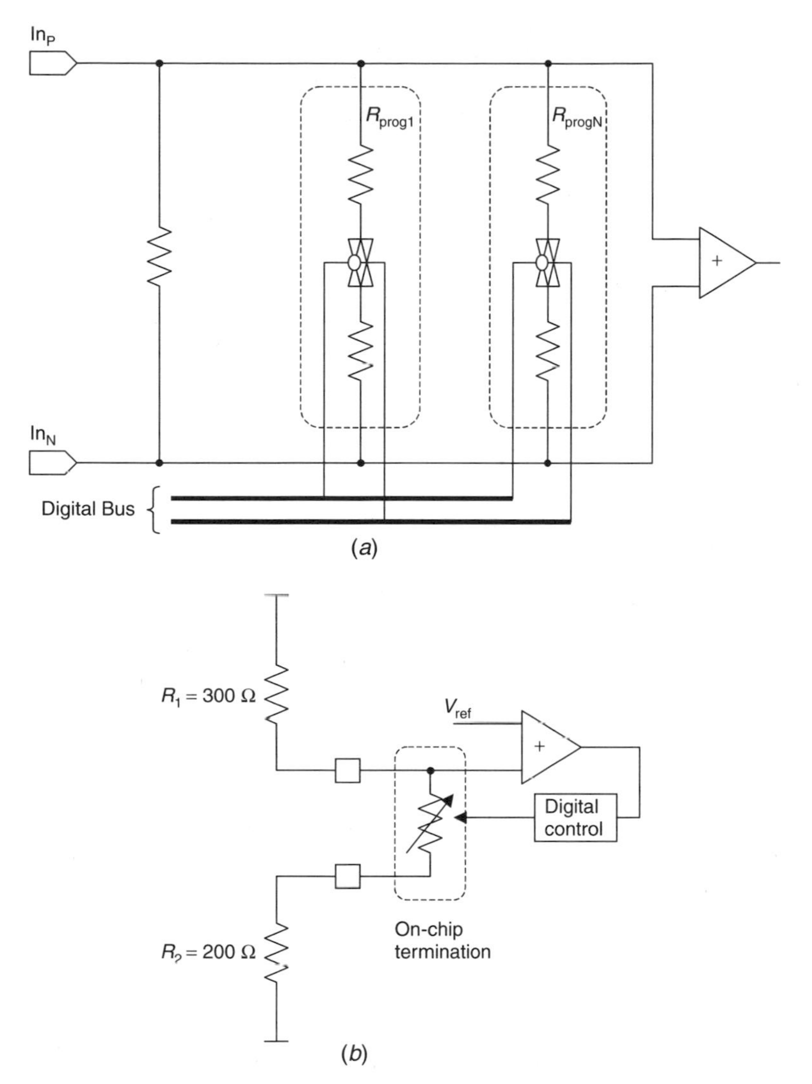

Figure 6.14 Simplified diagram of a variable input termination scheme (*a*) and the circuit required to calibrate the resistor setting (*b*).

termination on the board for these high-speed interfaces because of the congestion near these dense parts, high I/O speed, and remaining package and board stub, which creates another discontinuity. This constraint has forced IC designers to incorporate the termination on-die, especially for very high speed interfaces. This trend is accelerating at each technology node. Figure 6.14(*a*) is a simplified diagram of a potential input termination scheme for a differential input buffer. The termination resistor is located across the inputs to the differential input buffer. A digital bus is used to set the appropriate number of resistors in parallel to achieve the desired termination resistor value. Unsilicided poly is selected for the resistors because of their low sensitivity to voltage, manageable parasitic capacitance, and reasonable accuracy, although they can still vary by ±20%. For this reason, parallel resistors are used to compensate out the process and temperature variation. This implementation implies that some type of circuit must be used to sense and compensate out these factors. One scheme commonly used is to utilize external resistors to create a reference voltage that includes an on-chip version of the termination circuit shown in Figure 6.14(*b*). Although this scheme works, it does require two additional pins and couples the accuracy of external components to the overall circuit performance. A second approach is to use trimming to produce accurate references. This approach can allow removal of the extra pins but adds considerable complexity and additional cost since a trim scheme must be implemented using some type of fuse technology (metal or poly). Again, the critical items to be monitored when adding on-chip termination are:

1. *Process, supply, and temperature variation.* Sensitivity to these parameters must be minimized by at least ±10%.
2. *Additive pin capacitance.* This must be minimized; otherwise, it will limit the maximum operating speed of the interface.
3. *Capacitance as a function of voltage.* This must be minimized and should be less than 5% of the total pin capacitance to ensure that the effects do not significantly affect interface performance.

It is critical to provide an accurate termination impedance to maintain matching with the driving source and transmission line. It is also becoming important to have the termination resistance configurable to allow the end user to select the appropriate termination for their particular application.

6.7 IMPEDANCE MATCHING

Impedance matching is critical to ensure that good signal integrity is maintained when an output buffer is driving a transmission line. If the impedances do not match, the reflection that will occur will degrade the overall system performance. This issue has only become important for advanced processes because of the higher data rates and reduced I/O timing budgets. Several schemes have been used to provide a programmable output impedance [7,8]. An example of one of these circuits is shown in Figure 6.15. Six control signals are used to set the output impedance of the buffer, ϕ_1 through ϕ_6. This same scheme can be used to control the slew rate as well as by staging when the individual devices turn on.

This circuit is not without its problems since the output impedance of the driver can vary significantly over process, temperature, and supply voltage. Again, the only way to counteract this variation is with some type of calibration circuit to offset these effects. The effects on the overall signal integrity of process,

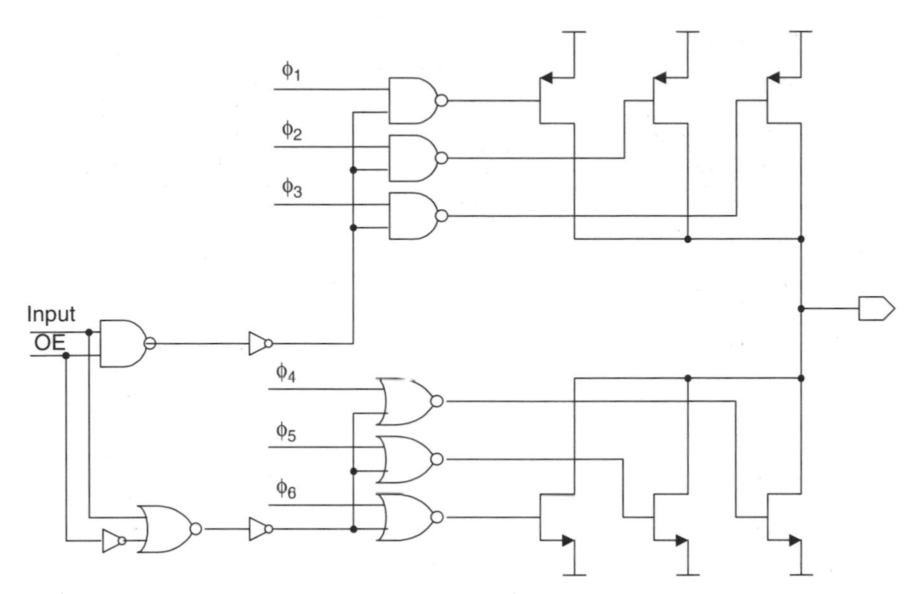

Figure 6.15 Simplified diagram of an impedance control circuit for an output buffer. (From Ref. 7.)

temperature, and supply should be accessed to determine if calibration is required to meet system timing specifications.

6.8 PREEMPHASIS

Two techniques are used to increase the I/O operating frequency, preemphasis, and equalization. For frequencies above 1 Gbps, both techniques are becoming important to include, and for 1.5 Gbps and above they are absolutely required. Preemphasis has moved from an academic topic to the mainstream product in the last few years, due mainly to the increased I/O bandwidth requirements dictated by the higher performance of core logic. Preemphasis is basically the concept of providing an overdrive of the current any time the data signal makes a transition.

The concept of preemphasis is illustrated in Figures 6.16 and 6.17. Figure 6.16 is a simplified diagram of a transmitter, a receiver, and an interconnect between the two that could consist of the PC board, connectors, or cables. The transmit chip does not have preemphasis and transmits the waveform shown below the transmitter. The resulting waveform at the receiver end is shown below the receiver chip. The waveform received illustrates the effects of the losses through a PC board or cables since they act as a low-pass filter because of the greater losses at the higher frequencies.

Figure 6.17 shows the same system configuration except that the transmitting chip includes preemphasis. The additional drive current occurs at each point where the data make a high → low or low → high transition. This added drive

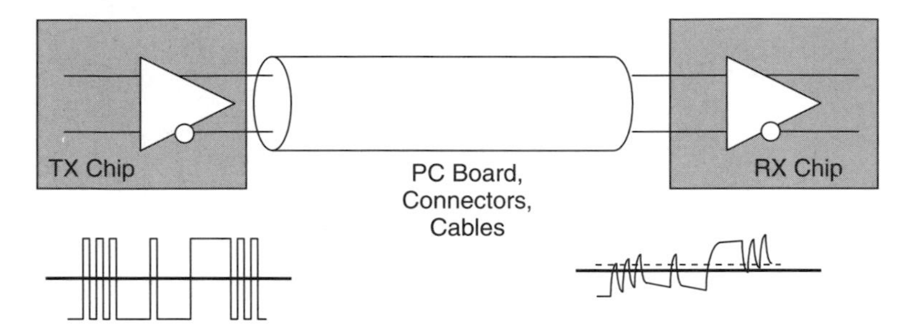

Figure 6.16 High-speed interface between two chips without preemphasis.

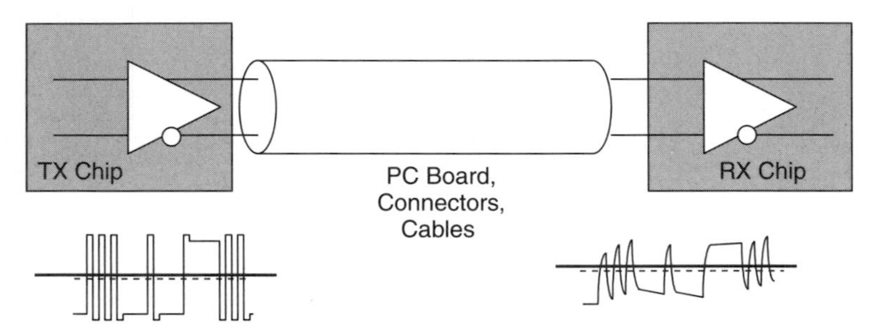

Figure 6.17 High-speed interface between two chips with preemphasis.

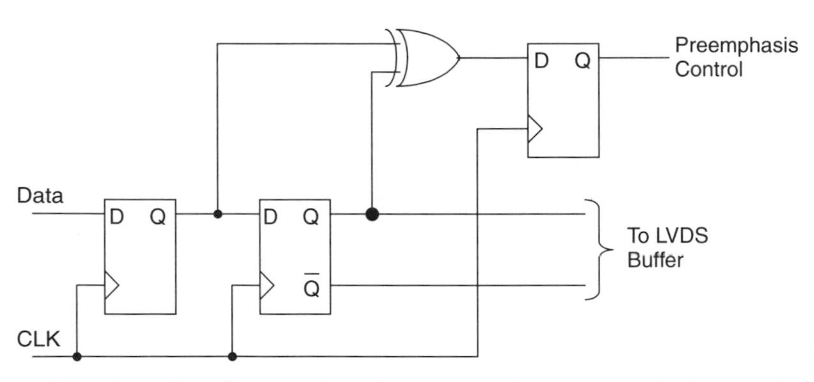

Figure 6.18 Simplified diagram of circuit for generating the preemphasis control signal.

current helps to increase the voltage swing at the receiver, making it easier for the receiver to recover the data.

Preemphasis control can be achieved by using a circuit similar to the one shown in Figure 6.18. The data are sampled in the serializer to determine when a transition occurs and a control signal is produced for enabling the additional current drive to the output buffer that is shown in Figure 6.19. The two transistors, MP_2 and MN_2, are enabled only when a transition occurs on the data to effectively

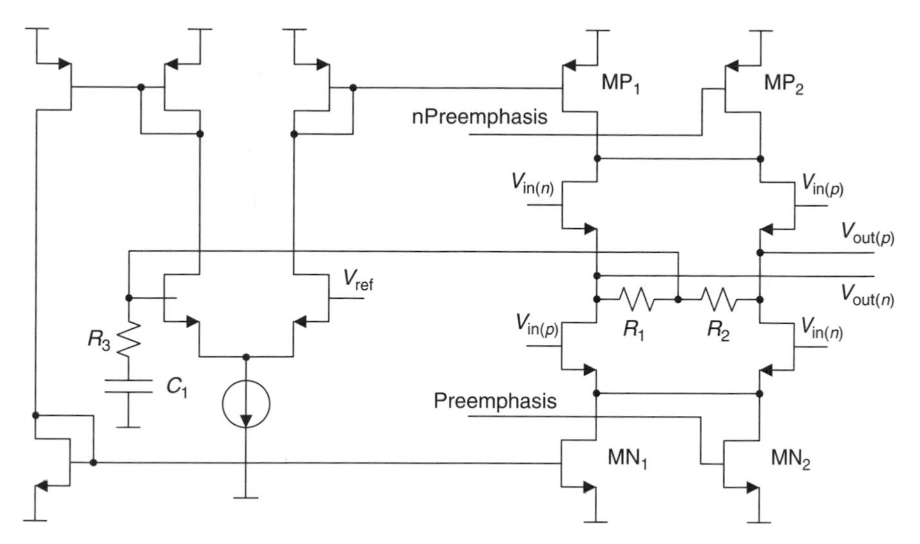

Figure 6.19 Simplified diagram of an LVDS output buffer with preemphasis added.

increase the current drive of the buffer, which increases the voltage swing. This simple example shows a single device being used to control the preemphasis, but in reality, multiple devices are used to allow the amount of preemphasis to be configured based on the specific application the devices go into. Process, supply, and temperature effects must also be considered when designing a preemphasis scheme to ensure that this circuit provides proper tracking to maintain good signal integrity.

6.9 EQUALIZATION

Whereas preemphasis is used for the transmit side of the data link, equalization is used on the receiver side of the link. Equalization has also become a necessary part of any high-speed input buffer design. Use of preemphasis with equalization provides the best overall solution for meeting high-speed timing requirements. Figure 6.20 is a simplified circuit for implementing equalization. The equalization scheme functions by monitoring the present data ($V_{\text{in(n)}}$) and the previously data ($V_{\text{in(n-1)}}$) through a sample-and-hold circuit. The drain current through MN_1 shown in Figure 6.20 can be expressed as

$$I_{D\text{MN1}} = g_{m\text{MP1}} \lfloor V_{\text{in(n)}} - V_{\text{ref}} \rfloor - g_{m\text{MP3}} \lfloor V_{\text{in(n-1)}} - V_{\text{ref}} \rfloor$$

that can be simplified to

$$I_{D\text{MN1}} = g_{m\text{MP1}} \{ \lfloor V_{\text{in(n)}} - V_{\text{ref}} \rfloor - \alpha \lfloor V_{\text{in(n-1)}} - V_{\text{ref}} \rfloor \}$$

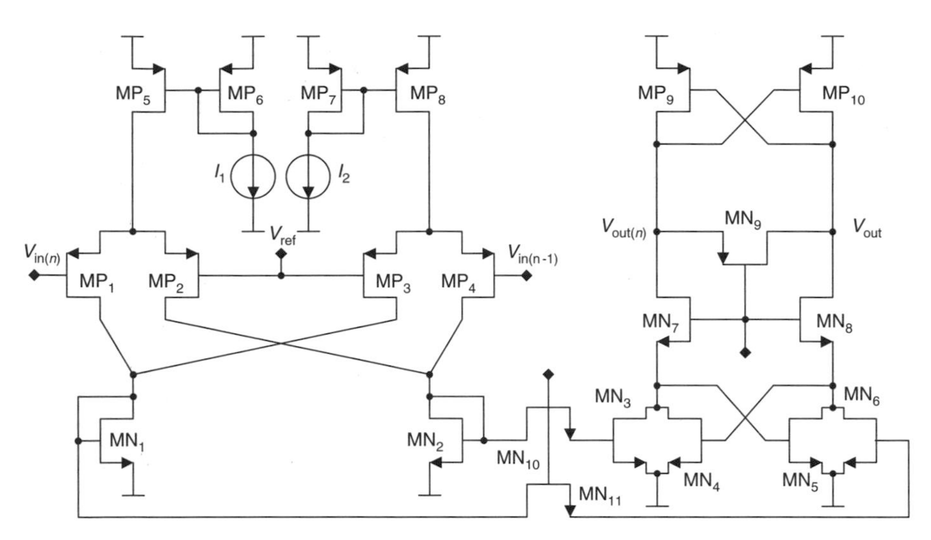

Figure 6.20 Simplified diagram of an equalizer. (From Ref. 1.)

by making $\alpha = g_{m\text{MP3}}/g_{m\text{MP1}}$. The α parameter can be varied by scaling the tail currents, I_1 and I_2, or adjusting the device sizes of MP_3 and MP_4 relative to MP_1 and MP_2. It is best to make the α-parameter variable so that it can be optimized for a particular application. The addition of equalization has been shown to increase the data rate by approximately 10% [2].

6.10 CONCLUSION

I/O design on the more advanced processes has become progressively more difficult given the ever-increasing data rates required by the overall system. Designers must have a complete understanding of the end-application to ensure that stringent requirements are met. New circuit techniques must be employed to meet the higher IO bandwidth requirements such as pre-emphasis and equalization. Inductive peaking may be used in CML buffers to increase further the bandwidth. All of these techniques add to the complexity of the I/O buffer design.

To meet successfully the higher data rates requires faster edge rates that create more simultaneous switching noise. Designers are now forced to include on-die decoupling to help reduce supply droop and ground bounce. Die integration within the package must be considered carefully to minimize package parasitics. I/O buffers cannot be designed as a single entity. The entire I/O system must be considered to ensure it is fully optimized to provide the required performance while maintaining a small overall die area. Initial macro modeling should be done prior to starting detailed design to develop overall requirements for the entire system from the transistor level through the system level.

Future trends with I/O design show no sign of changing. Data rates will only increase. Serial interfaces are becoming more prevalent given their ability to deliver high-speed operation using less overall power and I/O pin counts. The higher data rates of these serial interfaces will require an even greater level of integration to ensure performance targets can be met. Additionally, modeling will become imperative as the end customers for the IC will be required to run system level simulations to verify their board and system level design.

REFERENCES

[1] J. Y. Sim, J. J. Nam, Y. S. Sohn, H. J. Park, C. H. Kim, and S. I. Cho, A CMOS transceiver for DRAM bus system with a demultiplexed equalization scheme, *IEEE J. Solid-State Circuits*, Vol. 37, pp. 245–250, Feb. 2002.

[2] B. S. Song and D. C. Soo, NRZ timing recovery technique for band-limited channels, *IEEE J. Solid-State Circuits*, Vol. 32, pp. 514–520, Apr. 1997.

[3] M. W. Allam and M. I. Elmasry, Dynamic current mode logic (DyCML): a new low-power high-performance logic style, *IEEE J. Solid-State Circuits*, Vol. 36, pp. 550–558, Mar. 2001.

[4] H. Sanchez, J. Siegel, C. Nicoletta, J. P. Nissen, and J. Alvarez, A versatile 3.3/2.5/1.8-V CMOS I/O driver built in a 0.2-μm, 3.5 nm Tox, 1.8 V CMOS technology, *IEEE J. Solid-State Circuits*, Vol. 34, pp. 1501–1511, Nov. 1999.

[5] R. Mooney, C. Dike, and S. Borkar, A 900 Mb/s bidirectional signaling scheme, *IEEE J. Solid-State Circuits*, Vol. 30, pp. 1538–1543, Dec 1995.

[6] H.-M. Rein and M. Moller, Design considerations for very-high-speed Si bipolar IC's operating up to 50 Gb/s, *IEEE J. Solid-State Circuits*, Vol. 31, pp. 1076–1090, Aug. 1996.

[7] T. Matano, Y. Takai, T. Takahashi, Y. Sakito, I. Fujii, Y. Takaishi, H. Fujisawa, S. Kubouchi, S. Narui, K. Arai, M. Morino, M. Nakamura, S. Miyatake, T. Sekiguchi, and K. Koyama, A 1-Gb/s/pin 512-Mb DDRII SDRAM using a digital DLL and a slew-rate-controlled output buffer, *IEEE J. Solid-State Circuits*, Vol. 38, pp. 762–768, May 2003.

[8] T. Takahashi, M. Uchida, T. Takahashi, R. Yoshino, M. Yamamoto, and N. Kitamura, A CMOS gate array with 600 Mb/s simultaneous bidirectional I/O circuits, *IEEE J. Solid-State Circuits*, Vol. 30, pp. 1544–1546, Dec. 1995.

[9] M. D. Ker and T. S. Wu, Novel octagonal device structure for output transistors in deep-submicron low-voltage CMOS technology, *IEEE International Electron Devices Meeting*, pp. 889–892, 1996.

[10] F. A. Musa and A. C. Carusone, Clock recovery in high-speed multilevel serial links, *IEEE International Symposium on Circuits and Systems*, Vol. 5, pp. 449–452, May 2003.

[11] P. Heydari, Design and analysis of low-voltage current-mode logic buffers, *IEEE International Symposium on Quality Electronic Design*, pp. 293–298, Mar. 2003.

[12] P. Heydari and R. Mohavavelu, Design of ultra high-speed CMOS CML buffers and latches, *IEEE International Symposium on Circuits and Systems*, Vol. 5, pp. 208–211, May 2003.

[13] A. Boni, A. Pierazzi, and D. Vecchi, LVDS I/O interface for Gb/s-per-pin operation in 0.35-μm CMOS, *IEEE J. Solid-State Circuits*, Vol. 36, pp. 706–711, Apr. 2001.

[14] M. D. Ker and C. H. Chuang, Electrostatic discharge protection design for mixed-voltage CMOS I/O buffers, *IEEE J. Solid-State Circuits*, Vol. 37, pp. 1046–1055, Aug. 2002.

[15] T. L. Polgreen and A. Chatterjee, Improving the ESD failure threshold of silicided NMOS output transistor by ensuring uniform current flow, *IEEE Trans. Electron Devices*, Vol. 39, No. 2, p. 379, 1992.

[16] R. Senthinathan and J. L. Prince, Simultaneous switching ground noise calculation for packaged CMOS devices, *IEEE J. Solid-State Circuits*, Vol. 26, No. 11, pp. 1724–1728, Nov. 1991.

CHAPTER 7

DRAM

7.1 INTRODUCTION

Over the last decades, manufacturers succeeded in optimizing and scaling down dynamic memories (DRAMs). Key enablers of the successful DRAM shrink path were the three-dimensional one-transistor/one-capacitor cell and the latching cross-coupled sense amplifier. These DRAM-specific circuits, however, are facing specific challenges when crossing into the 100-nm regime, and as a consequence, well-known principles may have to be abandoned. In this chapter we discuss these specific challenges. To set the groundwork, we start by touching briefly on the most important DRAM basics.

7.2 DRAM BASICS

In dynamic memories, data are stored as a volatile charge on a capacitor. Due to the volatile nature, the stored charge would leak away unless each memory cell gets refreshed regularly. Following early developments with 3T1C cells [1], the optimized 1T1C cell emerged [2]. The acronym 1T1C is an abbreviation for a memory cell consisting of one capacitor used to store the information and one transistor used to access the stored information.

Figure 7.1 is a schematic representation of the arrangement of memory cells into a memory array. Word lines are connecting the gates of the access transistors along a column; bit lines are running perpendicular and are connecting the sources along a row. An idealized layout with a generic capacitor is shown in Figure 7.2.

Nano-CMOS Circuit and Physical Design, by Ban P. Wong, Anurag Mittal, Yu Cao, and Greg Starr
ISBN 0-471-46610-7

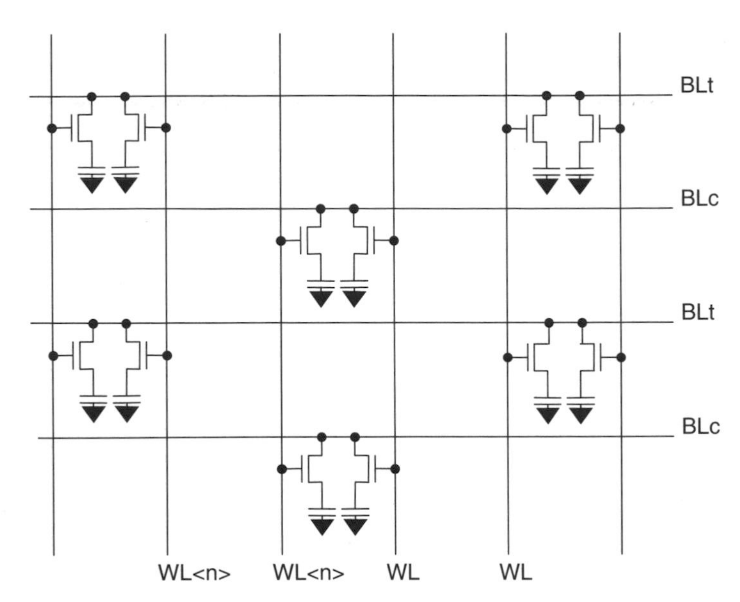

Figure 7.1 Schematic representation of a DRAM array. Word lines (WLs) are running vertically, bit lines (BLs) horizontally. Logically, bit lines are organized into true/complement pairs.

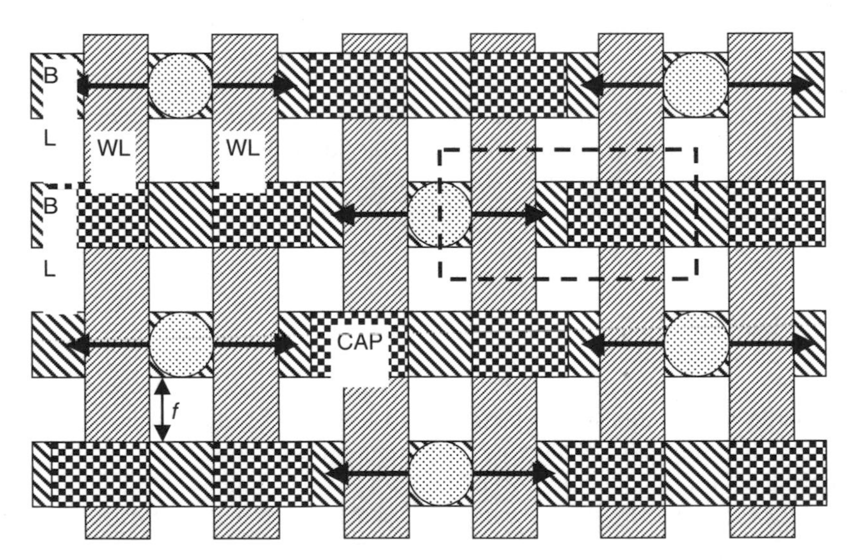

Figure 7.2 Idealized layout of an $8f^2$ DRAM array. The memory cell outline is given by dashed lines. Arrows indicate the direction of current flow through the transfer device. Circles indicate contact from bit line to diffusion of the transfer device. Word lines (WLs) run vertically; bit lines (BLs) are oriented perpendicularly. A storage capacitor (CAP; checkerboard pattern) may occupy more than $1f^2$ area.

No details are included as to how this structure can be realized on silicon. The only relevant information is that this cell layout requires an area of $8f^2$ per cell, with f denoting the minimum feature size of the process. A cell of $8f^2$ area with the given arrangement permits an efficient implementation of the folded bit-line architecture [3]. In this architecture, bit lines are arranged as complementary pairs, alternating between true bit lines (t) and complement bit lines (c). Word lines are connecting to cells either on the true or on the complement bit lines (Figure 7.1).

The folded bit line is the architecture of choice for most, if not all, modern DRAMs. In folded-bit-line architecture, a precise differential amplifier, the sense amplifier, is attached to each complementary pair of bit lines. As the complementary pair is laid out in closest proximity, most array noise appears as common-mode noise and does not degrade the differential amplification.

Figure 7.3 shows the schematics of a typical sense amplifier. To minimize the area, one sense amplifier is shared between two memory arrays, one to its left side and one to its right side. The array to be sensed is selected through multiplexer devices. Equalizer transistors allow biasing of the sense amplifier at the desired precharge level of VAA/2, with VAA denoting the high level of the bit-line voltage. Finally, the cross-coupled pair in the center implements the sense-amplifier core.

A typical memory access cycle is shown in Figure 7.4. Initially, the array is held in the precharge state. All multiplexer and equalizer gates are biased high and the bit lines are held at VAA/2. As the first step in the activation sequence, between either the right and left array for sensing has to be selected. In the example, the right array is to be read out. This is achieved by turning the left

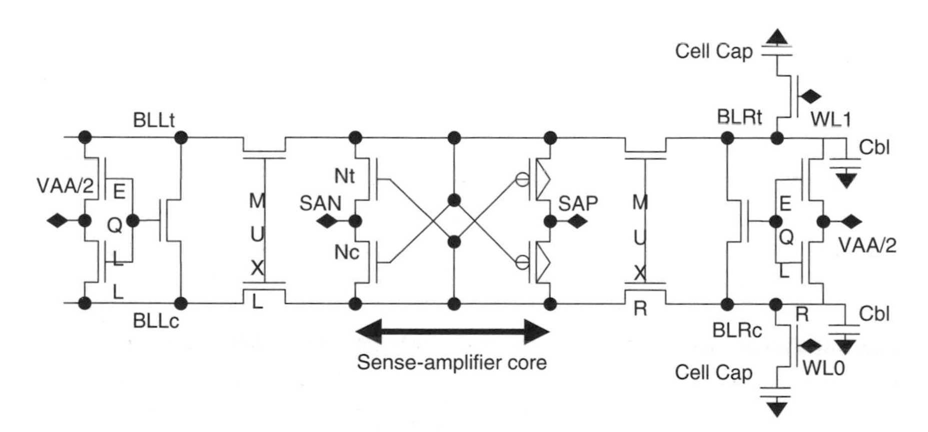

Figure 7.3 Schematics of a classical shared DRAM sense amplifier. For clarity, we have included only one of the data path access devices, which are not limiting in the scaling context. Multiplex and precharge devices permit selection between the left (L) and right (R) array. Thus, one sense amplifier core may be used for two adjacent memory arrays. N_t and N_c denotes n-sense amplifier transistors that pulls down the true or complementary branch.

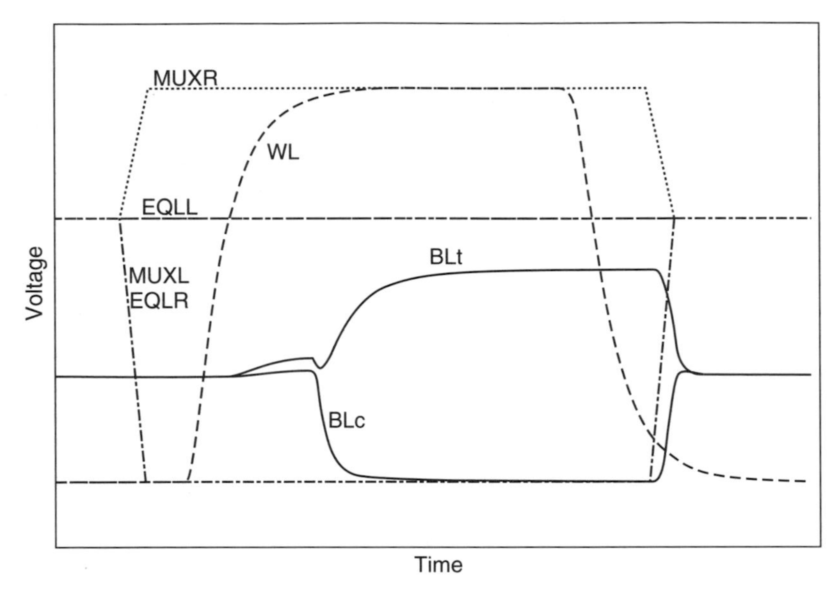

Figure 7.4 Example of a typical time development of signals in the array. Shown is the sensing of a 1. Solid lines indicate true and complementary bit lines. The dashed line is the word line. The dotted and dash-dotted lines indicate equalizer and multiplexer signals. MUXL and EQLR are lying on top of each other.

multiplexer device (MUXL) and the right equalizer (EQLR) off. Thereafter, the word line can be raised, which transfers the stored charge of the memory cell onto the bit line. Charge sharing between bit lines at equalization voltage and high (low) potential in the cell causes the bit-line voltage to increase (decrease). After read-out, the sense amplifier can be enabled to amplify the differential signal on the complementary bit-line pair. Enabling of the sense amplifier is performed by driving SAN from VAA/2 to 0 V and SAP from VAA/2 to VAA. In this fashion, the restored signal is automatically written back correctly into the cell. To end the cycle, signals return to their precharge levels.

This sensing scheme has multiple advantages. The number of devices is minimal for a circuit providing equalizing, left–right selection, and amplification. In addition, the interconnect scheme is relatively simple, permitting a very compact layout. This fact is of great importance for the chip size of DRAMs, due to the huge number of sense amplifiers required. Besides being advantageous for area, this circuit is also beneficial from a power perspective. During amplification, only the high-going bit line needs to be charged from VAA/2 to VAA, consuming a charge of CBL $\times$ VAA/2 in the process (CBL: bit-line capacitance). At the end of the cycle, the high bit line at VAA and the low bit line at 0 V are simply shorted by the equalizer transistor, creating the natural precharge level of VAA/2. Thus, equalization is fast and requires no additional current.

The 1T1C cell with the cross-coupled sense amplifier is highly optimized for area. Thus, it is very desirable to keep this basic structure while shrinking into the sub-100-nm regime. However, success in doing so will require solution of multiple puzzles. Practically any piece has its own associated specific problem. These are discussed in the next few sections.

7.3 SCALING THE CAPACITOR

The storage capacitor is the first key element. The signal available to the sense amplifier, V_{signal}, is related to the storage capacitance, C_{cell}, through simple charge sharing with the bit-line capacitance C_{BL}:

$$V_{\text{signal}} = \frac{C_{\text{cell}}}{C_{\text{cell}} + C_{BL}}\left(V_{\text{cell}} - \frac{V_{AA}}{2}\right)$$

Typically, the bit-line capacitance is five to 10 times the capacitance of the memory cell. The bit-line capacitance includes the total capacitance of the memory plus the capacitance of the sense amplifier. DRAM cell designers strive to maintain the same level of cell capacitance over the generations. In this way, sensitivity to leakage losses for the stored charge will remain comparable. Also, the capacitive divider ratio will stay constant or even improve if the bit-line capacitance drops over the generations.

A key enabler of large capacitances in feature sizes down to the 100-nm regime was the three-dimensional capacitor. Extension into the third dimension can be either above the silicon (stacked capacitor) or down into the silicon (trench capacitor). Examples of both these technologies are shown in Figures 7.5 and 7.6.

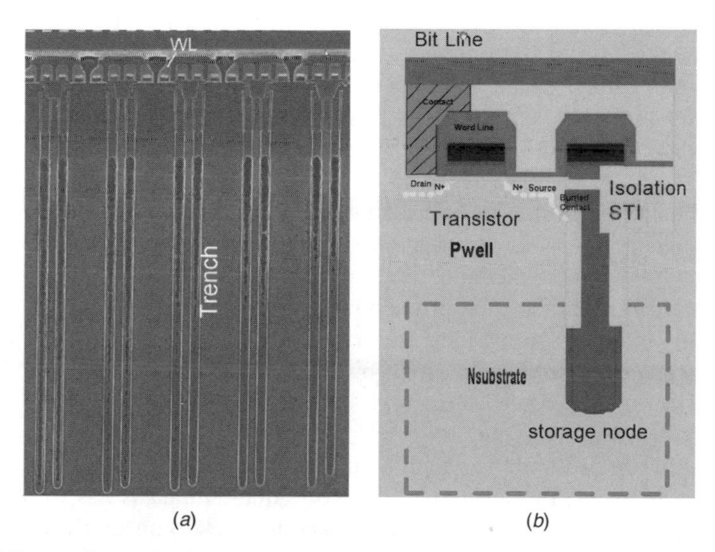

Figure 7.5 (*a*) Example of a trench capacitor cell. The trench etches deep into the silicon. Note that the trench depth is factors larger than the length of the word line defining the feature size f. (*b*) Simplified diagram of the word line and associated memory cell.

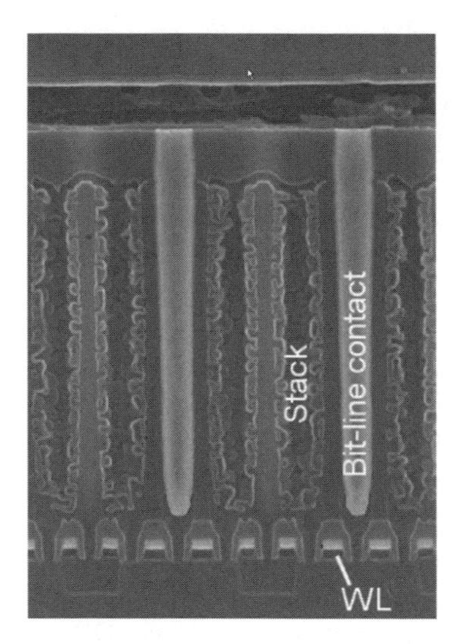

Figure 7.6 Example of a stacked-capacitor cell. Note the corrugated surface of the stacked capacitor employed for additional surface enhancement.

Figure 7.5(*a*) shows a trench capacitor memory cell, and Figure 7.5(*b*) is a schematic drawing of the trench cell. The trench etches deep into the silicon, which drastically increases the capacitor area available above the footprint. Figure 7.6 shows a stacked-capacitor cell. Here, area enhancement is smaller than in the trench approach, as the stack cannot be made too high. Otherwise, it would be difficult to planarize the dielectric layer above to enable processing additional metal layers on top of the stacked capacitor. To compensate for this disadvantage, further capacitance enhancement techniques may be employed. As shown in Figure 7.6, the capacitors are implemented with a roughened surface in order to boost the area. Even further improvement of the capacitance is obtained when a conventional dielectric layer is replaced with one having higher ε_r materials—Ta_2O_5 is in widespread use today.

Through continuous improvement, this approach has been successful over the most recent generations. However, another scaling problem will be how to charge and discharge the cell capacitance through the series resistance of the capacitor and/or contacts. Etching a deeper trench (see Figure 7.5) at a smaller feature size automatically implies a strong increase in total resistance. For the stack approach, a similar problem is seen in the long-bit-line contact dropping down from the bit line to the silicon. The bit-line and stack contacts both need to be realized between the word lines of the array.

Thus, until the 100-nm generation, major emphasis was put on achieving sufficient capacitance. In the future, however, equal emphasis has to be put into guaranteeing a small enough resistance to keep the *RC* constant low. This will have to be achieved through low-ohmic-contact materials and shallow gate stacks.

Scaling of the capacitor is expected to be carried on with an additional focus on keeping the total contact resistance at a level such that the capacitance can be read out. If, however, cell designers manage to keep the capacitance roughly constant, they need to provide a memory cell transistor capable of driving sufficient current to read this cell fast enough. The transistor is therefore the subject of the next section.

7.4 SCALING THE ARRAY TRANSISTOR

Our main assumption has been to maintain an $8f^2$ memory cell as shown in Figure 7.2. In this cell, both the width and length of the array device are identical to the feature size f. Therefore, unlike devices for logic, geometrical scaling, is not a free parameter for DRAM device design. DRAM device scaling has to be achieved through alternative means. To fully comprehend this notion, it is helpful to go over the boundary conditions for the array device. To allow a more general discussion, we introduce some new abbreviations for voltages. These are VPP for the word-line voltage of a selected cell, VLL for the word-line voltage of a deselected cell, and VBB for the voltage of the p-well of the array device (p-type array devices are no longer in use). A triple-well process is used, where the array p-well is inside an n-well; this allows backbias on the transfer transistors that raises V_{th} and also isolates the arrays from substrate noise and to some extent ionizing particle upsets.

The tightest constraint for the array device is given by the DRAM leakage requirements. The time period of the DRAM refresh is standardized at a typical refresh period of 64 ms. Using a memory cell capacitance of 20 fF and a sustainable leakage loss of 300 mV, we end up with a total leakage budget of 100 fA. If we assign 10% of the total budget to the device, device leakage has to be limited to 10 fA. As leakage is increasing with temperature, this condition has to be met at the highest operation temperature. Recognizing that the worst off-state condition is encountered for cells that are connected to a bit line where the associated sense amplifiers are active and pull the bit lines low, we have

$$I_{\text{leakage}}(V_d = \text{VAA};\ V_g = \text{VLL};\ V_s = 0;\ V_{\text{bulk}} = \text{VBB};\ \text{high temperature}) = 10\ \text{fA}$$

However, the device cannot be optimized solely for leakage. On top of that, the current during read-out needs to be sufficiently high. Simplifying, we require the initial read-out slope of the voltage across the cell capacitor to be 750 mV/2n. The value has to be achieved in the read-out condition where the bit lines are precharged at VAA/2, pushing the device into source-follower mode. As the threshold voltage increases with decreasing temperature, this condition typically is tightest at low temperature. The fact that device mobility degrades with increasing temperature does not compensate for this threshold effect. Therefore;

$$I_{\text{read}}(V_d = \text{VAA};\ V_g = \text{VPP};\ V_s = \text{VAA}/2;\ V_{\text{bulk}} = \text{VBB};\ \text{low temperature}) = 10\ \mu\text{A}$$

Let us first summarize means to guarantee the leakage requirements. Figure 7.7 illustrates the improvements. A negative VBB has been used in most standard DRAMs for several generations. Biasing VBB negative shifts the threshold of the

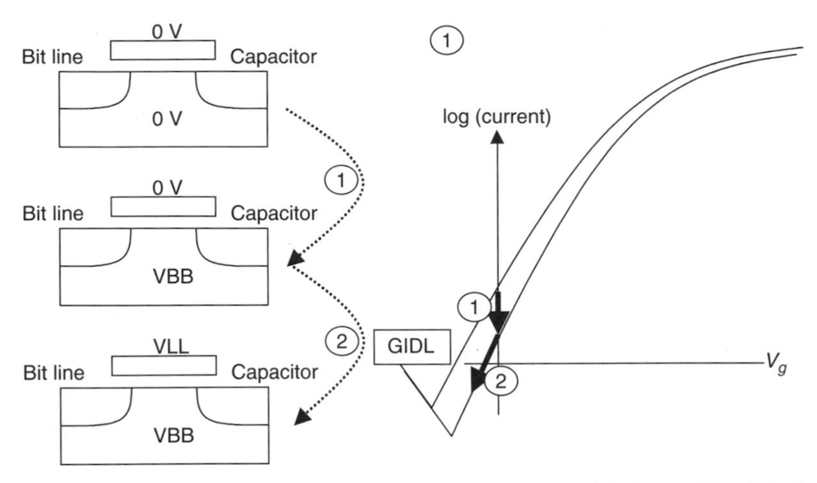

Figure 7.7 Voltage engineering of the array device to control leakage. The right half of the figure is a logarithmic representation of the transfer device current to emphasize the subthreshold regime. The left half shows the history of newly introduced voltages. At first, the negative p-well voltage VBB was introduced, which modified the IV–curve. Around the 100-nm node, the word line voltage for nonselected devices was set negative to achieve a lower subthreshold leakage but not too negative or it will be in the GIDL regime.

device up, thereby reducing the off-current leakage. However, we have to keep in mind that VBB also acts as the p-well voltage for n-diffusion of the storage node. Increasing VBB will therefore increase diode leakage across the junction. An optimum design point must be carefully chosen.

In the 100 nm regime, this rather classical measure turned out to be insufficient. As an improvement, a new voltage VLL was introduced [4]. VLL, the word-line voltage for deselected cells, is set slightly negative to decrease the subthreshold current of the array device. Again, an optimum value has to be chosen carefully, as VLL reduces subthreshold leakage, but at the same time it increases gate-induced drain leakage (GIDL) at the node junction. Thus, retention is at first expected to improve through improved subthreshold behavior, but for more negative values of VLL, the trend reverses due to GIDL.

Choosing optimum values for VBB, VLL, and VAA, an array device can be engineered that meets the leakage criteria for data retention. The most critical device parameter is the threshold of the device, which needs to be set as high as necessary. However, the subthreshold slope is not scaling with geometry. Thus, to guarantee retention, the threshold has to be kept relatively unchanged over the generations, yet on-current requirements of the array device must support the ever-increasing read-out speed of the cell, needed to scale the access time. This scenario is totally different from scaling in logic processes where performance through scaling is gained at the expense of incurring ever-increasing off-state leakages.

The straightforward solution to this problem is to employ a boosted gate voltage VPP. A boosted VPP has been employed in DRAMs for several generations at least since 1980. The main reason was the desire to write the full VAA level into the memory cell, which required the gate voltage of the transfer device to fulfill: VPP > (VAA + V_{th}). In the context of our scaling discussion, a boosted VPP additionally increases the drive current in the read-out condition. Therefore, it is advantageous to choose VPP as high as possible while technology boundary conditions are still being met. A hard limit is set through the oxide thickness—the electrical fields may not exceed breakdown (GOI) limits. The limit is time dependent dielectric breakdown (TDDB), representing a reliability issue. The maximum field in use is limited to about 5 mV/cm although the breakdown voltage is greater than 10 mV/cm. To push this limit higher, modern DRAM processes employ dual-gate oxide thickness devices. In peripheral circuits, devices use thin gate oxides optimized for performance, while the array device is build with a thicker oxide, allowing a boosted VPP. An additional, though softer limit is set through the design of the VPP charge pump. To keep pump efficiency high, VPP cannot be set too high. Previously, a boosted word line was of minor importance, due to the high supply voltage. However, it is expected to gain relevance in the future as the external supply voltage of the memory is decreasing rapidly. Next-generation devices are expected to operate at 1.5 V or less. The principles described above are expected to enable scaling below 100 nm. Yet at some point in the scaling road map, even these measures will no longer be adequate. More radical approaches have to be considered. A straightforward, though technically challenging solution may be found by browsing through the historical records of the DRAM capacitor. Some generations ago, the planar, two-dimensional capacitor was replaced by a three-dimensional vertical capacitor. Along the same lines, we must look for further innovations to continue scaling into the sub-100-nm technologies. The access transistor is the next component that will be the scaling limiter and will require innovations to overcome the scaling limitations. Experimental DRAMs where the transistor is folded vertically into a trench cell has already been demonstrated [5,6]. The advantage of this topology is the decoupling of the array device's channel length from the target feature size, f. Figure 7.8 illustrates the technique. The transistor no longer is a planar device but lies vertically in between the bit-line contact diffusion at the silicon surface and the trench deeper in the silicon.

Although challenging, ideas remain to be discovered and implemented to scale the 1T1C cell below the 100-nm feature size. Building the array once we have the capacitor and access transistor taken care of does not impose a significant additional problem if care is taken that all resistance and (coupling) capacitances are modeled correctly. The next challenge is the design of the sense amplifier.

7.5 SCALING THE SENSE AMPLIFIER

An optimized circuit for sense amplification has evolved over the generations. Scaling imposes two systematic problems onto a sense amplifier: decrease of the operating voltage of the array (VAA) and decrease of the device area. We discuss the effect of the voltage first.

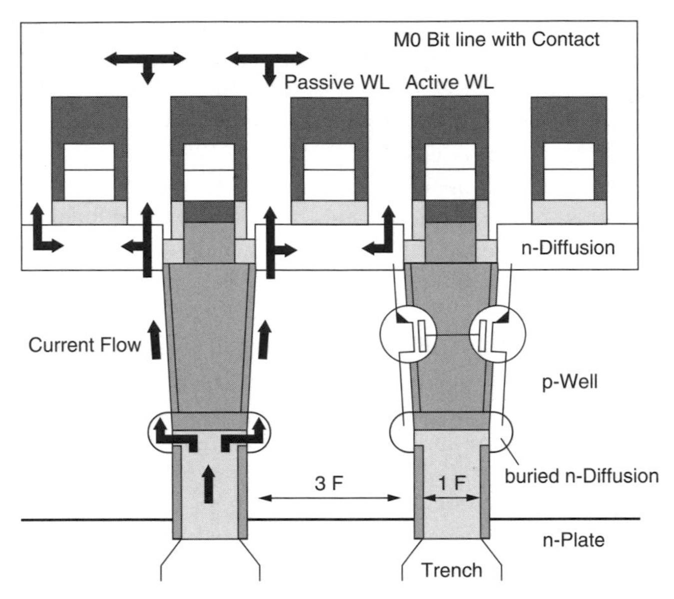

Figure 7.8 Memory cell with a vertical access transistor. Vertical orientation of the transistor extends the three-dimensional integration of the memory cell capacitor. A major advantage is the decoupling of the gate length of the access transistor from the feature size, f.

Similar to the access transistor, the sense amplifier has to meet constraints in speed and leakage. Sensing speed is most critical during sensing of the 0 (see Figure 7.9). Prior to the read-out of the cell, the bit lines are precharged at VAA/2. After the word line is asserted, the voltage of BLt has dropped due to the charge sharing between BLt and the memory cell. BLt needs to be discharged through the n-sense-amplifier transistor Nt, which is gated by BLc (cf. Figure 7.3). The voltage on BLc after read-out, however, will typically drop below VAA/2 through line–line coupling between the closely spaced bit lines. Thus, sensing of cells storing a 1 starts at a slightly higher level on SAN than sensing of cells storing a 0. As, typically, thousands of sense amplifiers are activated in parallel to read out all the bits along a word line, this fact may lead to a significant delay in sensing the 0 if most of the other cells store a 1. In this case, SAN cannot be pulled down fast enough against the load of all bit lines sensing a 1. Sensing of the 0 is therefore the unfavored state, particularly at an ever-lowered VAA. As shown in Figure 7.9, with diminishing VAA, sensing speed is degrading rapidly and ultimately, will not be fast enough to permit column access without disturbing the information developing on the bit line. As the threshold increases with decreasing temperature, this effect is most pronounced at low temperature, due to the higher threshold. A common technique to increase sensing speed is by lowering the threshold of the sense device. Therefore, advanced DRAM processes employ low-threshold n-transistors in the sense amplifier. Lowering the device threshold increases leakage through the

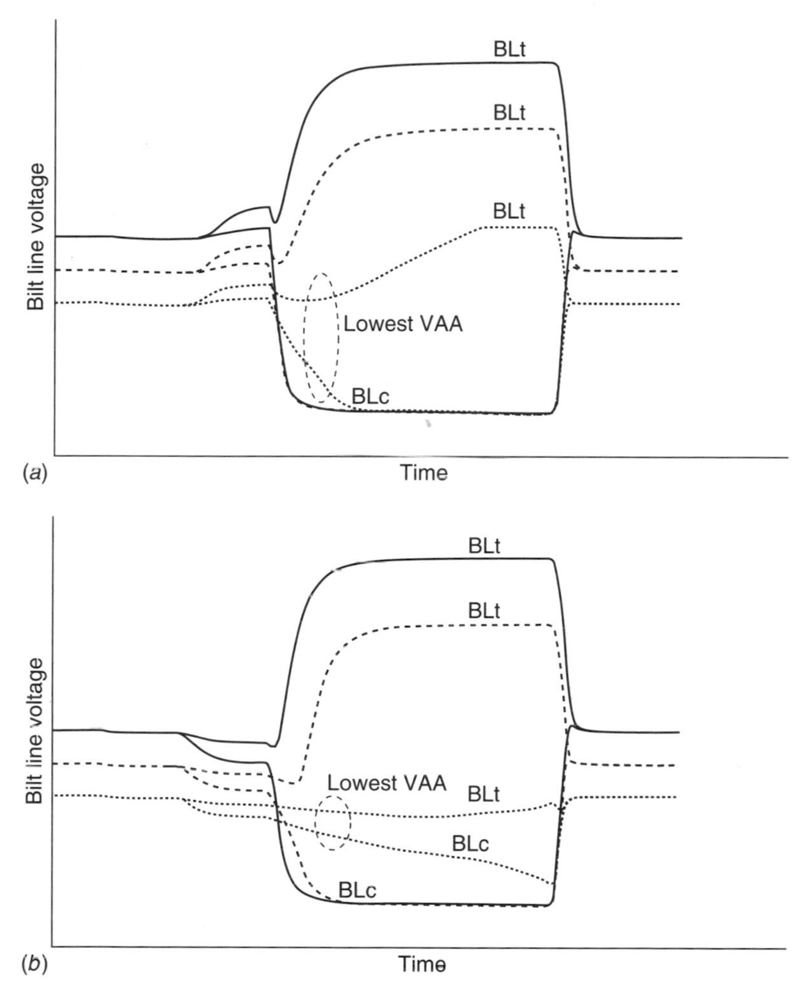

Figure 7.9 Typical bit-line voltage development for sensing at low temperature at various levels of VAA (solid/dashed/dotted lines): (*a*) sensing of a 1; (*b*) sensing of a 0. Sensing speed for a 0 deteriorates much faster with lowered VAA.

sense-amplifier pair once the sense-amplifier transistors have latched. In earlier generations we could easily arrive at a design point for the sense amplifier where both constraints could be met (leakage and speed). However, below 100 nm the picture starts to change to the extent that a device design point may no longer exist. In this regime, the array voltage VAA needs to be reduced to the order of 1 V. Thus, VAA/2, which sets the order of the overdrive at sensing, is only about 500 mV. Therefore, very low threshold devices would be required to maintain sufficient sense speed. As the subthreshold slope is limited, these low thresholds would result in unmanageable subthreshold leakage after latching of the sense amplifier.

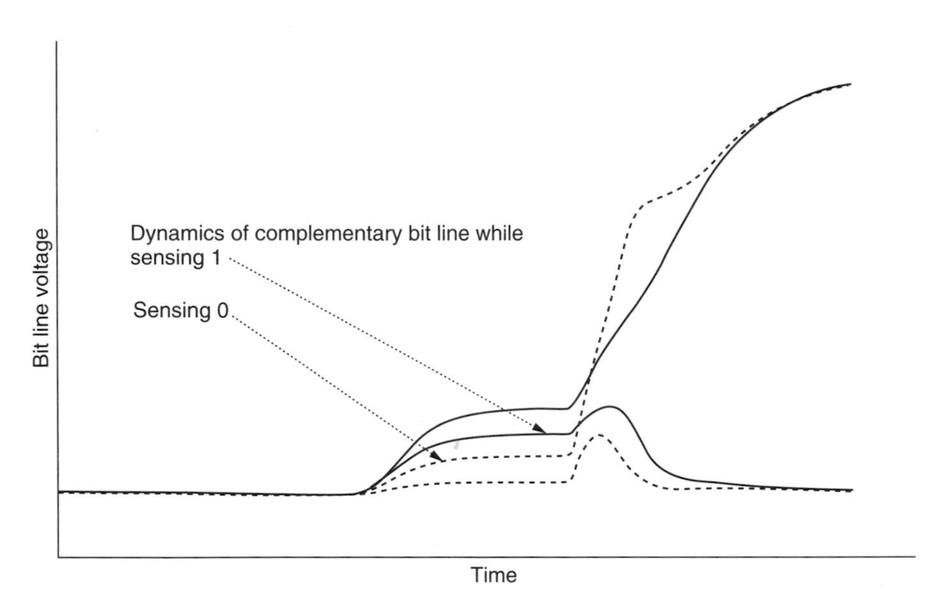

Figure 7.10 Typical bit-line voltage development for precharge ground sensing on true and complementary bit lines. Dashed (solid) lines show sensing of 0 (1). A reference cell on the complementary bit line needs to be read out to establish a midpoint. Reading out a 0 (1) on the true bit line causes a voltage increase through charge sharing, which is smaller (larger) than the reference bit line. Difference on the reference cell bit line is created by line–line coupling.

Multiple approaches are being investigated to improve DRAM sensing performance at lower array voltages. The simplest one is to precharge the bit lines at 0 V [7]. Under this condition, the p-sense-amplifier transistors will be the amplifying devices. However, as SAP is pulled up from the precharge state of 0 V to VAA, these devices can operate under full VAA such that sufficient gate overdrive can easily be maintained. Even though this simple solution appears to be attractive at first glance, there are many disadvantages. The most notable is the need to introduce a reference cell for sensing. The reference cell needs to have the same capacitance as the cell and has to be precharged to the midlevel of VAA/2. To read out a cell on a true bit line, the reference cell needs to be connected to the complementary bit line, and vice versa. Reading out this cell causes an increase in the voltage on the complementary bit line, which is equivalent to half of the full signal (Figure 7.10). Under this condition a correct differential signal between the bit line and the complementary bit line can be obtained for both a 0 and a 1 stored in the memory cell. Reference cells require additional area and a special layout, as they require a port to precharge the cell to VAA/2. Additionally, ground-level precharge doubles the operating current of the array. During amplification, one bit line has to be pulled fully from ground to VAA. After sensing, this charge is dumped to ground and lost. Ground-level precharge may therefore, be a possible solution, but has a significant power penalty.

Another approach is discussed in [8]. Here, the authors propose to tackle the problem at its root cause: the variation of the device threshold with the temperature. To compensate for this effect, active regulation of the well-bias is employed to lower the threshold at low temperature while maintaining a sufficiently high threshold at high temperature. In this way, both high sensing speed at low temperature and low array leakage at high temperature can be achieved.

An alternative technique has been proposed where the precharge levels of the bit line and sense amplifier are different [9]. The bit lines are precharged to VAA/2 to keep the advantage of the midlevel precharge array, whereas the sense amplifier is precharged to the full VAA level to have enough headroom for the sense-amplifier devices. The circuit promises good functionality, but the drawback is an increased device count and wiring complexity compared to the traditional sense amplifier.

Voltage scaling certainly is imposing a directly visible hard limit on sense-amplifier operation. A more subtle limit is set by shrinking of the device area. With decreasing area, random mismatches of the threshold voltage due to random dopant concentration will become significant. High-performance SRAM and analog designers have been battling this problem since the 250-nm node. One technique that may be implemented by analog designers is by increasing the device area since ∂V_{th} is roughly proportional to $\sqrt{1/\text{device area}}$ [10]. Unfortunately, this approach is not reasonable within an area- and pitch-constrained sense amplifier. Recognizing this problem, several solutions have been proposed in the current literature where active offset-correcting sense amplifiers are implemented [11–13]. A common principle in these solutions is to predistort the precharge level on the bit lines after the end of the equalization period such that sense-amplifier offset is canceled out. Again, as a negative side effect, the device count is increased, and more complicated wiring schemes have to be implemented in the sense amplifier.

7.6 SUMMARY

In summary, for both lowered operation voltages and sense-amplifier device mismatch, no imminent limit is seen for scaling below 100 nm. The proposed solutions suggested promise to enable continued scaling of the DRAM. There are, however, problems in topology and circuit designs that will require new solutions in order to continue scaling a manufacturable and robust design.

REFERENCES

[1] W. M. Regitz and J. Karp, A three transistor-cell, 1024-bit, 500 NS MOS RAM, *IEEE International Solid-State Circuits Conference*, Vol. XIII, pp. 42–43, Feb. 1970.

[2] C. N. Ahlquist, J. R. Breivogel, J. T. Koo, J. L. McCollum, W. G. Oldham, and A. L. Renninger, A 16 384-bit dynamic RAM, *IEEE J. Solid-State Circuits*, Vol. 11, pp. 570–574, Oct. 1976.

[3] K. Itoh, R. Hori, H. Masuda, Y. Kamigaki, H. Kawamoto, and H. Katto, A single 5 V 64K dynamic RAM, *IEEE International Solid-State Circuits Conference*, Vol. XXIII, pp. 228–229, Feb. 1980.

[4] H. Tanaka, M. Aoki, T. Sakata, S. Kimura, N. Sakashita, H. Hidaka, T. Tachibana, and K. Kimura, A precise on-chip voltage generator for a gigascale DRAM with a negative word-line scheme, *IEEE J. Solid-State Circuits*, Vol. 34, pp. 1084–1090, Aug. 1999.

[5] S. Wuensche, M. Jacunski, H. Streif, A. Sturm, J. Morrish, M. Roberge, M. Clark, T. Nostrand, E. Stahl, S. Lewis, J. Heath, M. Wood, T. Vogelsang, E. Thoma, J. Gabric, M. Kleiner, M. Killian, P. Poechmueller, W. Mueller, and G. Bronner, A 110 nm 512 Mb DDR DRAM with vertical transistor trench cell, *Symposium on VLSI Circuits Digest 16*, pp. 114–115, June 2002.

[6] T. Kirihata, G. Mueller, M. Clinton, S. Loeffler, B. Ji, H. Terletzki, D. Hanson, C. Hwang, G. Lehmann, D. Storaska, G. Daniel, L. Hsu, O. Weinfurtner, T. Boehler, J. Schnell, G. Frankowsky, D. Netis, J. Ross, A. Reith, O. Kiehl, and M. Wordeman, A 113 $mm^2$600 Mb/s/pin 512 Mb DDR2 SDRAM with vertically-folded bitline architecture, *IEEE International Solid-State Circuits Conference*, Vol. XLIV, pp. 382–383, Feb. 2001.

[7] J. Barth, D. Anand, J. Dreibelbis, and E. Nelson, A 300 MHz multi-banked eDRAM macro featuring GND sense, bit-line twisting and direct reference cell write, *IEEE International Solid-State Circuits Conference*, Vol. XLV, pp. 156–157, Feb. 2002.

[8] K. Hardee, F. Jones, D. Butler, M. Parris, M. Mound, H. Calendar, G. Jones, L. Aldrich, C. Gruenschlager, M. Miyabashi, K. Taniguchi, T. Arakawa, A 0.6 V 205 MHz 19.5 ns tRC 16 Mb Embedded DRAM, *IEEE International Solid-State Circuits Conference*, Vol. XLVII, pp. 200–201, Feb. 2004.

[9] J.-Y. Sim, Y.-G. Gang, K.-N. Lim, J.-Y. Choi, S.-K. Kwak, K.-C. Chun, J.-H. Yoo, D.-I. Seo, and S.-I. Cho, Charge-transferred presensing and efficiently precharged negative word-line schemes for low-voltage DRAMs, *IEEE Symposium on VLSI Circuits*, Session 22–4, 2003.

[10] M. J. M. Pelgrom, A. C. J. Duinmaijer, and A. P. G. Welbers, Matching properties of MOS transistors, *IEEE J. Solid-State Circuits*, Vol. 24, pp. 1433–1439, Oct. 1989.

[11] T. Furuyama et al., A new sense amplifier technique for VLSI dynamic RAM's, pp. 47–44, *Proceedings of the IEEE International Election Devices Meeting*, 1981.

[12] S. H. Hong, S. H. Kim, S. J. Kim, J. Wee, and J. Y. Chung, An offset cancellation bit-line sensing scheme for low-voltage DRAM applications, *IEEE International Solid-State Circuits Conference*, Vol. XLV, pp. 154–155, Feb. 2002.

[13] J.-Y. Sim, K.-W. Kwon, J.-H. Choi, S.-H. Lee, D.-M. Kim, H.-R. Hwang, K.-C. Chun, Y.-H. Seo, H.-S. Hwang, D.-I. Seo, C. Kim, and S.-I. Cho, A 1.0 V 256 Mb SDRAM with offset-compensated direct sensing and charge-recycled precharge, *IEEE International Solid-State Circuits Conference*, Vol. XLVI, Feb. 2003.

CHAPTER 8

SIGNAL INTEGRITY PROBLEMS IN ON-CHIP INTERCONNECTS

8.1 INTRODUCTION

Decades of remarkable technology scaling have resulted in the fabrication of integrated circuits (ICs) with smaller feature sizes, higher levels of integration, and faster operating frequencies. Although these advances largely benefit IC performance, they also lead to complications that pose significant challenges to on-chip interconnect design, such as inductance effect, crosstalk noise, and power supply stability. In this chapter the major concerns of signal integrity are addressed from both a design exploration and performance analysis perspective. First, the scaling trend and figures of merit for high-speed interconnect are introduced. Then, various approaches and design guidelines for signal integrity analysis are discussed, ranging from the extraction of line parasitics and representation of the circuit model, to the analysis of timing and noise. Finally, we present practical solutions to improve the quality of signal routing in a nanometer design, using both physical design optimization and circuit design techniques.

In Chapter 2 the trends and challenges of interconnect scaling were discussed, mainly from the standpoint of process technology. Figure 8.1 illustrates the multiple-level on-chip interconnect structure in modern microfabrication. Interconnect can be divided into two main categories, depending on the scale of the blocks it connects: local layers, which link transistors and small circuit blocks

Nano-CMOS Circuit and Physical Design, by Ban P. Wong, Anurag Mittal, Yu Cao, and Greg Starr
ISBN 0-471-46610-7

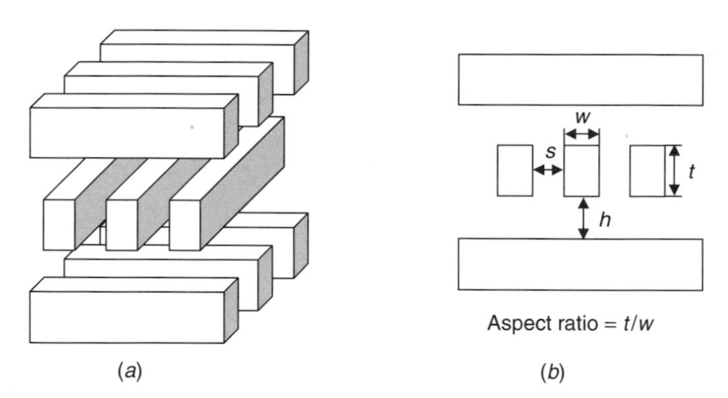

Figure 8.1 Contemporary interconnect structure and geometric parameter definitions: (*a*) three-dimensional view; (*b*) cross-sectional view.

(of length about 10 to 500 μm), and global layers, which link the large function blocks in a microprocessor (length usually > 300 μm). Typical applications of global interconnect include a data bus, clock distribution, power supply, and even spiral inductors in radio-frequency design. Whereas local interconnect is relatively short and scales with technology thus has less of an impact on circuit performance, the length of global interconnect can extend up to the chip-edge size and can form a significant performance bottleneck in nanometer design, especially for signal integrity issues. Hence, the focus of this chapter is the analysis of signal integrity problems in global interconnect at both the design exploration and verification stages (note that the principles are applicable to local interconnect as well).

Signal integrity refers to the quality of signal transportation during circuit operation. Depending on the functionality of the signal (e.g., logic switching, data transferring, power supply, or clock), it includes a broad set of design issues, such as crosstalk noise, delay uncertainty, signal overshoot, finite slew rate, *IR* drop, dI/dt noise, and clock skew and jitter. In this chapter we address general concerns in signal timing and waveform distortion for high-speed IC design. In the nanometer regime, the deteriorating nature of signal integrity is an inevitable result of technology scaling: To support the increasing device density, on-chip interconnect has a smaller pitch at advanced technology nodes, as shown in Figure 8.2(*a*) [1,2]. On the other hand, the demand of delay reduction requires the scaling of cross-sectional area of metal wires, which is the product of width (w) and thickness (t). As a consequence, metal wires become narrower and taller, leading to a larger aspect ratio (defined as t/w), as shown in Figure 8.2(*a*). This approach alleviates the interconnect delay problem, however, causing stronger coupling between neighboring lines. Figure 8.2(*b*) illustrates the ratio of coupling capacitance (C_c) to metal-to-ground capacitance (C_g) and the ratio of mutual inductance (L_m) to self-inductance (L_s), assuming that line spacing (s) is equal to line width. Both capacitive and inductive coupling between adjacent lines increases dramatically: For example, the C_c/C_g ratio will increase by more than 70% from the 180-nm node to the 65-nm node. Hence, signal

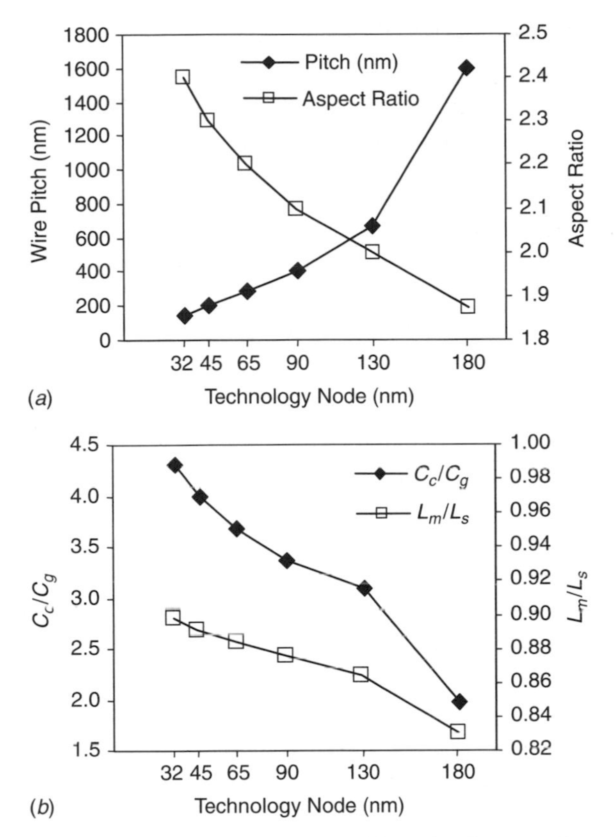

Figure 8.2 Interconnect coupling intensifies with technology scaling, as a result of smaller wire pitch and larger aspect ratio: (*a*) trend of global wire pitch and aspect ratio scaling; (*b*) trend of capacitive and inductive coupling at global layer.

transportation suffers a stronger impact from neighboring line switching at future technology nodes.

Another important factor that aggravates crosstalk noise is faster signal switching. Today's high-performance microprocessor design has achieved multiple-gigahertz on-chip clock speed. Since both capacitive coupling $[C(dV/dt)]$ and inductive coupling $[L(dI/dt)]$ are ac phenomena, faster signal switching (i.e., shorter dt) inherently causes larger noise magnitudes, which is a problem because the noise tolerance of circuits worsens with supply voltage (V_{dd}) scaling. Due to considerations of device reliability and chip power consumption, supply voltage has been scaled from 1.8 V at the 180-nm node to about 0.7 V at the 65-nm node. Meanwhile, the device threshold voltage (V_{th}) is reduced (although at a relatively slower rate due to leakage concerns) in order to achieve sufficient driving capability. As a result, the noise margin of digital circuits, especially dynamic circuits, becomes much lower than before, thereby imposing extremely demanding noise control and signal integrity solutions for robust nanometer design.

8.1.1 Interconnect Figures of Merit

Before discussing the design methodologies for solving signal integrity problems, we would like to introduce the major figures of merit (FOMs) that are generally used to evaluate on-chip interconnect performance:

1. *Signal delay and energy consumption.* Signal delay (timing) is the primary FOM in synchronous design. It is usually measured at the $50\% V_{dd}$ point, from the input of the driver to the end of the line (i.e., the input of receiver), as shown in Figure 8.3. Delay is a function of driver strength gate and wire loading. At 0.25 μm and above, the primary wire loading is due to metal-to-ground capacitance and is thus largely proportional to wire length. As technology scales, the majority of wire loading has shifted to the coupling capacitance. Hence, signal delay is not only a function of the local wire geometry but also depends on the switching activities of its neighboring lines. Another issue in future designs is interconnect energy consumption ($C \cdot V_{dd}^2$); to support higher levels of integration, the total length of wires in a chip increases dramatically (e.g., at 45 nm node, it will be more than 5 km), leading to a larger total capacitance and thus, to more energy dissipation. Note that interconnect energy consumption is proportional to total capacitance and the square of V_{dd}, and to the first order is independent of line resistance and inductance.
2. *Crosstalk noise.* In general, noise describes voltage glitches that distort a given signal's desired value. Crosstalk noise originates from coupling of a given line to its neighboring lines, through either coupling capacitance or mutual inductance. At the 180-nm node, cross coupling affects only high-performance designs significantly. Entering the sub-100-nm regime, however, it becomes a greater concern, of comparable importance to timing, power, and area cost concerns in all IC designs. Crosstalk noise profoundly affects the performance of a circuit, particularly in two aspects: First, a large noise may trigger false logic switching and thus, cause malfunctions; second and more generally, crosstalk noise distorts the original switching waveform, and depending on the time it switches, can lead to a significant delay

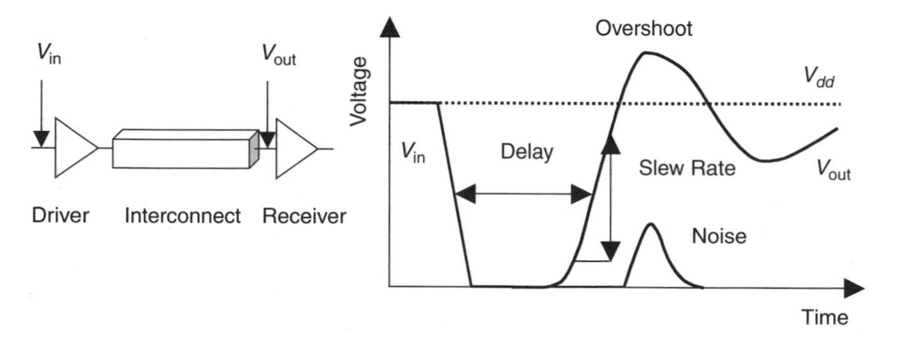

Figure 8.3 Definitions of major FOMs for on-chip interconnect analysis.

variation (as much as 80% at the 65-nm node). Thus, if designers continue to use the current worst-case design approach, leaving adequate margins to tolerate such timing uncertainties, they will suffer from an excessive timing budget and severely reduced performance. We discuss noise-aware timing analysis in more detail in Section 8.3.4.

3. *Waveform integrity (slew rate, overshoot, and signal ringing).* In digital design, an ideal signal is a step function switching instantaneously between 0 and V_{dd}. However, this model is only an approximation to the actual switching waveform, which in reality has a nonzero slew rate and possible signal overshoots and ringing. *Slew rate* (or *rise time*) is defined as the time required for the signal to change from 10% to 90% (or 30% to 70% for more recent designs) of its final value. Due to limited driver strength and lossy wire conditions, switching in reality has a finite slew rate, which delays signal stabilization and transportation. As circuit speed increases rapidly, this portion of the timing budget needs to be well controlled. Faster circuit speeds also result in other high-frequency electromagnetic effects, such as undesired transmission line behavior and increased difficulty in achieving impedance matching. Furthermore, increased circuit speeds may also lead to signal overshoot (i.e., above V_{dd} or below ground) and ringing, as shown in Figure 8.3. While signal ringing disrupts the settling time, overshoot can cause reliability problems with the gate oxide of the receiver, especially for pass-gate circuits. As the scaling trend continues, waveform distortion in current interconnect architectures eventually limits the bandwidth as the rate of data communication approaches terabits per second [3,4]. New interconnection technology and design techniques are required to overcome this limitation.

From a design perspective, these FOMs need to be considered carefully to avoid signal integrity degradation. They are applied throughout this chapter for both performance analysis and design solutions. Driven by recent process-design cooptimizations, these performance FOMs can be further integrated with technology considerations to maximize the throughput of on-chip communication.

8.2 INTERCONNECT PARASITICS EXTRACTION

To perform timing and signal integrity analyses, it is necessary to translate interconnect layout and technology information, such as the width and length of the wire, neighboring line conditions, and related dielectrics, into electrical parameters, so that they can be combined with other circuit components to evaluate performance. This is achieved through parasitics extraction. Based on the design and technology specifications, a physical line is usually converted into a netlist composed of resistors (R), capacitors (C), and if necessary, inductors (L). The conventional wire modeling approach is to use a lumped RC segment; however, with circuit operation frequency on the rise, this model lacks the accuracy to model a high-performance interconnect sufficiently. As technology scales, more physical concerns should be reflected into the extraction methodology during both early design planning and postdesign verification stages. For instance, line inductance, which previously was neglected in the model, is a crucial factor that affects the performance of some global interconnects in modern designs. In this section we discuss guidelines and feasible solutions for interconnect *RC* and *RLC* extractions. Furthermore, we also present techniques for the characterization of on-chip parasitics, which are used in design calibration and model verification.

8.2.1 Circuit Representation of Interconnects

The fundamental electrical behavior of a metal wire can be fully determined using Maxwell's equations:

$$\nabla \cdot D = \rho \qquad \nabla \times E = -\frac{\partial B}{\partial t} \qquad \nabla \cdot B = 0 \qquad \nabla \times H = J + \frac{\partial D}{\partial t} \tag{8.2.1}$$

in conjunction with the rule of charge conservation:

$$\nabla \cdot J + \frac{\partial \rho}{\partial t} = 0 \tag{8.2.2}$$

However, solving these equations directly requires a prohibitive amount of computation. Hence, depending on the range of interested frequencies and the length of a line, these equations are usually simplified to improve the computation efficiency. Figure 8.4 illustrates several circuit models for on-chip interconnect, based on various approximations.

The behavior of a wire is frequency dependent. At dc, it behaves as a resistor, causing both losses in the voltage supply (*IR* drop) and static power consumption

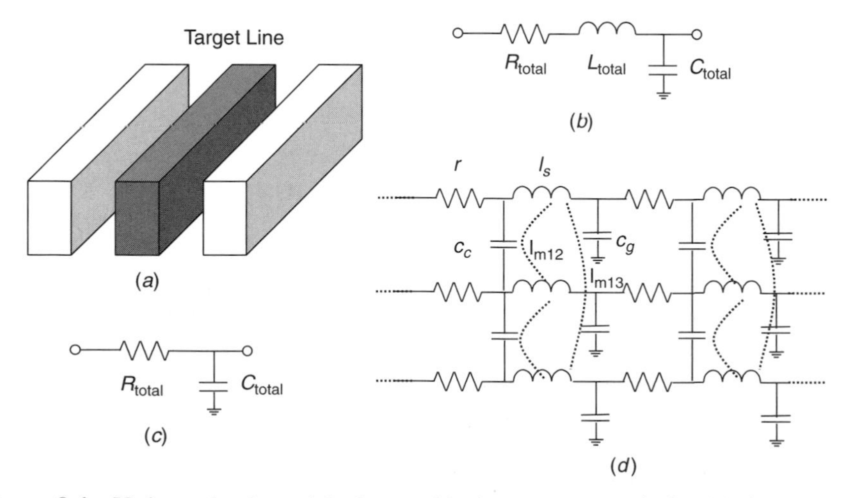

Figure 8.4 Various circuit models for on-chip interconnect analysis: (*a*) three parallel lines; (*b*) lumped *RLC* model; (*c*) lumped *RC* model; (*d*) distributed *RLC* model.

(IR^2). In ac operation, line activities are affected by the interaction between electric and magnetic fields. In current IC designs, if the signal frequency is relatively low and the line length is much shorter than the wavelength of the signal (e.g., at 10 GHz, the wavelength is about 17 cm for $\kappa = 3.0$ dielectrics), the quasistatic assumption is usually applicable. Under the quasistatic assumption, the displacement current in equation (8.2.1), $\partial D/\partial t$, is ignored, decoupling electric and magnetic fields. Hence, line capacitance C and line inductance L can be defined and extracted independently, and the resulting line is represented by an *RC* or *RLC* equivalent circuit, as shown in Figure 8.4. For a uniform line, Maxwell's equations can be reduced to the telegraph equation (transmission line theory) to solve the electrical response:

$$\frac{\partial^2 V}{\partial x^2} = RC\frac{\partial V}{\partial t} + LC\frac{\partial^2 V}{\partial t^2} \tag{8.2.3}$$

where x is the length dimension, t the time, and V the voltage. Moreover, if the wire is significantly shorter than the signal wavelength, it is a safe approximation to treat the entire line as a lumped segment because the signal level is approximately constant along the entire length of the wire. The target line in Figure 8.4(a) can be represented by either a lumped *RC* or a lumped *RLC* model, as shown in Figure 8.4(b) and (c), respectively. A rule of thumb to determine whether or not a wire can be represented by a lumped circuit is to test its length against the following criterion [5]:

$$\text{length} \leq \lambda/20 \tag{8.2.4}$$

where λ is the signal wavelength. Note that the frequency spectrum that a digital signal contains is more closely related to its rise time, t_r (e.g., $\frac{1}{3.14t_r}$), than to the signal frequency itself. Therefore, λ should be estimated from the rise time of the signal. For example, for a wire with a rise time of 20 ps implemented with a $\kappa = 3.0$ dielectric material, the lumped model can be used as long as the line length is less than 500 μm. Global interconnect lengths are usually too long to satisfy the foregoing criterion, and thus need to be partitioned into segments with equal lengths satisfying equation (8.2.4). Each segment can be a ladder, T-circuit, or Π-circuit composed of scaled *RC* or *RLC* elements. A distributed *RLC* representation for three coupled lines [Figure 8.4(a)] is shown in Figure 8.4(d). For performance analyses, the lumped model can provide scalable and physical insights and thus is suitable for early stages of the design exploration. However, in further design stages when more accuracy is needed, the distributed model should be applied.

When and Where to Consider Inductance Inductance is a single measure of the distribution of the magnetic field created by a current. Since line inductance is negligible at low clock speeds, on-chip interconnect is conventionally modeled only with *RC* components; inductance is a concern for off-chip package designers.

However, when the clock frequency enters the gigahertz regime, the impedance contributed by the line inductance ($j\omega L$) becomes comparable to the line resistance R, and can even dominate the total metal impedance ($Z = R + j\omega L$) of global lines. In addition, the introduction of low-resistivity copper and low-κ dielectrics also increases the relative importance of L. The impact of inductance manifests in many important circuit design issues; it not only increases the signal delay but may also cause voltage overshoot and reduced slew rate, which can increase crosstalk noise on neighboring lines. Thus, the inductance effect becomes an increasingly important issue in signal integrity analysis. On the other hand, since the consideration of L usually requires very expensive computation, it is desirable to include L only when it is necessary. From the standpoint of signal timing, the importance of L can be determined by evaluating the following three time constants:

1. Driver input signal slew rate: t_r
2. Time of flight for a transmission line: $t_f = \text{length} \cdot \sqrt{LC}$
3. Elmore delay for an RC line: $t_e = RC/2 \cdot (\text{length})^2$

Here R, L, and C are the resistance, inductance, and capacitance per unit length, respectively. As a figure of merit, interconnects should be modeled as RLC lines if they satisfy the following two conditions [5,6]:

1. $t_f > t_e$. Under this condition, the inductance effect leads to a longer signal delay. Since the RLC model represents a transmission line, comparing the time constants of the RLC and RC models is equivalent to evaluating the impedance of a transmission line ($Z = \sqrt{L/C}$) with total line resistance $R \cdot \text{length}$. The RLC effect is critical for evaluating the line performance when Z is significantly large compared to the total resistance.
2. $t_r < 2t_f$. Since $2t_f$ is the time required for a signal to travel round-trip from the driver to the end of a line, this condition implies that when the switching is fast enough, the signal transportation is affected by the reflected wave, which exhibits transmission line characteristics.

Translated into relationships between interconnect length and RLC components, these two conditions can be summarized as

$$\frac{t_r}{2\sqrt{LC}} < \text{length} < \frac{2}{R}\sqrt{\frac{L}{C}} \tag{8.2.4}$$

In case the constraint on the left-hand side is larger than that on right-hand side:

$$t_r > 4L/R \text{ (input signal is not fast enough)} \tag{8.2.5}$$

the inductance effect can be ignored regardless of the line length. As an example, Figure 8.5 plots the range in which RLC is critical for a single global line. The inductance effect is more prominent for wires with intermediate lengths (e.g., about 2 to 10 mm for a typical 130-nm technology with a clock frequency of 1.2 GHz). Most global interconnects in advanced technologies fall into this range of lengths and thus need to be represented by *RLC* lines. Local and intermediate levels of metal lines can be modeled as *RC* circuits because they are highly resistive. Figure 8.5 also shows that progressively shorter interconnects will be dominated by inductance effects as well, due to the increase of frequency. It should be pointed out that the discussions above are based on a single interconnect. In reality, there are multiple neighboring lines at the same layer, and thus the inductance in equations (8.2.4) and (8.2.5) should include the total inductance in the current return loop. In addition to the transmission line behavior being affected by the interconnect layout, it also depends on the boundary conditions at both ends of the line, which can be changed by tuning the driver resistance and loading capacitance.

As signal frequencies continue to increase beyond 10 GHz and line lengths continue to increase past the signal wavelengths, the quasistatic assumptions cease to be effective. With the inclusion of displacement current, interactions between pairs of elements are not instantaneous and signal degradation becomes the key for correct analysis. Hence, the extraction of interconnect parasitics should be combined with electrical analyses to solve time- or frequency-domain responses. Although this is not an issue for either current or near-term designs, full-wave methods such as that in [7] will eventually be necessary for interconnect analysis.

8.2.2 *RC* Extraction

Although inductance effects are increasingly important for modeling on-chip interconnects, the *RC* equivalent circuit is still sufficiently accurate to model the majority of on-chip interconnects, particularly at local and intermediate layers. This parasitics extraction is necessary at both early design and postlayout stages. In early top-down design flows, the inclusion of interconnect parasitics is critical

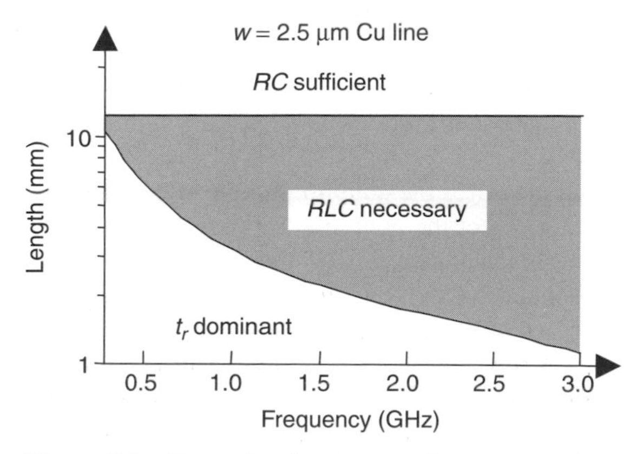

Figure 8.5 Example of inductance-important region.

for performing design synthesis and capturing major timing and signal integrity problems. Later, after the design is flattened, parasitic values are extracted from the layout to verify the design specifications. In general, for a well-defined layout pattern, *RC* and even *L* values can be obtained by applying three-dimensional electromagnetic field solving techniques. These numerical methods (e.g., the finite difference method, finite element method, method of moments) achieve very high accuracies but face two significant limitations in realistic design cycles. First, during early design planning, detailed layout information is not available for field solvers, and thus the flexibility of such an approach is limited. Second, as the total number of transistors on a chip exceeds several million, computation of a full capacitance matrix is prohibitively expensive. For these reasons it is more common in design to employ analytical models or lookup tables in conjunction with layout pattern recognition algorithms to achieve efficient run-time extraction. The accuracy of these analytical or table-lookup models is ensured against golden data obtained from either three-dimensional field solvers or test structure measurement results.

Modeling Approaches for RC Extraction On-chip interconnect structures are usually composed of metal lines with rectangular cross sections using a Manhattan layout across layers (Figure 8.1). This architecture not only simplifies both the manufacturing process and routing algorithms, but also greatly reduces the complexity of *RC* modeling efforts. For a uniform metal line of width w and thickness t [Figure 8.1(*b*)] its dc resistance per unit length, R, can be calculated as

$$R = \frac{\rho}{wt} \tag{8.2.6}$$

where ρ is the metal resistivity ($\rho = 2.2\ \mu\Omega\cdot\text{cm}$ for copper and $3.3\ \mu\Omega\cdot\text{cm}$ for aluminum). For instance, a typical 3-mm global copper line, which has $w = 0.8\ \mu\text{m}$ and $t = 0.8\ \mu\text{m}$, can be modeled as a 103-Ω resistor at the dc condition. In addition to metal lines, vias, which connect multiple layers vertically, contribute to path resistance as well. Via resistance per area is about $10^{-9}\ \mu\Omega\cdot\text{cm}^2$ at the 90-nm technology node [1]. Therefore, a $0.25\ \mu\text{m} \times 0.25\ \mu\text{m}$ via can be modeled as an equivalent 1.6-Ω resistor. With the steady increase in metal layers and shrinking of via size, the effect of via resistance can no longer be neglected in timing models; it can contribute as much as an additional 10% to the total critical path delay [8].

Metal capacitance measures the coupling between lines through electric fields. Depending on whether or not the coupling line is grounded, it is usually referred to as metal-to-ground capacitance, C_g (if the coupling line is ac grounded), or metal-to-metal capacitance, C_c (if the coupling line is a signal line). Examples of cross-sectional views of C_g and C_c are shown in Figure 8.6. From Maxwell's equations (8.2.1) we know that an electric field can be fully shielded by metal lines. Thus, capacitive coupling is a short-range effect: When there are multiple lines on the same layer, capacitive coupling decays rapidly with the increase in neighboring orders. For example, C_c between second- or even higher-order neighbors (i.e., there is at least one line inserted between coupling lines) is usually less than 10% of C_c between nearest neighbors. To achieve simplicity in modeling while maintaining sufficient analysis accuracy, only the nearest C_c is considered in extraction and performance analyses; the higher-order C_c values can be neglected. This results in the following matrixes, which are generally used in RC analysis of coplanar interconnects (Figure 8.6):

$$\mathbf{R} = \begin{pmatrix} \ddots & 0 & 0 & 0 & 0 \\ 0 & r_{i-1} & 0 & 0 & 0 \\ 0 & 0 & r_i & 0 & 0 \\ 0 & 0 & 0 & r_{i+1} & 0 \\ 0 & 0 & 0 & 0 & \ddots \end{pmatrix} \quad \mathbf{C} = \begin{pmatrix} \cdots & \cdots & 0 & 0 & 0 \\ \cdots & \cdots & \cdots & 0 & 0 \\ 0 & -c_{ci,i-1} & c_{gi} + c_{ci,i-1} + c_{ci,i+1} & -c_{ci,i+1} & 0 \\ 0 & 0 & \cdots & \cdots & \cdots \\ 0 & 0 & 0 & \cdots & \cdots \end{pmatrix} \tag{8.2.7}$$

Each value of r_i can be calculated from equation (8.2.6) while C_g and C_c elements can be generated from analytical models or using table-lookup approaches such as those described below.

To calculate capacitance efficiently at both stages of design synthesis and postlayout verification, multiple-layer three-dimensional interconnects are usually simplified to two-dimensional [9–11] or quasi-three-dimensional structures [12], based on layout patterns. If the layers above and below a line are routed densely, they can be approximated as a ground plane, leading to two-dimensional models, as shown in Figure 8.6. Under this condition, C_g and C_c becomes scalable functions of line cross-sectional dimensions. Analytical models for C_g and C_c are listed below [11].

For top-layer interconnects [metal line above one ground plane, as shown in Figure 8.6(a)]

$$\frac{C_g}{\varepsilon} = \frac{w}{h} + 2.217\left(\frac{s}{s + 0.702h}\right)^{3.193} + 1.171\left(\frac{s}{s + 1.510h}\right)^{0.7642}\left(\frac{t}{t + 4.532h}\right)^{0.1204}$$

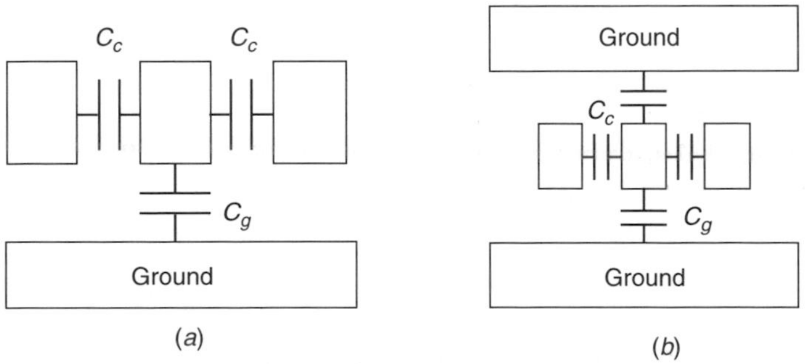

Figure 8.6 Two-dimensional capacitance modeling for local and global interconnects (cross-sectional view): (a) top-layer interconnects; (b) local-layer interconnects.

$$\frac{C_c}{\varepsilon} = 1.144\frac{t}{s}\left(\frac{h}{h+2.059s}\right)^{0.0944} + 0.7428\left(\frac{w}{w+1.592s}\right)^{1.144}$$
$$+ 1.158\left(\frac{w}{w+1.874s}\right)^{0.1612}\left(\frac{h}{h+0.9801s}\right)^{1.179} \tag{8.2.8}$$

For local-layer interconnects [metal line between two ground planes, as shown in Figure 8.6(*b*)]

$$\frac{C_g}{\varepsilon} = \left(\frac{w}{h_1}+\frac{w}{h_2}\right) + 2.04\left(\frac{t}{t+4.5311h_1}\right)^{0.071}\left(\frac{s}{s+0.5355h_1}\right)^{1.773}$$
$$+ 2.04\left(\frac{t}{t+4.5311h_2}\right)^{0.071}\left(\frac{s}{s+0.5355h_2}\right)^{1.773}$$
$$\frac{C_c}{\varepsilon} = 1.4116\frac{t}{s}\exp\left(-\frac{2s}{s+8.014h_1}-\frac{2s}{s+8.014h_2}\right)$$
$$+ 1.1852\left(\frac{w}{w+0.3078s}\right)^{0.25724} \tag{8.2.9}$$
$$\cdot\left[\left(\frac{h_1}{h_1+8.961s}\right)^{0.7571}+\left(\frac{h_{21}}{h_2+8.961s}\right)^{0.7571}\right]\exp\left(-\frac{2s}{s+3h_1+3h_2}\right)$$

where ε is the dielectric constant and dimension variables are as defined in Figure 8.1 [in equation (8.2.9), h_1 and h_2 refer to dielectric thickness above and below the metal line, respectively]. These models are generated from physical considerations and the coefficient values are fitted from field solver results. Therefore, they are highly scalable and achieve an accuracy of within 5 to 10%. After two-dimensional values of C_g and C_c per unit length are obtained, the total capacitance can be calculated by multiplying them by the length of line.

In more general cases, a long interconnect can first be partitioned into several segments, where each segment can be matched to a predefined layout pattern, depending on line conditions at the same layer and layers above or below [12,13]. Then, to build the **C** matrix, an analytical model or lookup table, which is verified with field solver or silicon measurements, is applied to calculate C_g and C_c for each segment [13]. Combined with the analogous **R** matrix, *RC* timing and noise characteristics can be examined further using analysis tools.

On-chip Parasitics Characterization Techniques for characterizing parasitics are necessary not only for model verification, but more importantly, for the generation of direct models from silicon measurements. For example, with an elegant and simple technique available for capacitance measurements, capacitance values for typical layout patterns can be extracted directly from test structures and then used to build lookup tables, thereby reducing process uncertainties and modeling errors. Because on-chip interconnects are relatively narrower and shorter than off-chip metal lines in packages, they have smaller capacitances (<100 fF/mm) and inductances (<1 nH) and thus require a high-resolution characterization that is easy to implement. The general relationships between *RLC* and electrical parameters (voltage and current) are

$$R = \frac{V}{I} \qquad I = C\frac{dV}{dt} \qquad V = L\frac{dI}{dt} \tag{8.2.10}$$

From the first equation, metal resistance values can easily be extracted from dc $I-V$ measurements. Whereas off-chip capacitances can be characterized by *LCR* meters, it is desirable to perform capacitance extractions on-chip for interconnects, due to their small values. Based on equation (8.2.10), the charge-based capacitance measurement (CBCM) technique has been developed and achieves a resolution of sub-0.1 fF [14]. The accuracy of this technique is limited by transistor mismatch in the test circuits. As circuit frequencies exceed the gigahertz range, techniques for characterizing high-frequency impedances in either the frequency (e.g., S-parameter characterization) or time domain are necessary [15,16]. These techniques are generally applicable to both on- and off-chip *RLC* extractions. However, the testing procedure is much more complicated than in-situ measurements and usually requires special test structures.

8.2.3 Inductance Extraction

Inductance effects become significant in the nanometer regime, particularly for global interconnects. In contrast to *RC* extraction, in which only a line itself and its nearest neighbors should be included, inductance coupling is a long-range effect and, since magnetic fields penetrate well beyond the metal surface, decays very slowly with increasing line spacing. The fundamental definition of inductance is

$$L = \frac{\oint_A B\,ds}{I} \tag{8.2.11}$$

where I is the current, B the magnetic field induced from I, and s the integration loop. If s is the same as I, L is self-inductance, whereas if s follows a different conducting path, equation (8.2.11) defines mutual inductance. This definition indicates that inductance calculations follow a loop property, and thus the determination of inductive behavior must consider the entire current loop. However, in modern interconnect structures [driver-line-loading capacitance, as shown in Figure 8.3(a)], there are no dc paths to form a well-defined loop. As a result, return current usually spreads over a long range, which complicates the analysis. Consequently, the extraction analysis should include all neighboring lines that are possibly involved in the current loop.

Partial and Loop Inductance Because of the uncertainty of the return current path, it is difficult to calculate loop inductance [defined in equation (8.2.11)] in realistic designs. To overcome this difficulty, the concept of partial inductance is introduced, in which the induced current is assumed to return at infinity, avoiding the need to define the return loop. This inductance calculation technique, known as the partial element equivalent circuit (PEEC) method [17], is very suitable for design automation since it depends only on the geometry of the lines. Based on PEEC, a general partial inductance matrix can be composed from L calculations of each line segment:

$$\mathbf{L} = \begin{pmatrix} \cdots & \cdots & \cdots & \cdots \\ \cdots & L_{si} & L_{mij} & \cdots \\ \cdots & L_{mji} & L_{sj} & \cdots \\ \cdots & \cdots & \cdots & \cdots \end{pmatrix} \tag{8.2.12}$$

This matrix is then combined with the RC matrix [equation (8.2.7)], and with the aid of circuit simulators, the result is used to determine the current loop. In this method, each inductance element (i.e., self-inductance L_s or mutual inductance L_m in equation (8.2.12) can be calculated using field solvers, such as FastHenry and Raphael, or closed-form solutions. For rectangular cross-sectional wires, the closed-form solutions are first derived by Rosa and Grover [18] and then simplified to the following relationships when $1 >> w$, t, and d [19]:

$$L_s = \frac{\mu_0}{2\pi}\left[l \ln \frac{2l}{w+t} + \frac{l}{2} + 0.2235(w+t)\right] \tag{8.2.13}$$

$$L_m = \frac{\mu_0}{2\pi}\left(l \ln \frac{2l}{d} - 1 + d\right)$$

Here μ_0 is the magnetic permittivity of the dielectrics; w, t, and l are the width, thickness, and length of the segment, respectively; d is the center-to-center distance between two lines; and L_m is the mutual inductance of two equal-length lines (a more general solution for L_m of non-equal-length lines is also provided in Ref. 19). These expressions indicate that inductance has a nonlinear dependence on segment length. Therefore, in contrast to RC extraction, which is scalable with length, L must be calculated over the entire length of the wire. Furthermore, the logarithmic function in equation (8.2.13) implies that L has a weaker dependence on line geometry than do R and C. Note that only lines on the same layer, which are parallel to each other, contribute to inductive coupling; lines on neighboring layers do not influence the coupling, due to their orthogonal layout.

Although PEEC can deal with general inductance extractions without a priori knowledge of the current return loop, the nonsparsity of the inductance matrix (caused by the long-range inductive coupling) leads to expensive computations in further analyses [20]. Unlike the **C** matrix, in which it is sufficient to keep only the short-range coupling values, the **L** matrix cannot be truncated for simplicity; simply discarding the L_m values of distant neighboring lines causes model instability [21]. Many efforts have been employed to improve the computation efficiency of this complex matrix. One example is the **L** matrix truncation method, which uses the power grid as the boundary for the susceptance matrix extraction, which is the inverse matrix of **L** and has the desirable property of sparsity [22,23]. Although a number of computationally efficient techniques have been developed thus far, there is still not one solution that is both SPICE compatible and simple and general enough for on-chip interconnect structures. An approximated loop inductance model is desirable for evaluating the physical definition of inductance [equation (8.2.11)] especially during the early stages of design exploration. This model can be described using lookup tables [24] or analytical models. The latter approach is particularly suitable for the specialized global clock structure, which is well shielded from neighboring wires by the power and ground lines [25,26].

Frequency-Dependent R(f)L(f) The phenomenon of frequency-dependent R and L $[R(f),L(f)]$ has previously been a concern only for package and microwave design due to their large wire sizes. However, as the chip operating frequency increases into the gigahertz regime, these effects migrate to on-chip interconnects as well. This is because at high frequencies the depth of current penetrating the metal (skin depth) becomes comparable to or even smaller than the cross-sectional dimensions of the global interconnect. For example, at 1 GHz the skin depth of copper is about 2 μm; as the frequency increases, this depth decreases with the square root of the frequency. As a result, the conducting cur rent density in the metal line is no longer uniform and the metal impedance becomes dependent on the operating frequency. Therefore, the conventional representation of an on-chip wire using a constant R and L (dc RL) is no longer adequate because it does not model this frequency dependence. Figure 8.7(a) illustrates the distribution of the cross-sectional current density for three parallel lines using Raphael, the PEEC-based RLC extraction tool. As the frequency rises, the current moves toward the surface of the wire and away from neighboring lines conducting current in the same direction. This nonuniform current distribution is referred to as the *skin effect* for a single-line case and the *proximity effect* when neighboring lines are inductively coupled, and leads to significantly larger resistances at higher frequency, as shown in Figure 8.7(b) (note that line inductance drops only slightly and eventually saturates).

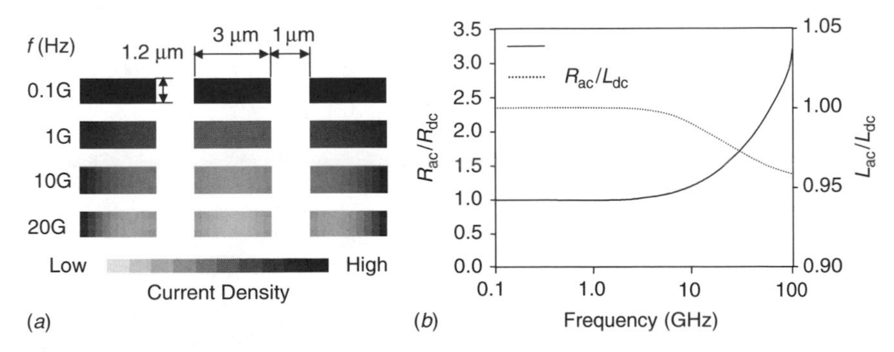

Figure 8.7 PEEC simulation results of $R(f)$ and $L(f)$: (a) cross-sectional current density distribution (copper line thickness = 1.2 μm); (b) frequency dependence of R and L.

Publications have proposed a number of ways to accurately analyze the effect of frequency-dependent R and L: the use of high-frequency RL [26] values, predetermined loop RL values [25], or the analytical equivalent-circuit model [27] for timing estimation in the gigahertz regime, which increases the extraction and analysis complexity. However, a waveform comparison between dc RL and $R(f)L(f)$ shows that the difference in predicted delay is very small: in Figure 8.8(*a*), for the projected 90-nm CMOS technology [2], the delay and rise times found using the dc RL model match well with those given by the $R(f)L(f)$ model [28]. This phenomenon can be explained by the dominance of the inductive impedance ωL in the voltage response at the rising edge: In current copper wire technology, when the switching frequency exceeds multiple gigahertz and the skin effect becomes pronounced, ωL is usually much larger than R. Thus, in the gigahertz regime, the delay is more sensitive to changes in L than to changes in R. As frequency increases further, however, L decreases only slightly [Figure 8.7(*b*)]; in addition, R and L have opposing dependencies on frequency, which further reduces the overall impact of $R(f)L(f)$ on signal delay. In conclusion, dc RL values are sufficient for delay analysis. After the rising edge, the output signal slows down and resistance dominates in the overshoot and ringing portions of the waveforms, so that there are differences in the amplitude and period of ringing.

In contrast to the insensitivity of delay to $R(f)L(f)$, $L(di/dt)$ noise on power supply is strongly suppressed by $R(f)L(f)$, due to the larger resistance values at higher frequencies. As shown in Figure 8.8(*b*), using $R(f)L(f)$ predicts smaller peaking and faster damping. This implies that less decoupling capacitance is required to stabilize the power supply when considering the frequency dependence of metal impedance. In our example, to reduce the peak noise below 10% of V_{dd}, dc RL predicts that 134 pF of decoupling capacitance is required, while

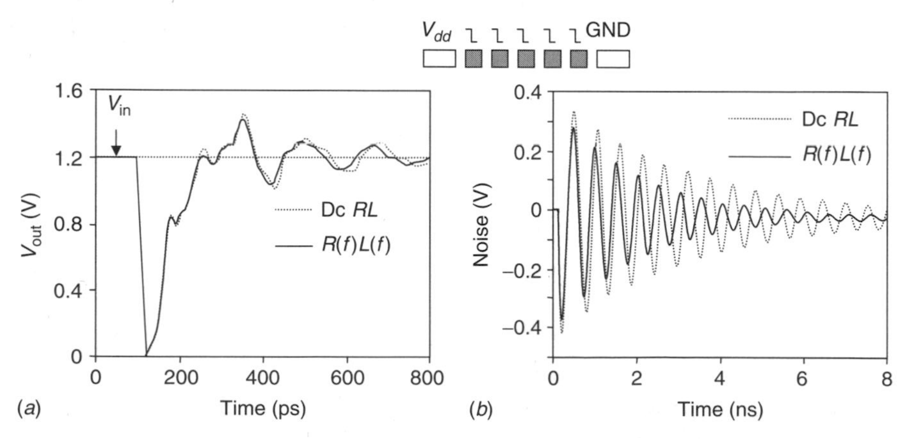

Figure 8.8 Impact of $R(f)L(f)$ on circuit performance: (*a*) Output waveforms with ramp input ($w = 5$ μm, $s = 2$ μm, length = 3 mm; $w_{Vdd} = 10$ μm); (*b*) Noise on the ground line (for power lines, $w = 50$ μm, length = 500 μm; $C_{\text{decoupling}} = 50$ pF; $I = 100$ mA).

$R(f)L(f)$ predicts 115 pF. By considering the frequency dependence, more than 15% of the area can be saved from this smaller decoupling capacitance requirement. This effect can be significant since the increasing need for power supply stability has led to rapidly rising area costs of decoupling capacitors. Therefore, correct consideration of $R(f)L(f)$ at the multiple-gigahertz regime can help alleviate such concerns while providing sufficient power supply stability.

Difference between On- and Off-chip Inductance For several decades, inductance has been a concern for the design of off-chip interconnects such as those on board-level package designs. Although a great deal of inductance modeling work has been developed for off-chip designs, these efforts cannot be adopted directly for on-chip interconnects because of the more complicated wiring environment and different geometries of on-chip interconnects. The major differences between on- and off-chip interconnect are summarized as follows:

1. *Return path.* In off-chip designs, ground planes are usually placed generously in the layout to reduce the inductance (e.g., the stripe-line structure); these additional wires do not add significant overhead to the design. The resulting current-return paths are well defined, and therefore approximate formulas can be derived for inductance analyses. In contrast, on-chip interconnects usually do not have well-defined return current paths because of the limited routing resources.
2. *Resistive loss.* Off-chip interconnects have larger cross sections than those on-chip. They are much less lossy than on-chip interconnects and thus suffer from more prominent transmission-line behavior, such as wave reflections. Low-loss transmission-line theory can be applied in off-chip interconnect analysis. However, on-chip transmission lines usually suffer from high loss, and the analysis is more complicated;
3. *Routing complexity.* On-chip interconnects are routed much more densely than those off-chip, and a significantly larger number of neighboring wires need to be included in the analysis to estimate the current return path and interconnect behavior correctly.
4. *Termination.* It is relatively easy to terminate off-chip interconnects by using a resistor to match the characteristic impedance of the line [$Z_0 = (L/C)^{1/2}$]. In contrast, it is very challenging to terminate an on-chip interconnect ideally, because the characteristic impedance of on-chip interconnects is not purely resistive ($Z_0 = [(R + j\omega L)/j\omega C]^{1/2}$), and the driver size is typically optimized for delay minimization. Its output impedance may not be equal to Z_0, and the input impedance of on-chip loads is almost exclusively capacitive.

8.3 SIGNAL INTEGRITY ANALYSIS

After the physical layout information is converted into equivalent electrical *RC* or *RLC* components, depending on the frequency and accuracy of interests, the interconnect performance can be analyzed using either generic circuit simulators (e.g., HSPICE, SPECTRE) or analytical modeling approaches. In conventional performance-driven designs, signal path delay is the sole focus of design exploration and optimization. However, signal integrity problems that were previously negligible, including crosstalk noise, signal slew rate, and voltage overshoot, emerge to the center of wire-centric design in the nanometer regime as a result of both rapid technology scaling and an increasingly tighter timing budget. Correct and adequate consideration of these issues is crucial to successful chip design and implementation, and has been learned for many design cases at 130 nm and succeeding technology generations. In particular, it is essential for timing analysis tools to capture degradations in signal integrity, which have become comparable in magnitude to the nominal timing. In this section we first present practical and efficient techniques to analyze signal integrity concerns in physical design for both *RC* and *RLC* interconnects. Then, to incorporate the impact of crosstalk noise into early timing analyses (e.g., at global routing and synthesis stages) and thus achieve fast timing closure, design methodologies for noise-aware timing analysis are discussed in further detail.

8.3.1 Interconnect Driver Models

For design simplicity, signal transportation along an on-chip wire can conveniently be partitioned into two parts [Figure 8.9(*a*)]: from the input of the gate (driver) to the output of the gate, and from the near end of a line to the far end (i.e., input of the receiver). The signal delay is thus decoupled into gate delay and line delay, and each part is first analyzed individually and then summed together to calculate the overall timing. The performance of local circuits is usually dominated by gate delay, due to the short length of interconnects, whereas for global signaling, line delay is at least as important as gate delay, even sometimes accounting for the majority of signal timing due to the long wire length. To mitigate this effect, the size of global interconnect drivers, including both logic gates and inserted repeaters, should be optimized to minimize total path delay. Furthermore, even if a driver does not contribute unwanted noise, its size strongly affects the magnitude of crosstalk noise: A large driver in the static state provides a better dc connection to ground and as a result, suppresses both capacitive and inductive coupling-induced noise. For these purposes, it is important to have proper driver and gate loading models, for efficient analysis and optimization.

A switching driver can be modeled as a either time-variant voltage source [29] or a current source [30], as shown in Figure 8.9(*b*) and (*e*) (note that the receiver is simply modeled as a loading capacitor at the far end of the line). The Thévenin equivalent model [Figure 8.9(*b*)], which is comprised of a ramping voltage source and a linear resistor (R_{dr}), naturally captures the interaction of the gate with interconnect loading. For instance, in typical gate delay analyses, *RC* or *RLC* interconnects are usually approximated as an effective capacitance (C_{eff}) or single-Π circuit [Figure 8.9(*c*) and (*d*), respectively], using model order reduction techniques (e.g., moment-matching-based asymptotic waveform evaluation [31–33].

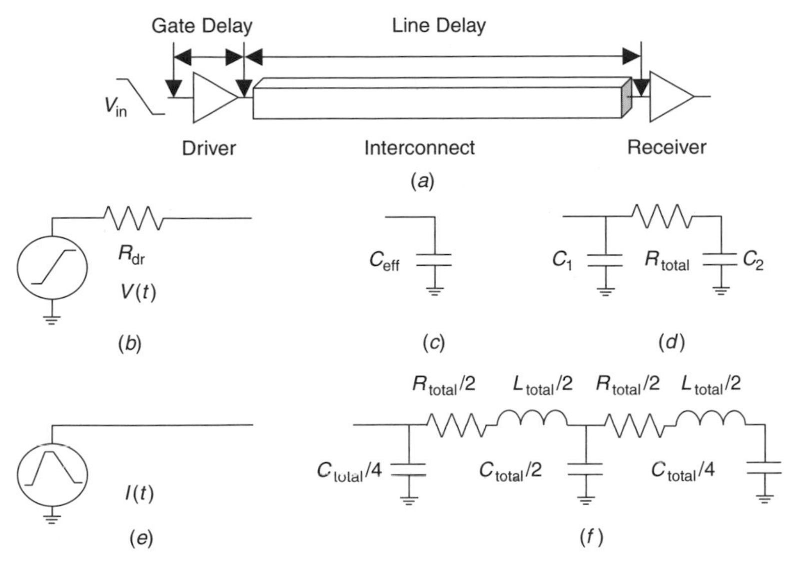

Figure 8.9 Various driver [(*b*) and (*e*)] and line loading [(*c*), (*d*), and (*f*)] models for gate delay calculation: (*a*) definitions of gate delay and line delay; (*b*) Thévenin model; (*c*) Effective capacitance; (*d*) single-Π model; (*e*) Norton model; and (*f*) two-Π model for *RLC* lines.

Under this approximation, the gate delay can easily be calculated from the $R_{dr}C_{eff}$ product. On the other hand, using a single resistor to model a switching gate can lead to inaccuracies in the prediction of the slew rate of gate output signal, especially when its input slew rate and loading capacitance vary significantly over a wide range. To overcome this problem in practical *RC* analyses, the values of R_{dr} and C_{eff} are fit by iteratively matching two points of the gate output waveform (e.g., 50% and 90%). For instance, in characterizations of cell library components, the procedure generates a lookup table of R_{dr} as a function of loading capacitance and input slew rate. Besides the ability to make gate delay predictions, the Thévenin model also provides the basis for optimizing the driver size for overall path delay minimization. A rule of thumb for this purpose in *RC* analysis is the condition that [34]

$$\text{gate delay} = \text{line delay} \tag{8.3.1}$$

Beyond conventional *RC* timing analysis, the value of R_{dr} in the Thévenin model is not applicable to crosstalk noise predictions because a static gate is always in its linear operation condition, whereas a switching gate operates in both linear and saturation regimes. A different resistance, whose value is usually smaller than R_{dr}, should be used to model a static driver for signal integrity analyses. Furthermore, a single linear resistor is too simple to predict the full-waveform characteristics in *RLC* analyses and cannot fully capture higher-order phenomena such as voltage overshoot and signal ringing. More accurate waveform integrity analyses can be performed by employing a time-varying current source model

[Figure 8.9(*e*)], which physically captures transistor behavior over the entire switching range [30]. For *RLC* interconnects, it is also necessary to use more complicated equivalent loading models, such as the symmetrical 2-Π *RLC* circuit [Figure 8.9(*f*)], in order to match the waveform reflections at both the near and far ends of the wire [30]. (Note that in the single-Π model, C_1 and C_2 are not equal, due to the resistance shielding effect.)

8.3.2 *RC* Interconnect Analysis

The model created from interconnect performance analyses can be a line tree, which contains branching segments with different *RC* (or *RLC*) elements, loading capacitances, and neighbor coupling conditions, but not floating capacitors and resistor loops. General solutions to this linear system rely on a variety of numerical techniques performed in either the time or frequency domain. For instance, one such approach combines *RC* (or *RLC*) matrixes [equations (8.2.7) and (8.2.12), respectively] with Kirchoff's voltage and current laws [e.g., equation (8.2.10)] and solves the output voltage using a matrix approximation technique [27]; another approach utilizes the transfer function from the input to output and then predicts the signal delay and output waveform by matching a number of moments [27,31]. Not only are these approaches capable of handling various layout configurations and switching patterns, but they also can provide very accurate timing and noise information for design verifications. However, their role is very limited in the placement and routing stages, because of the difficulty involved in explicitly relating the line performance to physical layout using numerical solutions. Furthermore, to achieve high accuracies, these numerical techniques usually require an expensive computation time, which restricts their use in advanced full-chip analysis. In contrast to numerical solutions, analytical performance metrics have excellent model scalability and simplicity, making them suitable for purposes of design optimizations; however, they have trade-offs between accuracy and model generality. To obtain insights into signal integrity issues and to further investigate circuit and physical design solutions, our discussions in this section are focused on analytical modeling efforts.

In local and intermediate layers of on-chip interconnects, resistive and capacitive effects dominate the response of the line to voltage switching, although it is also necessary to consider inductance effects for some global interconnects. Within the accuracy requirements, *RC* analyses are preferred over *RLC* analyses in practice because of their simplicity and efficiency, advantages that originate from the nature of short-range capacitance coupling. Therefore, before performing analyses of timing and signal integrity, a screening process based on criteria similar to those described in Section 8.2.1 is usually performed to identify and limit the use of *RLC* modeling. Even with rapid technology scaling, *RC* analyses are still advantageous and therefore are used for the majority of interconnect timing and crosstalk noise estimates.

***RC* Interconnect Timing Analysis** Much effort has been made to develop analytical *RC* interconnect timing metrics because of their ability to link line performance easily with physical layout definitions (e.g., line widths, line lengths, spaces). The most popular metric is the Elmore delay, which describes the first moment of an impulse response, because it is suitable for all levels of *RC* tree analysis [35]. As proven in Ref. 36, the simple Elmore delay is the upper bound on the actual 50% V_{dd} delay of an *RC* tree with a ramp input applied and hence it is a safe choice for *RC* delay estimates. To further improve the accuracy of the Elmore delay metric as well as to extend prediction to include more characteristics of switching (such as the slew rate), the full output waveform of a single *RC* line can be solved in closed form by asymptotically matching higher orders of moments from the transfer function [37]. The accuracy of these analytical metrics is usually within 10% of the numerical results, which is sufficiently accurate for early design stages. However, it should be noted that these metrics handle only the case of a single line or line tree and do not consider the impact of neighboring line switching, an issue that becomes increasingly important as technology scales down.

The timing analysis of a line is complicated by the presence of neighboring lines, whose electrical behavior couple into that of the target line via C_c (Figure 8.10). To simplify this coupling scenario, the target line can first be decoupled into an equivalent single line and then the analytical metrics (e.g., the Elmore delay) can be applied to calculate timing. In this approach, C_c is converted to an effective ground capacitance using the concept of switching factors (SFs), and then merged with C_g to separate a pair of *RC* lines, as shown in Figure 8.10. The idea of the switching factor is based on the *Miller effect* across the coupling capacitance C_c. This effect can be understood by considering the following scenario. If the neighbor line (i.e., line B in Figure 8.10) is in its static state, the voltage swing on C_c is V_{dd}; however, when the voltages at both nodes of C_c (i.e., V_A and V_B) switch simultaneously, C_c experiences a different voltage swing. In this situation, to approximate C_c as a ground capacitance with only one switching node, the effective C_c should be calculated as

$$C_{c\,\text{effective}} = \text{SF} \cdot C_c \quad \text{and} \quad \text{SF} = 1 - \frac{\Delta V_B}{\Delta V_A} \tag{8.3.2}$$

where ΔV is the voltage change during the overlapping period of voltage switching. According to this formula, SF equals 0 and 2 for the in-phase (i.e., V_A and V_B switch in the same direction) and out-of-phase cases, respectively, if V_A and

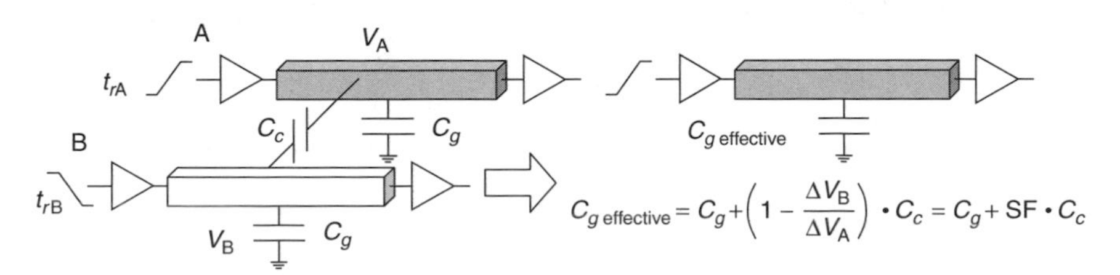

Figure 8.10 Switch factor–based *RC* line decoupling.

V_B are both step inputs. However, in the nanometer regime the finite slew rate is no longer negligible, and the signal switching can no longer be modeled as step input. As a result, the bound of SF depends further on the ratio of the slew rates of V_A and V_B (t_{rA} and t_{rB}) and can be as large as [−1, 3], assuming that the switching threshold of the receiver is 50% V_{dd} [38,39]. The worst-case delay scenario for V_A, in which SF = 3 instead of 2, occurs when V_A and V_B are out of phase and t_{rB} is at least twice as small as t_{rA}. Consequently, the equivalent total ground capacitance is $C_g + 3C_c$, which is larger than $C_g + 2C_c$ for step inputs. Since C_c usually dominates C_g as a result of technology scaling [Figure 8.2(*b*)], this modification in SF is important for correct estimates of the timing bound.

Capacitive Coupling Noise In switching factor-based timing analysis, SF is 0 if line B does not switch. This approximation is appropriate when the nonswitching line can be treated as a ground node, which is true only if the coupling noise is negligible. However, due to higher C_c/C_g ratios and larger line resistances in advanced technology, crosstalk noise has become so pronounced that this assumption no longer holds. Figure 8.11(*a*) is a representation of two coupled *RC* lines using a lumped-circuit model to evaluate the resulting noise (line is modeled as 2-Π). The switching line that induces noise is usually called the *aggressor*, and the line that suffers from the interfering noise is termed the *victim*. Note that for capacitive coupling, only adjacent lines affect the victim; the effect from higher-order neighbors is negligible. The capacitive noise always appears in the same direction as the voltage switching of the aggressor line. A large noise on the victim line not only leads to excessive delay uncertainty but also introduces potential logic malfunctions. The latter problem is especially serious for designs with lower noise margins, such as those with higher clock frequencies, lower supply voltages, and those implemented using dynamic logic. Because high-speed circuits have many of these noise-susceptible properties, the effects of crosstalk noise are considered at nearly every stage of their design in order to reduce the number of expensive design iterations and ultimately ensure success of the design.

Two major metrics are typically employed to evaluate the impact of noise: noise peak (V_{peak}) and noise width, as illustrated in Figure 8.11(*b*). V_{peak} describes

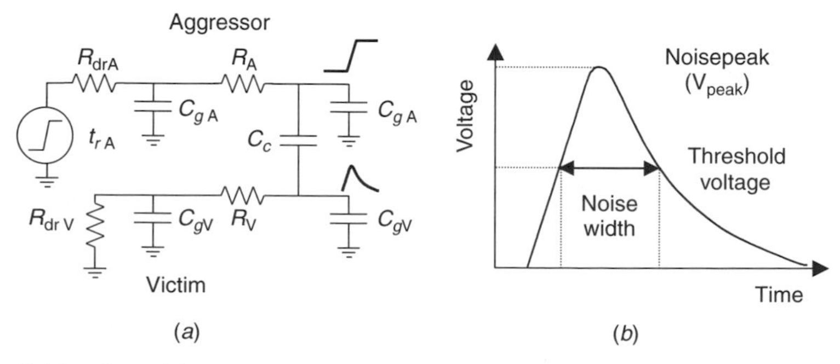

Figure 8.11 Capacitive coupling noise in *RC* analysis: (*a*) lumped model for a pair of coupled *RC* lines; (*b*) major noise characteristics.

the maximum amount of crosstalk noise between two nets, and its value depends on the coupling capacitance, other loading capacitances and parasitic resistances, the switching slew rate of the aggressor, and the victim driver strength. Using the dominant-pole method [40], V_{peak} can be approximated as

$$\frac{V_{\text{peak}}}{V_{dd}} = \frac{t_x}{t_{r\text{A}}}(1 - e^{-t_{r\text{A}}/t_v}) \tag{8.3.3}$$

where t_x and t_v are the settling times for the aggressor and victim lines, respectively, whose values can be calculated in closed form from other *RC* parasitics [41,42]. Similar analytical solutions are provided in Refs. 43 and 44. According to these theoretical results and experiments using actual circuits, V_{peak} has been found to be more sensitive to the ratio $C_c/C_{g\text{V}}$ than to other parameters [42]. In fact, if the victim line is highly resistive and the aggressor switching is very fast, V_{peak} approaches the upper limit of charge sharing:

$$\frac{V_{\text{peak}}}{V_{dd}} = \frac{C_c}{C_c + C_{g\text{V}}} \tag{8.3.4}$$

In addition to $C_c/C_{g\text{V}}$, the resistance of the victim driver (R_{drV}) also plays an important role in determining the value of V_{peak}. Incorporating these observations into the design techniques helps to improve both optimization and suppression of the undesired coupling.

The peak noise amplitude V_{peak} is not the only metric used to characterize noise. Even if V_{peak} exceeds a certain threshold, the receiver may still be immune to noise in certain cases: for instance, if the noise has a very narrow width and the receiver capacitance is large (i.e., the noise is too fast to trigger a low-bandwidth receiver). For this reason, noise width, which describes the length of time that the value of the noise is larger than a given threshold, is generally used to represent the speed of the noise. One advantage of this metric in practical design is that it can be solved in closed form and thus fits well in routing and screening algorithms [41]. To predict the effect of noise on timing more accurately, we need to have a representation of the entire noise waveform. Capacitive crosstalk noise, such as the one whose characteristic is illustrated in Figure 8.11(*b*), can be modeled as a linearly rising edge that reaches the V_{peak} value and decays exponentially after that peak [44]. More details of this model are presented in Section 8.3.4.

8.3.3 *RLC* Interconnect Analysis

While *RC* analyses are most applicable to highly resistive nets at local and intermediate layers, inductance effects are frequently encountered in wide global wires, which transport signals between functional blocks, distribute clock references, and supply power to logic gates. To design these wires properly in high-performance circuits, *RLC* models and techniques for characterizing the signal transportation are required. However, the inclusion of inductance effects increases the complexity of timing and noise analyses significantly, for two main reasons. First, unlike capacitive coupling, in which a line is affected only by its nearest neighbors, the effect of inductive coupling extends for a longer range. In contrast to the behavior of electric fields, magnetic fields are nonzero at the surface of a metal, and therefore mutual inductance decays very slowly with distance. Second, there is uncertainty associated with the inductive current return path in a circuit because on-chip interconnect structures do not provide a clear dc path to form the current loop. Therefore, *RLC* analysis is not a local problem, meaning a sufficiently large number of neighbors must be considered to obtain the correct solution. It is also more difficult to obtain simple analytical solutions for major performance metrics because inductance induces nonmonotonic behavior (e.g., ringing and overshoot), as indicated by equation (8.2.3), and thus more moments need to be matched to approximate the output characteristics.

***RLC* Interconnect Timing Analysis** In contrast to *RC* lines, *RLC* interconnects behave differently during the propagation of voltage switching; they have an increased delay, faster slew rate, and ringing as well as overshoot. Figure 8.12(*a*) is an example of the impact of inductance on the ramp response waveform of a typical global line in 180-nm technology. While the 50% V_{dd} delay increases due to the inductive impedance, the signal slew rate, which is especially critical for clock edge and crosstalk noise, is reduced. When evaluating this effect, two factors should be considered: On one hand, a sharper signal edge is preferred in digital design because of the shorter period needed for the state transition; on the other hand, the faster a signal switches, the larger the crosstalk noise (from both capacitive and inductive coupling). Therefore, an optimal design should achieve the smallest slew rate within the noise constraint specified. Voltage ringing and overshoot exist in *RLC* but not in *RC* lines, as a result of the transmission line behavior of the inductive element. These undesired characteristics may cause further undesired effects: Ringing affects the signal stability of clocks since large oscillations can be sensed erroneously as

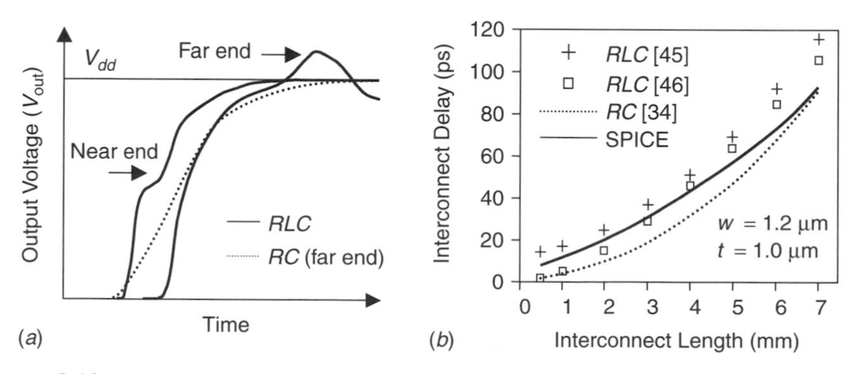

Figure 8.12 Comparisons of switching behavior: (*a*) output waveform comparison; (*b*) comparisons of delay prediction.

a transition, thus causing a logic fault, while voltage overshoots may increase power consumption and degrade the reliability of the gate oxide as well as of the overall device. In addition to waveform characteristics that exist at the far end of a *RLC* line, other undesired behavior can occur at various points along the line. For example, Figure 8.12(*a*) illustrates that at the near end of a line, the voltage waveform can have a plateau in the middle of the transition edge due to the impedance mismatch of the driver and line. Voltage plateaus such as this that occur near the threshold voltage may exacerbate the driver delay, but this effect can be optimized via driver sizing.

Similar to the analysis of *RC* lines, *RLC* timing analysis can be performed using either general numerical techniques [27,45] or analytical solutions. The latter is especially desirable for global *RLC* interconnect routing and optimization because of its efficiency. To simplify the complexity of the inductive coupling for purposes of modeling, the target line is first decoupled from its neighbors using the concept of equivalent loop inductance (L_{loop}). For a well-shielded structure such as a clock, L_{loop} can be calculated in closed form [25,26], whereas for a multibit data bus structure, it is easy to build a lookup table in which the L_{loop} values are functions of both line configurations and input switching patterns [24]. After the L_{loop} value is calculated, the output waveform for a single *RLC* line is solved using a moment-matching technique [30,46,47], and the delay metric can then be approximated as

$$\text{delay} = \frac{e^{-2.9\zeta^{1.35}} + 1.48\zeta}{\omega_n} \tag{8.3.5}$$

where ζ and ω_n are functions of the line parasitics [46]. Similar results are also provided by [47,48]. Note that signal ringing and overshoot occur when $4L/R^2C > 1$ (where R, L, and C are the total values of line parasitics over the entire line length). Based on analytical delay metrics, Figure 8.12(*b*) evaluates the line delay as predicted by various *RC* and *RLC* models. Overall, signal delay increases by about 15% when considering inductance effects, and *RLC* models match SPICE results well within the length of 2 to 5 mm [8].

Besides the property of the line itself, the signal delay also depends on the input vector of a multibit data bus, when neighboring lines are switching simultaneously. Understanding the aggressors' switching directions in the worst case is important when designing a verification tool to identify potential signal integrity problems. If the simple *RC* model is applied, it is known that a victim line suffers from the largest effective coupling capacitance when the direction of its switching patterns is exactly opposite to that of its adjacent lines, thus generating the worst-case delay. With the inclusion of long-range inductance coupling, it is necessary to determine the complete input switching pattern for worst-case estimates. The two candidates for the worst-case input vector are related by symmetry, as shown in Figure 8.13. The first case occurs when all neighboring lines switch in the direction opposite to the target line [Figure 8.13(*a*)]. This delay is largest when the lines are *RC*-dominated. The second input pattern that leads to

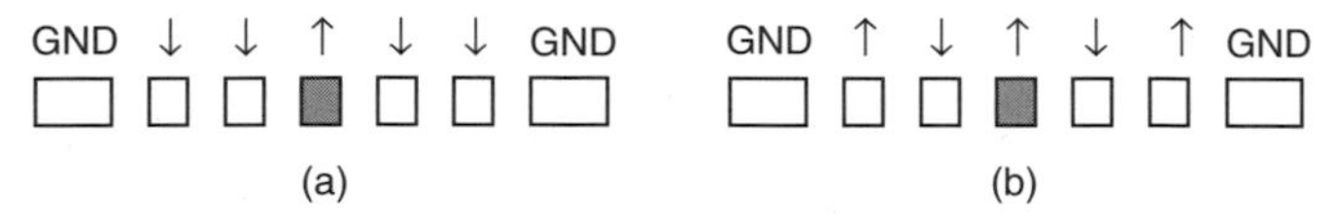

Figure 8.13 Worst-case input vector candidates (↑, switch up; ↓, switch down): (*a*) Pattern 1: capacitive coupling prone; (*b*) pattern 2: inductive coupling prone.

a worst-case delay occurs when all higher-order neighbors switch in the same direction as the target line; this occurs when the inductance effect is dominant [illustrated in Figure 8.13(*b*)] because in-phase switching generates the largest loop inductance. At the 180-nm technology node, the second input pattern accounts for the worst delay in most circuit examples [24], but in reality the worst-case switching pattern depends on the technology in addition to the *RLC* parameters.

In conclusion, with a proper input vector, the *RLC* model generates the upper bound for signal delay estimates, and the *RC* model usually provides the lower bound for slew rate calculations.

Inductive Coupling Noise Continuous increases in operating frequency and global line length not only lead to pronounced inductance effects, but also exacerbate crosstalk noise in the nanometer regime. There are two fundamental differences between capacitive crosstalk and inductive crosstalk:

1. *Polarity of noise.* In capacitive coupling, crosstalk noise [$C(dV/dt)$] always occurs in the same direction as the aggressor switches. However, inductive coupling induces noise [$L(dI/dt)$] through the return current, which opposes the direction of the aggressor switching and occurs more instantaneously than does $C(dV/dt)$. Hence, given an aggressor switching, inductive noise generally has the opposite polarity of capacitive noise and appears earlier in time, as illustrated in Figure 8.14(*a*). For first- and second-order neighboring lines, both positive (i.e., capacitive coupling) and negative (i.e., inductive coupling) noise peaks can be seen to match *RLC* predictions. However, these opposing factors suppress the overall amplitude of the coupling noise at adjacent lines.
2. *Coupling range.* Since the return current induced by inductive coupling spreads over a long range, even higher-order victim lines may suffer from *RLC* crosstalk noise, as shown in Figure 8.14(*a*). While capacitive coupling decays rapidly with increasing distance, inductive noise cannot be ignored for nonadjacent lines. In fact, without the opposing capacitive noise, the maximum inductive noise (i.e., negative peak) is larger for second-order than for first-order neighbors [Figure 8.14(*a*)].

Due to the competing nature of these two coupling mechanisms, it is difficult to predict the overall behavior of *RLC* crosstalk noise accurately without the aid of circuit simulators. Figure 8.14(*b*) shows the complexity of the relationship between peak noise and line length. V_{peak} values in wider lines are more

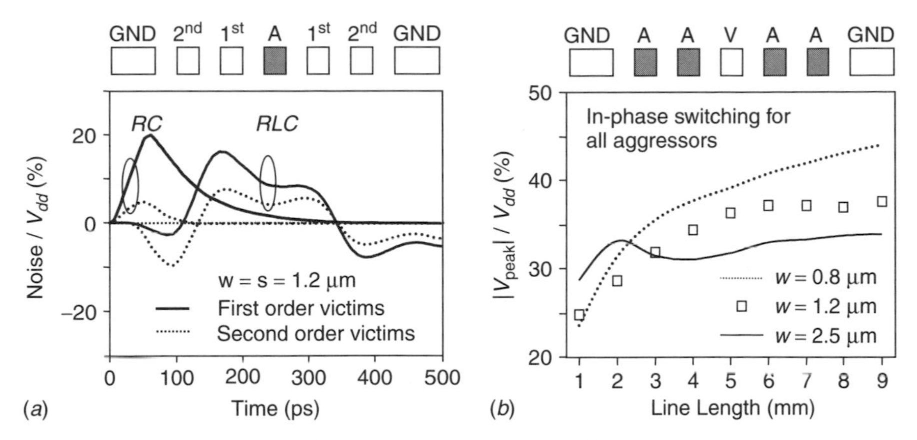

Figure 8.14 Comparisons of crosstalk noise with a 180-nm copper technology (line length = 3 mm; A, aggressor; V, victim): (*a*) noise waveform comparison; (*b*) peak noise at different width.

prone to inductance effects and exhibit a nonmonotonic dependence, whereas in *RC*-dominated, narrower lines, V_{peak} values increase with greater length. Furthermore, inductively coupling noise is more severe and difficult to control than that caused by capacitance coupling, especially in long parallel data bus structures. In practice, power and ground lines are inserted for every two to four signal lines in order to restrict the return current loop. However, even with this preventive measure, inductive noise can still attack victims across the shielded region [49]. For these reasons, it is preferred to apply layout (e.g., [50]) or circuit techniques to prevent dramatic inductance effects at early design stages, rather than relying on expensive analysis tools in later verification steps.

8.3.4 Noise-Aware Timing Analysis

Designs with tighter metal pitches, larger aspect ratios, and increasing operating frequencies are affected more significantly affected by interconnect coupling effects. Excess crosstalk noise may cause false switching on the victim net, but even a small amount of noise can change the victim delay significantly, resulting in a dynamic delay. This undesired effect occurs when the timing of a stage (i.e., gate and interconnect) becomes uncertain due to coupling from the switching activity of neighboring gates, resulting in dynamic delay, Due to the restoring nature of CMOS logic, only noise glitches exceeding the receiver's switching threshold can induce functional failures. In contrast, dynamic delays are more general and can easily be larger than 20 to 30% of nominal delays for short wires ($<$ 500 μm) [51]. Figure 8.15(*a*) shows the increase in delay uncertainty for a 3-mm global wire through a number of technology generations. In this example, the worst-case, normalized dynamic delay approaches 80% in the nanometer regime (note that the low-to-high transition experiences more of a dynamic delay because the PMOS victim is weaker than the NMOS aggressor in

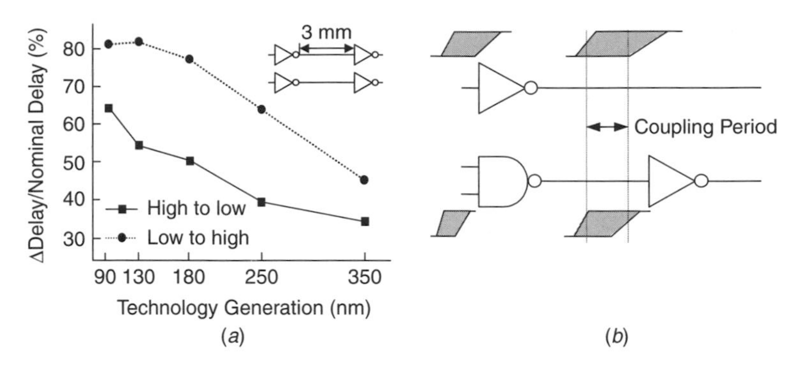

Figure 8.15 Noise-aware timing analysis using the switching window method: (*a*) trend in noise-induced dynamic delay; (*b*) switching window-based timing analysis (shadow areas represent the time window of a possible switching event).

this scenario) [51]. Large delay uncertainties such as this pose severe challenges for high-performance designs with very tight timing budgets.

To avoid chip timing failures in the worst-case coupling scenario, it is important to consider dynamic delays in static timing analysis (STA) and leave a large enough margin to tolerate delay fluctuations. An approach that is commonly used in *RC* timing analysis to compute the earliest and latest crosstalk delays involves scaling the coupling capacitances on critical paths by the switching factor, which is bounded by [0,2] or more accurately, [−1, 3], and then modeling them as an equivalent grounded capacitance for delay calculations. This technique is conservative and easily implemented and does not require information from neighboring nets. However, although this simplification reduces the complexity of analysis, it can nonetheless result in an estimate that is overly pessimistic or a routing space that is unnecessarily restricted, since crosstalk noise affects signal delay only when the aggressor and victim switch at the same time. If the aggressor and victim do not have switching overlap, there is no need to consider dynamic delay. Therefore, the conventional approach overestimates the timing bound and wastes computation time.

The key to improving the accuracy of noise-aware STA is to include temporal and functional information regarding the signal nets. This is realized by introducing the concept of a *switching window* (also called a *timing window*), which is the period of time within which a node makes transitions, as shown in Figure 8.15(*b*) [52,53]. The signal delay of two coupled nodes may change due to the crosstalk noise only when they have overlapping switching windows; otherwise, the signal timing is immune to dynamic delay [Figure 8.15(*b*)]. The remaining problem is to determine the switching window for each node. Although the timing information of the aggressor can be employed for this purpose, the aggressor's switching window may depend on the victim's switching window, resulting in a typical "chicken-and-egg" problem. A more general answer relies on iterative computations, although there are several approaches that can resolve the cycles [53,54]. First, we can assume an initial coupling scenario for the nets (e.g., worst-case coupling), run the delay engine to estimate the delay bound (i.e., the switching window) for each node, and reevaluate the coupling scenario depending on the relationship of switching windows. This process is repeated until the timing windows converge [52].

Using noise-aware timing analyses that are based on switching windows can significantly reduce the pessimism of delay-bound estimates. Within this framework, further progress has been made to improve the efficiency and accuracy of timing window calculations, recognizing the fundamental relationship between crosstalk and dynamic delay. As illustrated in Figure 8.16(*a*), the signal delay of a victim line changes in the presence of crosstalk noise, because the noise waveform is induced to the victim and distorts the original voltage propagation. Depending on the position at which the noise is injected, different changes in delay are observed. Therefore, with knowledge of the nominal characteristics of the switching voltage and noise glitches, along with the timing information at the inputs of both aggressor and victim lines, dynamic delay at the stage outputs can be predicted using waveform superposition. This idea is captured in the delay change curve (DCC), which represents the delay as a function of relative signal arrival time between the aggressor and victim inputs [51,55,56]. Figure 8.16(*b*) shows the measurement results of DCC from a 6-mm global wire in a 0.35-μm technology [51]. By using DCC, the output timing window is accurately scaled down compared to the traditional estimate of the delay bound using switching factors, and the result matches the peak-to-peak magnitude in the DCC. In practice, the DCC can be generated efficiently from analytical waveform superposition [51].

8.4 DESIGN SOLUTIONS FOR SIGNAL INTEGRITY

Timing-critical interconnects, such as the clock and global signal bus, are usually designed for optimal delay, rise time, and noise. This optimization process includes the prevention, analysis, and repair of signal integrity problems, at different design stages. In the early stages of physical design, extra routing constraints can be added where possible to avoid excessive noise. Circuit

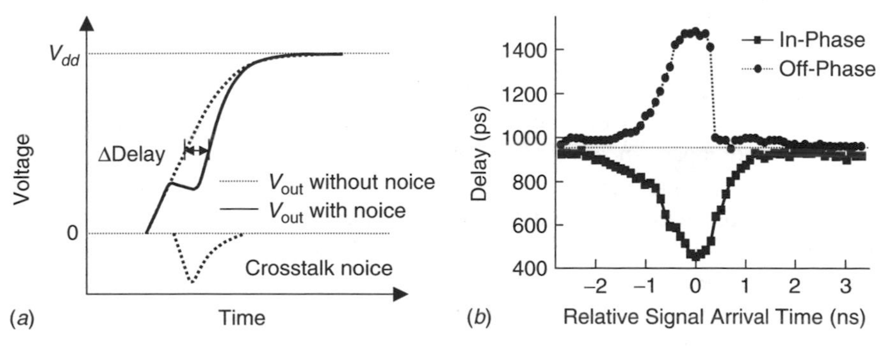

Figure 8.16 DCC for timing window estimation: (*a*) waveform superposition–caused delay change; (*b*) delay change curve from measurement.

design techniques such as repeater insertion are effective in improving the quality of the signal and interconnect performance. At the postrouting stage, it is usually necessary to fix signal integrity problems using interconnect tuning. Each of these techniques has their advantages and drawbacks. Preventing signal integrity problems by adding routing constraints is easy to implement but is costly in terms of chip size and power consumption. Also, it cannot fix all errors and may make unnecessary changes due to the rough nature of crosstalk noise estimates at this stage of design. Interconnect tuning at the postrouting stage is more precise but occurs very late in the design flow and can suffer from convergence problems. In general, all of the above approaches must be used at different design stages if signal integrity is to be kept under control with minimal impact on cost and productivity. In this section we discuss pre- and post-layout strategies for optimizing this physical interconnect structure as well as design techniques for signal integrity-aware design.

8.4.1 Physical Design Techniques

The delay and rise times are strong functions of the physical interconnect structure (i.e., length, width, spacing, driver size, etc). The magnitude of the coupling noise is strongly dependent on how close together the wires are placed, the distance for which they neighbor each other, and the neighboring transition activity, which is in turn determined by the drive strength and load capacitance. Various techniques can be used to optimize each of these properties, such as use of noise-constrained routing, net reordering, gate sizing, or interconnect geometry optimization.

Noise-Constrained Routing Crosstalk is highly dependent on routing. In recent years, CAD tools have begun to include signal integrity prevention and correction measures during the routing stage, which is called *noise-constrained* or *noise-immune/avoidance routing*. This problem is NP-hard and cannot be solved rigorously, and thus a heuristic approach is needed. First, an initial solution is constructed based on conventional routing solutions. After that, the crosstalk on each net is estimated. An example of a crosstalk noise estimate is as follows: A predefined boundary (e.g., prerouted power/ground grid) is used to divide the design into different regions. The coupling between different regions is assumed to be zero. For capacitive coupling, use the assumption that only the coupling capacitance (spacing) is controlled by layout design. Other parameters (driver strengths, load capacitance, input waveforms, etc.) either cannot be modified, or a modification is undesirable. The capacitive coupling noise is assumed to be proportional to the coupling capacitance. For each of the nets within a region, calculate the sensitivity of the net to the capacitance coupling noise, measured by the *capacitive crosstalk coefficient*:

$$C_i = \sum_{j \neq i} c_{ij} \tag{8.4.1}$$

where c_{ij} is the coupling capacitance between nets i and j. Coupling capacitance decreases rapidly beyond the first neighbor, and usually only the first-order neighbors need to be included in the summation above. The metric above neglects the fact that if two nets switch at different times, their crosstalk noise may not affect circuit performance. However, characterizing all possible switching cases requires exhaustive timing analyses, which in turn depend on crosstalk. Therefore, in the worst case, we can use the summation of all coupling capacitances from neighboring wires to represent the total crosstalk. An alternative is to add preliminary timing information in the crosstalk estimate to avoid a prohibitively large overestimation. A noise constraint can be set as

$$C_i < \overline{C_{\max}} \tag{8.4.2}$$

If a violation is found, compensation techniques such as spacing increase, shield insertion, and net reordering can be used to improve the design. If a region continues to have violations after these changes, some nets in the region may be removed and rerouted through other regions. Because of the rough estimates of the crosstalk noise, noise-constrained routing can easily lead to over- or under-design.

As discussed in Section 8.3, inductive coupling becomes more important with increased clock frequency and technology scaling. A similar model can be applied to estimate the sensitivity of a net to inductive noise using the *inductive crosstalk coefficient* K_i [57]:

$$K_i = \sum_{j \neq i} l k_{i,j} \quad \text{and} \quad K_i \leq \overline{K_{\max}} \tag{8.4.3}$$

where $k_{i,j}$ is the inductive coupling coefficient between nets i and j, and l is the region length. Inductive noise has a much longer coupling range than capacitive coupling, and thus more nets need to be included in (8.4.3) than in the capacitive coupling case. Again, a boundary such as an existing power/ground grid is usually needed to constrain the problem. However, for inductive coupling, this power/ground screening rule can sometimes underestimate the inductive noise.

Another technique for reducing the neighboring line activity is to use intentional skewing of the driver. This technique makes the wires within a bus have both normal and shifted timing. As a result, no adjacent wires will switch simultaneously, and both the normal and delayed signals will experience less of a coupling effect from their neighbors. The shifted timing of a wire can be established with either an inverter chain or a two-phase clocking scheme. Although this technique introduces a delay on the time-shifted wires, the overall bus delay for the shifted wires is reduced because the dominant crosstalk delay is suppressed.

Driver Sizing We now look at the impact of driver sizing on signal integrity from the point of view of both the victim and aggressor drivers. Intuitively, if the victim driver is sized up, its effective conductance increases, allowing it to hold a signal on a net more steadily. On the other hand, if an aggressor driver is sized up, the amount of noise it can induce on a victim is increased. Therefore, increasing the driver size has a twofold impact on crosstalk. The noise on the wire with the sized driver is decreased, but its induced noise neighbor lines will increase. A more quantitative view of how much the driving sizing and various other interconnect parameters (C_c, C_{a1}, etc.) will affect the crosstalk noise for a specific design is shown in Figure 8.17 [58]. Figure 8.17(*a*) illustrates the coupling noise model in which both aggressor and victim lines are divided into three regions: the interconnect segment before the coupling location, the coupling location, and the segment after the coupling location. The noise sensitivity of each model parameter for a practical circuit is shown in Figure 8.17(*b*).

Interconnect Tuning Interconnect tuning should be carried out simultaneously with transistor sizing. For *RC* lines, the most effective way to reduce interconnect delay through tuning is to increase the wire width. Wider lines generally have less delay because when the width is increased, the reduction of resistance occurs faster than the increase in total capacitance (as it is dominated by coupling capacitance). Because the two dominating considerations of capacitive

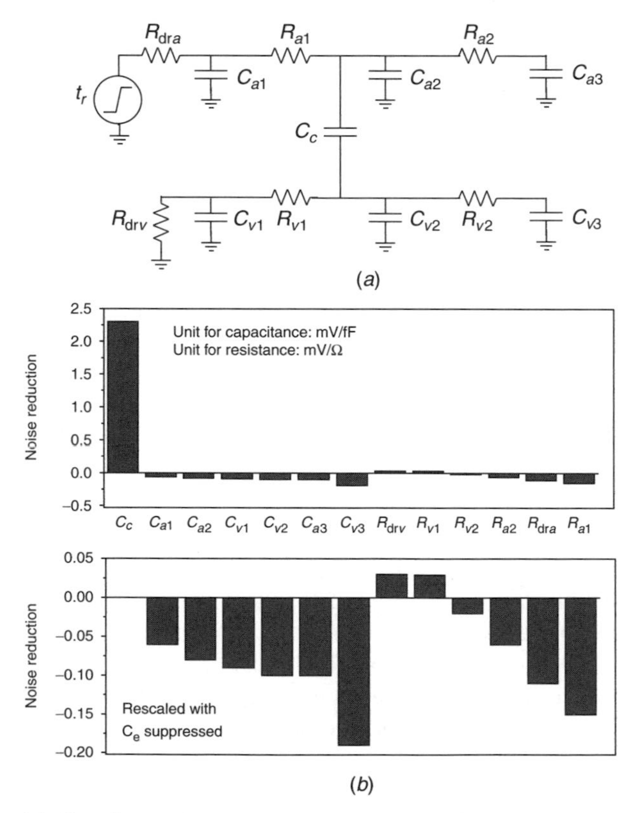

Figure 8.17 (*a*) Coupling noise model; (*b*) sensitivity of peak noise on model parameters [58].

crosstalk are coupling capacitance and neighbor switching conditions, the most effective way of reducing noise is to increase the spacing and number of bus order permutations. Some simple rules of thumb for *RC* net tuning are:

1. To reduce delay, increase interconnect width (more effective than increase spacing) or insert repeaters.
2. To reduce crosstalk, increase spacing (more effective than increase width), reorder nets, or insert repeaters.

For inductive nets, interconnect tuning becomes more tricky. Widening the wire can lead to more inductance on the dominant line, which can exhibit inductive ringing and extra delay. Furthermore, the increase of loop inductance caused by the increase of wire spacing will offset some benefits of the reduction in coupling capacitance. When wide lines are needed to drive a large load, they may need to be divided into small fingers interspersed with VDD/GND shields. Figure 8.18 is an example of a global clock structure in which interconnects are split into three wires and fully shielded. It can be seen from the graph that a significant performance gain (here represented by delay and rise time) can be achieved using the same routing area by optimizing the interconnect geometry. We define the interconnect signal-to-return ratio as the ratio of the total clock width ($T_{CLK} = NW_{CLK}$) to the total ground shield width [$T_{GND} = (N+1)W_{CLK}$], and observe the following rule for a fully shielded clock structure at a clock frequency of 2 GHz with the noise constrained $\leq 10\%$ [26]:

$$\text{Optimal delay}: \quad T_{CLK} : T_{GND} \approx 0.9 \text{ to } 1 \quad S : W_{GND} \approx 0.4 \text{ to } 0.5 \tag{8.4.4}$$

$$\text{Optimal power}: \quad T_{CLK} : T_{GND} \approx 0.8 \quad S : W_{GND} \approx 0.7 \tag{8.4.5}$$

These ratios will decrease (implying increased W_{GND}) with an increased frequency and line splitting number because of the increased importance of ground return resistance. The simple rules of thumb for designing *RLC* interconnect are:

1. Provide at least as much close return path as the signal ($T_{GND} \geq T_{CLK}$).
2. Use larger than minimal spacing, because the resulting reduction in coupling capacitance is greater than the increase in loop inductance.

The use of VDD/GND as shield wires within high-speed buses is the most common design technique to limit signal-line coupling, but at the cost of an increased routing area. It effectively eliminates capacitive coupling and the associated delay uncertainty. For *RLC* nets, ground shields provide close current-return paths and reduce the loop inductance. They also reduce the inductive noise generation because the magnetic field outside the pair occurs in opposite directions and cancel each other. Figure 8.19 shows the impact of shielding density on signal noise and delay. As shown in Figure 8.19, the noise of an inductance-dominated line ($W = 2.5$ μm, $S = 1.25$ μm) exhibits a linear dependence on the number of signal lines, because shielding is a less effective technique for

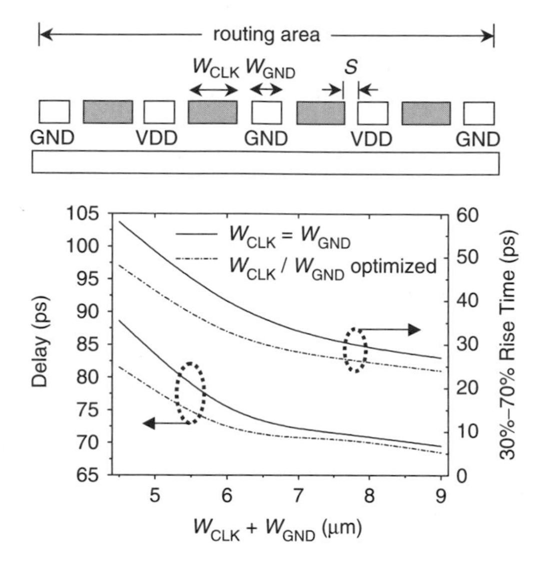

Figure 8.18 Improved performance can be achieved from interconnect geometry optimization.

controlling noise for inductive coupling than for capacitive coupling. For delay optimization, the delay curve plateaus beyond the point at which a shield line is placed between every three wires for $W = 0.8$ μm. For a more inductive line with $W = 2.5$ μm, the delay continues to degrade with an increased number of lines between the shields. In general, the optimal area efficiency for shielding is realized when a shield line is placed between about every two to four signal lines. For future technologies with higher operating frequencies, dedicated ground planes may be needed to reduce the inductive coupling. This technique is often used by PCB and package designers but is too expensive for on-chip designers in the current technology. It is important to notice that in actual designs, other considerations, such as wire congestion, noise, and power line *IR* drop, are also important considerations for deciding interconnect geometries.

A practical example for the design of a high-performance microprocessor applies various noise avoidance techniques to all victim nets and shows the resulting average percentage of noise reduction on the 48,000 longer interconnects [58]. From this example, it is observed that wire spacing is the most effective noise avoidance technique but it is also costly. Furthermore, while victim driver sizing is also comparably effective, wire sizing proved to be the technique that is the least effective for noise avoidance. Of course, the effectiveness of a particular noise avoidance technique depends on the particular interconnect/driver characteristics of a net.

8.4.2 Circuit Techniques

Repeater Insertion Repeater (buffer) insertion is a key solution for reducing the large delay of long interconnects, but with the penalty of increased chip

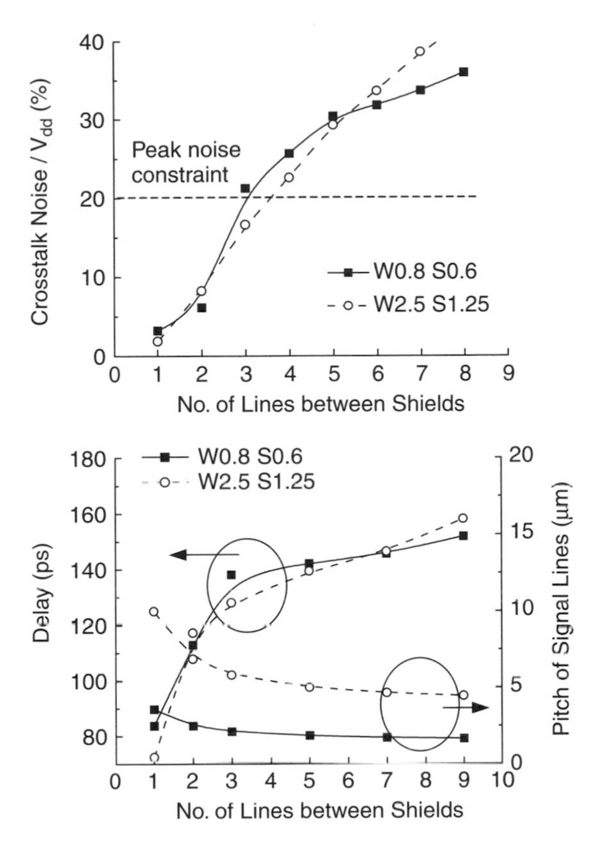

Figure 8.19 Optimal shielding is a shield line between every two to four signal lines.

area and power consumption. This technique breaks down long interconnects and inserts drivers (repeaters) in between the resulting segments(Figure 8.20), essentially reducing the delay dependence on wire length from quadratic to linear and thus greatly alleviating the delay problem of long interconnects. It also vastly improves signal slew rate at the far-end receiver because of the regenerative nature of CMOS drivers. With the exception of the first and last segments of the path, the repeaters are usually inserted at uniform intervals, because in practice the driver and receiver sizes may not be the same as the repeater size. Also in practice, the repeaters are usually implemented by a cascaded inverter pair to achieve the best delay reduction.

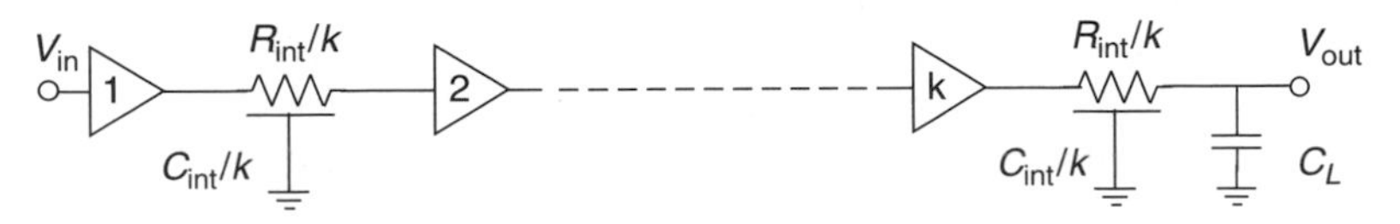

Figure 8.20 Reducing *RC* interconnect delay by repeater insertion.

In the *RC* regime, the most commonly cited optimal buffer sizing expression is that of Bakoglu [34]. The optimal number of repeaters is

$$k_{\text{opt}} = \sqrt{\frac{0.4 R_{\text{int}} C_{\text{int}}}{0.7 C_0 R_0}} \tag{8.4.6}$$

where R_0 and C_0 are the output resistance and input capacitance of a minimum-size repeater. The size of the repeaters is

$$h_{\text{opt}} = \sqrt{\frac{R_0 C_{\text{int}}}{R_{\text{int}} C_0}} \tag{8.4.7}$$

However, results obtained from equation (8.4.7) are often unrealistically large; typical standard cell libraries may include inverters or buffers up to 50 to 100 times the minimum size, whereas (8.4.7) can give results in the range of 400 to 700 times minimum. In practice, a larger delay is usually tolerated by adequate repeater insertion rather than by optimal repeater insertion. An expression was derived in [59] to optimize a weighted delay-area product rather than a pure delay metric. The results were on the order of 50 to 60% smaller than (8.4.7):

$$\begin{aligned} W_{\text{optarea}} = {} & \frac{0.541}{R_{\text{int}} C_{\text{int}}} (-0.231 R_D C_{\text{in}} - 0.126 R_{\text{int}} C_{\text{int}} \\ & + \sqrt{0.053 R_D^2 C_{\text{in}}^2 + 0.058 R_D C_{\text{int}} R_{\text{int}}^2 C_{\text{int}}^2 + 1.708 R_D C_{\text{in}} R_{\text{int}} C_{\text{int}}}) \end{aligned} \tag{8.4.8}$$

The delay based on (8.4.8) is higher, but the area and power costs are considerably smaller. If optimizing for energy-delay product, the value is even smaller.

As the line is broken down into shorter segments, the net is also more immune to noise. Repeater insertion reduces the parallel length of interconnects, which strongly affects the crosstalk noise. Figure 8.21 illustrates the effect of noise on a victim net, with and without a repeater. The top wire is the aggressor net and the bottom is the victim. As shown in part (*b*), inserting a buffer results in a smaller noise pulse at the input of the inserted buffer than at the input of the receiver in part (*a*). This small noise is easily suppressed by the regenerative nature of the buffer. For inductive noise coupling, the length of the original current return path is now shortened by returning the current through the repeaters, resulting in a smaller current loop and hence smaller inductive coupling. However, repeater insertion is less effective for *RLC* line delay reduction, because the time constant for an *LC* line ($\sqrt{LC}$) is approximately linearly proportional to the wire length, instead of quadratically proportional as in the *RC* case.

The placement of repeaters on adjacent lines can be staggered to minimize the impact of coupling capacitance on delay and crosstalk noise (Figure 8.22) [8]. The repeaters are offset so that each gate is placed in the middle of its neighboring gates' interconnect loads. The effective switching factor is limited to one, because potential worst-case simultaneous switching on adjacent wires is present for only

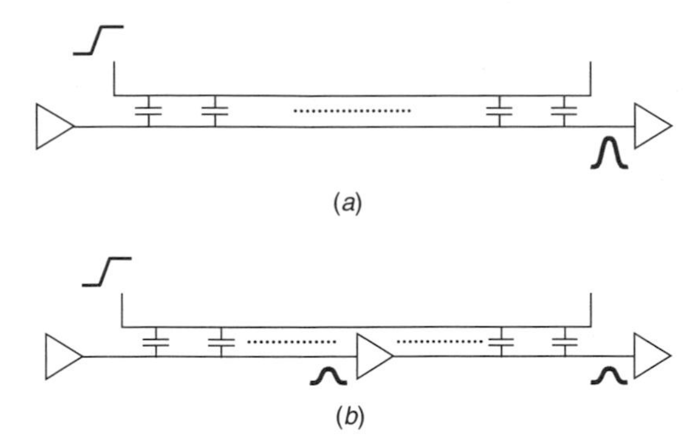

Figure 8.21 Repeater helps to suppress coupling noise on victim nets: (*a*) without repeater; (*b*) with repeater [63].

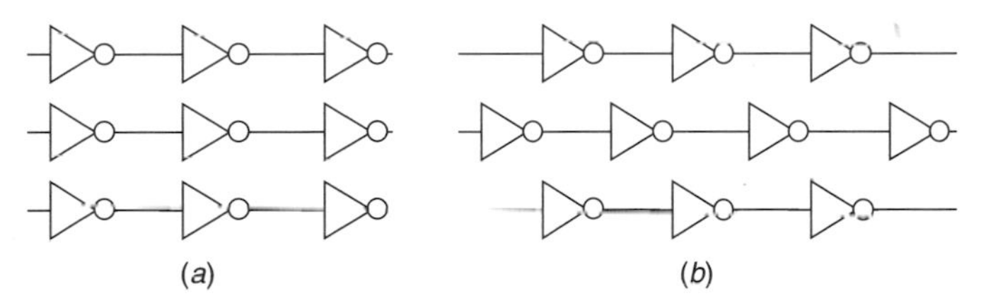

Figure 8.22 Staggered repeaters to reduce delay variation from switching pattern.

half the length of the victim line, while the other half of the victim line will experience the best-case neighboring switching activity, due to symmetry. With staggered repeaters, the delay uncertainty due to neighboring wire switching condition can be greatly reduced.

Recent studies indicate that repeaters use increasingly larger area, power, and design resources and are inherently limited in how much they can improve performance. New ideas have emerged in recent years to drive a long interconnect more efficiently, including the regenerative *booster* [60,61]. Unlike repeaters, the booster is attached along the wire to locally enhance the transmitted signal and does not intrude on the interconnect routing. It senses when a voltage transition is occurring on interconnect and provides an additional current boost to speed up the transition. An example of a booster circuit schematic and timing diagram is shown is Figure 8.23. This booster has two skewed inverters to detect a transition before it reaches the normal inverters. They drive a feedback path to locally accelerate the switching signal. In addition, a Muller-C element is included to prevent a direct path between VDD and GND. One advantage of the booster circuit is that its performance is insensitive to placement variations and can be placed at almost any point along the bus; thus they are affected less by the underlying signal routing constraints. Furthermore, it does not affect the polarity of the signal and supports both bidirectional and multisource configurations. Some experiments using this technique have demonstrated that boosters can drive longer interconnects, and cost less area and power than repeaters. However, a

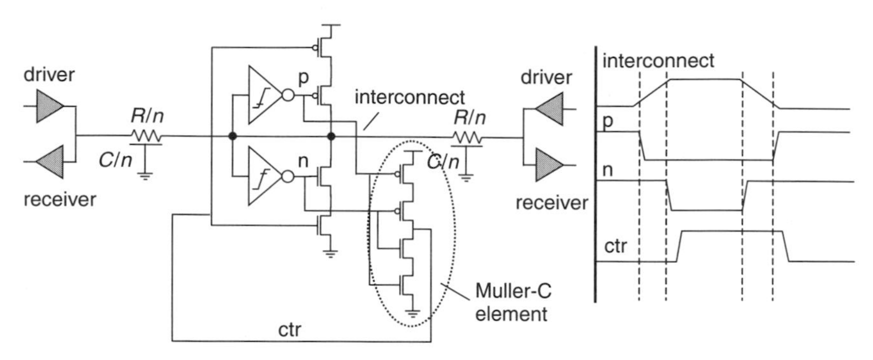

Figure 8.23 Example of booster circuit.

major design issue for these regenerative boosters is the potential metastability problems that are inherent in positive-feedback circuits, prohibiting signals with arbitrary pulse width to propagate through.

Keeper Circuit Dynamic gates are often used in performance-critical units of microprocessors and other high-performance VLSI circuits. Unlike static CMOS gates, the charge lost from a dynamic node due to noise cannot be restored, and as a result, dynamic gates are more vulnerable to noise than static CMOS gates. Dynamic floating nodes can be avoided by employing a static path through a pull-up and/or pull-down device referred to as a *keeper* (Figure 8.24) [62]. The keeper circuit restores the lost charge due to coupling noise, charge sharing, and subthreshold leakage current. However, with increasingly large noise and leakage current, the keepers much be sized up accordingly, which can significantly degrade the performance of dynamic circuits.

Differential Signaling Differential signals are inherently more robust to noisy environments than single-ended signals. The basic idea behind differential signaling is illustrated in Figure 8.25, in which two tightly coupled lines are used to transmit the data differentially. At the receiving location, these two signals are compared to determine their logic polarity. Differential signaling can be implemented in both voltage and current modes.

The differential signaling approach offers a high rejection of common-mode interferences such as crosstalk noise and supply-rail variations. It also provides

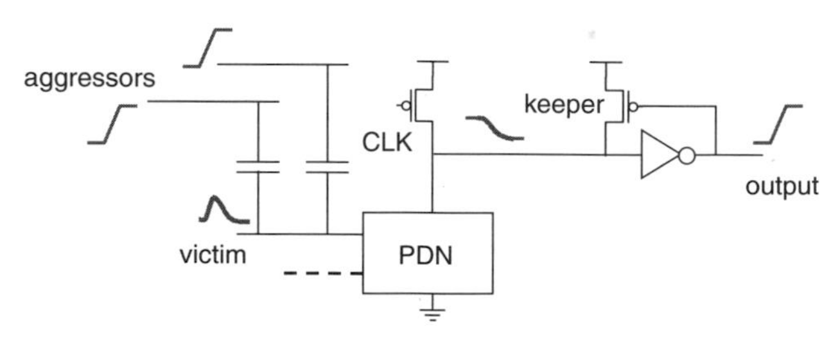

Figure 8.24 Keeper can restore lost charge in dynamic circuits.

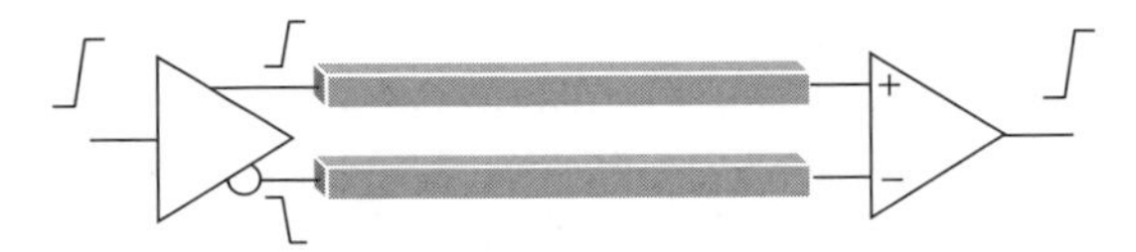

Figure 8.25 Differential signaling.

other advantages compared with single-ended signaling, including: (1) it has a built-in nearby return path for every signal wire, so less noise is coupled to other nets; (2) because of its high noise immunity, a low signal swing can be used to reduce power consumption—operation with swings as low as 200 mV has been demonstrated; and (3) the signal is isolated from the supply rails and the associated noise, making all supply noise occur in common mode to the differential receiver, which is usually designed to have excellent common-mode rejection. For these reasons, differential signaling can survive in much noisier environments and operate at much higher signaling rates than its single-ended counterparts. However, implementation of this technique comes with significant costs because it requires $2 \times N$ routing tracks for N signals. Also, the transmitter and receiver require extensive design management and may still be vulnerable to clock skew and jitter variations.

The technique of combining differential signaling with current mode logic has been widely adopted for off-chip interconnections. An example of current-mode bidirectional differential signaling that can operate up to 6.4 GHz is shown in Figure 8.26. For future generations of high-performance circuits, when the inductive noise become a significant issue on-chip as the chip operation frequency increases or when the skew generated by the power noise is too high, this technique may become a promising solution for on-chip interconnections.

8.5 SUMMARY

Signal integrity issues, including crosstalk noise, signal overshoot, and even power supply noise, have become important concerns in contemporary design. To guarantee sufficient chip yield and to avoid high design costs, it is preferable

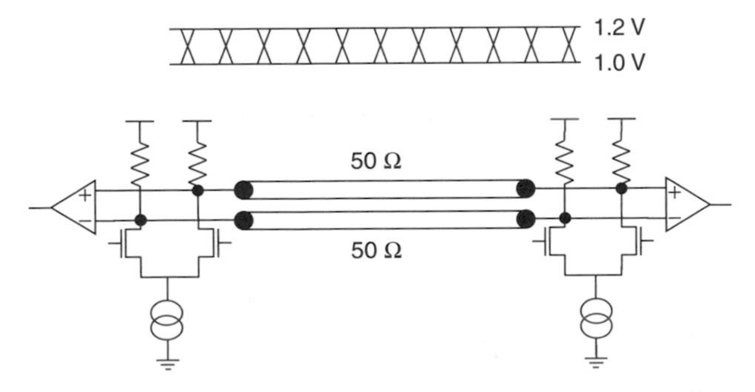

Figure 8.26 Example of bidirectional differential signaling.

to consider these problems as early as possible in the design flow rather than only relying on expensive full-chip verification and correction techniques at post-layout stage. For this purpose, conventional timing-driven design methodologies have evolved to accommodate signal integrity constraints into every design stage. These constraints can be formulated using analytical models, which physically link the design space with circuit performance. Various circuit and physical design techniques presented in this chapter provide further options to help improve the quality of signals. Most important, for a designer in practice, it is the constant awareness of possible signal integrity problems that builds a firewall to prevent potential hazards.

REFERENCES

[1] International Technology Roadmap for Semiconductors, *http://public.itrs.net.*

[2] Berkeley Predictive Technology Models, *http://www-device.eecs.berkeley.edu/~ptm.*

[3] D. A. B. Miller and H. M. Ozaktas, Limit to the bit-rate capacity of electrical interconnects from the aspect ratio of the system architecture, *J. Parallel Distribut. Comput.*, Vol. 41, No. 1, pp. 42–52, Feb. 1997.

[4] A. Deutsch et al., Bandwidth prediction for high-performance interconnections, *IEEE 50th Electronic Components and Technology Conference*, pp. 256–266, 2000.

[5] A. Deutsch et al., When are transmission-line effects important for on-chip interconnections, *IEEE Trans. Microwave Theory Tech.*, Vol. 45, No. 10, pp. 1836–1846, Oct. 1997.

[6] Y. I. Ismail, E. G. Friedman, and J. L. Neves, Figures of merit to characterize the importance of on-chip inductance, *IEEE Trans. VLSI Syst.*, Vol. 7, No. 4, pp. 442–449, Dec. 1999.

[7] P. J. Restle, A. E. Ruehli, S. G. Walker, and G. Papadopoulos, Full-wave PEEC time-domain method for the modeling of on-chip interconnects, *IEEE Trans. Comput. Aided Des. Integrated Circuits Syst.*, Vol. 20, No. 7, pp. 877–887, July 2001.

[8] Y. Cao et al., Effects of global interconnect optimizations on performance estimation of deep submicron designs, *Proceedings of the International Conference on Computer Aided Design*, pp. 56–61, Nov. 2000.

[9] T. Sakurai and K. Tamaru, Simple formulas for two- and three-dimensional capacitances, *IEEE Trans. Electron Devices*, Vol. 30, pp. 183–185, 1983.

[10] J.-H. Chern, J. Huang, L. Arledge, P. -C. Li, and P. Yang, Multilevel metal capacitance models for CAD design synthesis systems, *IEEE Electron Device Lett.*, Vol. 13, pp. 32–34, 1992.

[11] S.-C. Wong, G.-Y. Lee, and D.-J. Ma, Modeling of interconnect capacitance, delay, and crosstalk in VLSI, *IEEE Trans. Semicond. Manuf.*, Vol. 13, No.1, pp. 108–111, Feb. 2000.

[12] W. Jin, Y. Eo, W. R. Eisenstadt, and J. Shim, Fast and accurate quasi-three-dimensional capacitance determination of multilayer VLSI interconnects, *IEEE Trans. VLSI Syst.*, Vol. 9, No. 3, pp. 450–460, June 2001.

[13] E. You et al., Parasitic extraction for multimillion-transistor integrated circuits: methodology and design experience, *IEEE Custom Integrated Circuits Conference*, pp. 491–494, 2000.

[14] D. Sylvester, J. C. Chen, and C. Hu, Investigation of interconnect capacitance characterization using charge-based capacitance measurement (CBCM) technique and three-dimensional simulation, *IEEE J. Solid-State Circuits*, Vol. 33, No. 3, pp. 449–453, Mar. 1998.

[15] A. Deutsch, Electrical characteristics of interconnections for high-performance systems, *Proc. IEEE*, Vol. 86, No. 2, pp. 315–355, Feb. 1998.

[16] Y. Eo, W. R. Eisenstadt, and J. Shim, *S*-parameter-measurement-based high-speed signal transient characterization of VLSI interconnects on SiO_2–Si substrate, *IEEE Trans. Adv. Packag.*, Vol. 23, No. 3, pp. 470–479, Aug. 2000.

[17] A. E. Ruehli, Inductance calculations in a complex integrated circuit environment, *IBM J. Res. Dev.*, pp. 470–481, Sept. 1972.

[18] E. B. Rosa and F. W. Grover, *Formulas and Tables for the Calculation of Mutual and Self-Inductance*, U.S. Government Printing Office, Washington, DC, 1916.

[19] X. Qi et al., On-chip inductance modeling and *RLC* extraction of VLSI interconnects for circuit simulation, *Proceedings of Custom Integrated Circuits Design Conference*, pp. 487–490, 2000.

[20] K. Gala et al., On-chip inductance modeling and analysis, *Proceedings of Design Automation Conference*, pp. 63–68, 2000.

[21] Z. He, M. Celik, and L. Pileggi, SPIE: sparse partial inductance extraction, *IEEE Design Automation Conference*, pp. 137–140, 1997.

[22] A. Devgan, J. Hao, and W. Dai, How to efficiently capture on-chip inductance effects: introducing a new circuit element *K*, *IEEE International Conference on Computer Aided Design*, pp. 150–155, Nov. 2000.

[23] M. W. Beattie and L. T. Pileggi, On-chip induction modeling: basics and advanced methods, *IEEE Trans. VLSI Syst.*, Vol. 10, No. 6, pp. 712–729, Dec. 2002.

[24] Y. Cao et al., Effective on-chip inductance modeling for multiple signal lines and application on repeater insertion, *IEEE Trans. VLSI Syst.*, Vol. 10, No. 6, pp. 799–805, Dec. 2002.

[25] B. Krauter and S. Mehrotra, Layout based frequency dependent inductance and resistance extraction for on-chip interconnect timing analysis, *IEEE Design Automation Conference*, pp. 303–308, 1998.

[26] X. Huang, P. Restle, T. Bucelot, Y. Cao, and T. -J. King, Loop-based interconnect modeling and optimization approach for multi-GHz clock network design, *IEEE J. Solid-State Circuits*, Vol. 38, No. 3, p. 457–463, Mar., 2003.

[27] C.-K. Cheng, J. Lillis, S. Lin, and N. Chang, *Interconnect Analysis and Synthesis*, Wiley, New York, 2000.

[28] Y. Cao, X. Huang, D. Sylvester, T. King, and C. Hu, Impact of frequency-dependent interconnect impedance on digital and RF design, *IEEE International ASIC/SoC Conference*, pp. 438–442, Sept. 2002.

[29] F. Dartu, N. Menezes, and L. T. Pilleggi, Performance computation for precharacterized CMOS gates with *RC* loads, *IEEE Trans. Comput. Aided Des. Integrated Circuits Syst.*, Vol. 15, No. 5, pp. 544–553, May 1996.

[30] X. Huang, Y. Cao, D. Sylvester, T. King, and C. Hu, Analytical performance models for *RLC* interconnects and applications to clock optimization, *IEEE International ASIC/SoC Conference*, pp. 353–357, Sept. 2002.

[31] L. T. Pilleggi and R. A. Rohrer, Asymptotic waveform evaluation for timing analysis, *IEEE Trans. Comput. Aided Des.*, Vol. 9, No. 4, pp. 352–366, Apr. 1990.

[32] J. Qian, S. Pullela, and L. Pilleggi, Modeling the "effective capacitance" for the *RC* interconnect of CMOS gates, *IEEE Trans. Comput. Aided Des. Integrated Circuits Syst.*, Vol. 13, No. 12, pp. 1526–1535, Dec. 1994.

[33] A. B. Kahng and S. Muddu, New efficient algorithms for computing effective capacitance, *International Symposium on Physical Design*, pp. 147–151, 1998.

[34] H. B. Bakoglu, *Circuit, Interconnections, and Packaging for VLSI*, Addison-Wesley, Reading, MA, 1990.

[35] W. C. Elmore, The transient analysis of damped linear networks with particular regard to wideband amplifiers, *J. Appl. Phys.*, Vol. 19, No. 1, pp. 55–63, 1948.

[36] R. Gupta, B. Tutuianu, and L. T. Pileggi, The Elmore delay as a bound for *RC* trees with generalized input signals, *IEEE Trans. Comput. Aided Des. Integrated Circuits Syst.*, Vol. 16, No. 1, pp. 95–104, Jan. 1997.

[37] T. Sakurai, Closed-form expressions for interconnection delay, coupling, and crosstalk in VLSIs, *IEEE Trans. Electron Devices*, Vol. 40, No. 1, pp. 118–124, Jan. 1993.

[38] P. Chen, D. A. Kirkpatrick, and K. Keutzer, Miller factor for gate-level coupling delay calculation, *Proceedings of the International Conference on Computer Aided Design*, pp. 68–74, Nov. 2000.

[39] A. B. Kahng, S. Muddu, and E. Sarto, On switch factor based analysis of coupled *RC* interconnects, *IEEE Design Automation Conference*, pp. 79–84, 2000.

[40] M. Kuhlmann and S. S. Sapatnekar, Exact and efficient crosstalk estimation, *IEEE Trans. Comput. Aided Des. Integrated Circuits Syst.*, Vol. 20, No. 7, pp. 858–866, July 2001.

[41] J. Cong, D. Z. Pan, and P. V. Srinivas, Improved crosstalk modeling for noise constrained interconnection optimization, *Asia and South Pacific Design Automation Conference*, pp. 373–378, 2001.

[42] M. R. Becer et al., Analysis of noise avoidance techniques in DSM interconnects using a complete crosstalk noise model, *IEEE Proceedings of Design, Automation and Test in Europe Conference and Exhibition*, pp. 456–463, 2002.

[43] D. Sylvester and C. Hu, Analytical modeling and characterization of deep-submicrometer interconnects, *Proc. IEEE*, Vol. 89, No. 5, pp. 634–664, May 2001.

[44] L. H. Chen and M. Marek-Sakowska, Closed-form crosstalk noise metrics for physical design applications, *IEEE Proceedings of Design, Automation and Test in Europe Conference and Exhibition*, pp. 812–819, 2002.

[45] A. Odabasioglu, M. Celik, and L. T. Pileggi, PRIMA: passive reduced-order interconnect macromodeling algorithm, *Proceedings of the International Conference on Computer Aided Design*, pp. 58–65, Nov. 1997.

[46] Y. I. Ismail, E. G. Friedman, and J. L. Neves, Equivalent Elmore delay for *RLC* trees, *IEEE Trans. Comput. Aided Des. Integrated Circuits Syst.*, Vol. 19, No. 1, pp. 83–97, Jan. 2000.

[47] A. B. Kahng and S. Muddu, An analytical delay model for *RLC* interconnects, *IEEE Trans. Comput. Aided Des. Integrated Circuits Syst.*, Vol. 16, No. 12, pp. 1507–1514, Dec. 1997.

[48] Y.-C. Lu, M. Celik, T. Young, and L. T. Pileggi, Min/max on-chip inductance models and delay metrics, *Proceedings of the Design Automation Conference*, pp. 341–346, 2001.

[49] X. Huang et al., *RLC* signal integrity analysis of high-speed global interconnect, *Technical Digest, International Electron Devices Meeting*, pp. 731–734, Dec. 2000.

[50] Y. Massoud, S. Majors, T. Bustami, and J. White, Layout techniques for minimizing on-chip interconnect self inductance, *Proceedings of the Design Automation Conference*, pp. 566–571, 1998.

[51] T. Sato et al., Bidirectional closed-form transformation between on-chip coupling noise waveforms and interconnect delay-change curves, *IEEE Trans. Comput. Aided Des. Integrated Circuits Syst.*, Vol. 22, No. 5, pp. 560–572, May 2003.

[52] R. Arunachalam, K. Rajagopal, and L. T. Pileggi, TACO: timing analysis with coupling, *Proceedings of the Design Automation Conference*, pp. 266–269, 2000.

[53] P. Chen, D. A. Kirkpatrick, and K. Keutzer, Switching window computation for static timing analysis in presence of crosstalk noise, *Proceedings of the International Conference on Computer Aided Design*, pp. 331–337, Nov. 2000.

[54] B. Thudi and D. Blaauw, Non-iterative switching window computation for delay-noise, *Proceedings of the Design Automation Conference*, pp. 390–395, 2003.

[55] Y. Sasaki and G. D. Micheli, Crosstalk delay analysis using relative window method, *IEEE International ASIC/SoC Conference*, pp. 9–13, Sept. 1999.

[56] Y. Sasaki and K. Yano, Multi-aggressor relative window method for timing analysis including crosstalk delay degradation, *Proceedings of Custom Integrated Circuits Design Conference*, pp. 495–498, 2000.

[57] J. D. Ma and L. He, Toward global routing with *RLC* crosstalk constraints, *IEEE/ACM Design Automation Conference*, June 2002, pp. 669–672.

[58] M. R. Becer, D. Blaauw, V. Zolotov, R. Panda, and I. N. Hajj, Analysis of noise avoidance techniques in DSM interconnects using a complete crosstalk noise model, *Proceedings of the Design, Automation and Test in Europe Conference and Exhibition*, pp. 456–463, 2002.

[59] D. Sylvester and K. Keutzer, System-level performance modeling with BACPAC: Berkeley advanced chip performance calculator, *Proc. SLIP*, pp. 109–114, 1999; *http://www.eecs.umich.edu/~dennis/bacpac/.*

[60] I. Dobbelaere, M. Horowitz, and A. El Gamal, Regenerative feedback repeaters for programmable interconnections, *IEEE J. Solid-State Circuits*, Vol. 30, No. 11, pp. 1246–1253, Nov. 1995.

[61] A. Nalamalpu, S. Srinivasan, and W. P. Burleson, Boosters for driving long onchip interconnects: design issues, interconnect synthesis, and comparison with repeaters, *IEEE Trans. Comput. Aided Des. Integrated Circuits Syst.*, Vol. 21, No. 1, pp. 50–62, Jan. 2002.

[62] R. Colwell and R. L. Steck, A 0.6 μm BiCMOS processor with dynamic execution, *IEEE International Solid-State Circuits Conference*, pp. 176–177, 1995.

[63] C. J. Alpert, A. Devgan and S. T. Quay, Buffer insertion for noise and delay optimization, *35th IEEE/ACM Design Automation Conference*, pp. 362–367, 1998.

CHAPTER 9

ULTRALOW POWER CIRCUIT DESIGN

9.1 INTRODUCTION

Throughout the past three decades the continuous technology scaling kept providing designers with faster devices, higher integration capacity, and less energy per transition. All these contributed to the five-order-of-magnitude improvement on microprocessor performance over this period. However, while the performance demand of future applications continues to grow, scaling beyond the 90-nm node became increasingly difficult. One of the main barriers in this trend is the excessive chip power consumption exacerbated by performance-driven scaling and integration. At the pace of current scaling trend, each process generation achieves 30% reduction in capacitance per node, a twofold increase in electrical node integration density, 14% of die size growth, 15% supply voltage reduction, and a twofold frequency increase. As a result, the CPU active power consumption increases nearly 2.7-fold every two years, according to industry data [1]. On the academic side, Figure 9.1(*a*) shows a survey of front-edge processor design published in ISSCC during the years 1980–2000, where the power consumption grows 1.4-fold every three years [2]. Furthermore, the reduced V_{th} in scaling leads to an excessive leakage increase of three- to fivefold per generation [3]. The estimated scaling trend of leakage power according to ITRS parameters is plotted in Figure 9.1(*b*).

High power consumption causes performance and reliability degradation in desktop computers and servers. With the increasing microarchitecture complexities, clock frequencies, and die sizes, next-generation multiprocessor server boxes

Nano-CMOS Circuit and Physical Design, by Ban P. Wong, Anurag Mittal, Yu Cao, and Greg Starr
ISBN 0-471-46610-7

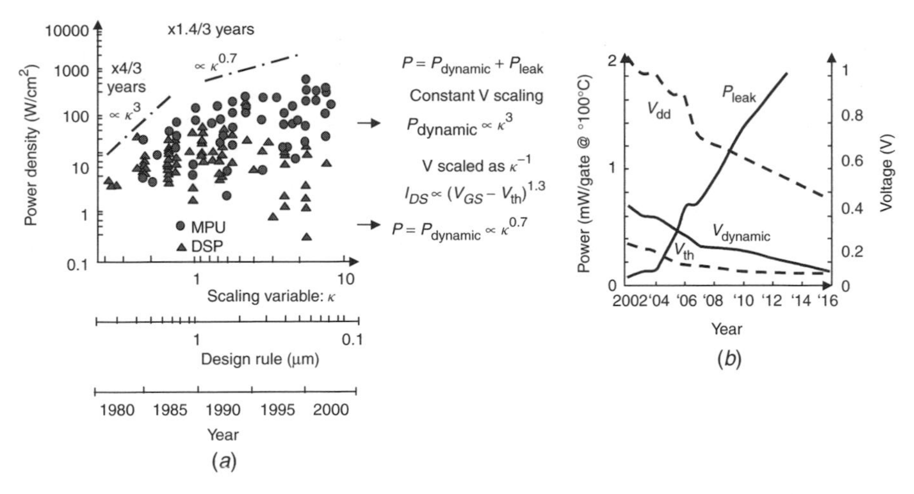

Figure 9.1 Dynamic and leakage power increase due to scaling: (*a*) power of processors published in ISSCC, 1980–2000; (*b*) estimated scaling trend of voltage and power per device according to ITRS. (From Ref. 2.)

may soon need budgets for liquid-cooling or refrigeration hardware. This transition is likely to cause a breakpoint with a step upward in the ever-decreasing price–performance ratio curve [74]. At the other end of the performance spectrum, excessive leakage power reduces the operation time in battery-supported applications such as laptop computers, cellular phones, and PDAs. Power dissipation limits have emerged as a major constraint for VLSI. Low-power design hence becomes the key challenge, especially for the future technology nodes beyond 90 nm, where the effective oxide thickness (EOT) is set to the range 1 to 1.6 nm [35]. With such thin EOT, the gate oxide tunneling leakage and gate-induced drain leakage (GIDL) becomes significant and comparable to the subthreshold current. Currently, standby leakage on a 130-nm technology is typically less than 10% of the total current (i.e., standby plus dynamic). On the 90-nm technology node, this ratio increases to 30% or more, while the projections for 65 nm are even greater.

As the technology scales in the inevitable trend toward ever-increasing power density, the task of minimizing system power consumption involves optimization at all levels of the design. From hardware architecture, software operating system down to the physical circuit design, a range of power reduction opportunities exist on all levels. The coordination of various power-aware design techniques across levels is the key in minimizing the power consumption in an ultralow-power application, such as battery-supported systems. Other computing-intensive designs with less stringent power budget may benefit from a certain subset of these techniques in achieving the optimum operation efficiency. As power has emerged as the performance limiter for designs in nano-CMOS regime, lower-power processors and servers will come out ahead in such applications as well.

Considering the phases in the design process when these power reduction techniques are applied, they can be divided into two categories as design-time and run-time techniques. The optimization of design-time techniques is finished and fixed during the design phase, while the run-time techniques apply different real-time control on the design for different periods of workload to optimize overall power consumption. The leakage suppression techniques are mainly in the run-time group since they kick in only during system idle periods. In Sections 9.2 and 9.3 we provide an overview of existing design-time and run-time power control methods on different levels of system design, with the focus on circuit-level logic and memory design techniques. Technology innovations for low-power design are introduced in Section 9.4. The perspective of ultralow-power design techniques for future technology nodes beyond 90 nm is discussed in Section 9.5.

9.2 DESIGN-TIME LOW-POWER TECHNIQUES

9.2.1 System- and Architecture-Level Design-Time Techniques

At the system level, the goal of power reduction techniques is to minimize unnecessary activity. The system partitioning technique partitions a system or algorithm into spatially local clusters by exploiting the locality of references, leading to shorter local buses, and less activity on the highly capacitive global buses. Optimizations are needed during chip assembly to limit the length of wide buses, allowing only longer, narrower buses. The floor plan must be adjusted to favor reducing the length of wide buses at the expense of lower-signal-count buses. Other techniques include event-driven design methodology, minimized data transfer, power-aware medium-access protocol, and network routing [4,5].

At the architecture level, designs implemented with parallel hardware allow reduction in supply voltage and clock frequency without degradation in system throughput. Optimized hierarchical memory system reduces memory accesses and applies a caching scheme to exploit the data locality in memory accesses. A power-aware compiler makes optimized trade-off between code size and speed in favor of energy reduction. Power-efficient I/O interconnect design reduces bus switching capacitance and applies data coding to minimize bus transitions [4].

9.2.2 Circuit-Level Design-Time Techniques

At the circuit level, numerous techniques are used to build power-optimized circuitry.

1. Exploiting stack effect at design time. By stacking two off transistors, the subthreshold leakage current is reduced significantly compared to a single off device, due to simultaneous reductions in gate–source, body bias, and drain–source voltages. This stack effect has been exploited extensively in various leakage reduction techniques. Most of these approaches apply run-time standby control using schemes of multiplexed low-leakage input vectors, gate modification [45],

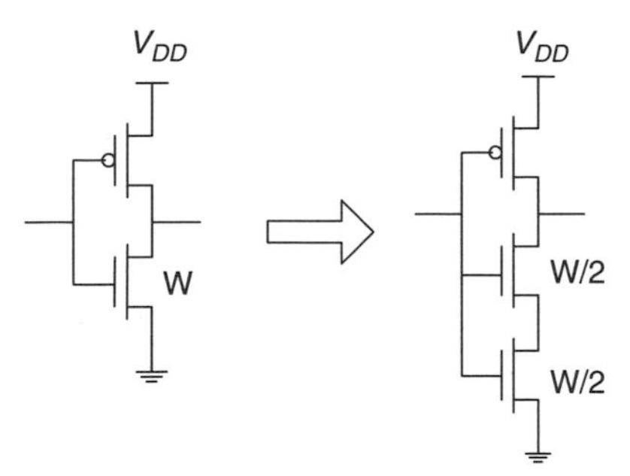

Figure 9.2 NMOS two-stack with stack forcing. (From Ref. 27.)

and series transistor insertion [44] to convert the standby circuits to stacked structure, introduced in Section 9.3.2. At design time a stack-forcing technique [27] forces a nonstack device to a stack of two devices without affecting the input load (as shown in Figure 9.2). With this method, leakage of a stack-forced logic gate is reduced by a factor of 9 at certain delay penalties, similar to the dual-V_{th} technique but without the process complexity of a second V_{th}. Stack forcing can be applied to noncritical paths, resulting in reduced standby and active leakage without affecting the speed of critical paths with normal gate design. In the same work it was shown that this stack technique for leakage reduction is expected to improve with technology scaling, which makes the leakage control technique that exploits the stack effect more effective in future technologies [27].

2. *Input reordering*. Input reordering is a gate-level technique that can be used to optimize circuit delay and capacitive power consumption. Appropriate input ordering minimizes the switching activity at internal nodes and reduces active power. Various algorithms based on analytical modeling of circuit structure, internal nodes capacitance, signal probability (of being logical one), and transition probability has been proposed to solve for optimum input order [47–49]. General rules for input reordering were summarized. Among those, one commonly recognized rule is to place signals with the largest switch probability closest to the output terminal [48,50], which minimizes connection activity to power rails and at the same time leads to optimized performance. Another study in this area further takes into account the optimization of the overall power consumption of the fan-in, fanout gates and the reordered gate [51]. All these input reordering algorithms produce average power reduction ranging from 3.6 to 12%. Compared to other low-power techniques, input reordering usually achieves a limited power saving ratio; however, it does not require any extra device and architectural modifications and therefore can easily be used in conjunction with other low-power techniques. These properties encourage its application.

3. *Transistor sizing*. Transistor sizing is an important knob in designing for desired trade-off between power, delay, and area concerns. Sizing optimization has been explored extensively with several optimization tools, such as TILOS [38] and EinsTuner [39]. These tools are capable of approximating the solution for minimizing

the overall power consumption of a circuit under given delay constraints. As the first synthesizer approach for sizing optimization, TILOS assumes a simple *RC* delay model in posynomial programming optimizations [38]. It handles circuits sized up to 250,000 transistors. Applying TILOS to a variety of high-performance chip designs provided 40 to 50% power reduction [7]. More than a decade later, the EinsTuner tool developed by IBM research improved the delay model in TILOS by accurately simulating channel-connected components, and implemented gradient-based nonlinear optimization. As the result, EinsTuner achieved better solutions with higher accuracy, but at the cost of reduced capability for resolving large-scale circuits (three days of computing time for a 2796-transistor adder circuit which has over 5600 variables and over 5600 constraints) [39]. In addition, a number of other works explored further improvements on sizing optimization by taking into account the short-circuit power dissipation [40] and rise/fall time delay elements [41].

4. Applying multiple supply and threshold voltages. Supply (V_{DD}) and threshold (V_{th}) voltages are the key factors in optimizing the balance among active power, leakage power consumption, and circuit performance. At run time, V_{DD} and V_{th} can be varied dynamically to enhance system power efficiency at different workloads, as will be introduced in Section 9.3.1. At design time, numerous works have been dedicated to solving for optimum V_{DD} and V_{th} for a high-speed energy-efficient design. Closed-form formulas were derived considering short-channel effects and V_{th} variation [58]. Sensitivity balanced analysis with variables of V_{DD}, V_{th}, and sizing suggested possible energy savings of 40 to 70% at 20% delay overhead [9]. On the other hand, the use of multiple V_{DD} and V_{th} variables is motivated by the observation that a circuit's overall performance is often limited by a few critical paths, while the path delay distribution of the entire circuit actually spreads widely [54]. As shown in Figure 9.3(*a*), a dual-V_{th} technique can be used to speed up critical paths with low-V_{th} devices while leaving noncritical paths with high-V_{th} leakage suppression. Figure 9.3(*b*) shows dual-V_{th} optimization effects on path delay distribution, where the goal is to balance the path delays and speed up critical paths. This technique has been used extensively in many implementations [55,56], while the optimization space can be expanded further by combining the multiple V_{DD} assignment and transistor sizing design techniques in path balancing. Throughout the study of this optimization scheme, algorithms were developed to select transistors in noncritical paths that can be assigned high V_{th} values without affecting system performance (by turning a noncritical path into a critical path) [57]. In another work it was concluded that no more than three discrete values for V_{DD}, V_{th}, or sizing are needed for an efficient design [8].

5. Nonminimum channel length. Typically, the smallest channel length permitted by a process is used within a design. This has been the traditional design practice up to the 130-nm technology node, but now, given the significant increase in leakage current, designers are being forced to partition devices into minimum and nonminimum channel lengths, to reduce the leakage current. Increasing the channel length can have a significant impact on the overall standby current of a

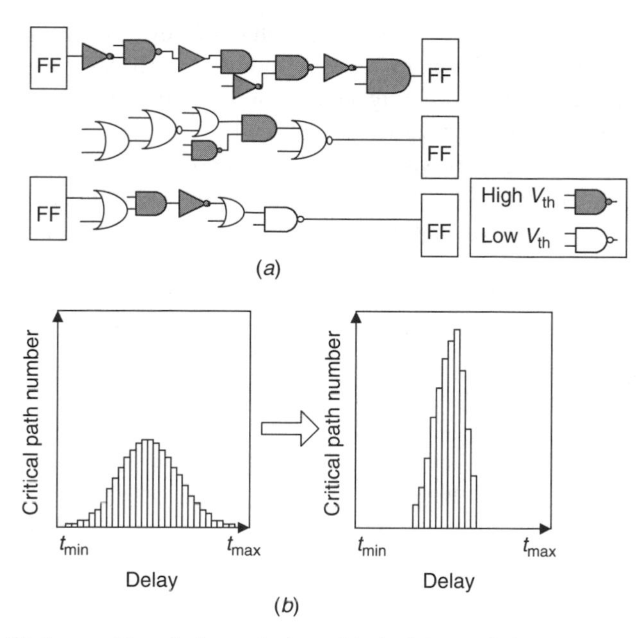

Figure 9.3 High-speed low-leakage design with dual-V_{th} technique: (*a*) applying low-V_{th} value to critical path transistors; (*b*) path delay distribution before and after dual-V_{th} optimization.

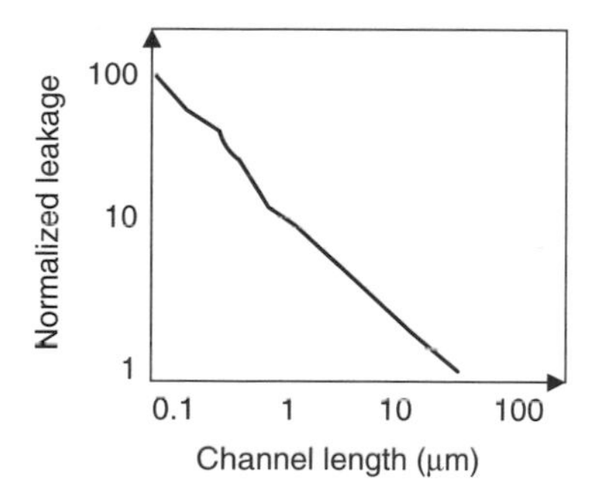

Figure 9.4 Normalized I_{ds} leakage as a function of channel length.

design, especially if it is widely applied to a large number of devices. Figure 9.4 illustrates how the source–drain leakage current changes as a function of the channel length for a 100-nm process. All leakage numbers have been normalized to the leakage for the channel length of a 15-μm device. Approximately 60% leakage reduction can be achieved by changing the channel length from 100 nm to 150 nm. Similar to the dual-V_{th} scheme, nonminimum channel lengths can be used on noncritical speed paths to balance the system delay distribution. By applying nonminimum channel lengths, the process variation effects on these paths are reduced as a secondary benefit, since the channel-length variance now becomes a smaller percentage of the longer-channel devices. Furthermore, while using multiple threshold devices involves additional process cost with requirements of more masks and processing steps, a nonminimum channel-length design approach can provide a lower-cost solution.

6. Low-power standard cell library and on-demand library generation. The availability of a standard cell library with power-minimized components facilitates low-power system implementation. Low-power library cells are implemented with energy-efficient logic style and customized sizing, as well as versions with different threshold voltages for design with different specifications. It was shown that a logic synthesizer using a low-power library designed with appropriate strategies produced designs with significant performance and power improvements [10] compared to designs with a general-purpose standard cell library. Furthermore, the approach of on-demand library generation [52] overcomes the limited-size problem of a customized low-power standard cell library, providing ultimate flexibility for implementation The ASIC design methodology with on-demand library generation is shown in Figure 9.5, where a tailored library is generated according to the performance estimation results and supplied to cell-based design tools. This design flow is featured by postlayout transistor sizing, which optimizes the library by downsizing the cells based on the information extracted from the preliminary layout. In this way the area and power redundancy in conventional fixed library design is eliminated, resulting in a fully optimized physical implementation. It was reported that the power dissipation of circuits implemented with an on-demand library is reduced by 77% maximum and 65% on average without an increase in delay [53].

7. Reducing interconnect power consumptions. Interconnects, including both on-chip lines and wires in a package, have become a major source of power consumption. As a result of an increasing number of layers and more compact line dimensions, metal line capacitance can take up to 70% of the total chip capacitance in contemporary design [2]. Moreover, a rapid increase in chip operating

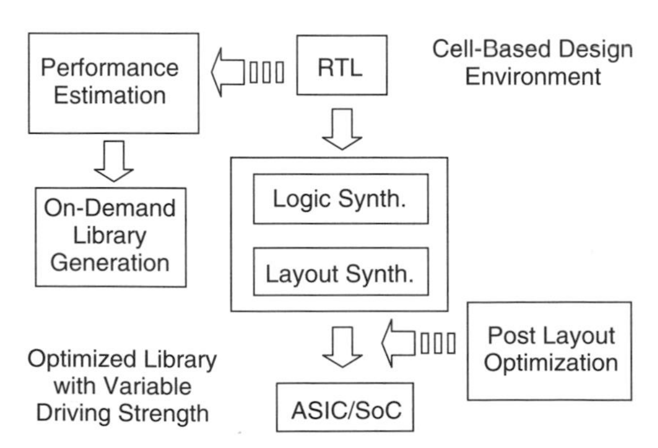

Figure 9.5 ASIC design methodology with on-demand library generation. (From Ref. 52.)

frequencies further exacerbates the amount of dynamic power dissipated in the interconnect system, in the format of $CV_{dd}^2 \times$ frequency. Note that line inductance does not consume power directly during voltage switching. Based on interconnect functionalities, three types of wires have been recognized as the dominating factors in power consumption: on-chip signal lines, interconnects in I/O systems, and clock distribution networks. To improve their power efficiencies, a number of innovations have been explored from both technology and design perspectives. A general approach to reducing interconnect power consumption is to apply low-voltage swing. This technique has been used widely in I/O systems, such as low-voltage differential signaling (LVDS). LVDS not only saves power but also enhances the speed of I/O signaling. Yet as signal-coupling noise increases significantly in the nanometer regime, concerns of signal integrity and design cost limit its application in on-chip signaling and clock networks. For global signal lines, bus shuffling or encoding has been demonstrated to reduce the power consumption on coupling capacitance by either shuffling the bus placement or coding the switching patterns, so that worst-case coupling capacitance can be minimized. Another power-aware approach for interconnect design is the introduction of nonorthogonal global layers to reduce the total signal line length. For instance, in X-architecture [97], 45° layout is allowed. In this case about 20% of the total wire length can be saved, cutting the interconnect power cost proportionally [97].

9.2.3 Memory Techniques at Design Time

The high-density, low-power, and low-cost features of stand-alone and embedded random-access memories (RAMs) have contributed to improvements in various electronic systems. Nowadays, microprocessor designs incorporate large memory components, which consume a significant portion of a systems power budget. For instance, 30% of Alpha 21264 and 60% of StrongARM are devoted to cache and memory structures [60]. For battery-supported applications with a low duty cycle, the memory leakage power can even dominate the overall system power consumption and determine battery life. Driven by a requirement for optimum system power efficiency in various applications, low-power memory design has been a major area that has experienced rapid and remarkable progress. In Sections 9.2.3 and 9.3.3, design- and run-time techniques for low-power SRAM and DRAM designs are introduced [6,31].

Low-Power SRAM at Design Time

1. Partial activation of multi-divided word line and bit line. Partial activation is an effective approach in reducing the charging capacitance of heavily loaded word and bit lines in SRAM. Simply by dividing the memory array into subblocks [14], word- and bit-line loads can be reduced significantly. However, this technique carries a large penalty, due to additional control logic and routing. Other techniques keep the integrity of the memory array and focus on decoding logic restructuring. As shown in Figure 9.6(a), the divided-word-line (DWL)

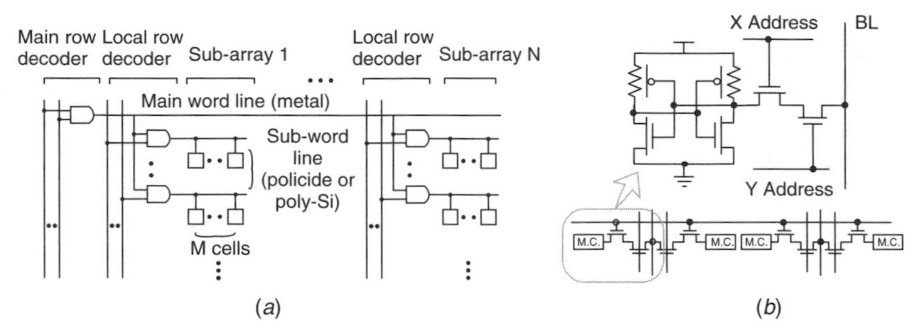

Figure 9.6 Schemes for partial activation of a multi-divided word line: (*a*) DWL structure; (*b*) SRAM cell used for SCPA architecture. [Part (*a*) from Ref. 15; part (*b*) from Ref. 59.]

scheme [15] adopts a two-stage hierarchical row decoder structure. During each memory access only one sub-word line is activated, which typically carries 10 to 25% capacitance compared to the undivided main word line. Subsequently, both the capacitive power consumption and the word-line delay are reduced substantially. The DWL scheme has been used extensively in most high-density SRAMs of 1 Mb and greater [15]. To further increase the capacitance reduction ratio, other approaches used a combination of DWL and a multiple-row decoder and a three-stage hierarchical row decoder scheme. Single-bit-line cross-point cell activation (SCPA) architecture [59] is another scheme aiming for bit-line current minimization by single-cell activation. As shown in Figure 9.6(*b*), memory access activates only one SRAM cell on the cross-point of the X and Y address controls. A 16-Mb SRAM implemented in SCPA achieved 36% active current reduction and 10% area reduction compared to conventional DWL structure as reported.

2. *Pulse operation.* Pulsed word line (PWL) operation can shorten the active duty cycle to the minimum time required for reading and writing operations [61]. As a result, active power consumption during memory access is reduced. Figure 9.7(*a*)

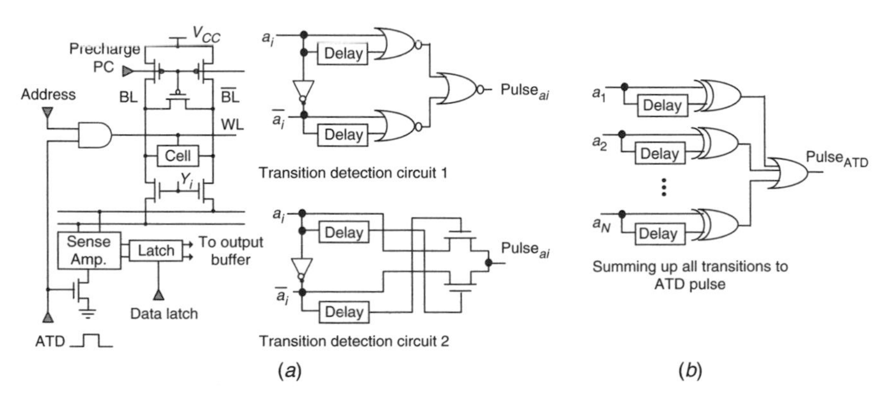

Figure 9.7 PWL operation: (*a*) partial schematic and timing diagram; (*b*) ATD pulse generation circuits. [Part (*a*) from Ref. 61; part (*b*) from Ref. 6.]

shows a PWL partial schematic and timing diagram. In this scheme an address transition detection (ATD) [6] unit was used to generate pulses from address and control signal transition detection. The XD pulse shown on the schematic is formed as ATD rises and then controls word-line operation through the X decoder and sense amplifiers. The ATD pulse generation circuits are shown in Figure 9.7(*b*). This pulsing scheme can also be applied to highly capacitive predecode lines, write-bus lines, bit lines, and sense circuitry [16–20].

3. Cell driving schemes at reduced V_{DD}. As reducing supply voltage effectively suppresses both active and standby SRAM power consumption, many low-power SRAM designs with sub-1-V operation have been implemented during the past decade. To achieve 100-MHz operation at a 0.5- to 0.8-V supply voltage, these designs employ various cell driving schemes, such as a driving source line (DSL) [11], negative word-line driving (NWD) [12], and boosted offset-grounded data storage (BOGS) [13]. Figure 9.8 shows the cell schematic and operation waveforms of DSL and NWD. The DSL scheme connects a source line of cross-coupled inverters to negative voltage V_{BB} during the read cycle, and leaves the source line floating during the write cycle. As a result, the cell read access time

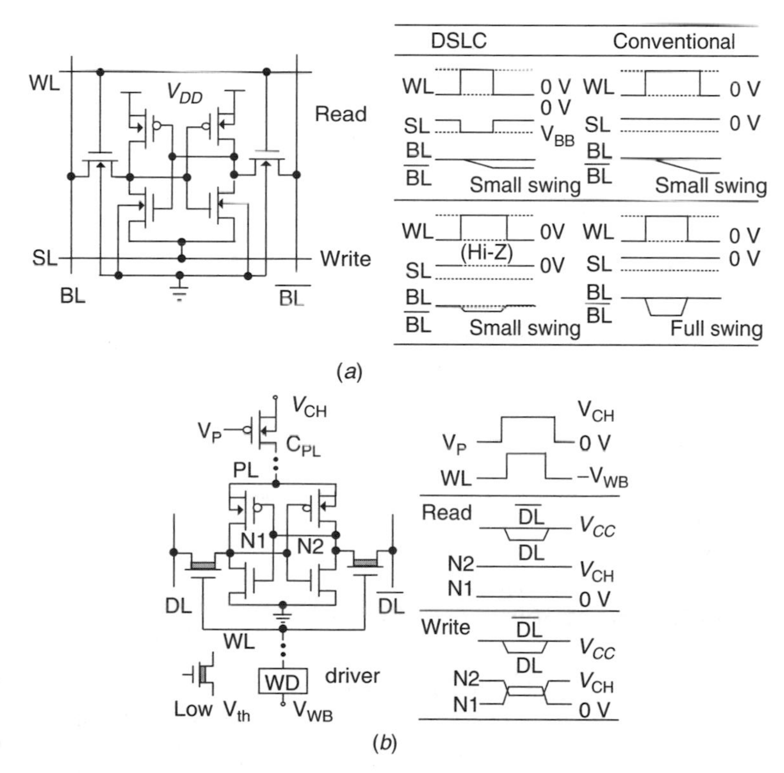

Figure 9.8 DSL and NWD schemes: (*a*) DSL cell and read/write cycle timing diagram; (*b*) NWD cell and read/write cycle timing diagram. [Part (*a*) from Ref. 11; part (*b*) from Ref. 12.]

is improved with boosted gate-to-source voltage and forward bias at the body source–substrate junction of the transistors. The write cycle is also improved since the NMOS transistors in the cross-coupled inverter pair are inactive. The NWD scheme uses low-V_{th} access transistors (Q_{t1} and Q_{t2}) with negative cutoff gate voltage, and a high-V_{th} cross-coupled inverter pair with boosted gate voltage ($V_{\mathrm{CH}} > V_{CC}$) to achieve both improved access time and reduced standby leakage. By exploiting gate–source bias and V_{th} control, both the DSL and NWD schemes enhanced the memory operation speed at sub-1-V supply voltage compared to conventional cell implementation, and suppressed standby leakage current. However, there are several overheads involved with the application of these schemes, such as low-efficiency charge pump operation in generating negative source voltage (DSL) and high leakage flowing from the boosted storage node to the bit line (NWD). BOGS is another cell-driving scheme aimed at solving these problems. Here, shifting the voltage potential of the data storage node pairs from 0.5 V/0 V to 1.3 V/0.65 V eliminates the need for negative source-line voltage generation. Equalizing the boosted potential level between bit-line precharging and word-line driving avoids leakage from the boosted storage node to the bit line. The scheme also applies a charge-recycling method to save power in source-line voltage control.

4. *Low-power sense amplifier designs.* A sense amplifier on an I/O line usually consumes dc current of 1 to 5 mA [6]. When the number of I/O lines on a high-speed processor increases to obtain higher data throughput, the power consumption of the sense amplifier becomes an even larger portion of the total chip power. As shown in Figure 9.7, the pulsed operation scheme efficiently reduces sense amplifier power consumption by switching it on only during the pulse active period. Figure 9.9(*a*) shows a latch-type PMOS cross-coupled sense amplifier design proposed in 1989 [62]. Compared to a conventional paired current-mirror amplifier, this design achieves 50% reduction in sense delay and 80% reduction in dc current with full output swing. The equalizer used to equilibrate the paired outputs of this amplifier requires accurate timing control for stable operation. Figure 9.9(*b*) shows another high-speed sense amplifier design [63]. This amplifier senses bit-line current difference instead of voltage difference. With this design the detectable data-line voltage swing is reduced to less than 30 mV. Compared to conventional voltage sense amplifiers, which require a voltage swing of 100 to 300 mV, this current sense amplifier design saves 60% power consumption with a fixed delay of 1.2 ns [64]. Since the bit-line voltages are kept equal in this design, the sense amplifier possesses an intrinsic equalizing function, which simplifies timing control of the operation.

Low-Power DRAM at Design Time Over the last decade successive circuit advancements have produced a power reduction of two to three orders of magnitude for a fixed-capacity DRAM chip. Similar to the case for SRAM, reduced active current in DRAM helps to achieve low power consumption, low junction temperature, and low-cost packaging. Reductions in charging capacitance and operation voltage have been exploited as the main techniques in DRAM active

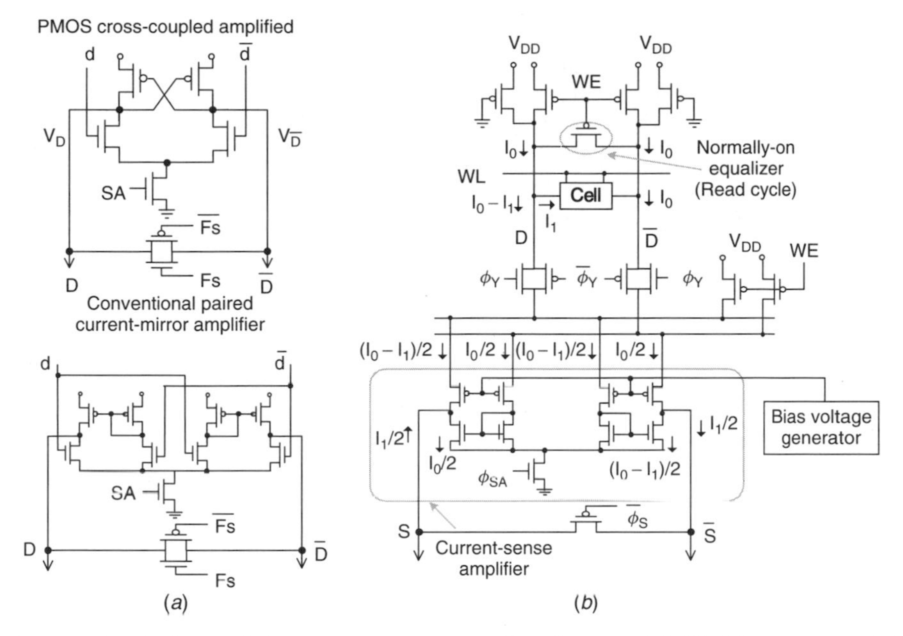

Figure 9.9 High-speed low-power sense amplifier designs: (*a*) PMOS cross-coupled amplifier design; (*b*) current sense amplifier design. [Part (*a*) from Ref. 62; part (*b*) from Ref. 63.]

power control. Meanwhile, the application of subthreshold current suppression schemes such as standby negative gate-to-source bias is indispensable for future battery-supported DRAM systems [6].

Both being key VLSI memory system components, DRAM and SRAM are similar in operation, architecture, and power consumption sources. Therefore, they share many similar power reduction techniques. Since low-power SRAM design is discussed elsewhere in the chapter, DRAM design-time and run-time power control schemes are discussed briefly below, with a focus on techniques designed specifically for DRAM structures.

1. Charging capacitance reduction and increased refresh time. Similar to DWL [15] and SCPA [59] in SRAM, partial activation schemes for multi-divided data and word lines were used to minimize the charging capacitances. As a result, the active power consumption is reduced and the signal-to-noise ratio for memory access operation is improved. Figure 9.10(*a*) and (*b*) show two schemes applying DRAM data- and word-line partial activation, respectively [87,88]. In these approaches the data and word lines are partitioned into multiple sections. These sublines are activated with additional control logic, such as Y decoders in partial data-line activation and row select lines (RX) in partial word-line activation. Shared I/O, sense amplifiers and decoding logic can help reduce the control circuitry overhead [87]. Another static current reduction technique that has been

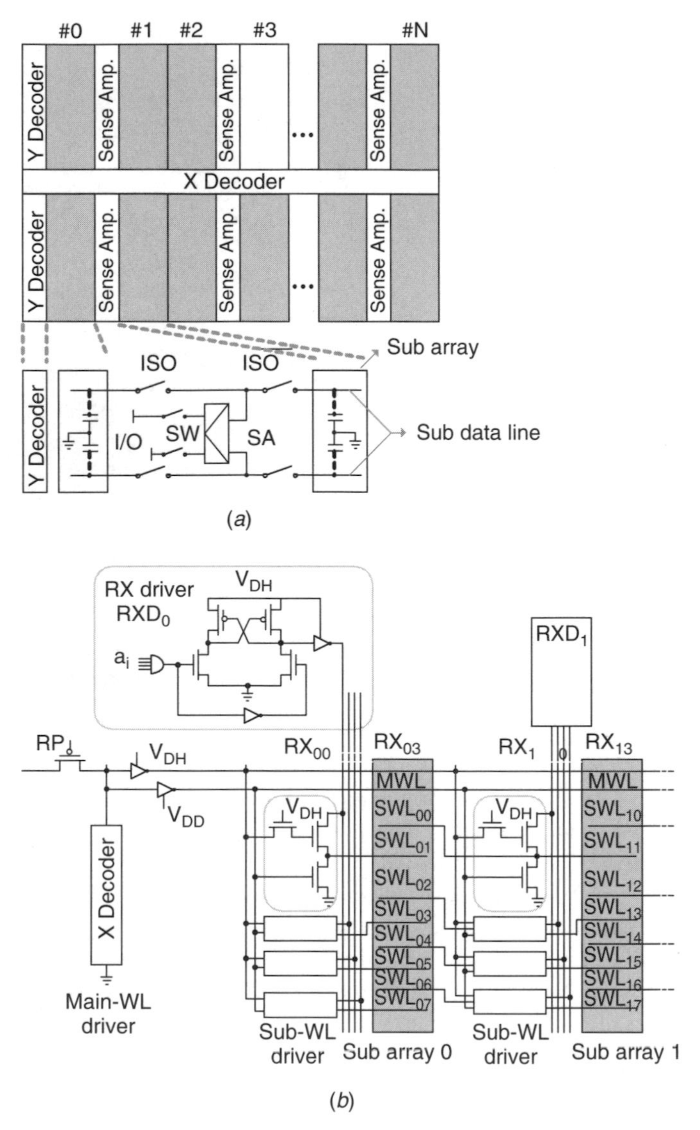

Figure 9.10 Partial activation schemes for DRAM power reduction: (*a*) partial activation of multi-divided data line; (*b*) partial activation of multi-divided word line. [Part (*a*) from Ref. 87; part (*b*) from Ref. 88.]

used together with partial activation schemes is refresh time increase [87]. With flexible control of subsections of the data lines, memory refresh time can be extended without affecting normal operation. This is accomplished by controlling multiple times the number of arrays during refreshing compared to the number of arrays activated concurrently in the normal cycle. The increased self-refresh time leads to a reduction in refreshing current and DRAM static power.

2. Operating voltage reduction. Driven by the scaling and low-power requirements, the supply voltage of DRAM has been reduced from 12 V about two decades ago to the current level-of-approach 1-V operation. Further voltage scaling into the sub-1-V region presents a significant challenge due to the operation-speed degradation and exacerbated leakage power dissipation caused by V_{th} scaling. The key to overcome these difficulties lies in fast sense amplifier and memory operation designs, and effective subthreshold leakage suppression schemes, which will be introduced in a run-time low-power DRAM section. Furthermore, the half-V_{DD} data-line precharge scheme [89] halves the data-line power with reduced voltage swing. The large spike current caused during restoring or precharging periods is also halved, leading to quieter operation with less noise. Finally, the other indispensable contributor in the application of various memory power control techniques is the on-chip voltage down converter, which generates different voltage levels required, such as the precharging voltage in a half-V_{DD} data-line operation scheme. These converters provide stable and accurate output voltage under rapidly changing load current [6].

9.3 RUN-TIME LOW-POWER TECHNIQUES

9.3.1 System- and Architecture-Level Run-Time Techniques

System level run-time techniques can be applied to optimize system management strategy based on real-time operation information such as workload. These techniques include various dynamic-power-aware scheduling schemes that arrange tasks according to estimated execution time [21], dynamic power management (DPM) that dynamically reconfigures an electronic system to provide the service requested with a minimum number of active components [22], and energy-aware routing in communication network applications [23].

At the architecture level the dynamic voltage and frequency scaling technique (DVS, also called DFS or DVFS) [24] is a well-known method used to reduce power consumption when executing a certain task with feedback loop control on V_{DD} and system clock frequency. To reduce further leakage during the idle period, the dynamic V_{th} scaling (DVTS) [25] scheme is used to adjust the threshold voltage adaptively by means of body bias control. Both forward body bias (FBB) and directional adaptive body bias (ABB) have been used as enhancements of conventional reverse body bias (RBB) control. FBB has the desirable result of improving the short-channel effects of a transistor, thus reducing sensitivity to critical-dimension variations [85]. To compensate the within-die variation effect, the within-die ABB (WID-ABB) technique was proposed, which integrates phase detectors and generates appropriate body bias for each circuit block. The effects of ABB and WID-ABB are shown in Figure 9.11. Here sevenfold σ reduction of die frequency distribution is achieved by ABB alone, while WID-ABB reduces the variation an additional threefold, allowing virtually 100% of the dies to be accepted in the highest frequency bin [86]. V_{DD} and V_{th} hopping is a scheme

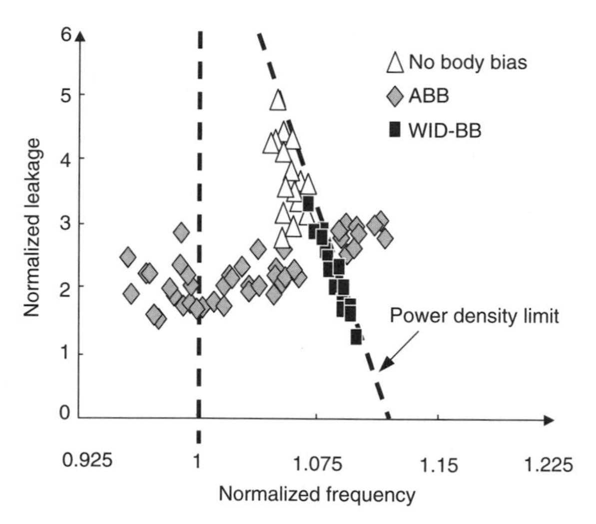

Figure 9.11 ABB and WID-BB control effect on leakage versus frequency distribution. (From Ref. 86.)

similar to DVS, with V_{DD} and V_{th} adjusted at discrete levels controlled by a software feedback loop [26].

FBB and ABB techniques are not without problems. Substrate noise can modulate circuit performance unless the body bias supply is well decoupled and distributed like another supply, which takes away route resources as well as requiring chip area for the decoupling capacitors. There is always a danger of latch-up if the FBB pushes the body too high and forward biases the junction of the transistors. FBB will also increase the junction capacitance and thus increase the dynamic power of the chip. Adaptive negative body bias is the preferred implementation for low-power design. Negative body bias reduces junction capacitance and subthreshold leakage, therefore improving dynamic power as well as standby power.

For body bias to work, the transistors must be tuned for higher body effect so that the negative body bias will raise the V_{th}, thus reducing the leakage current. Gate-induced drain leakage (GIDL) has surfaced in the 90-nm nodes and beyond. GIDL alters the subthreshold curve of a device when the leakage current increases with reduced gate drive. Body bias exacerbates this effect and can negate the effect of body bias (see Figure 9.12).

For a high-reliability server-class high-performance microprocessor, burn-in may be required for the reliability screen. In most cases the power during burn-in is prohibitive, to the extent that only one part can be burned in at a time, due to the power supply limitation. This severely limits productivity in burn-in, so it forces designers to sacrifice performance to reduce power so that more parts can be burned in simultaneously in the same oven. Negative body bias can be used to reduce subthreshold power in burn-in ovens, so that designers do not need to trade performance just to facilitate burn-in of the microprocessor [104].

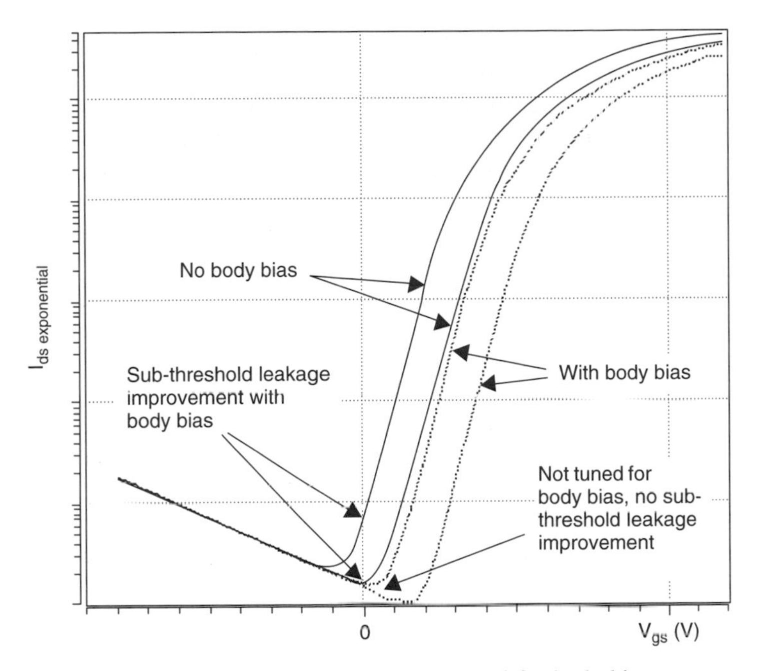

Figure 9.12 Transistors must be tuned for body bias.

Clock distribution networks represent another major source of power consumption, particularly in high-performance microprocessor designs. In a 72-W 600-MHz Alpha processor, half of the power is dissipated in the clock distribution [93]. Among the various approaches used to mitigate this problem, clock gating [75] is an important architectural run-time technique which effectively minimizes clock-involved active power consumption by preventing unneeded activities in logic modules as well as eliminating unnecessary power dissipation in the clock distribution network. While clock gating can be rather easily implemented on the circuit level by switch insertion, it poses many optimization problems on the architectural level. These problems include clock tree construction with minimal total wire length, timing constrains management for clock domains, gated clock nets skew minimization, and others. Besides clock gating, new resonant clock structures have been proposed, using either coupled traveling- or standing-wave oscillators or spiral inductor-based resonant grids [94–96] In these approaches, electromagnetic energy is oscillating in the *LC* system rather than being dissipated as heat in *RC*. Therefore, power loss is reduced. More than an 80% clock power saving has been shown at a resonance frequency of 1.1 GHz [96].

9.3.2 Circuit-Level Run-Time Techniques

1. Exploiting stack effect at run time. Similar to the stack-forcing method [27] that transforms a single device into stacks at design time, various other run-time techniques exploit the stack effect by converting circuits to a stacked structure in standby mode. Motivated by the wide variation in leakage power of a circuit according to the input vectors [42], these techniques aim at reducing the leakage of gates by applying their low-leakage inputs during the standby period. To find an input vector corresponding to minimum leakage power, a number of algorithms have been proposed, including random sampling with a given confidence level [42], genetic estimation [43], and heuristic search based on leakage observability measures [44] and Boolean network modeling [45]. By assigning the specified input vector to a 32-bit static CMOS Kogg–Stone adder, up to a twofold leakage reduction can be achieved [46]. For circuits with a large logic depth, multiplexer insertion- and gate modification-based schemes were used to apply control to internal nodes [45]. As shown in Figure 9.13(*a*), insertion of a multiplexer enables the access to the internal node X. Here the multiplexer is implemented as an AND gate since one of the inputs to the multiplexer is fixed. Figure 9.13(*b*) shows two ways to modify a fully complementary CMOS gate in order to connect its output to 1 or 0 during the standby period. With this scheme the leakage in both the modified gate and its fanout gate are reduced due to the stack effect. Benchmark circuits applying input vector control implemented with these two schemes achieved a 10 to 70% leakage power saving in speed and area costs [45]. Besides these internal node access schemes, another approach is to insert a series low-V_{th} switch to those internal gates in a high-leakage state [44]. During the standby period the series leakage control switches turns off the leaky gates that are not accessible from an external input vector control.

2. *MTCMOS.* At the circuit level, representative run-time techniques are multi-threshold CMOS (MTCMOS) [64], variable-threshold CMOS (VTCMOS) [65], dynamic-threshold CMOS (DTCMOS) [66], and their derivatives. These techniques reduce standby leakage current by inserting series resistance or increase device V_{th} in standby mode. As shown in Figure 9.14(*a*), MTCMOS turns off a low-V_{th} logic block with a series high-V_{th} power switch. As proper sizing of the

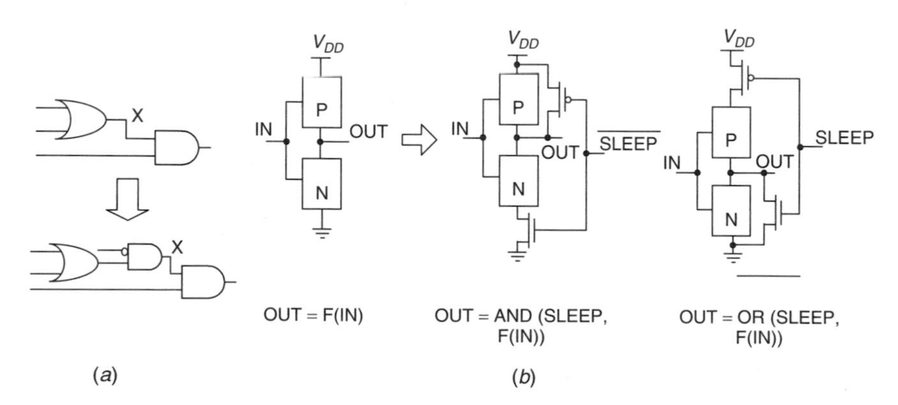

Figure 9.13 Methods to apply input vector control to circuit internal node: (*a*) multiplexer insertion (simplified to AND gate); (*b*) gate modification enables output control. (From Ref. 45.)

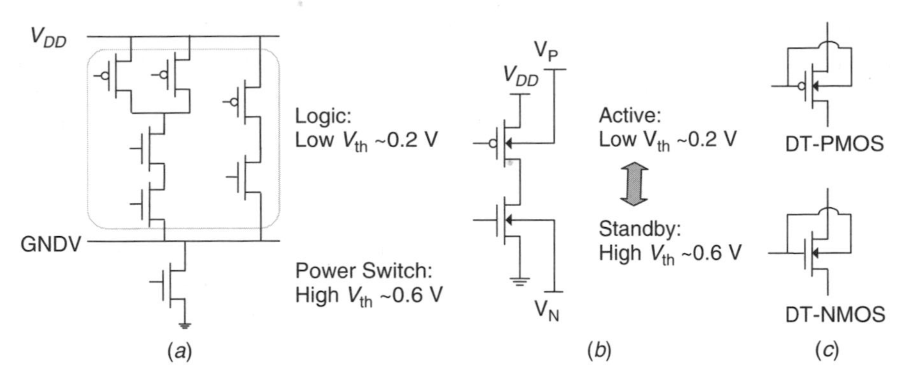

Figure 9.14 Run-time circuit-level schemes for low-power operation: (*a*) MTCMOS; (*b*) VTCMOS; (*c*) DTMOS.

high-V_{th} switch is required to balance the operation delay and area overheads, a hierarchical sizing algorithm was developed to minimize the overall silicon area at a given delay constraint [67]. The MTCMOS technique has been an effective technique in many low-power designs. However, as V_{DD} scales down into the sub-1-V regime, MTCMOS will experience reduced efficiency and eventually, failed functionality, due to the turn-on voltage requirement of the high-V_{th} device. For future low-voltage operations, improved structures, including super cutoff CMOS (SCCMOS) [68] and boosted-gate MOS (BGMOS) [69] were proposed to continue the effectiveness of the power switch–based leakage suppression scheme. The SCCMOS scheme applies negative gate to source bias voltage to a low-V_{th} switch in the standby mode, while BGMOS uses boosted gate-to-source overdrive voltage to speed up operation with high-V_{th} switch. Both of these two schemes effectively suppress leakage current at low V_{DD} but at the expense of extra voltage-level design cost. Furthermore, the zigzag super cutoff CMOS (ZCCMOS) and zigzag boosted gate CMOS (ZBGMOS) schemes were proposed as derivatives of SCCMOS and BGMOS that can improve wake-up time [76].

3. VTCMOS. Figure 9.14(*b*) shows the VTCMOS scheme, where the body bias of circuit in operation is adjusted during different operation modes to achieve the desired threshold voltages. Compared to MTCMOS, the implementation area overhead of VTCMOS is smaller, with transient current flow in the substrate much smaller than the active current pulled from power supplies. The application of VTCMOS is not limited by supply voltage scaling since active operation of VTCMOS is not affected by leakage control. However, as the technology scales toward shorter channel length, the body bias control effect on V_{th} becomes weaker [70]. The increase of within-die V_{th} variation due to reversed body bias effect on short-channel devices also reduces the efficiency of this scheme [71]. As a result, the use of forward bias becomes a more favorable design choice for future VTCMOS implementation [6].

4. DTCMOS. As conventional low-power circuit techniques such as MTCMOS and VTCMOS evolves to satisfy requirements of future design, the DTCMOS scheme has waited for the past decade to embrace the sub-1-V design era. As shown in Figure 9.14(*c*), DTCMOS was proposed at 1994 as a novel operation of MOSFET with gate-to-body connections [72]. With these connections the device V_{th} becomes a function of its gate voltage. As V_{gs} increases during active operation, V_{th} drops to provide a much higher current drive than that of a standard MOSFET with low V_{DD}. On the other hand, zero V_{gs} in the idle mode leads to high V_{th}, which suppresses leakage current effectively. To prevent excessive substrate capacitances and currents, the gate voltage of DTMOS has to be smaller than approximately one diode voltage (about 0.7 V at room temperature), which limits its application in designs with higher V_{DD}. SOI implementation helps reduce the junction cross sections and alleviates the forward-bias hazards. Several proposals were made to eliminate the low-voltage operation limit by using auxiliary MOSFETS or diodes to clamp the body–source forward-bias value or restrict it to a transient effect [73]. As future low-power design requires low-voltage operation, DTCMOS becomes a compelling candidate.

9.3.3 Memory Techniques at Run Time

Low-Power SRAM at Run Time

1. Peripheral circuit leakage suppression by SSI. Memory peripheral circuits are comprised of multiple iterative circuit blocks, which become leakage-intensive paths during the standby period with large total-channel width. Even in active operation mode, most of these circuits stay inactive except for a small portion of selected modules. These features enable simple and effective subthreshold current control. Many logic circuit leakage reduction techniques, such as gate-to-source back biasing, substrate-to-source back biasing, multi-V_{th}, and power switch schemes have been used for memory peripheral circuit leakage suppression [91]. As an example of the application of these techniques, Figure 9.15 shows the switched-source impedance (SSI) scheme [77], which turns off control circuitry

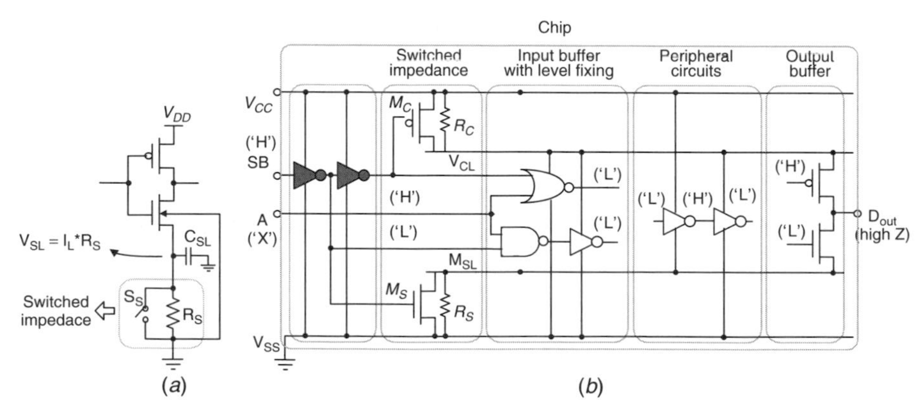

Figure 9.15 SSI scheme and its application to memory: (*a*) SSI circuit structure; (*b*) SSI applied to memory peripheral circuit leakage suppression. (From Ref. 77.)

leakage paths during the idle period. Level-fixing input buffers in Figure 9.13(*b*) are used to force internal nodes to predetermined levels. M_c, M_s, and the shaded inverters in this figure are high-V_{th} switches.

2. *Variable-threshold leakage suppression schemes.* Among the circuit-level leakage suppression schemes, adjustable body bias control has the property of preserving data stored in latch circuits. Therefore, it has been applied on memory arrays to achieve subthreshold leakage reduction. Similar to the VTCMOS technique, here substrate voltages of the nonselected memory cells are reverse biased to obtain high-V_{th} standby operation. Figure 9.16(*a*) shows the circuit schematics and timing diagram of a dynamic leakage cutoff (DLC) scheme [78]. DLC

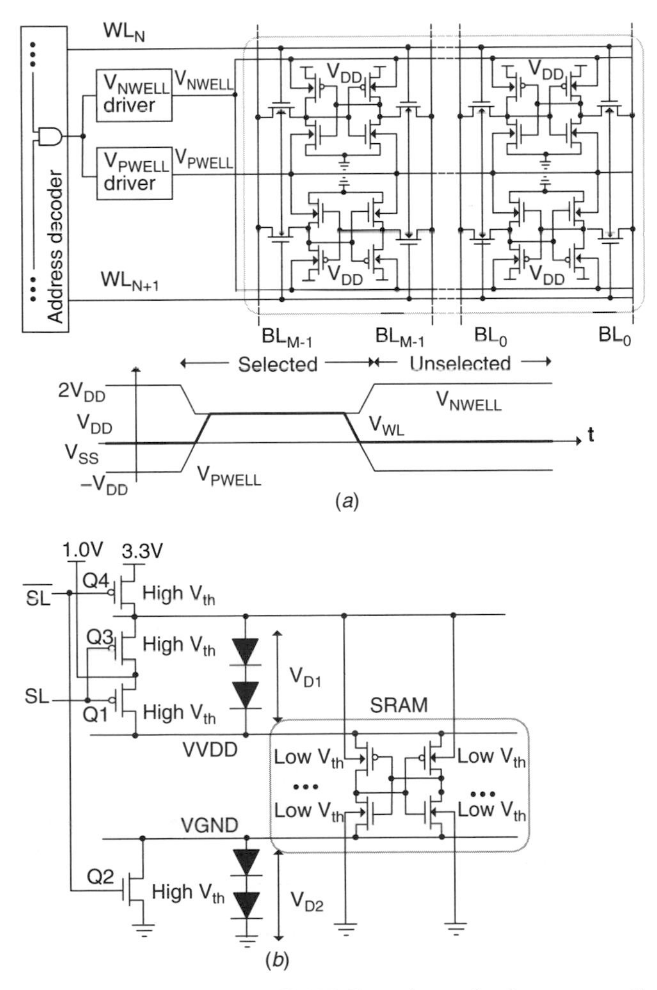

Figure 9.16 Dynamic leakage cutoff (DLC) and auto-backgate-controlled MT-CMOS (ABC-MT-CMOS): (*a*) schematics and timing diagram of DLC scheme; (*b*) configuration of ABC-MT-CMOS circuit. [Part (*a*) from Ref. 78; part (*b*) from Ref. 79.]

biases the substrate voltages of nonselected SRAM cells at about $2V_{DD}$ for V_{NWELL} and about $-V_{DD}$ for V_{PWELL}. Figure 9.16(*b*) shows the configuration of auto-backgate-controlled MT-CMOS (ABC-MT-CMOS) scheme [79]. The active operation voltage is 1 V with Q1, Q2, and Q3 turned on and Q4 off. During standby mode, Q4 is switched on while other transistors are turned off. The virtual V_{dd} and ground rails of ABC-MT-CMOS are clamped by diodes D1 and D2. With the reverse bias voltages $V_{\text{D1}} = V_{\text{D2}} = 1.15$ V, the leakage current is reduced to 20 pA/cell.

3. Gated supply and ultralow standby supply voltage schemes. At the architectural level, SRAM run-time leakage reduction techniques include gating off the supply voltage of the idle memory sections or putting the less frequently used sections into a drowsy standby mode. These approaches exploited the quadratic reduction of leakage power with V_{dd}, and achieved optimal power–performance trade-offs with the assistance of compiler-level cache activity analysis. Cache delay technique applied adaptive timing policies in cache-line gating, achieving 70% leakage saving at a modest performance penalty [28]. As shown in Figure 9.17(*a*), to further exploit the leakage control on caches with a large utilization ratio, the drowsy cache scheme allocated inactive cache lines to a low-power mode, where V_{dd} was lowered but still preserving the memory data [29].

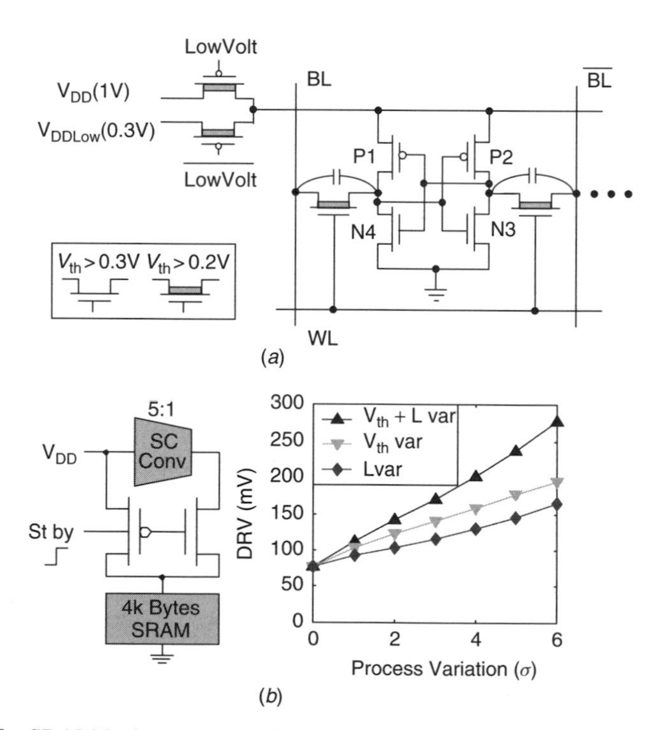

Figure 9.17 SRAM leakage suppression schemes applying ultralow standby supply voltage: (*a*) drowsy memory circuit; (*b*) dual-rail SRAM standby scheme and process effect on DRV. [Part (*a*) from Ref. 29; part (*b*) from Ref. 30.]

The dual-rail standby scheme shown in Figure 9.17(*b*) was proposed for ultralow power application, where the entire SRAM module was pushed into a deep sleep with a 300-mV standby supply voltage during the standby period. Over a 90% leakage power saving was achieved at this ultralow data retention voltage [30]. In this work it was also shown that stable SRAM data preservation for a 0.13-μm process is achievable in the 300-mV region, while the data retention voltage (DRV) increases about linearly with process variations in threshold voltage and channel length.

Low-Power DRAM at Run Time

1. Peripheral circuit power reduction. As noted earlier, logic circuit leakage reduction techniques such as gate-to-source back biasing, substrate to source back biasing, multi-V_{th}, and power switch schemes are effective methods for memory peripheral circuit leakage suppression. In addition, enhancing the conversion power efficiency of on-chip voltage converters and minimizing their standby current have been particularly important issues for low-power DRAM design. This is because low-voltage DRAM design has relied heavily on various current drive boosting or subthreshold current suppression schemes to achieve high-speed, low-power operation. These schemes usually require various voltage levels, such as back bias, reference, and precharge voltages [6,90,91].

2. Refresh time extension and charge recycling. As mentioned in Section 9.2.3, an increasing refresh time interval reduces the refresh current. This scheme was accompanied by the partial data-line activation technique, which provides flexible control on array operations [87]. Charge recycling is another scheme that reduces capacitive data-line power consumption during refreshing. Here the charge used in one array, conventionally poured out in every cycle, is transferred to another array and gets reutilized [90].

3. Gate–source offset driving schemes for DRAM cell. The boosted sense ground (BSG) and negative word-line (NWL) schemes has been well known as gate–source offset driving schemes applied on a DRAM cell. Both of these schemes vary the V_{th} of a DRAM cell transistor dynamically to achieve the desired active drive current and small subthreshold current. As shown in Figure 9.18, BSG raises the data-line voltage by ΔV_{DL} and NWL reduces the gate

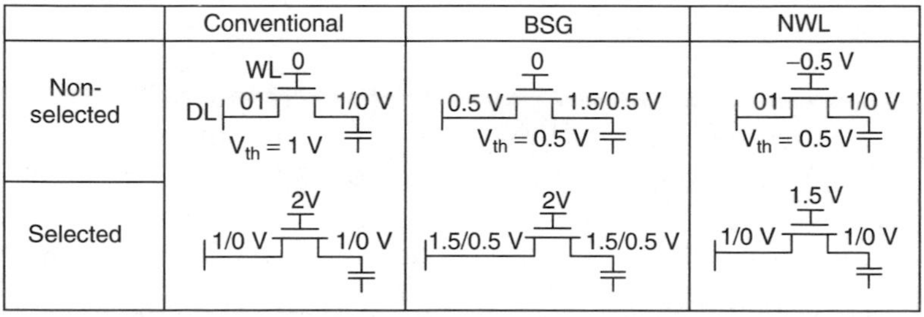

	Conventional	BSG	NWL
Non-selected	WL 0; DL 01; 1/0 V; V_{th} = 1 V	0; 0.5 V; 1.5/0.5 V; V_{th} = 0.5 V	−0.5 V; 01; 1/0 V; V_{th} = 0.5 V
Selected	2V; 1/0 V; 1/0 V	2V; 1.5/0.5 V; 1.5/0.5 V	1.5 V; 1/0 V; 1/0 V

Figure 9.18 Comparison of DRAM cell driving schemes (assuming that $V_{th0} = 1$ V and 1 V storage voltage). (From Ref. 92.)

voltage by ΔV_{WL} for nonselected cells during the standby period. The standby V_{th} subsequently increased suppresses cell leakage and enables the application of lower-V_{th} transistors for a DRAM cell. In this comparison, V_{th} is chosen to be 1 V for the conventional scheme and 0.5 V for BSG and NWL. The ΔV_{WL} generator implementation in NWL is comparatively easier, due to fact that there is the word-line discharging current than data-line sinking in BSG. Both of these two schemes, however, increase gate oxide stress for the nonselected cells [92].

9.4 TECHNOLOGY INNOVATIONS FOR LOW-POWER DESIGN

While scaling exacerbates the leakage power crisis for future designs, the evolving CMOS technology also provides designers with low-power process features, including choices of high V_{th}, thick gate oxide, and access to multiple wells, which facilitates the application of adaptive back-bias schemes. Beyond conventional CMOS, technology innovations have brought many novel devices and fabrication processes onto the stage, including SOI, double-gate devices, and strain Si. New assembly technologies such as system-in-a-package (SiP) help reduce package- and board-level capacitances and achieve low-power system integration.

9.4.1 Novel Device Technologies

The primary difficulty of transistor scaling lies in the control of off-state leakage. To solve this problem, advanced device structures have been proposed, such as fully depleted SOI (e.g., ultrathin-body (UTB) device [100]) and double-gate structure (e.g., FinFET [101]). Among them, FinFET has been considered the foremost candidate for ~10-nm-gate-length device technology, due to its superior scalability and a process flow and layout similar to that of a conventional MOSFET. Figure 9.19(a) illustrates the FinFET structure. It is usually made from

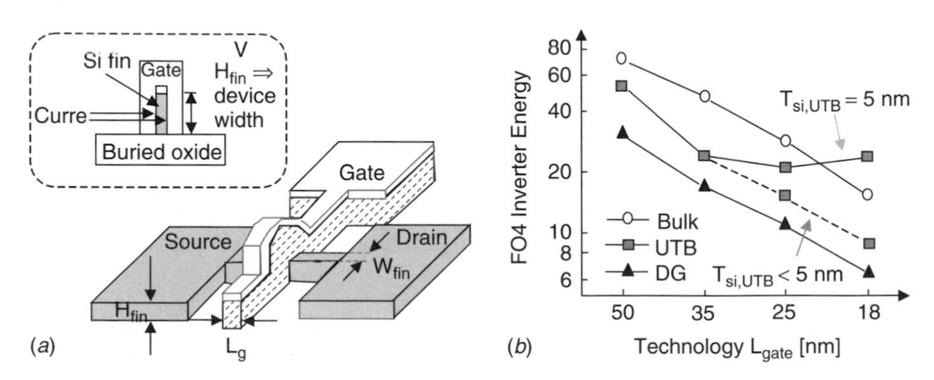

Figure 9.19 FinFET structure and potential active power saving benefit: (a) FinFET three-dimensional schematic; (b) performance comparison between technologies. (From Ref. 102.)

SOI substrate, with a gate straddling a thin, fin-shaped body, which forms two self-aligned channels along the sidewalls of the fin. The top of the fin is usually covered by a hard mask and not part of the channel. The device width is defined by the fin height (H_{fin}) and multiple fins can be used to realize different device widths. With two gates controlling the thin channel, the short-channel effect of the device is suppressed efficiently. Other advantages of FinFET include larger current due to higher carrier mobility in the intrinsic doped channel, and less capacitance because both depletion and junction capacitances are eliminated. As a result, FinFET offers excellent standby power reduction because of the suppression of subthreshold leakage from its double-gate nature. Furthermore, since FinFET devices can have higher drive current than traditional bulk silicon, power supply voltage can be reduced to match the same performance as bulk CMOS; thus the active power of the circuit can also be greatly reduced. Figure 9.19(*b*) shows the energy saving of a FO4 inverter using FinFET and UTB devices compared to that of classic bulk silicon. An energy consumption reduction of up to 60% can be achieved using FinFET. With the excellent scalability and significant circuit performance advantages offered by FinFET structure, it may be adopted for IC production as early as the 65-nm technology node (about 25 nm physical gate length) [102].

In addition to SOI and double-gate devices other technology innovations include low-κ dielectric (air-gap Cu technology) and the newly developed strain silicon with a nitride cap for 90 nm and below.

9.4.2 Assembly Technology Innovations

I/O systems consume a large amount of power due to the intrinsic large capacitance. Wires in a package substrate and printed-circuit board are much longer than on-chip lines, and the capacitance associated with them can be two times larger than that in a chip. New types of assembly technology and interface signaling design techniques provide promises for low-power system integration, such as system-in-a-package (SiP), three-dimensional integration, *RF* wireless interconnects, and optical interconnections [98]. Among these explorations, the SiP approach can integrate multiple heterogeneous chips and *RF* passive components in one package [99]. It reduces I/O power dissipation considerably without requiring substantial development for new design principles.

9.5 PERSPECTIVES FOR FUTURE ULTRALOW-POWER DESIGN

The trend toward prosperity of future pervasive computation/communication applications with superior portability and intelligence will keep pushing system designs into a lower power regime. Besides the evolution of the foregoing techniques that can be scaled into next-generation applications, the potential areas and techniques having the most impact on future ultralow power design include:

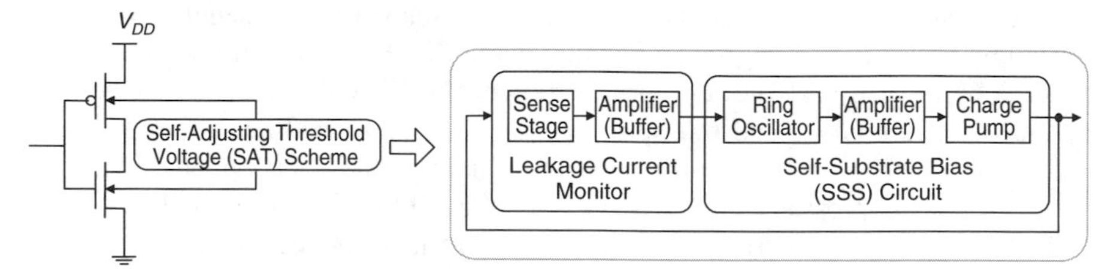

Figure 9.20 VT-sub-CMOS logic with a stabilization scheme. (From Ref. 32.)

9.5.1 Subthreshold Circuit Operation

With power consumption orders of magnitude lower than that of a normal strong inversion circuit, subthreshold circuit operation is a strong candidate for future applications with self-sustainable energy scavenging design. Compared to conventional CMOS logic, subthreshold circuits also have advantages of increased transconductance gain and near-ideal static noise margin. However, due to the absence of conducting inversion channels, its sensitivity to power supply, temperature, and process variations can be prohibitively high without proper control, which limits the near-term application of this scheme. During the effort to overcome these difficulties, some subthreshold logic families have been proposed, including variable V_{th} subthreshold CMOS (VT-sub-CMOS) and subthreshold dynamic V_{th} MOS (sub-DTMOS) logic. As shown in Figure 9.20, VT-sub-CMOS logic applies an additional stabilization scheme, where a stabilization circuit monitors any change in the transistor current due to temperature and process variations and transmit an appropriate bias to the substrate. Both logic and stabilization circuits of VT-sub-CMOS operate in the subthreshold region. The DTMOS logic introduced in Section 9.3.2 is a compelling candidate for low-voltage operation. Compared to subthreshold CMOS logic, sub-DTCMOS has a larger gate capacitance but provides much higher active current. With the power-delay product (PDP) similar between these two subthreshold logic families, sub-DTCMOS can be operated at a much higher switching frequency while maintaining the same energy/switching ratio. Both VT-sub-CMOS and sub-DTMOS achieve desired robustness and tolerance to process and temperature variations but at the penalty of additional stabilization circuitry or process complexities [32].

9.5.2 Fault-Tolerant Design

On the front end of technology and voltage scaling, future devices and interconnects are subject to larger process variations and higher vulnerability to external interference such as natural radiation and electrical noise. Relaxing the requirement from 100% correctness in operation to a reasonable error rate can drastically reduce the design costs, but at the same time requires a reliable ultralow power design to be equipped with a certain degree of fault tolerance. Up to now, numerous fault-tolerant schemes on various design levels and application fields have been implemented, such as error correction codes (ECCs) in DRAM design and communication processes, computer architecture verification schemes with redundancy solutions in hardware (a triplication and voting scheme (TMR), a watch dog processor design, and a dynamic implementation verification architecture (DIVA) [33]) or software approaches (simultaneously and redundantly threaded processor (SRT) [34]), and so on. The future robust ultralow power system will be an integration of cross-level cooperative fault-tolerant schemes, just as today's systematic low-power design approach.

9.5.3 Asynchronous versus Synchronous Design

The synchronous timing scheme has carried VLSI design successfully through the past two decades of exponential growth, resulting in masterpieces of modern processor design, well-established design methodology, and advanced computer-aided design tools resolving and optimizing the synchronization. However, as the designer's goal evolves toward multi-GHz frequency operation, a larger system with increased complexity, and at the same time a limited budget of power consumption, the conventional synchronous scheme inevitably runs into significant problems. Clock uncertainty control and the power consumed by the clock distribution network are major barriers in reducing system design cost. Increased process variations also hurt the synchronous system performance, severely where worst-case timing characterizations have to be enforced over all other circumstances. As a returning rival with considerable potential for power-efficient design, the asynchronous design methodology has been investigated extensively in recent years [80–83]. Compared to the synchronous scheme, asynchronous design has the advantages of using power only for useful work, optimizing subcomponents for typical instead of worst-case conditions, lower noise and electromagnetic emission, and reduced difficulty in global timing coordination [80,81]. Asynchronous design is especially well suited to applications where the computation load fluctuates unpredictably and when a large discrepancy exists between worst case and typical condition performance [80]. The existing possible solutions range from complete asynchronous circuits to globally asynchronous, locally synchronous (GALS) [84] systems. GALS results from the evolution of synchronous architecture. As an intermediate stage, it reduces the difficulty in design methodology transition, takes less area, but consumes more power than fully asynchronous implementation [82].

9.5.4 Gate-Induced Leakage Suppression Schemes

Gate leakage currents are still not the dominant leakage components in 130-nm technology. But with the rapid scaling trend of thinner t_{ox}, gate oxide tunnel leakage and gate-induced drain leakage (GIDL) will soon be comparable to the subthreshold current, and effective techniques will be required to get it under control. Traditional leakage reduction techniques such as reducing supply voltage and shutting off the unused sections are still effective methods treating the new leakage elements. Other existing approaches dedicated to gate leakage suppression include pin-reordering [36] and electrical field relaxation (EFR) [37] schemes. The pin-reordering technique exploits the gate leakage dependence on the location of "off" devices within a nonconducting stack. Pin-reordering optimization results show a 22 to 82% reduction in standby gate leakage and up to 25% in run-time gate leakage [36]. The EFR scheme achieves 90% reduction in GIDL current by relaxing the gate-to-drain voltage of SRAM cell transistors from 1.5 V to 1 V [37]. Application of dual-t_{ox} is another technique suggested for future high-speed, low-power DRAM designs, where thin t_{ox} on the periphery helps achieve faster operation, and thick t_{ox} of the core cells ensures stable operation and suppressed gate tunneling leakage current. Similarly, dual-V_{th} and dual-V_{DD} can be applied to satisfy different requirements between RAM cells and peripheral circuits, leading to an optimized memory design for both performance and power concerns. Besides circuit improvements, future innovations at the technology level, such as the development of new gate-dielectric materials with low leakage and a high dielectric constant, may be the most desirable solution [103].

REFERENCES

[1] V. De and S. Borkar, Technology and design challenges for low power and high performance, *International Symposium on Low Power Electronics and Design,* pp. 163–168, Aug. 1999.

[2] T. Sakurai, Perspectives on power-aware electronics, *IEEE International Solid-State Circuits Conference*, pp. 1–16, Feb. 2003.

[3] B. Chatterjee et al., Effectiveness and scaling trends of leakage control techniques for sub-130 nm CMOS technologies, *International Symposium on Low Power Electronics and Design,* pp. 122–127, Aug. 2003.

[4] P. J. M. Havinga and G. J. M. Smit, Design techniques for low power systems, *J. Syst. Archit.*, Vol. 46, No. 1, 2000.

[5] M. Sheets et al., Power management for PicoRadio, *Gigascale Systems Research Center Workshop*, June 2002.

[6] K. Itoh, K. Sasaki, and Y. Nakagome, Trends in low-power RAM circuit technologies, *Proc. IEEE*, pp. 524–543, Apr. 1995.

[7] J. P Fishburn and S. Taneja, Transistor sizing for high performance and low power, *Custom Integrated Circuits Conference*, pp. 591–594, May, 1997.

[8] M. Hamada, Y. Ootaguro, and T. Kuroda, Utilizing surplus timing for power reduction, *Custom Integrated Circuits Conference*, pp. 89–92, May, 2001.

[9] R. Brodersen et al., Methods for true power minimization, *International Conference on Computer Aided Design*, Nov. 2002.

[10] C. Piguet et al., Low-power low-voltage library cells and memories, *IEEE International Conference on Electronics, Circuits and Systems*, Vol. 3, pp. 1521–1524, Sept. 2001.

[11] H. Mizuno and T. Nagano, Driving source-line (DSL) cell architecture for sub-1-V high-speed low-power applications, *Digest of Technical Papers, Symposium on VLSI Circuits*, pp. 25–26, June 1995.

[12] K. Itoh, A. R. Fridi, A. Bellaouar, and M. I. Elmasry, A deep sub-V, single power-supply, SRAM cell with multi-Vt, boosted storage node and dynamic load, *Digest of Technical Papers, Symposium on VLSI Circuits,* pp. 132–133, June 1996.

[13] H. Yamauchi, T. Iwata, H. Akamatsu, and A. Matsuzawa, A 0.5 V single power supply operated high-speed boosted and offset-grounded data storage (BOGS) SRAM cell architecture, *IEEE Trans. VLSI Syst.*, Vol. 5, No. 4, pp. 377–387, Dec. 1997.

[14] O. Minato et al., A 20 ns 64 K CMOS RAM, *IEEE International Solid-State Circuits Conference*, pp. 222–223, Feb. 1984.

[15] J. S. Caravella, A low voltage SRAM for embedded applications, *IEEE J. Solid-State Circuits*, Vol. 32, No. 3, pp. 428–432, Mar. 1997.

[16] M. Yoshimoto et al., A 64 Kb full CMOS RAM with divided word line structure, *IEEE International Solid-State Circuits Conference*, Vol. XXVI, pp. 58–59, Feb. 1983.

[17] B. S. Amrutur and M. A. Horowitz, Techniques to reduce power in fast wide memories, *Proc. SLPE'94*, pp. 92–93, 1994.

[18] T. Mori et al., A 1 V 0.9 mW at 100 MHz 2 k × 16 b SRAM utilizing a half-swing pulsed-decoder and write-bus architecture in 0.25 μm dual-Vt CMOS, *IEEE International Solid-State Circuits Conference*, pp. 22.4-1–22.4-2, Feb. 1998.

[19] K. W. Mai et al., Low-power SRAM design using half-swing pulse-mode techniques, *IEEE J. Solid-State Circuits*, Vol. 33, No. 11, pp. 1659–1671, Nov. 1998.

[20] S. Flannagan et al., Two 64 K CMOS SRAMs with 13 ns access time, *IEEE International Solid-State Circuits Conference*, Vol. XXIX, pp. 208–209, Feb. 1986.

[21] H. Aydin et al., Dynamic and aggressive scheduling techniques for power-aware real-time systems, *Real-Time Systems Symposium*, London, Dec. 2001.

[22] L. Benini, A. Bogliolo, and G. De Micheli, A survey of design techniques for system-level dynamic power management, *IEEE Trans. VLSI Syst.*, Vol. 8, No. 3, pp. 299–316, June 2000.

[23] J. Gomez, A. T. Campbell, M. Naghshineh, and C. Bisdikian, Power-aware routing in wireless packet networks, *IEEE International Workshop on Mobile Multimedia Communications*, pp. 380–383, Nov. 1999.

[24] T. D. Burd, T. A. Pering, A. J. Stratakos, and R. W. Brodersen, A dynamic voltage scaled microprocessor system, *IEEE J. Solid-State Circuits*, Vol. 35, No. 11, pp. 1571–1580, Nov. 2000.

[25] C. H. Kim and K. Roy, Dynamic V_{TH} scaling scheme for active leakage power reduction design, *Proceedings of Design, Automation and Test in Europe Conference and Exhibition*, pp. 163–167, Mar. 2002.

[26] S. Lee and T. Sakurai, Run-time voltage hopping for low-power real-time systems, *Design Automation Conference*, pp. 806–809, June 2000.

[27] S. Narendra et al., Scaling of stack effect and its application for leakage reduction, *International Symposium on Low Power Electronics and Design*, pp. 195–200, Aug. 2001.

[28] S. Kaxiras, Z. Hu, and M. Martonosi, Cache decay: exploiting generational behavior to reduce cache leakage power, *International Symposium on Computer Architecture*, pp. 240–251, June–July 2001.

[29] K. Flautner et al., Drowsy caches: simple techniques for reducing leakage power, *International Symposium on Computer Architecture*, pp. 148–157, May 2002.

[30] H. Qin et al., SRAM leakage suppression by minimizing standby supply voltage, *IEEE International Symposium on Quality Electronic Design*, Mar. 2004.

[31] M. Margala, Low-power SRAM circuit design, *IEEE International Workshop on Memory Technology, Design and Testing*, pp. 115–122, Aug. 1999.

[32] H. Soeleman, K. Roy, and B. C. Paul, Robust subthreshold logic for ultra-low power operation, *IEEE Trans. VLSI Syst.*, Vol. 9, No. 1, pp. 90–99, Feb. 2001.

[33] T. M. Austin, DIVA: a reliable substrate for deep submicron microarchitecture design, *ACM/IEEE International Symposium on Microarchitecture*, 1999.

[34] S. K. Reinhardt and S. S. Mukherjeem, Transient fault detection via simultaneous multithreading, *International Symposium on Computer Architecture*, 2000.

[35] A. Ono et al., A 100 nm node CMOS technology for practical SOC application requirement, *IEEE International Electron Devices Meeting*, pp. 511–514, 2001.

[36] D. Lee, W. Kwong, D. Blaauw, and D. Sylvester, Analysis and minimization techniques for total leakage considering gate oxide leakage, *Design Automation Conference*, pp. 175–180, June 2003.

[37] K. Osada, Y. Saitoh, E. Ibe, and K. Ishibashi, 16.7fA/cell tunnel-leakage-suppressed 16-Mbit SRAM based on electric-field-relaxed scheme and alternate ECC for handling cosmic-ray-induced multi-errors, *IEEE International Solid-State Circuits Conference*, pp. 260–261, Feb. 1996.

[38] J. P. Fishburn and A. E. Dunlop, TILOS: a posynomial programming approach to transistor sizing, *International Conference on Computer-Aided Design*, pp. 326–328, Nov. 1985.

[39] A. R. Conn et al., Gradient-based optimization of custom circuits using a static-timing formulation, *Design Automation Conference*, pp. 452–459, June 1999.

[40] M. Borah, R. Owens, and M. Irwin, Transistor sizing for low power CMOS circuits, *IEEE Trans. Comput. Aided Des. Integrated Circuits Syst.*, Vol. 15, No. 6, 665–671, 1996.

[41] S. Ma and P. Franzon, Energy control and accurate delay estimation in the design of CMOS buffers, *IEEE J. Solid-State Circuits,* Vol. 29, No. 9, pp. 1150–1153, Sept. 1994.

[42] J. P. Halter and F. N. Najm, A gate-level leakage power reduction method for ultra-low-power CMOS circuits, *IEEE Custom Integrated Circuits Conference*, pp. 475–478, May 1997.

[43] Z. Chen, M. Johnson, L. Wei, and W. Roy, Estimation of standby leakage power in CMOS circuit considering accurate modeling of transistor stacks, *International Symposium on Low Power Electronics and Design*, pp. 239–244, Aug. 1998.

[44] M. C. Johnson, D. Somasekhar, L. Chiou, and K. Roy, Leakage control with efficient use of transistor stacks in single threshold CMOS, *IEEE Trans. VLSI Syst.*, Vol. 10, No. 1, pp. 1–5, Feb. 2002.

[45] A. Abdollahi, F. Fallah, and M. Pedram, Runtime mechanisms for leakage current reduction in CMOS VLSI circuits, *International Symposium on Low Power Electronics and Design*, pp. 213–218, Aug. 2002.

[46] Y. Ye, S. Borkar, and V. De, A new technique for standby leakage reduction in high-performance circuits, *Digest of Technical Papers, Symposium on VLSI Circuits*, pp. 40–41, 1998.

[47] E. Musoll and J. Cortadella, Optimizing CMOS circuits for low power using transistor reordering, *European Design and Test Conference*, pp. 219–223, Mar. 1996.

[48] S. C. Prasad and K. Roy, Circuit optimization for minimization of power consumption under delay constraint, *International Conference on VLSI Design*, pp. 305–309, Jan. 1995.

[49] R. Hossain, M. Zheng, and A. Albicki, Reducing power dissipation in CMOS circuits by signal probability based transistor reordering, *IEEE Trans. Comput. Aided Des. Integrated Circuits Syst.*, Vol. 15, No. 3, pp. 361–368, Mar. 1996.

[50] W. Z. Shen, J. Y. Lin, and F. W. Wang, Transistor reordering rules for power reduction in CMOS gates, *Asian and South Pacific Design Automation Conference*, pp. 1–6, Aug. 1995.

[51] M. Hashimoto, H. Onodera, and K. Tamaru, Input reordering for power and delay optimization, *IEEE International ASIC Conference and Exhibit*, pp. 194–199, Sept. 1997.

[52] H. Onodera, M. Hashimoto, and T. Hashimoto, ASIC design methodology with on-demand library generation, *Digest of Technical Papers, Symposium on VLSI Circuits*, pp. 57–60, June 2001.

[53] M. Hashimoto and H. Onodera, Post-layout transistor sizing for power reduction in cell-based design, *Asia and South Pacific Design Automation Conference*, pp. 359–365, Feb. 2001.

[54] S. Sirichotiyakul et al., Duet: an accurate leakage estimation and optimization tool for dual-V_t circuits, *IEEE Trans. VLSI Syst.*, Vol. 10, No. 2, pp. 79–90, Apr. 2002.

[55] Z. Chen et al., 0.18 μm dual V_t MOSFET process and energy-delay measurement, *International Electron Devices Meeting*, pp. 851–854, Dec. 1996.

[56] K. Fujii, T. Douseki, and M. Harada, A sub-1 V triple-threshold CMOS/SIMOX circuit for active power reduction, *IEEE International Solid-State Circuits Conference*, pp. 190–191, Feb. 1998.

[57] L. Wei et al., Design and optimization of low voltage high performance dual threshold CMOS circuits, *Design Automation Conference*, pp. 489–494, June 1998.

[58] K. Nose and T. Sakurai, Optimization of V_{DD} and V_{TH} for low-power and high-speed applications, *Asia and South Pacific Design Automation Conference*, pp. 469–474, Jan. 2000.

[59] M. Ukita et al., A single-bit-line cross-point cell activation (SCPA) architecture for ultra-low-power SRAM's, *IEEE J. Solid-State Circuits*, Vol. 28, No. 11, pp. 1114–1118, Nov. 1993.

[60] S. Manne, A. Klauser, and D. Grunwald, Pipeline gating: speculation control for energy reduction, *International Symposium on Computer Architecture*, pp. 132–141, July 1998.

[61] O. Minato et al., A 20 ns 64 K CMOS static RAM, *IEEE J. Solid-State Circuits*, Vol. 19, No. 6, pp. 1008–1013, Dec. 1984.

[62] K. Ishibashi et al., A 9-ns 1-Mbit CMOS SRAM, *IEEE J. Solid-State Circuits*, Vol. 24, No. 5, pp. 1219–1225, Oct. 1989.

[63] E. Seevinck, A current sense-amplifier for fast CMOS SRAMs VLSI circuits, *Digest of Technical Papers, Symposium on VLSI Circuits*, pp. 71–72, June 1990.

[64] K. Sasaki et al., 7-ns 140-mW 1-Mb CMOS SRAM with current sense amplifier, *IEEE J. Solid-State Circuits*, Vol. 27, No. 11, pp. 1511–1518, Nov. 1992.

[65] S. Douseki et al., 1-V power supply high-speed digital circuit technology with multithreshold-voltage CMOS, *IEEE J. Solid-State Circuits*, Vol. 30, No. 8, pp. 847–854, Aug. 1995.

[66] T. Kuroda et al., A 0.9-V, 150-MHz, 10-mW, 4 mm^2, 2-D discrete cosine transform core processor with variable threshold-voltage (VT) scheme, *IEEE J. Solid-State Circuits*, Vol. 31, No. 11, pp. 1770–1779, Nov. 1996.

[67] J. Kao, S. Narendra, and A. Chandrakasan, MTCMOS hierarchical sizing based on mutual exclusive discharge patterns, *Design Automation Conference*, pp. 495–500, June 1998.

[68] H. Kawaguchi, K. Nose, and T. Sakurai, A super cut-off CMOS (SCCMOS) scheme for 0.5-V supply voltage with picoampere stand-by current, *IEEE J. Solid-State Circuits*, Vol. 35, No. 10, pp. 1498–1501, Oct. 2000.

[69] T. Inukai et al., Boosted gate MOS (BGMOS): device/circuit cooperation scheme to achieve leakage-free giga-scale integration, *IEEE Custom Integrated Circuits Conference*, pp. 409–412, May 2000.

[70] T. Kuroda, Low power CMOS digital design for multimedia processors, *International Conference on VLSI and CAD*, pp. 359–367, Oct. 1999.

[71] K. Kanda, K. Nose, H. Kawaguchi, and T. Sakurai, Design impact of positive temperature dependence on drain current in sub-1-V CMOS, *IEEE J. Solid-State Circuits*, Vol. 36, No. 10, pp. 1559–1564, Oct. 2001.

[72] F. Assaderaghi et al., A dynamic threshold voltage MOSFET (DTMOS) for ultra-low voltage operation, *International Electron Devices Meeting*, pp. 809–812, Dec. 1994.

[73] F. Assaderaghi, DTMOS: its derivatives and variations, and their potential applications in microelectronics, *International Conference on Microelectronics*, pp. 9–10, Oct. 2000.

[74] D. M. Brooks et al., Power-aware microarchitecture: design and modeling challenges for next-generation microprocessors, *IEEE Micro Mag.*, Vol. 20, No. 6, pp. 26–44, Nov. 2000.

[75] G. E. Tellez, A. Farrahi, and M. Sarrafzadeh, Activity-driven clock design for low power circuits, *IEEE/ACM International Conference on Computer-Aided Design*, pp. 62–65, Nov. 1995.

[76] K. S. Min, H. Kawaguchi, and T. Sakurai, Zigzag super cut-off CMOS (ZSCCMOS) block activation with self-adaptive voltage level controller: an alternative to clock-gating scheme in leakage dominant era, *IEEE International Solid-State Circuits Conference*, pp. 1–10, Feb. 2003.

[77] M. Horiguchi, T. Sakata, and K. Itoh, Switched-source-impedance CMOS circuit for low standby subthreshold current giga-scale LSI's, *IEEE J. Solid-State Circuits*, Vol. 28, No. 11, pp. 1131–1135, Nov. 1993.

[78] H. Kawaguchi et al., Dynamic leakage cut-off scheme for low-voltage SRAMs, *Digest of Technical Papers, Symposium on VLSI Circuits*, pp. 140–141, June 1998.

[79] K. Nii et al., A low power SRAM using auto-backgate-controlled MT-CMOS, *International Symposium on Low Power Electronics and Design*, pp. 293–298, Aug. 1998.

[80] N. C. Paver and D. A. Edwards, Is asynchronous logic good for low-power? *IEE Colloquium on Low Power Analogue and Digital VLSI: ASICS, Techniques and Applications*, pp. 4/1–4/5, June 1995.

[81] C. H. Van Berkel, M. B. Josephs, and S. M. Nowick, Applications of asynchronous circuits, *Proc. IEEE*, Vol. 87, No. 2, pp. 223–233, Feb. 1999.

[82] C. Piguet, M. Renaudin, and T. J.-F. Omnes, Special session on low-power systems on chips (SOCs), *Conference and Exhibition on Design, Automation and Test in Europe*, pp. 488–494, Mar. 2001.

[83] V. G. Oklobdzija and J. Sparso, Future directions in clocking multi-GHz systems, *International Symposium on Low Power Electronics and Design*, p. 219, Aug. 2002.

[84] D. M. Chapiro, Globally-asynchronous locally-synchronous systems, Ph.D dissertation, Stanford University, Oct. 1984.

[85] M. Miyazaki et al., A 1000-MIPS/W microprocessor using speed-adaptive threshold-voltage CMOS with forward bias, *IEEE International Solid-State Circuits Conference*, pp. 420–421, Feb. 2000.

[86] J. Tschanz, Adaptive body bias for reducing impacts of die-to-die and within-die parameter variations on microprocessor frequency and leakage, *IEEE International Solid-State Circuits Conference*, pp. 422–423, Feb. 2002.

[87] T. Sugibayashi et al., A 30 ns 256 Mb DRAM with multi-divided array structure, *IEEE International Solid-State Circuits Conference*, pp. 24–26, Feb. 1993.

[88] K. Kennnizaki et al., A 36/spl mu/A 4 Mb PSRAM with quadruple array operation, *Digest of Technical Papers, Symposium on VLSI Circuits*, pp. 79–80, May 1989.

[89] N. C.-C. Lu and H. H. Chao, Half-V/SUB DD/bit-line sensing scheme in CMOS DRAMs, *IEEE International Solid-State Circuits Conference*, Vol. 19, No. 4, pp. 451–454, Aug. 1984.

[90] T. Kawahara et al., A charge recycle refresh for Gb-scale DRAM's in file applications, *IEEE International Solid-State Circuits Conference*, Vol. 29, No. 6, pp. 715–722, June 1994.

[91] K. Itoh, Low-voltage memories for power-aware systems, *International Symposium on Low Power Electronics and Design*, pp. 1–6, Aug. 2002.

[92] K. Itoh, *VLSI Memory Chip Design*, Springer-Verlag, New York, 2001.

[93] D. W. Bailey and B. J. Benschneider, Clocking design and analysis for a 600-MHz alpha microprocessor, *IEEE J. Solid-State Circuits*, Vol. 33, pp. 1627–1633, Nov. 1998.

[94] J. Wood, T. C. Edwards, and S. Lipa, Rotary traveling-wave oscillator arrays: a new clock technology, *IEEE J. Solid-State Circuits*, Vol. 36, pp. 1654–1665, Nov. 2001.

[95] F. O'Mahony, C. P. Yue, M. Horowitz, and S. S. Wong, 10 GHz clock distribution using coupled standing-wave oscillators, *International Solid-State Circuits Conference*, pp. 1–4, 2003.

[96] S. C. Chan, K. L. Shepard, and P. J. Restle, Design of resonant global clock distributions, *International Conference on Computer Design*, pp. 248–253, 2003.

[97] M. Igarashi et al., A diagonal-interconnect architecture and its application to RISC core design, *International Solid-State Circuits Conference*, pp. 272–273, 2002.

[98] J. D. Meindl et al., Interconnecting device opportunities for gigascale integration (GSI), *IEEE International Electron Devices Meeting*, pp. 525–528, 2001.

[99] K. L. Tai, System-in-package (SIP): challenges and opportunities, *IEEE Asia and South Pacific Design Automation Conference*, pp. 191–196, Jan. 2000.

[100] Y.-K. Choi, K. Asano, N. Lindert, V. Subramanian, T.-J. King, J. Bokor, and C. Hu, Ultra-thin body SOI MOSFET for deep-subtenth micron era, *Technical Digest, IEEE International Electron Devices Meeting*, pp. 919–921, 1999.

[101] X. Huang, W.-C. Lee, C. Kuo, D. Hisamoto, L. Chang, J. Kedzierski, E. Anderson, H. Takeuchi, Y.-K. Choi, K. Asano, V. Subramanian, T.-J. King, J. Bokor, and C. Hu, Sub-50 nm FinFET: PMOS, *Technical Digest, IEEE International Electron Devices Meeting*, pp. 67–70, 1999.

[102] L. Chang, Y.-K. Choi, D. Ha, P. Ranade, S. Xiong, J. Bokor, C. Hu, and T.-J. King, Extremely scaled silicon nano-CMOS devices, *Proc. IEEE*, Vol. 91, No. 11, pp. 1860–1873, Nov. 2003.

[103] Y. Nakagome, M. Horiguchi, T. Kawahara, and K. Itoh, Review and future prospects of low-voltage RAM circuits, *IBM J. Res. Dev.*, Vol. 47, No. 5/6, 2003.

[104] B. Wong, Method to reduce leakage during a semiconductor burn-in procedure, U.S. patent 6,649,425, Nov. 18, 2003.

CHAPTER 10

DESIGN FOR MANUFACTURABILITY

10.1 INTRODUCTION

As feature sizes shrink in nano-CMOS technologies, the process capabilities are not keeping up with the scaling requirements. The subwavelength gap widens, making it harder to print most structures [4]. Some structures are even harder to print, leading to lithographical distortions which in some cases result in yield loss as well as performance degradation [2]. The industry at large has relied on optical proximity correction (OPC) and other resolution extension technologies (RETs) to cope with the subwavelength gap (see Chapter 3 for a full discussion of OPC and RET). However, OPC correction ability is limited, leaving an important role for designers in the chip development process to enhance the yield of the design. Designers must understand the lithography step well enough to create layouts that would result in the least distortion and apply this knowledge to the design.

Interconnect manufacturing issues represent the largest yield detractor in nano-CMOS processing. A design put together without design for manufacturability (DFM) in mind can result in copper erosion and dishing, changing the designed characteristics affecting electromigration and timing. This will result in speed down bin and hence reduction in the average selling price of a product. Certain wiring patterns can result in high yield loss due to shorts. Open via is another major yield detractor in copper technology. Interconnect density variation causes interlayer dielectric (ILD) thickness variation (see Figure 11.15), resulting in yield loss due to underpolish metal shorts as well as unexpected timing due to variation of capacitive parasitics [2].

Nano-CMOS Circuit and Physical Design, by Ban P. Wong, Anurag Mittal, Yu Cao, and Greg Starr
ISBN 0-471-46610-7

Poly critical dimension (poly-CD) is affected by poly density and pitch and can lead to unexpected timing [1,3]. Failing maximum time as a result of longer poly-CD will result in speed down bin. Failing minimum time due to narrow poly-CD in minimum time paths will result in a nonfunctional part and should be considered seriously in the design. Since this poly-CD difference is systematic, it can be corrected to some extent in the lithographical step (see Chapter 3 for details). This correction will require pattern density analysis and a mask change. If the transistors are all oriented in the same direction in the design, this correction is much easier. If not, the fabrication engineers may need several iterations to determine the optimum correction for the two orientations of the transistors. Such corrections in the fabrication could result in a delay in product introduction.

Antenna problems can lead to yield loss due to gate damage and in some cases, degrade transistor performance by inducing early negative bias temperature instability (NBTI) V_{th} shifts. As a result, a nano-CMOS process needs a paradigm shift in the design methodology to enable the manufacturing of circuits with high yield while introducing the least amount of variation and parasitics in the design. In the next three sections we go through some design case studies to illustrate this need.

10.2 COMPARISON OF OPTIMAL AND SUBOPTIMAL LAYOUTS

The following case studies illustrate both good and bad layout practices. In Chapter 11 we explore the effects of layout on device parameter variability and how to avoid it.

Figure 10.1(*a*) shows two versions of the layout view of a library gate. There are several areas that would render this gate unscalable and increase its sensitivity

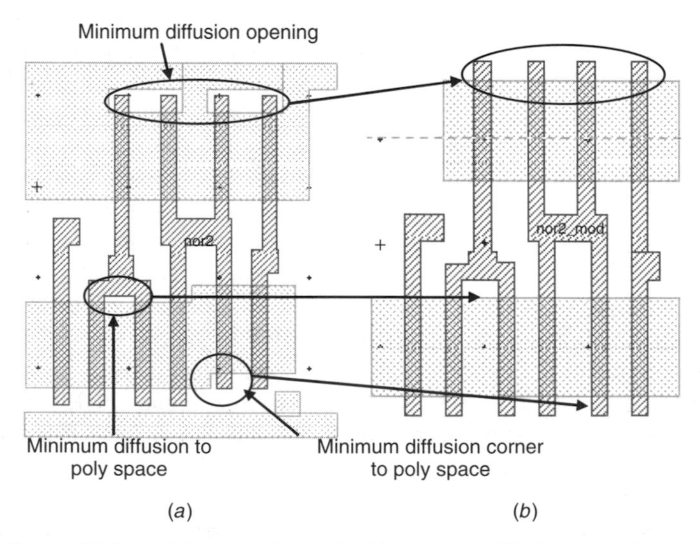

Figure 10.1 (*a*) Layout view of a library gate. (*b*) Improved layout.

to process variation and transistor end leakage. Process distortion of the minimum diffusion openings causes loss of end-cap coverage. This in turn results in severe transistor leakage.

The minimum diffusion space to the inverted U-shaped poly resulted in a higher average transistor length and hence, lower drive current. The minimum diffusion corner to poly space resulted in marginal end-cap coverage in perfect alignment and would fail with alignment offset, even though the diffusion corner in this case is a small jog. If the inner corner of this jog were longer, it would be more serious. The cells were designed for 130-nm node. The printed aerial image of the cells scaled to 90 nm is shown in Figure 10.2. An improved layout of the same cell is shown in Figure 10.1(*b*), which shows improvement in all the areas in which the first cell was marginal or failing.

Contact and via open is another major yield detractor. For this reason it would be prudent to use two contacts where there is space to land two contacts. Figure 10.3 shows an example layout and how that can be improved. Figure 10.4 shows a cell with a single via; redrawing the metal one landing pads to accommodate more than one via would improve the yield of the product.

For nano-CMOS technologies it would be easier to control poly-CD if all the transistors of the entire design are oriented in the same direction. If biases are needed to correct for lithographical and etch distortions, it would be easier to implement on a design with all transistor poly aligned in the same direction. This is especially important for analog circuits, memory bit cells, sense amplifiers, and other critical circuits. Figure 10.5(*a*) shows a layout where transistors are oriented vertically as well as horizontally. Figure 10.5(*b*) shows an implementation where all the transistors are oriented in the same direction. This layout also includes other improvements, which we discussed earlier.

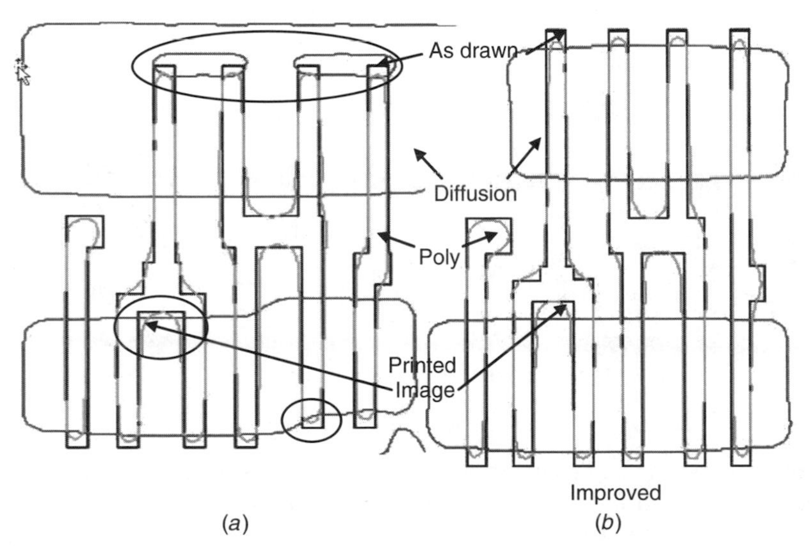

Figure 10.2 (*a*) Printed aerial image of the cells in Figure 10.1. (*b*) Improved version.

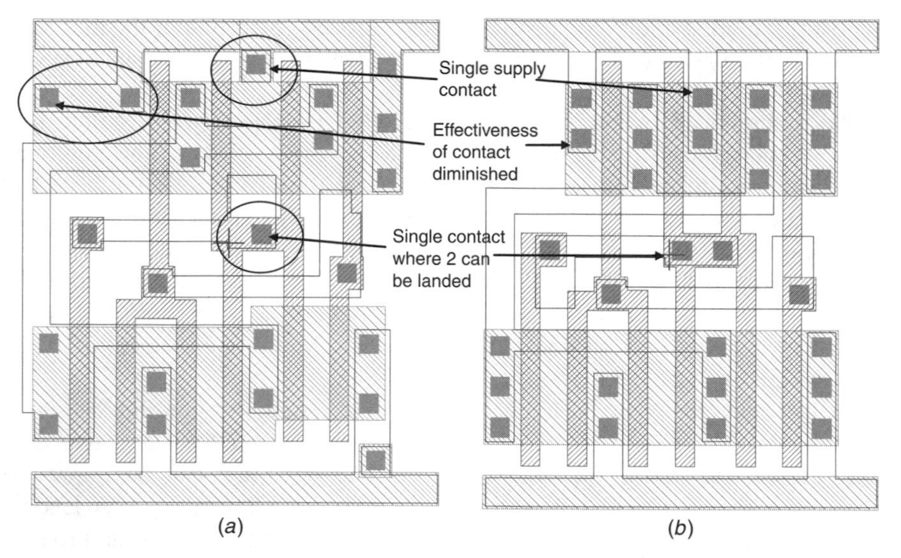

Figure 10.3 (*a*) Layout with two contacts. (*b*) Improved version.

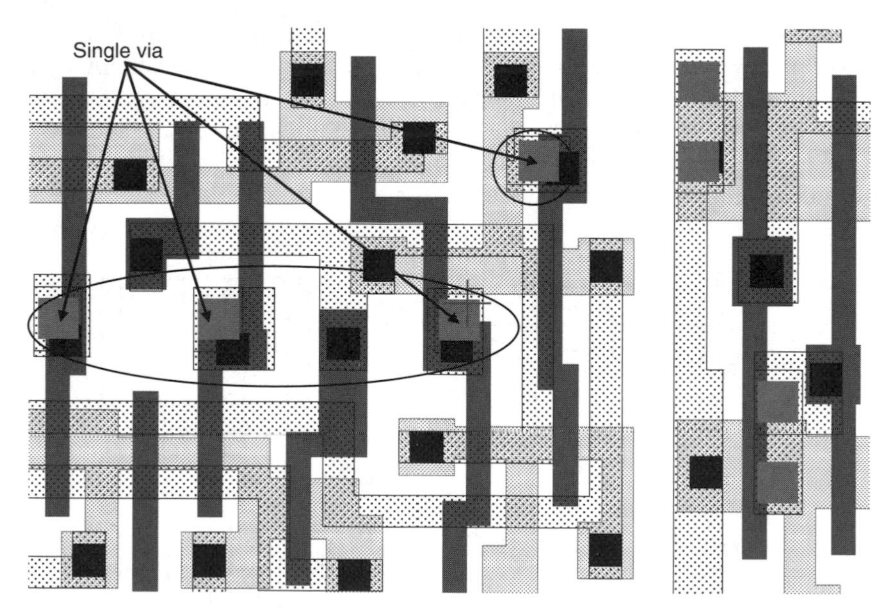

Figure 10.4 Cell with a single via.

Figure 10.6 shows a layout where misalignment and diffusion flaring can result in shorts. Diffusion flaring at node x coupled with misalignment such that poly y overlaps diffusion at node x will cause node A to short to node B. There are two locations in this layout where this can occur, shown circled in Figure 10.6.

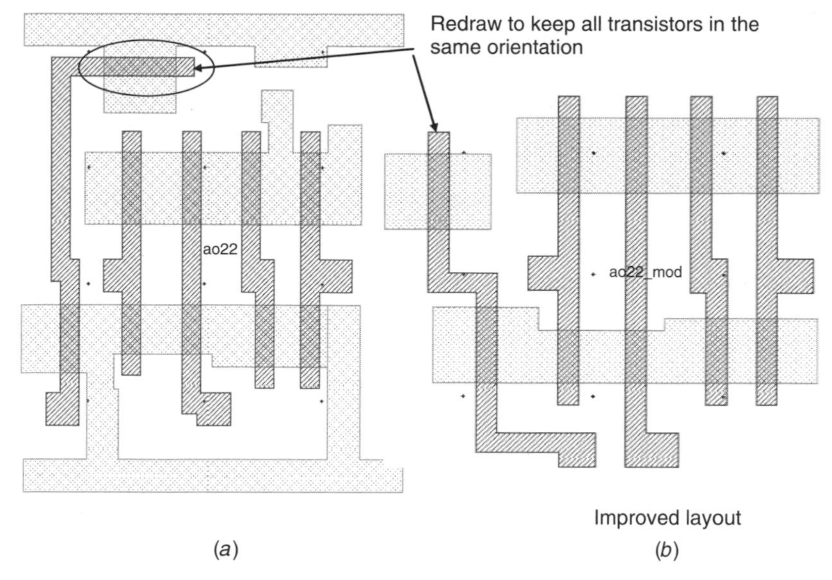

Figure 10.5 (*a*) Layout with transistors drawn both vertically and horizontally. (*b*) Improved layout.

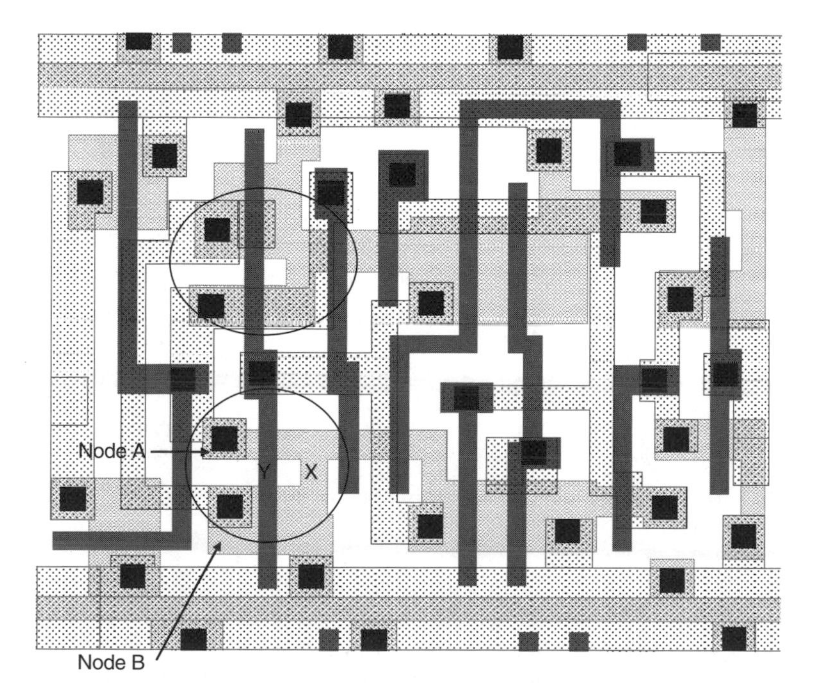

Figure 10.6 Layout that results in shorts.

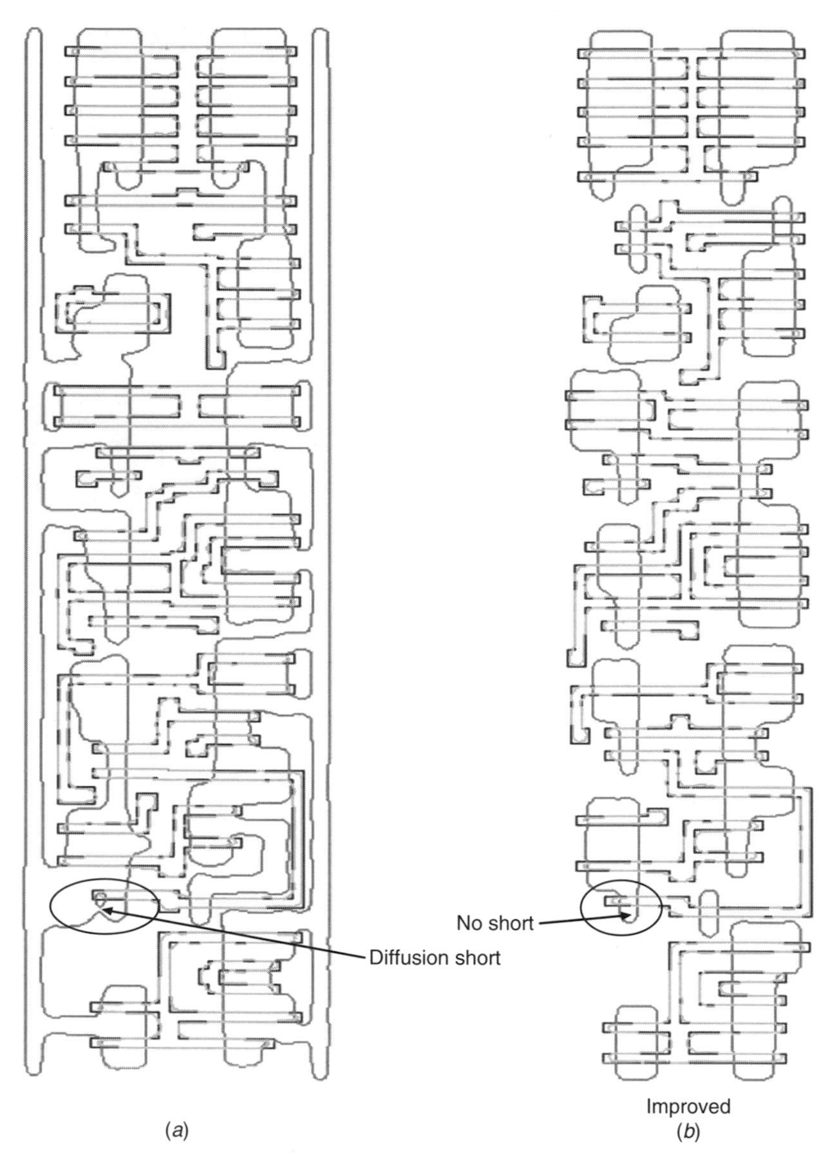

Figure 10.7 (*a*) Print simulation aerial image of a poorly designed flip-flop. (*b*) Improved version.

Figure 10.7(*a*) is a print simulation aerial image of a poorly designed flip-flop. The layout of this flip-flop is shown in Figure 10.8. The C-shaped diffusion with minimum diffusion space shown in Figure 10.8 will result in a diffusion short, due to diffusion flaring. The short is very evident in the image shown in Figure 10.7(*a*). The improved layout, where there is no C-shaped diffusion, did not short [Figure 10.7(*b*)].

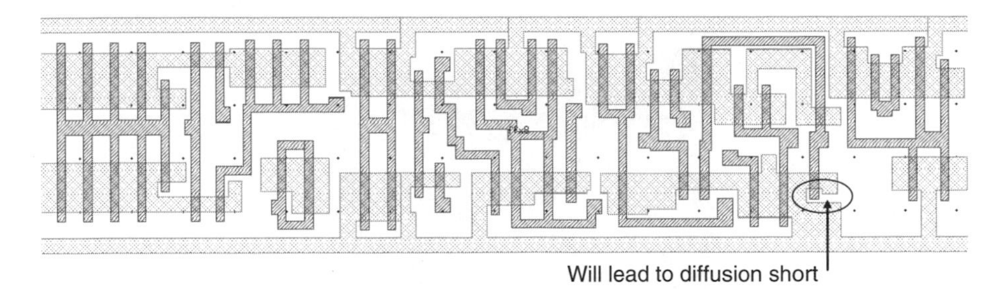

Figure 10.8 Layout of the flip-flop in Figure 10.7.

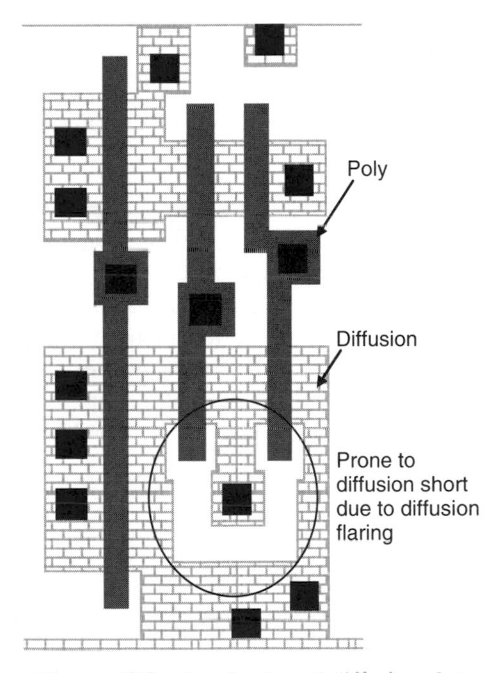

Figure 10.9 Poorly shaped diffusion slot.

Figure 10.9 shows another layout that will be a challenge for lithographic engineers. The minimum U-shaped diffusion slot is not only difficult to print but will also be difficult to scale to the next node. It is also prone to shorting due to diffusion flaring, as in the case of C-shaped diffusion in the flip-flop example (Figure 10.8).

Figure 10.10 shows a layout with the contact drawn too close to the edge of the diffusion that is minimum spaced to a T-shaped poly. Poly flaring in conjunction with misalignment could result in poly-to-contact short. The solution is to increase the poly-to-diffusion edge space so that the poly flare does not encroach onto the diffusion.

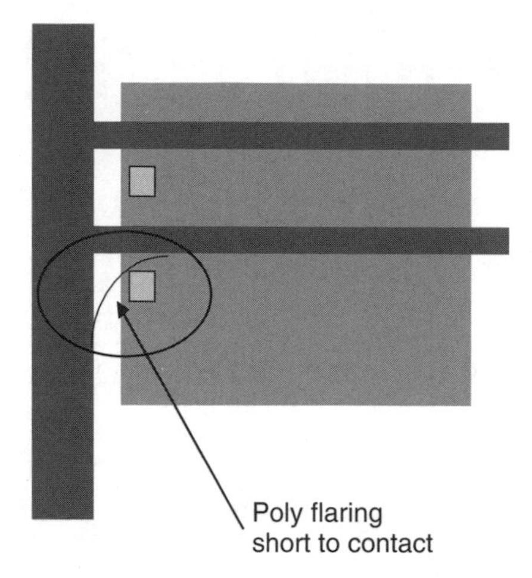

Figure 10.10 Layout with contact drawn close to the edge of the diffusion.

For nano-CMOS technology nodes, lithographic simulations are a must even for engineers who are very experienced with layout–lithographic interactions. This will aid in layout of the critical layers as well as proper placement of contacts, poly, diffusion, and via. In some subtle cases even metallization coverage of contact and via can be an issue, since the metal minimum width and space are getting small and are already requiring OPC as well as phase-shift masking (PSM) in the 90-nm node. The rule of thumb is to keep polygons simple without intricate jogs and keep poly bends as far away from the diffusion edge as possible without having to grow the cell.

10.3 GLOBAL ROUTE DFM

In Chapter 11 we will explore techniques to reduce variation as well as to implement correct by construction clock routes that will also be manufacturing friendly. In this section we will go over some of the techniques to improve other global route performance and their impact on yield.

Copper interconnect processing is still one of the main difficulties in the manufacture of nano-CMOS chips even when not using low-κ dielectric. First and foremost is the interconnect density across the chip. The ideal methodology is that the router maintain an almost uniform density across the entire chip. In reality, a router is still incapable of doing that. Tools are available to help achieve such density. Although not perfect, the goal of uniformity is at least headed in the right direction. Working in conjunction with metal fill and slotting, the metal density can be made quite consistent.

Another yield loss mechanism results when we have minimum width and spaced wires running in parallel for long distances. This yield loss is caused by the collapse of a resist due to the capillary forces acting on the walls of the resist

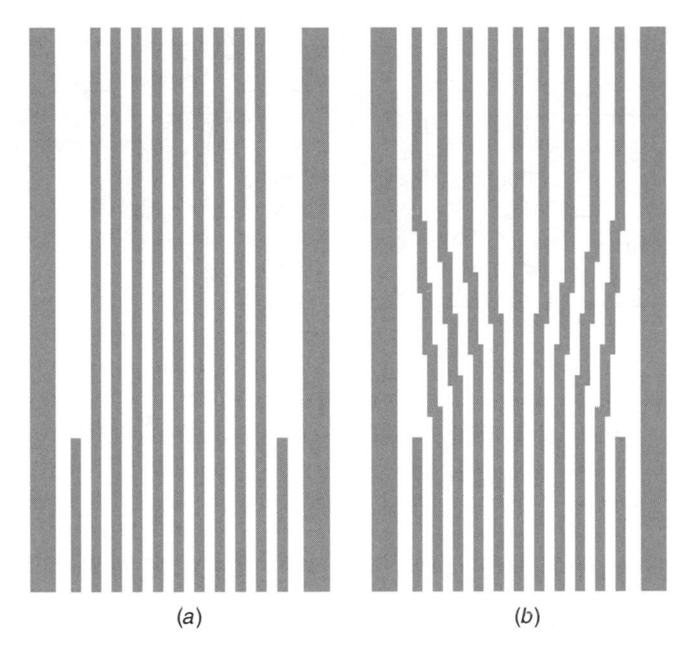

Figure 10.11 (*a*) Traditional and (*b*) wire spread routes.

during rinse. This capillary force pulls the resist walls together, especially when one trench is filled completely with surfactant while the neighboring trench is partially filled. The resulting difference in the force acting on the wall causes collapse of the resist wall.

If a router is capable of spreading wires apart whenever there is space, it will reduce the yield impact due to particulate foreign material, causing shorts or open as well as defects due to resist wall collapse. Spread in the wire of even a few micrometers can have a significant impact on yield. The other advantage of spreading the wire is the improvement in performance as well as signal integrity. Figure 10.11 shows an example of a route with wire spreading.

10.4 ANALOG DFM

In nano-CMOS processes, analog circuit design, suffers severely due to digital centric optimizations. Certain key aspects must be considered to ensure that analog circuits can be manufactured successfully. One issue that must be addressed early is whether to consider developing special design rule checks for analog circuits to improve their reproducibility in the presence of process variation. Some examples of the type of rules that should be considered are shown in Figure 10.12. This figure illustrates the layout difference between digital and analog devices. The analog device has less aggressive design rules, resulting in a larger device. The benefit of these less aggressive rules is that the analog device will be less

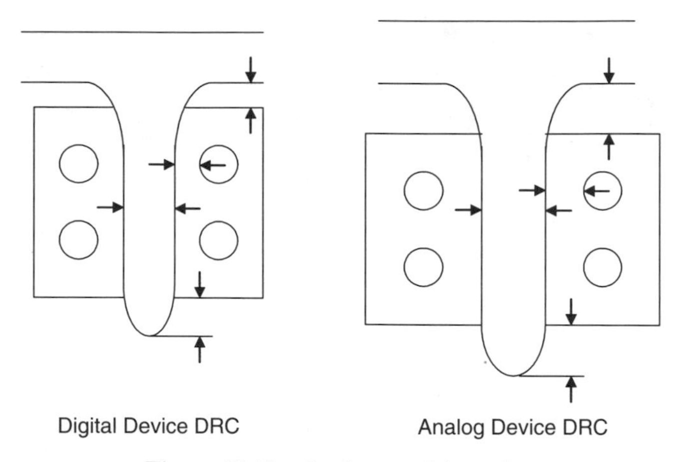

Figure 10.12 Analog precision rules.

sensitive to process variation than the digital cell. Mask alignment issues can affect the performance of an analog cell severely. Techniques to minimize the impact of misalignment on analog circuits are covered in depth in Chapter 11. The effective gate length variation can be reduced by increasing the spacing and overlap requirements so that misalignment of the poly mask will result in less variation due to poly gate flaring, as shown in Figure 10.12. Similarly, increasing the contact to gate and contact to diffusion edge will result in less resistance variation and contact-to-gate capacitance variation if the contact mask shifts. Avoiding use of the minimum gate length can also reduce the variation of analog devices. Some care must be taken when selecting the channel length, since most processes are tuned specifically for digital circuits, which demand minimum channel length driven pitch for density. It is possible for the pitch of an analog block to fall within a pitch range that has greater variability. If the OPC algorithm is not applied correctly, it is possible to get mask-generated artifacts in the physical design, creating yield issues. It is strongly recommended that analog designers consult their fabricator or foundry to determine the variability as a function of channel length so that "forbidden" pitches and channel lengths may be avoided. To better control poly-CD, subresolution assist features (SRAFs or scatter bars) are added to the design, which resulted in "forbidden" pitches. See Chapter 3 for a tutorial of the lithographical issues in nano-CMOS technologies and a detailed explanation of these effects. It is important to orient all analog transistors in the same direction for better poly-CD control as well as to minimize V_{th} variation. See Chapter 11 for further details.

Some analog circuits, especially phase-locked loops (PLLs), use a lot of capacitors as decoupling capacitors as well as loop filter capacitors. These capacitor banks increase diffusion density, which makes it difficult to clear the nitride layer over the diffusion after the shallow trench isolation (STI) etch. It would be necessary to break these banks apart to keep the diffusion density below the process threshold set by the fabricator.

10.5 SOME RULES OF THUMB

- Avoid minimum-spaced and minimum-width wires wherever possible to minimize erosion distortion of the signal lines, which increases resistivity and degrades timing that is not comprehended by the tools.
- Wide wires may require more space, since the walls of wide trenches have a tendency to collapse, causing shorts. The sidewall incline of wider wires is also greater and can result in shorts to neighboring wires.
- Diffusion flaring causes size variation of narrow-width devices, which is layout dependent. If a small transistor is needed, it should always be drawn without a dogbone-shaped diffusion (see Figure 11.17). The simpler the shape of a polygon, the easier it is to process, and the OPC is also simpler.
- STI stress causes mobility degradation and must be included in SPICE simulations. A better solution is to design out stress as much as possible. In Chapter 11 we discuss some strategies to design this effect out of a layout.
- N_{well} proximity effects can cause as much as a 50-mV V_{th} shift for NMOS and a 20-mV V_{th} shift for PMOS [Figure 11.23(*a*)]. Attention must be paid to the placement of matched devices where the orientation and space to the well are identical.
- Limiting the degrees of freedom in a layout, such as by having all transistors oriented the same way, can dramatically improve process control and optimization. Poly-direction alignment for all critical poly and memory devices is very important even if the logic transistors cannot conform to that general direction.
- Design uniformity and the use of tiled devices guarantee identical devices, which helps in device matching.
- Constraining poly pitch and the use of dummy devices to guarantee the neighborhood desired makes the lithographic processes easier and results in better poly-CD control. The use of SRAF requires poly-pitch constraints. Another side benefit is that of more uniform implant-poly proximity effects, which results in less variation.
- Symmetry in critical layout and the use of precision rules will help to ensure that the end caps have ample diffusion overlap (see Chapter 11 for a full discussion).
- The use of multiple contacts and vias has a major impact on yield.
- Use more structured design methodology where random layout patterns are not allowed. In Section 10.2 we have shown several examples that illustrate how random layout patterns can cause serious yield problems.
- Uniform polygon density should be maintained over the entire chip where possible, using tools to assist where needed. Fill and slot metals where needed; wire spreading is the preferred density normalizing technique. Break up capacitor arrays to reduce diffusion density.
- Precision or analog design rules should be used with analog cells.

10.6 SUMMARY

In this chapter we explored the physical design aspects for ease of manufacturing and better yield. In Chapter 11 we cover the circuit design aspects, including some physical design styles that will exacerbate process variability and the resulting impact on DFM as well as on circuit performance. Manufacturing yields are critical to the success of products and companies and can no longer be the sole responsibility of manufacturing engineers. Applying good design practices, especially in nano-CMOS technologies, will improve both yields and chip performance by reducing parasitics. As we have seen in the case studies, design has a tremendous impact on manufacturability and yield for nano-CMOS chips [5]. Variation-robust circuits and physical designs that are tailored for manufacturing and yield may result indirectly in dramatically cheaper process as well as better performance.

REFERENCES

[1] Future of semiconductor manufacturing, workshop, *IEEE International Electron Devices Meeting*, 2002.

[2] M. Orshansky, Computer-aided design for manufacturability, University of California–Berkeley, 2002.

[3] B. E. Stine, D. S. Boning, J. E. Chung, D. J. Ciplickas, and J. K. Kibarian, Simulating the impact of pattern-dependent poly-CD variation on circuit performance, *IEEE Trans. Semicond. Manuf.*, Vol. 11, No. 4, Nov. 1998.

[4] F. Schellenberg, Sub-wavelength lithography using OPC, *Semiconductor Fabtech*, 9th ed., MAR 1999.

[5] R. Radojcic, Old rules no longer apply: what's yield got to do with IC design? *EETiMES*, 2003.

CHAPTER 11

DESIGN FOR VARIABILITY

11.1 IMPACT OF VARIATIONS ON FUTURE DESIGN

The rapid scaling of silicon technology has enabled the dramatic success of integrated circuit (IC) design during the past few decades, allowing millions of transistors to be fully integrated onto a single chip. However, as the technology continues to shrink, precise control of chip manufacturing becomes increasingly difficult and expensive to maintain in the nanometer regime. Silicon processes such as lithography, oxidation, ion implantation, and chemical–mechanical planarization (CMP) suffer more severe variations as technology scaling continues. In addition, run-time environmental fluctuations [e.g., $L(di/dt)$ noise in V_{dd} and temperature change] also increase as chip operation frequency and power consumption escalate dramatically [1–3]. As a result, circuit performance exhibits much wider variability, leading to increasing yield degradation in successive technology generations. The robustness of circuits has emerged as a roadblock in advanced IC designs, and the integrated efforts of process and design engineers are required to mitigate its impact. We describe below some design techniques used to alleviate the effects of variation on the design.

11.1.1 Parametric Variations in Circuit Design

Circuit parametric variations refer to deviations in either the silicon process [e.g., effective channel length (L_{eff}), threshold voltage (V_{th}), metal width] or in circuit operation parameters (e.g., signal crosstalk, power supply noise, and temperature)

Nano-CMOS Circuit and Physical Design, by Ban P. Wong, Anurag Mittal, Yu Cao, and Greg Starr
ISBN 0-471-46610-7

MOSFET	
L_{eff}	16.7%
V_{th} (V)	30%
t_{ox} (Å)	10%
R_{ds} (Ω)	10%
Run-time	
V_{dd} (V)	10%
T (°C)	25-100

Interconnect	
ε	3%
ρ (Ω)	30%
w	20%
s (nm)	20%
t (nm)	10%
h	10%
R_{via} (Ω)	20%

(*a*) (*b*)

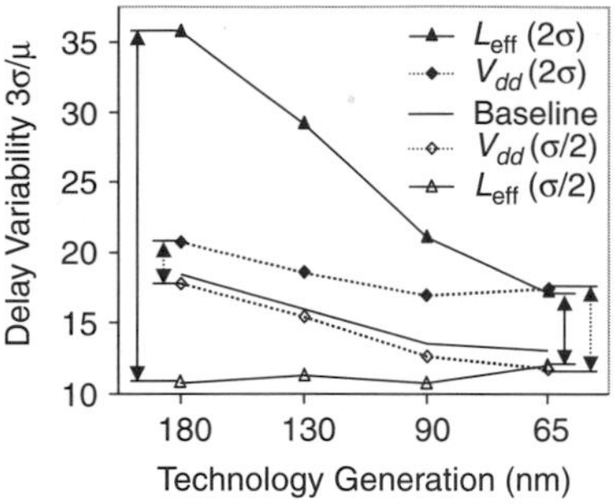

Figure 11.1 Circuit parametric variations and their impacts on delay variability: (*a*) $3\sigma/\mu$ of major variation sources at 130-nm node; (*b*) effect of L_{eff} and V_{dd}.

from nominal design values. They are introduced during either chip fabrication or run-time circuit operation. Assuming that these deviations are normally distributed, Figure 11.1(*a*) summarizes the 3σ/mean (μ) of major variation sources at the 130-nm technology node. The values are extracted from 2002 International Technology Roadmap for Semiconductors (ITRS) [1], with additional data from academic predictions [4–6]. According to the ITRS, similar or worse variations are expected at the 90-nm node and beyond, although augmented values are projected by industry [7]. Among these variation sources, circuit delay variability is the most sensitive to fluctuations in L_{eff}, V_{th}, metal dimensions, signal coupling, V_{dd} noise, and temperature [6]. Other parameters have either a weak impact on performance in current technology [e.g., parasitic source–drain resistance (R_{ds})] or they benefit from excellent variation control (e.g., dielectric constant); therefore, their impact is negligible in variation analysis. Furthermore, even the impact of first-order variations on variability changes with technology scaling. For instance, Figure 11.1(*b*) shows the effect of L_{eff} and V_{dd} control (i.e., reduce the parametric variation to σ/2 or relax it to 2σ) on the performance of a canonical critical path structure [6]. The critical path structure is based on the ITRS, and these projections are obtained by SPICE simulations using BPTM device and interconnect models [4,5]. Due to velocity saturation, L_{eff} has less of an impact on transistors with shorter channels, while the importance of V_{dd} increases as the channel length decreases [6]. Note that in previous technology generations, circuit performance variability was dominated by variations at the transistor and gate levels; but recent technology scaling has led to larger fluctuations in on-chip interconnect parameters, including line dimensions, resistivity (ρ), dielectric permittivity (ε), and via resistance (R_{via}) [8]. These interconnect variations are uncorrelated to variations in transistors and are relatively uncorrelated from one level to another, causing a corner model–based analysis of the overall variation to be prohibitively complex.

Parametric variations have an inherent spatial scale, and thus they are often characterized as either within-die (i.e., intradie) or die-to-die (i.e., interdie). *Die-to-die variation* affects each element of a chip equally and adds a random effect across the wafer. This variation determines the nominal value of each parameter on the die; these values differ among chips across the wafer as well as from wafer to wafer. Die-to-die variation comprises approximately 50% of the total critical dimension variance for today's technology [9]. Die-to-die variation is mostly design independent and is related to equipment properties, wafer placement, processing temperatures, and so on [9].

Within-die variation happens at the length scale of a die. In previous technologies, its effect was negligible, but in the nanometer regime it has become comparable to, and in some cases even substantially larger than, die-to-die variation [7]. For critical path delay variability, within-die variation affects the mean directly, whereas die-to-die fluctuation dominates the variance [9]. Within-die variation can be further divided into two contributors: systematic and random. *Systematic variations* can be predicted prior to fabrication; an example is layout-dependent channel length variation. Successful technology scaling relies on the effective compensation of systematic variation components in both the process and design phases. In contrast, *random variations* are due to the inherent unpredictability of the semiconductor technology itself. Examples of random variations include fluctuations in channel doping, gate oxide thickness, and dielectric permittivity, among others. Some run-time variations, such as V_{dd} noise, are also considered random components, due to the extreme difficulty of predicting their effects precisely. Since we cannot compensate for random phenomena, this type of variation may eventually pose the most significant challenge to nano-CMOS circuit design with satisfactory yield.

For a given operating temperature, random variations in L_{eff}, V_{th}, and V_{dd} are the most dominant variation sources of a logic gate. While fluctuations in L_{eff} and V_{dd} are relatively independent of each other, V_{th} is strongly correlated to the values of L_{eff}, V_{dd}, and transistor sizing. This is because the nominal V_{th} value of a short-channel MOSFET is affected directly by the DIBL effect, which is a function of L_{eff} and V_{dd}, while its variability, $\sigma_{V_{\text{th}}}$, depends on transistor size and is dominated by fluctuations in channel doping. The following relationship between V_{th} variation and transistor size holds in the nanometer regime [10,11] (also refer to Section 11.3):

$$\sigma_{V_{\text{th}}} \propto (W_{\text{eff}} L_{\text{eff}})^{-1/2} \tag{11.1}$$

It is necessary to consider these correlations for correct variation-aware design and optimization; their dependencies can be utilized to gain trade-offs among performance, power, and variability at the circuit level by tuning V_{dd}, V_{th}, and transistor size [11].

11.1.2 Impact on Circuit Performance

The increase in circuit parametric variations has been shown to cause wider performance distributions [6] and thus degrades chip yield, which refers to the percentage of total circuits whose propagation delays fall within a critical delay

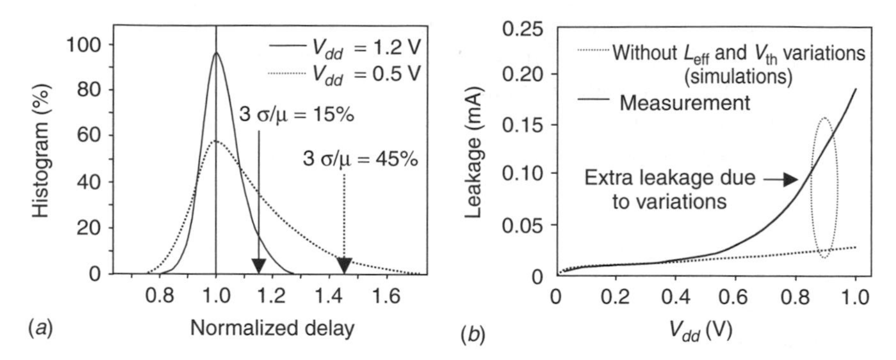

Figure 11.2 Impact of variations on delay variability and leakage power consumption in 130-nm technology: (*a*) Monte Carlo simulation results of a 4-bit adder; (*b*) measured leakage from a 4-KB SRAM.

cutoff. Figure 11.2(*a*) shows the delay histogram of a 4-bit adder from Monte Carlo simulations at the 130-nm technology node. Using the variation values in Figure 11.1(*a*), performance variability $3\sigma/\mu$ is as large as 15% at the nominal bias condition ($V_{dd} = 1.2$ V). In addition, it is observed that when V_{dd} is reduced to 0.5 V in order to save power consumption, variability worsens to 45% [Figure 11.2(*a*)]. Note that at low V_{dd}, the performance distribution becomes asymmetric, due to the nonlinear response of CMOS circuits to bias conditions [10]. In this situation, a lognormal distribution model is used to capture the statistical behavior because it is a better fit to the data than the more commonly used normal distribution model, especially for the extraction of mean values [12]. Besides the negative effect on variability, parametric variations also escalate the problem of power consumption, particularly in the context of leakage power. Figure 11.2(*b*) illustrates an experimental result from a 4-KB SRAM chip. In comparison to leakage values simulated without considering variations, the measured leakage current (I_{leak}) rises exponentially in the range of large V_{dd}, further threatening proper SRAM functionality and increasing pattern sensitivity failures. This dramatic leakage increase is caused by transistors with shorter L_{eff} and lower V_{th}: Those with small L_{eff} values suffer a severely degraded V_{th} value, due to drain-induced-barrier lowering (DIBL) and exhibit an exponential dependence of I_{leak} on V_{th}. Thus, they are very sensitive to variations [13,14]. Unfortunately, power consumption is already one of the main barriers to current high-performance design; the increasing variations cause further power concerns and therefore intensify this obstacle. Techniques to achieve robust design are thus a critical requirement for future IC success.

In Figure 11.2(*a*), we observe that variability worsens with lower V_{dd}, which implies that circuit yield degrades with power reduction. However, this phenomenon is not unique to the tuning of V_{dd}; it also occurs when tuning V_{dd} and transistor size. It has been shown that one of the most effective techniques for balancing power reduction with sacrifices in performance is to tune V_{dd}, V_{th}, and

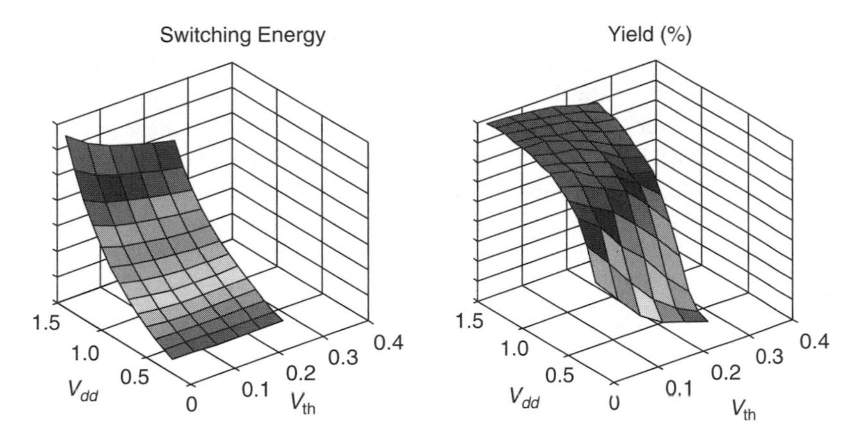

Figure 11.3 Yield degrades with power reduction.

transistor size, and exploit trade-offs among the three parameters. However, during optimization for power savings, delay variability increases at a rate similar to that of the nominal delay, and hence yield is reduced during this optimization [10,12,15]. Figure 11.3 demonstrates this point by plotting the switching energy and yield as functions of V_{dd} and V_{th} for an inverter chain sized for optimal delay at the 130-nm node [12]. As experiments and simulations show, it is desirable to use higher V_{dd} and lower V_{th} values in order to improve yield as long as energy and delay do not exceed their respective constraints [10,12]. Moreover, Figure 11.3 illustrates that while the switching energy exhibits a sharp reduction with decreasing V_{dd}, the yield actually degrades at a much slower rate. This relationship indicates that the energy–yield trade-off at the circuit level is favorable for low power design: a marginal sacrifice in yield can lead to a considerable reduction in energy consumption.

11.2 STRATEGIES TO MITIGATE IMPACT DUE TO VARIATIONS

11.2.1 Clock Distribution Strategies to Minimize Skew

As microprocessor clock frequency scales beyond 3 GHz, clock skew as a percentage of the clock period is getting to be substantial and needs to be minimized to enable further clock frequency scaling. Process variation is now an important part if the management of skew [16]. Process variations contribute to skews at the phase-locked loop (PLL) and at the clock distribution network. A brute-force method of minimizing clock skew at the clock distribution network is to use clock grids, as in the Alpha EV6 (21264) processor [17]. This method falls apart when the chip gets too large. When the *RC* delay of the shorting bars (grid) is equal to or more than the desired clock skew, having the grid will not improve skew; instead, it becomes parasitic capacitance and increases clock dynamic power.

As we add functionality to the chip at each subsequent node, chip size for most high-performance processors is not reduced with each node; instead, it increases in some cases. While gate delays are decreasing (see Chapter 1), interconnect delay is increasing at each subsequent node despite scaling and is even worse when unscaled (see Chapter 1, Figures 1.7 and 1.8 [19]). As a result, the line length, with *RC* delay equivalent to the gate delay, is getting shorter and shorter [see Figure 1.8(*c*)]. This is forcing designers to use finer grids to improve skew over a nongridded clock distribution. This will result in higher clock dynamic power. The grid on the EV6 processor cost quite a bit of power, where the total clock distribution power is at about 40% of the total chip power with a global grid capacitance of 2.5 nF and major grid capacitance of 3 and 6 nF in the local distribution, including the latches [22]. The grid itself consumes about 19% of the total chip power [17]. The finer the grid, the higher the power needed. In power-aware designs, a gridless clock tree is favored over gridded clock distribution. The motivation is due to the relatively large amount of power required [18] by a gridded clock distribution system to achieve a small skew reduction.

The use of balanced H-tree distribution is gaining popularity due not only to better power performance but also to the diminishing gains from gridded clock distribution, as we see the line length, with *RC* delay equivalent to the gate delay, shrinking in the nano-CMOS regime. However, the H-tree distribution system suffers from process variation skew and also requires load balancing to achieve low skew. This is because the load capacitance of the H-tree is about the same as the interconnect capacitance. Therefore, the load capacitance is proportionally larger than the total clock capacitance in the H-tree case. In the case of a gridded clock system, the interconnect capacitance dominates and is therefore very tolerant to load imbalances and does not require load balancing to guarantee low skew.

To minimize process variation skew, non-minimum-channel-length devices need to be used as clock drivers. This has to be traded off against area, power, and skew. It is important to understand the effects of increasing channel length of halo (pocket) implanted transistors and also at which channel-length setting the process will be at the sweet spot of critical dimensions (CD) control for the poly layer. All fabricators tend to optimize minimum poly length for the best CD control. They may not always succeed; hence we need to obtain those data from the fabricator and set the channel length at the lowest CD variation point or lowest channel length that offers the lowest CD variation. At a longer channel length, the CD variation as a percentage is still lower than even at the sweet spot, as described above, even though the absolute variation may be more for the longer-channel device. The problem with using the longer channel length is the increase in area, power, and number of stages required to buffer the phase-locked loop (PLL). The larger the number of stages needed to buffer the PLL, the higher will be the skew introduced, due to power supply and device variation.

The use of decoupling capacitors at the clock buffers is good insurance against power supply droops due to switching activity. Furthermore, when the clock buffer is built with an integrated supply decoupling capacitor, usually surrounding the buffer, a "moat" of nonswitching devices surrounds the buffer, thus reducing the power density around the clock buffer. This reduces the demand on the power distribution and also provides some isolation from heat sources when placed in a neighborhood of high-power-consuming circuits, such as the execution units. The skew of the resulting clock design will be lower as a result of lower power droop and temperature difference between clock buffers placed in the "colder" versus the "hotter" spots on the chip. It is also good design practice to minimize hot spots on the chip to improve performance. The high-power-consuming blocks need to be supplied with the power to maintain the edge rates; otherwise, this will self-limit the performance of the block and the chip. Therefore, breaking up such a high-power block to insert decoupling capacitors will not only maintain the supply integrity but also reduce the power density of the area in which this block resides. It is even more important when a clock buffer is placed in the vicinity of a hot spot since it will add to the skew, due to the temperature difference between a buffer in a hot spot compared to one located in a cold spot.

Clock buffer layout needs to be treated like analog layout and must be placed in one orientation throughout a chip, due to the horizontal and vertical CD differences. Off the mask itself there will be a 2-nm variation between the horizontal and vertical polygons at the 130-nm node, and not much less, if any, for the 90-nm node. This variation can be attributed to turning the writing beam on and off at different edges for horizontal versus vertical orientation. The CD variation in the resulting image after etch will be about two to three times worse. There is also a V_{th} variation introduced by the different times at which the halo implants are directed on the devices laid out horizontally versus vertically.

Since clock buffers are usually large, it is important to split them into smaller transistors so that the resistivity variation in the long narrow poly lines is mitigated. The devices should always be split into an even number of legs. When the transistors are folded, they become less sensitive to misalignment (Figure 11.4). Other misalignment effects are discussed later in the chapter.

Only one layout should be used for all clock buffers, and analog layout rules should be observed as far as possible (see Section 11.2.3 for analog variation strategies). The drive strength of the devices is maximized by minimizing the shallow trench isolation (STI) stress mobility degradation of NMOS devices (Figure 4.4). Dummy transistors are used to achieve this as well as to improve CD variation due to microloading effects during poly etch, lithographical effects, and proximity effects due to implant scattering on poly-gate sidewall [Figure 11.23(*b*)].

Shields for clock lines are imperative for gigahertz chips. Besides providing capacitive shielding, they act as inductive shield by providing a signal return path for the clock and other aggressor signals. Since the shields are placed manually, you can have a much more accurate extraction before the mask data-preparation step, so that the design is correct by construction before the tape is submitted for mask writing. During the mask data-preparation step, metal fills are added to areas where the metal density is below about 20%. The shields minimize the effect of the fringing capacitance change after the addition of metal fills to normalize

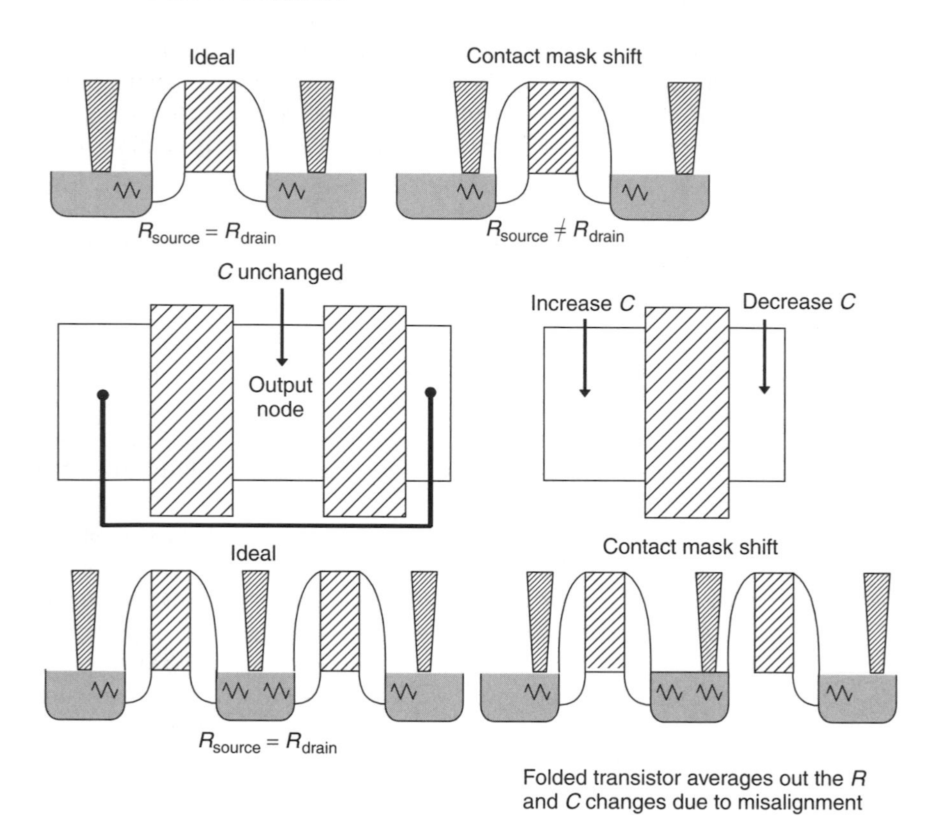

Figure 11.4 Misalignment effects on parasitic R and C; solution is folding the transistor.

metal density surrounding the clock and shields as well as above and below the clock routes. Observe the wide wire space and width rules to minimize yield loss as well as resistance variations due to chemical–mechanical planarization (CMP) effects such as erosion and dishing (see Chapter 2, Figure 2.26). On top of that, placing shields helps mitigate resistance variation due to CMP effects. It is very important to make sure that the wire density is uniform below and above the clock wires [21]. Nonuniform wire densities will result in interlayer dielectric (ILD) thickness variations, resulting in line delay variations and hence clock skews (see Figure 11.15) [20]. The ILD is thicker (t_1) over dense wiring areas and thinner in less dense areas (t_2). CMP will remove some of this variation but cannot completely eradicate serious ILD thickness variations. Tools are available to help normalize pattern density through fill and slotting. Also available are tools that will work with the router to spread wires apart to improve pattern density, but they do not resolve the density issues entirely. They need to work in conjunction with metal fill to normalize densities [21].

In a clock tree, the copper interconnect can be subjected to distortions due to subwavelength lithography, erosion, and dishing, which change the wire width and thickness, hence the *RC* delay, resulting in higher skew [25]. It is important for circuit designers to understand the impact of their layout on these effects and to work with process engineers to ensure that these newly exacerbated physical effects are not severely affecting the clock distribution wire delays of the various clock tree branches. Depending on the fabrication, using several narrower wires may work better for erosion and dishing. For others, slotting of the wide clock distribution wires is better. The use of several narrower wires has the advantage of having more skin surface for better high-frequency resistivity, depending on the frequency and skin depth. For most fabricators the use of several wires to form a wider wire works out better for erosion and dishing since the width of the dielectric between the wires is greater than that of slots and can resist erosion better. This is part of the correct-by-construction methodology, which will be the norm for future designs in the nano-CMOS regime.

Size variation introduced due to diffusion and poly flaring can influence clock skews if it is not taken into account [23]. This is especially problematic in cases where the clock line has to drive a large number of such transistors, especially in array designs. This can occur in the clocked sense amplifier, where the same clock line is connected to a number of similar devices, which could be 128 or even 256 instances. The devices as drawn could be small, but due to the diffusion flaring, the processed device width could increase by as much as 25%, depending on the layout, resulting in a corresponding increase in load as seen by the driver (see Figure 11.18). To minimize this effect it might be necessary to lay out the devices without dogbone-shaped diffusion wherever possible. It is always better to draw the devices with the minimum width without a dogbone-shaped diffusion and design for the load to avoid any surprise due to increased load after processing. This problem is going to be even worse as we go deeper into the subwavelength lithography regime. If the motivation for using a minimum-allowable sized transistor is layout area, a transistor without the "dogbone" may occupy equal or less layout area and is a better transistor in terms of variability and drive.

Another source of variation is the way that transistors are connected to clocks. The capacitance as seen by the clock differs depending on how the transistors are wired to the clock [22]. The worse configuration is feeding the clock through a pass transistor, as shown in Figure 11.5. Such a configuration is common in some cache memory designs to reduce the stack height in the decoder NMOS stack to improve speed. However, it causes the clock to see a load depending on the address pattern. If the address pattern turns on any of the pass gates M1 through Mn, the clock will see a higher load than if the pass gate is closed. The clock load is further dependent on the data pattern on the decoder N-tree. This creates an address pattern–dependent clock skew, and such a design should be avoided.

11.2.2 SRAM Techniques to Deal with Variations

The two most important components of an SRAM are the bit cell and the sense amplifier. These two components are also the ones most sensitive to process variations. As for the bit cell, its small size is the main reason it is prone to process

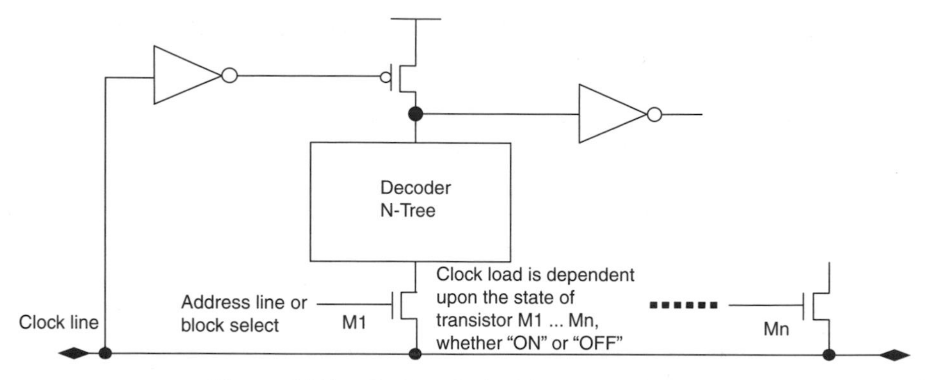

Figure 11.5 Clock as logical input to a decoder.

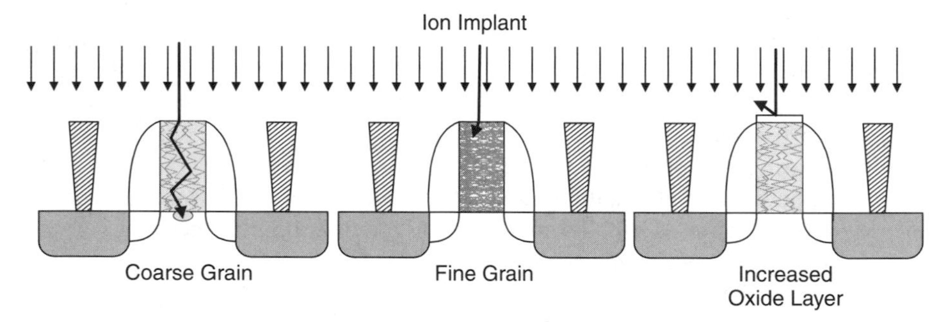

Figure 11.6 Dopant channeling.

variations, especially to V_{th} variations. Statistical implant variation, dopant channeling through the gate into the channel (see Figure 11.6), poly and diffusion CD variation, and dopant loss through the isolation oxide are the main causes of V_{th} variation [25,26]. Bit-cells with an area of less than 3 μm^2 will be particularly sensitive to V_{th} variation, due to dimensional variation modulating the V_{th}, due to roll-off characteristics and reverse short-channel (RSC) effects, including small width effects. The dimensional variation can be a result of process variation as well as process and layout interactions. Bit-cell layout will have a significant impact on these effects, which we will see in due course. Scaling the technology to sub-100-nm feature sizes will exacerbate these effects, and it is therefore important to understand how we can minimize these effects in our designs since most memory designs are handcrafted for high speed and density. Since it is handcrafted, it is even more important to ensure that such a design scales to minimize costly rework in future technology nodes.

Bit-Cell Designs to Illustrate Design Pitfalls Poly and diffusion CD and implant variation are the main causes of drive mismatch between two cross-coupled drivers. In Figure 11.7, M1 and M3 form one of the drivers, and M4 and

M6 comprise the other driver. The matched transistor pairs for a six-transistor bit cell are M1/M4, M3/M6, and M2/M5. The ratio between a pass (M2 or M5) and a pull-down (M1 or M4) NMOS must satisfy a certain value (1.8 to 2.2) to ensure cell stability to read/disturb in the intrusive read of single-port bit cells. A larger cell ratio comes at a price—that of cell size. The trade-off for bit-cell design is multifaceted and must be balanced for best performance at the lowest cell size yet the highest yield. Process variations and layout and process interactions can affect the device matching as well as the critical ratio, as mentioned above, resulting in reduced margins and tolerance to the process excursion window. A design is at risk of having low yield and poor performance if the problems described in this section are not dealt with in the design and layout stage.

As the nodal capacitance of a bit cell decreases with scaling, activating the word line (WL) during access can result in the coupling of a significant differential noise to the storage nodes of the bit cell formed by the drains of M1 and M3, or M4 and M6. The cause of this noise is due to the WL going high during access. Since one of the storage nodes of the bit cell is high while the other is low and the bit lines (BLs) are precharged high, one of the pass transistors is in saturation mode while the other is in the "off" state. The transistor in saturation will couple about two-thirds of its gate capacitance onto the source node, which in this case is the "low" storage node of the bit cell. The pass transistor connected to the high node of the cell will have its source and drain at V_{dd}. Since the WL goes from low to high during access, the gate-to-source voltage (V_{gs}) is initially negative for the pass gate connected to the high node of the cell. When the WL is high, the V_{gs} of this pass transistor is zero, which means that the transistor is still in the "off" state. The only capacitance coupling onto the high storage node due to the WL going from low to high is the overlap capacitance (C_{gd}) for

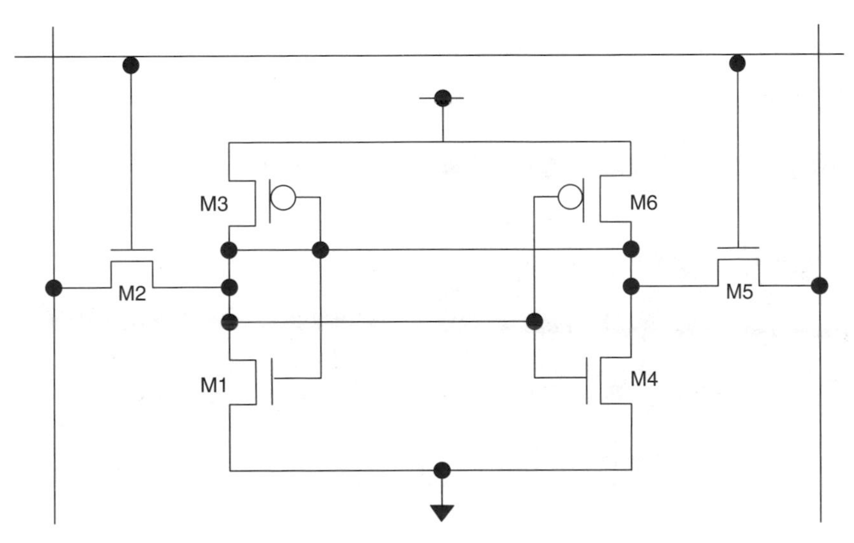

Figure 11.7 Schematic diagram of bit cell.

the "off" pass transistor. Since C_{gd} is about five times lower than two-thirds the gate capacitance, the low storage node will receive a relatively stronger coupling pulse than the high storage node; hence the differential noise is coupled onto the storage nodes, due to the mismatch of the coupling capacitance from the WL onto the low and high storage nodes. On top of that we have cell current flowing into the low node of the cell, raising the level of the low node. Therefore, a static noise margin alone can no longer guarantee a stable cell design unless one considers all the dynamic conditions as well, one of which is described above.

The ΔW value of transistors with STI has been improved greatly over the local oxidation of silicon (LOCOS) isolation. However, due to the extremely small width of the transistors (0.1 to 0.25 μm) used in the bit cell, even the comparatively small ΔW value is still quite significant (10 to 20%) for some bit cells and needs to be taken into consideration and certainly should be modeled in the bit-cell transistors.

Figure 11.8(*a*) shows a bit cell as drawn with model-based optical proximity corrections (MOPCs) applied, and Figure 11.8(*b*) is the bit cell after processing on a wafer. Figure 11.8(*a*) shows the bit-cell layout as drawn where the polygon corners are square with no end pullback. However, after processing on silicon, it does not look quite as drawn. The corners are rounded as a result of distortions due to subwavelength lithography and reactive ion etch (RIE) of the structures as shown in Figure 11.8(*b*). The distortions after processing can be minimized by applying the proper MOPCs, as can be seen in Figure 11.8(*a*). However, OPCs alone cannot compensate for all the distortions, especially as the lithography trend is showing that the subwavelength optical lithography gap is widening with each subsequent node (Figure 1.5). We will examine the influence of layout on the level of distortion and its effect on the design margin and performance as we go

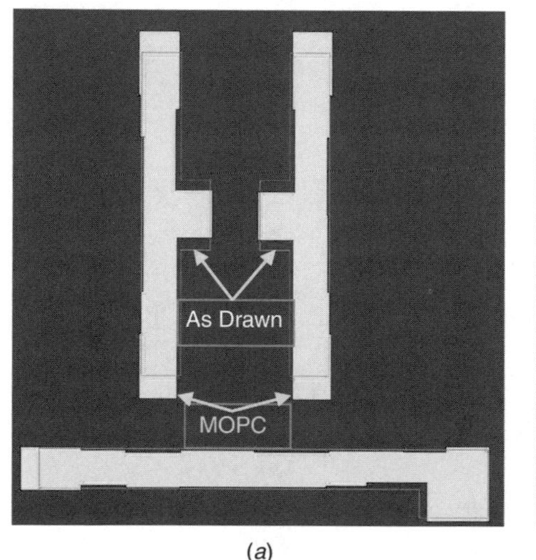

(*a*)

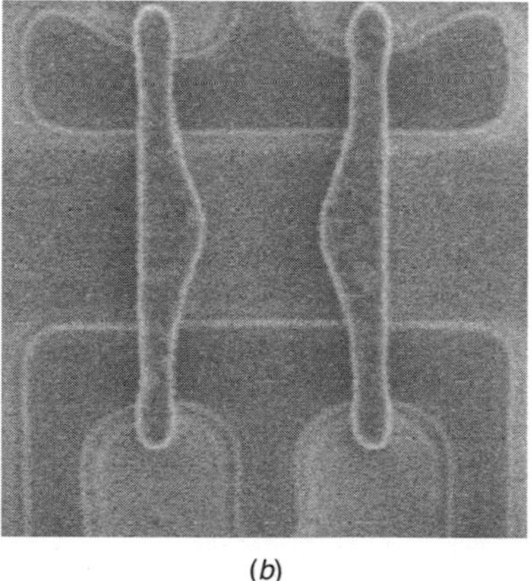

(*b*)

Figure 11.8 Bit-cell as drawn with (*a*) MOPC overlay and (*b*) after etch.

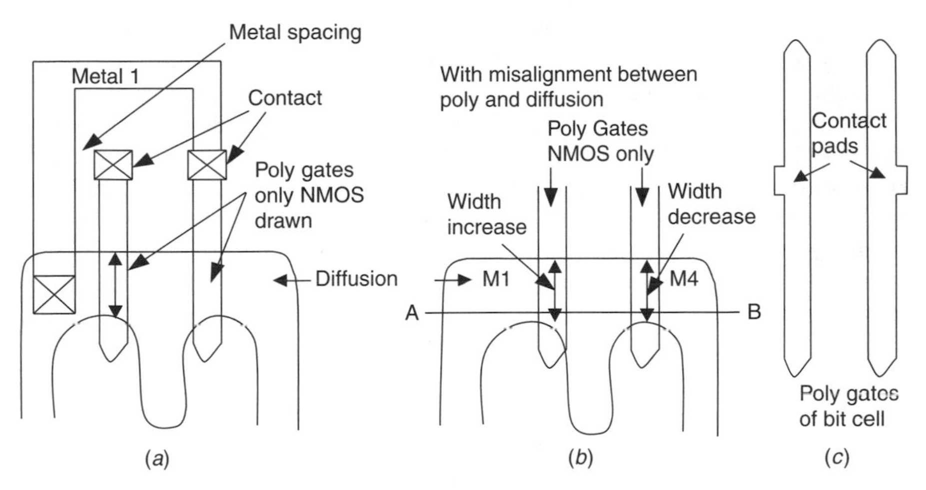

Figure 11.9 Bit-cell misalignment problems.

over the various bit-cell layouts. From this exercise it will be clear that we will have to rely more and more on layout to minimize effects due to lithography and etch distortions.

These distortions of the drawn shapes give rise to yet another source of variation in a bit cell due to the poly misalignment with respect to diffusion, as shown in Figure 11.9. The poly overdiffusion of some bit-cell design looks as shown in Figure 11.9(*a*). Such a bit-cell design is sensitive to misalignment because the poly is not centered over the diffusion curvature. The reason for the asymmetry in placement of the poly with respect to diffusion is due to the manner in which cross-coupling is achieved by metal 1 and the position of the contact pads on the poly gates. The contact pads are positioned as shown in Figure 11.9(*c*) in an attempt to minimize cell width. The contact pads in this design are flipped from the design shown in Figure 11.8, where the poly is centered over the diffusion curvature. This asymmetrical placement of poly gates [Figure 11.9(*c*)] over the diffusion curvature proves to be a bad trade for the small amount of area savings, as the resulting design is very sensitive to poly diffusion overlay misalignment. Since the contact pads on the poly are positioned away from the cell center, the poly contact would obstruct the metal 1 cross-couple interconnect, thus must be shifted toward the center of the bit cell to avoid shorting to the cross-coupling metal 1. This would require therefore, shifting the poly gates toward the center of the cell, thus causing asymmetry in the poly position over the diffusion, as shown in Figure 11.9(*a*).

When there is misalignment, in this example, to the right of the diffusion, the pull-down NMOS, M1, will increase in size while M4, the transistor, decreases as it rides down the tangent of the curvature. A better layout of such a bit-cell design is shown in Figure 11.8, where the poly is centered over the diffusion curvature to minimize or eliminate device-size changes due to horizontal misalignments. The only difference is that the contact pads on the poly gates are flipped so that they are by design moved away from the cross-coupling M1, thus allowing the poly to be positioned symmetrically over the diffusion curvature. Horizontal misalignment in such a design will not cause the mismatch that plagues the design with the asymmetrically placed poly gates over the diffusion curvature. With this design, the pull-down transistor width is narrowest with perfect alignment. Horizontal misalignment in such a cell will only increase the width of the pull-down transistor, resulting in an increase of the cell ratio. This is yet another advantage of placing the poly gates symmetrically with respect to the curvature of the diffusion. Unlike the case where poly placement over the diffusion curvature is asymmetrical as in Figure 11.9, this design actually improves cell stability with horizontal misalignment, whereas in the asymmetrical case, the cell stability is degraded.

Another bit-cell design is shown in Figure 11.10, where the contact pads on the poly gates are asymmetrical in the vertical axis. The bit cell after processing on silicon will look as shown in Figure 11.10(*b*). The poly gates will develop a bulging region where the contact pads are, and since they are asymmetrically spaced from the diffusion edge, vertical misalignment between poly and diffusion will result in device-size variation. If the poly misaligns downward, M1 will experience an increase in effective channel length due to the bulge on the poly and will therefore become weaker than M4 while M2 increases in width due to diffusion flaring around the contact. Since the diffusion contact pad is different for M2 and M4, flaring of M2 diffusion occurs closer to the poly edge and results in an asymmetrical size change between M2 and M4 as the poly misaligns downward. This results in a double whammy, with M2 increasing in strength while M1 decreases. This effectively reduces the cell ratio further and causes a mismatch in transistor strength between the match transistors as described earlier.

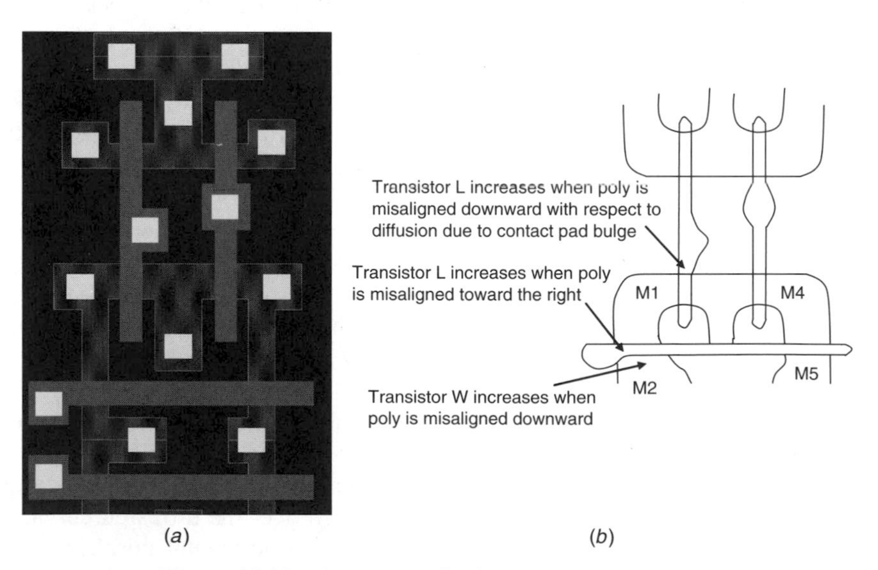

Figure 11.10 Asymmetry leads to process sensitivity.

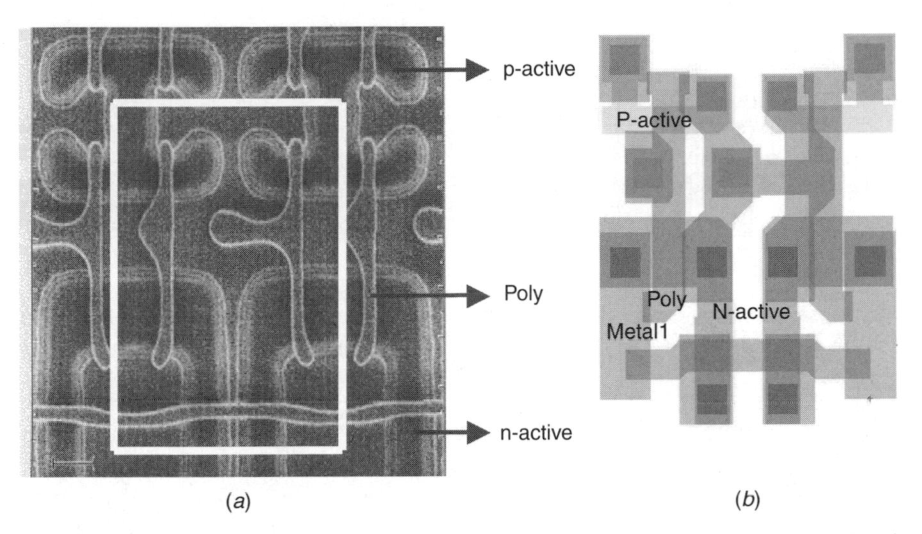

Figure 11.11 (*a*) Poly and diffusion patterns of the bit cell after lithography and etch. (*b*) Layout of the bit-cell.

When the poly misaligns horizontally to the right with respect to the diffusion, M2 becomes weaker with respect to M5, due to an increase in effective channel length as the contact pad bulge on the poly encroaches onto the diffusion of M2. Hence transistor matching in such a design is sensitive to misalignment as well. Applying the same deductions, one can see that the design in Figure 11.11 has the similar sensitivity to misalignment, due to the difference in the position of the poly flaring between the two poly gates and the diffusion flaring. This design would therefore be sensitive to horizontal as well as vertical misalignments.

The effects described in previous paragraphs are not evident from the structures drawn before the processing distortions. It is important for bit-cell designers to understand the types of distortions that a design goes through during fabrication. It is only by working with fabrication and process engineers that physical designers will see these effects and make corrections to a layout to avoid the pitfalls associated with these effects. We will describe a bit cell that applies techniques to counteract most of the issues associated with the processing distortions described earlier.

The bit-cell design that will thrive in the nano-CMOS regime is shown in Figure 11.12(*a*). The printing and etched image of the cell on silicon are shown in Figure 11.12(*c*). In this design all poly is placed in the same direction, which facilitates better poly CD control, easier for lithography and phase-shift masking (PSM) and in general better process control [24]. When the cell is arrayed, all transistors see the same poly patterns; hence, the poly proximity issues will be minimized. The poly proximity effect is yet another newly exacerbated effect that causes implant variation as a result of the difference in poly proximity [Figure 11.23(*b*)]. This effect is due primarily to the scattering of implants by

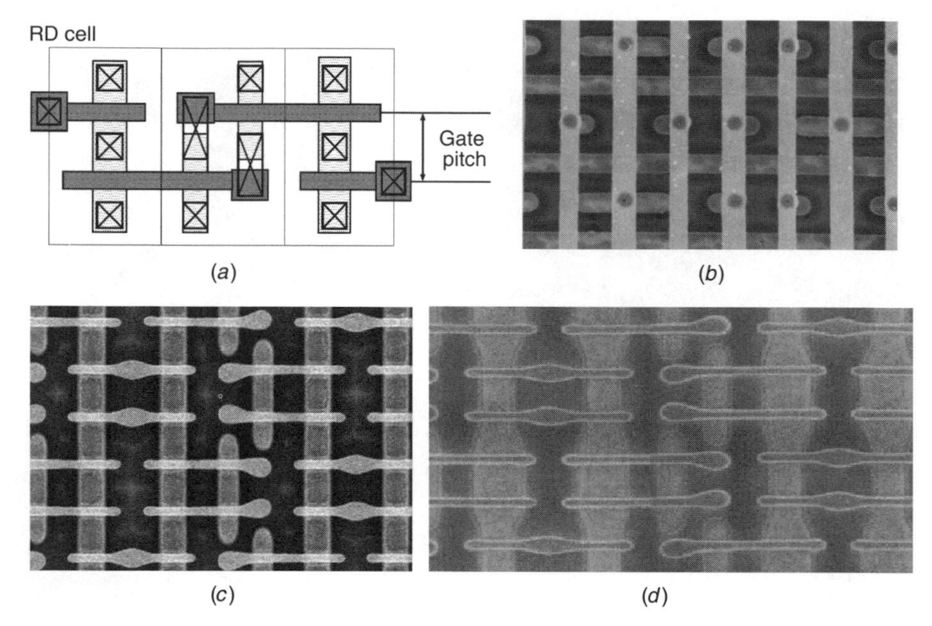

Figure 11.12 Structured design most tolerant to process excursions: (*a*) layout view; (*b*) cell metallization; (*c*) poly and diffusion; (*d*) earlier version of poly and diffusion. (SEM and layout courtesy of Trecenti/Hitachi.)

poly gates in the vicinity of the other transistors. When the transistor proximity changes, so does the effect. As laid out, this design guarantees that the proximity is consistent within the memory array.

This cell is also a lot less sensitive to misalignment than the designs shown in Figures 11.9, 11.10, and 11.11, due to the absence of bends in the diffusion and minimal poly bulges due to the contact pads plus the symmetry of the layout. The design of a similar cell is shown in Figure 11.12(*d*), where there is still a slight diffusion bend. As a result, this cell has some sensitivity to size change due to misalignment. The improved cell is shown in Figure 11.12(*a*) and (*c*), where all the diffusion edges are straight. This can be achieved through proper sizing of the length of the pass NMOS. To control cell leakage in the nano-CMOS regime, the pass NMOS length must be increased anyway. This increase in the length can be compensated by a corresponding increase in the width of the transistor in order to achieve the cell drive required. With proper sizing of the length of the pass NMOS, one can achieve pass NMOS to a pull-down NMOS width of 1:1, while the effective beta ratio of the two transistors is at a proper ratio of 1.8:1 to 2.2:1 to guarantee stability of the single-ported intrusively read cell. This design results in a diffusion strip without any bends, hence is a lithographically friendly design. Further, the STI stress effect is nonexistent as the diffusion end is at the end of the array. It is, however, important to take care of the end of the array to mitigate the difference as seen by the cell at the end versus those in the center of the array by including dummy transistors.

The aligned poly of this bit-cell design works well with the nano-CMOS design methodology of aligning all critical poly to minimize across-chip poly-CD variation. Notice also that the polygons, be they poly, diffusion, or metal, are drawn as straight as possible. The regularity of the polygons at any level makes it easy on lithography and process control. The pattern density within the structure is uniform, resulting in uniform proximity effect on the transistors. Other process controls, such as CMP, also benefit from uniform pattern density. We will certainly see more of such designs in the future.

There are many other advantages of this cell design, but they are beyond the scope of this book. Refer to Refs. 28 and 29 for a full discussion of other benefits of this cell design.

Characterization of Optical Proximity Correction for Design Most bit cells receive manual optical proximity correction (OPC) before committing to mask build. Here is yet another opportunity to correct lithographical distortions. The correction must be set at its optimum value for the appropriate scanner and the wavelength of the light source, or it could end up adding to distortions by over- or undercorrection. An example is shown in Figure 11.13, where the poly end-cap "hammer head" is too large, resulting in a bulging poly line end on the silicon. This, coupled with contact pad bulge, results in a poly gate shaped like a classic Coke bottle. The resulting cell design is again sensitive to misalignment in yet another direction, in this example, vertically. The effective channel length of transistors M1 and M4 is a function of poly-to-diffusion alignment. If the poly is shifted up with respect to the diffusion, the diffusion edge now encroaches on the bulging poly tip, resulting in an increase in the effective channel length

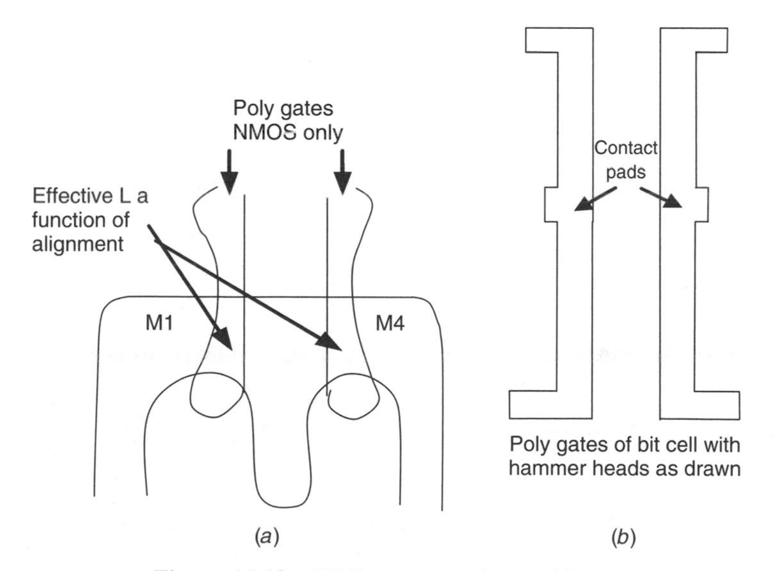

Figure 11.13 OPC overcorrection problems.

of the transistor. This reduces the drive of the pull-down transistors M1 and M4, thus reducing the cell ratio and its static noise margin, resulting in a less stable cell. It is imperative that we apply the proper OPC to the bitcell to avoid over- or undercorrecting for optical proximity effects. This may require several iterations to arrive at the optimum compensations and not result in the bulging tips (overcorrection) or too much poly-line end pullback (undercorrection). There are other effects due to over- or undercorrection; refer to Chapter 3 for a full discussion of OPC. When there is too much poly-line end pullback, we could end up in a situation where the transistor leaks due to insufficient end-cap coverage. This undercompensation causes cell stability problems as well as higher standby power consumption. Since OPC has a profound impact on the performance and yield of a bit cell, it is often applied manually and perfected through several iterations on silicon.

Figure 11.14(*b*) shows a 90-nm bit-cell with OPC applied. At 90 nm we can no longer rely on simple hammer heads as the only OPC. Notice the intricate OPC patterns that must be applied to properly correct for lithographical and etch distortions. The poly-line width is also biased up to achieve proper transistor channel length after lithography and etch. To arrive at the correct bias, several iterations and cell electrical characterization are required. The key is to correct for both lithographical and etch effects. Along the two sides of the cell we find the subresolution assist features (SRAFs) in the poly layer. Since they are subresolution, they do not print but do assist in maintaining uniformity during lithography when the poly pattern is nonuniform, especially at the array breaks for substrate and well taps. In some designs there is also a gap at the WL straps.

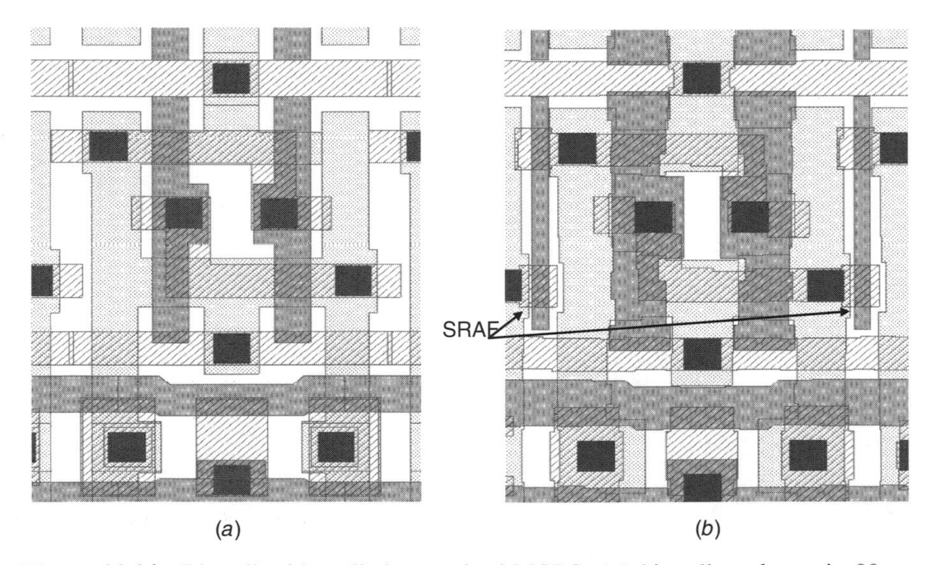

Figure 11.14 Bit cell with well-characterized MOPC: (*a*) bit cell as drawn in 90-nm rules; (*b*) with MOPC and SRAF. (© 1999 and 2002 Advanced Micro Devices, Inc.; reprinted with permission.)

This design avoids a lot of the drawbacks described earlier, especially in poly placement with respect to diffusion and the asymmetrically placed contact pads. Also, line end treatment in conjunction with the serifs and antiserifs on the poly has been optimized to achieve as straight a poly line over the diffusion as possible, thus avoiding misalignment sensitivities. The cell as drawn is shown in Figure 11.14(*a*) for comparison with the MOPC-corrected cell.

In the nano-CMOS regime, bit-cell sizes are getting to be small enough that the nodal capacitances fall within the error tolerance of fast, conventional extractions [23]. When we use a fast, conventionally extracted net list to build a simulation deck, the errors will add up and we will be surprised by the poor modeling when the silicon returns from the fabricator. Starting at the 90-nm node, it is prudent to start using field solvers for bit-cell parasitic extraction.

11.2.3 Analog Strategies to Deal with Variations

Having to deal with variability is not a new concept to analog designers. However, on the one hand, technology scaling, enables exponential improvement of digital circuit performance and functions on a chip, but on the other hand, has made analog design more challenging on many fronts. In this chapter we deal with the variability issues that confront analog designers. Although dealing with variability is not new, more issues have surfaced while others have become worse. Accurate modeling of these effects is very important for analog designers but will be a major challenge. Table 1.1 summarizes the modeling challenges that can affect analog designs.

Many analog circuits require good device matching. We will address some of the matching problems and the possible solutions to alleviate or minimize their impact on analog circuits. Listed below are the main sources of matching problems.

Design-Related Sources

- Asymmetry (leads to misalignment sensitivity)
- Small geometries (narrow-width effects; short-channel effects; larger V_{th} variation)
- Proximity effects [well proximity; poly proximity (linear proximity effects); microloading etch effects]
- Position of well and ground taps (body effect differential)
- Horizontal and vertical effects
- Temperature differential
- STI stress effects
- Diffusion and poly flaring (strong design influence in the nano-CMOS regime)
- Mirror layout effects (capacitance; R_{sd}; misalignment)

Process-, Device-, and Electrical Stress–Related Sources

- Random dopant fluctuation
- Dopant channeling through gate
- Poly-L variation; L_{eff} variation
- Degradation due to antenna effect
- Negative-bias temperature instability (NBTI)
- Hot-carrier injection (HCI)
- OPC (over- or undercorrection; poly line end pullback; poly necking and flaring; diffusion flaring)
- Metal density variation [ILD thickness variation (Figure 11.15); capacitance variation]

Techniques for Improving Matching

1. Increase input signal swings. Whenever possible, increase input signal swings. For instance, a sense amplifier flip-flop is a lot less sensitive to device matching than are sense amplifiers that need to detect low swing signals. Obviously, we cannot increase sense-amplifier input swings unless we are willing to sacrifice speed or use extremely large bit cells. Sense amplifiers must rely on other matching techniques.

2. Create a device layout library. This ensures the use of identical device geometry. When a larger size is needed, the devices are tiled. The input stage of an operational amplifier layout shown in Figure 11.16 illustrates the use of tiled devices.

3. Use dummy transistors. This minimizes proximity effect differences due to etch microloading effects and implant scattering by poly proximity. The use of dummy transistors can also alleviate STI stress effects [see Figure 4.20(c)].

4. Orient all analog transistors in the same direction. As discussed earlier, transistors laid out orthogonal to each other will result in higher CD variation as well

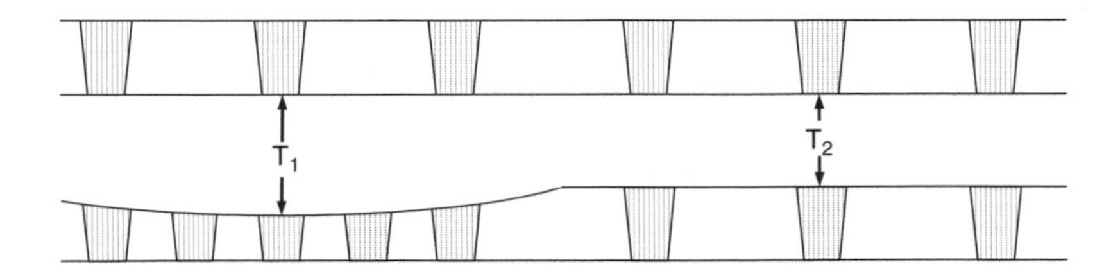

Figure 11.15 Wire density variation leads to ILD thickness variation.

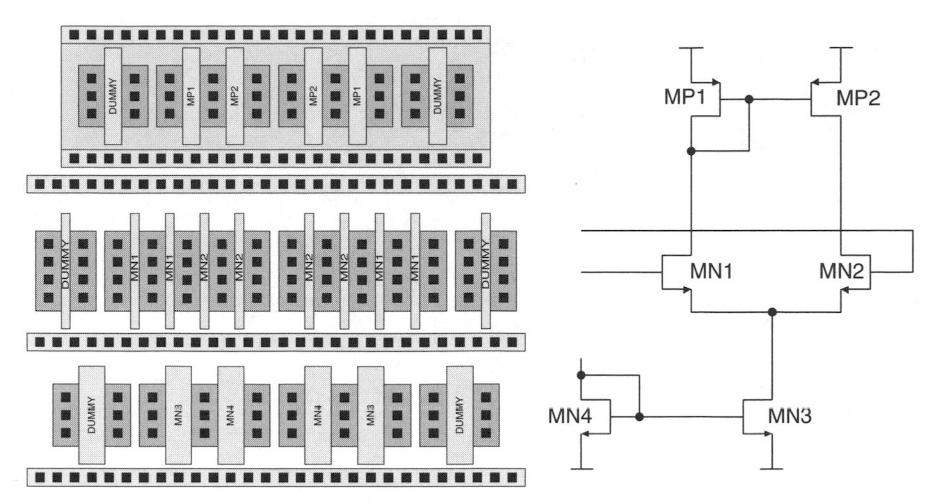

Figure 11.16 Layout using tiled devices.

as proximity effect variation. It is therefore important to orient all analog transistors in the same direction, along with the transistors of other critical circuits, such as clock buffers, sense amplifiers, and bit cells, to minimize CD variations that cannot easily be biased out.

Layouts to Avoid

- Avoid mirroring; instead, use step configurations and where possible, common centroid configurations (see Figure 11.17). A mirrored transistor layout in conjunction with misalignment will result in mismatch due to changes in parasitic capacitance and resistance of the drain and source of the transistors.
- Avoid minimum device size width or length. Minimum device width will result in higher threshold voltage (V_{th}) variation as a result of higher implant variation over a smaller area since dopant implant is a statistical event. Minimum-width transistors are also subjected to shape distortions, due to diffusion flaring as a result of a dogbone-shaped layout (see Figure 11.18). Minimum-channel-length transistors result in higher V_{th} variation, due to the steep roll-off of short-channel transistors in the nano-CMOS regime. A small change in the poly CD will have a large change in V_{th} (see Figure 11.19). Therefore, V_{th} matching for minimum-channel-length transistors is poor.
- Bent gate transistors result in size variation due to current mask making methods. It is also difficult to determine the transistor width and length of bent gate transistors, so their use should be avoided at all costs for analog design, especially if matching is important. They are also extremely sensitive to misalignment.

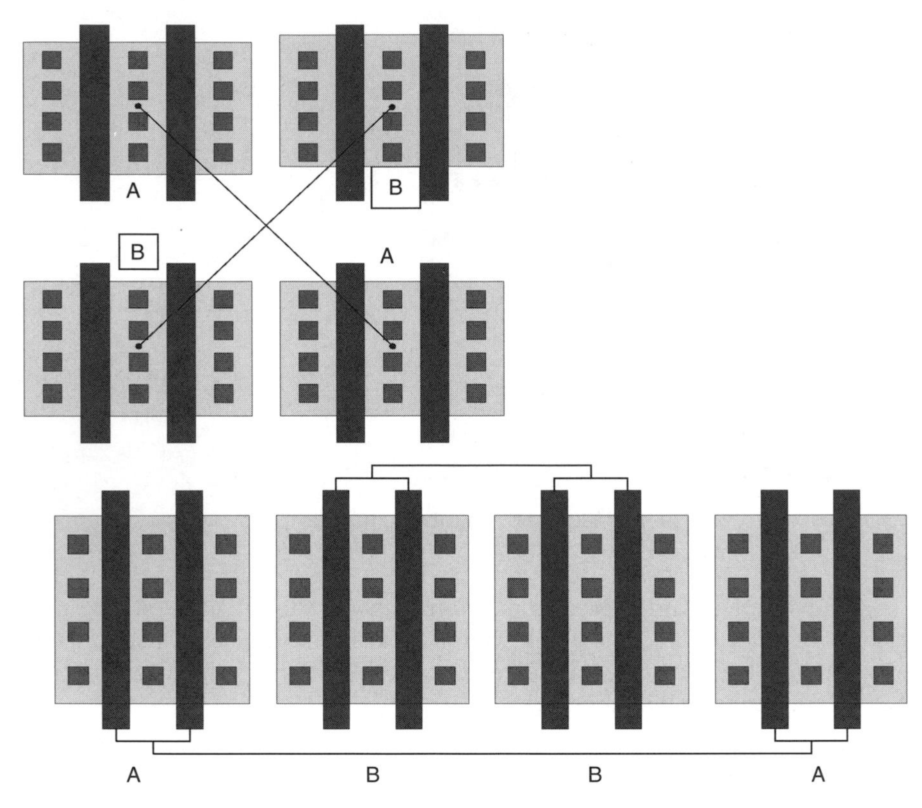

Figure 11.17 Common centroid layout for better matching; some possible configurations.

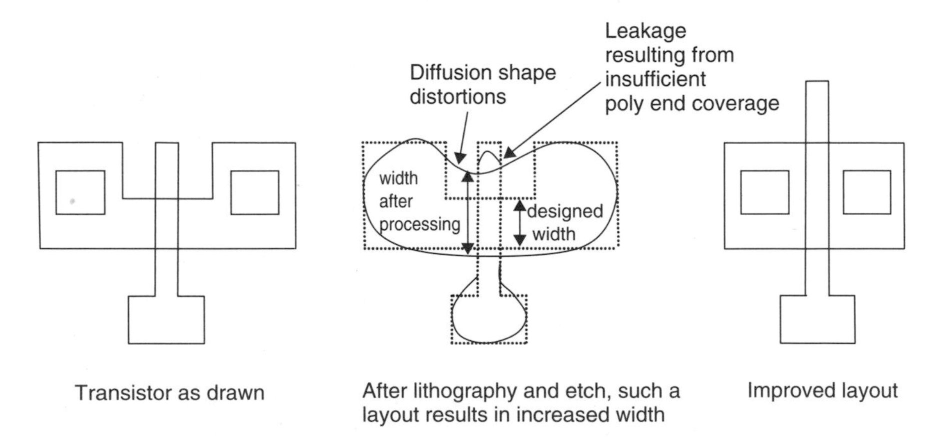

Figure 11.18 Diffusion flaring.

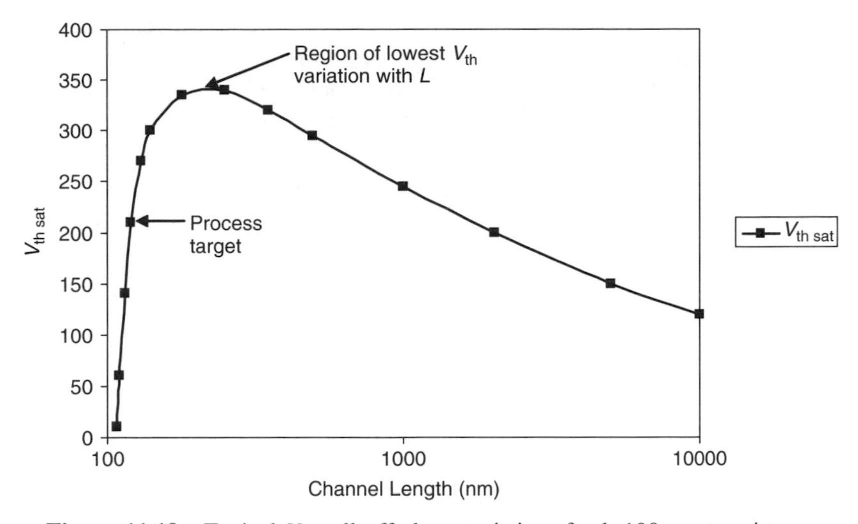

Figure 11.19 Typical V_{th} roll-off characteristics of sub-100-nm transistors.

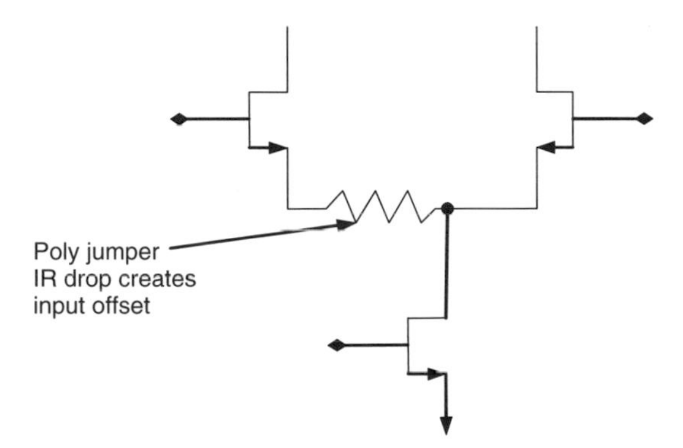

Figure 11.20 Current switch.

- Contact at a diffusion edge or corner should be avoided, due to lithographical distortion of corners and misalignment, which can result in contact resistance variation.
- Poly jumpers should never be used for any connection where there is dc current, especially for differential pairs. The higher sheet resistance (usually, two orders of magnitude higher even for silicided poly versus metal) can result in an offset due to *IR* drop (see Figure 11.20). Furthermore, resistance variability is greater for poly and silicided poly than for metal.

Layouts to Emphasize: Good Layout Practice

- Use larger-than-minimum poly end overlap of diffusion in all analog transistor layout as far as possible, especially for matched transistors. Some foundries have special rules and only allow certain discrete overlap; any other overlap will be reverted to the next minimum poly end overlap allowed by the design rules. This rule needs to be observed or you will not get the benefit of the extended overlap even though you have applied more than minimum poly overhang of diffusion to your layout.

- Increase the spacing of any bends in the poly from the diffusion edge to avoid misalignment-induced size variations due to poly flaring [see Figures 11.30 and 11.10(*b*)].
- Use only rectangular or square diffusion without dogbone-shaped diffusion or 90° corners close to poly edges (see Figures 11.18 and 11.31).
- Use a common centroid layout for low swing circuits where matching is very important (see Figure 11.17, where dummy transistors surrounding the active transistors are omitted for clarity).
- Use wide transistors, and fold large transistors so that misalignment will not affect parasitic resistance and capacitance matching (Figure 11.4). Use of wide transistors mitigates statistical implant variation on device parameters.
- Use longer-channel-length transistors (about five times the rule minimum) for better output impedance (the rate of change in $I_{d\ \mathrm{sat}}$ is low with respect to V_{ds}) and reduced CD variation impact.
- Use fully contacted diffusion, which reduces contact resistance variation. Use at least two vias—more if space and layout allow—for the same reason.
- Match interconnect parasitics, including Miller capacitance. Also, match the line length on the input as well as the output. The input lines must be matched at every level that is used, so that the matched transistors see the same stress during interconnect reactive ion etch as a result of antenna stress effects.

Circuit Techniques to Mitigate the Impact of Variability

- Keep the gate drive to the current sources as high as possible within the headroom constraint to minimize the impact of power supply variation and noise coupling on the gate of the current source. In current sources where the source node is connected to the supply or ground, it is a good idea to bypass the gate reference node to the supply or ground to improve power or ground noise rejection.
- Use thick oxide transistors as capacitors where charge leakage can cause failures, as in the loop filter capacitor of a PLL, since the charge on the loop filter represents the voltage-controlled oscillator (VCO) frequency. Charge leakage on the loop filter capacitor will result in static phase offset and can cause failure if the leakage exceeds the charge pump current. In high-multiplier PLLs the loop filter capacitor leakage will cause the VCO frequency to drift and result in higher jitter. For example, a multiply $\times$ 20 PLL will receive charge from the charge pump only every 20 clock cycles. The higher the multiplier, the greater the number of VCO output cycles

that elapse between charge pump updates. Use of metal capacitors is not recommended for PLL loop filters because of the greater area, and the capacitance is not well controlled. When the process is upgraded to a low-κ dielectric, this could change the loop bandwidth. Metal capacitors should be reserved for applications that require linear capacitance from rail to rail. Charge pump transistor subthreshold leakage will also cause very similar problems. Whereas gate leakage does not increase much with temperature, subthreshold leakage is a strong function of temperature.

Newly Exacerbated Physical Effects That Can Affect Analog Circuits RSC effects cause long-channel-device V_{th} to be lower than that of short-channel devices. As shown in Figure 11.19, as L increases, V_{th} increases until the peak and declines thereafter. When the channel length is longer than 1 μm, the V_{th} will be lower than the process target for the example technology. To achieve the least V_{th} variation it would be necessary to set the channel length at the peak of the curve, where the V_{th} variation with channel length is lowest.

Another effect that has become important as a result of pushing V_{th} is that under certain bias conditions the drive current increases with temperature, whereas at other bias conditions the drive current decreases with temperature. Figure 11.21 shows that effect, where the drive current of the transistor increases when its V_{gs} value is less that about 630 mV (below the crossover point) but decreases with temperature when its V_{gs} value is greater than about 630 mV. This can have a pronounced effect on the open-loop gain of an operational amplifier. It will also affect sense-amplifier gain, current switches or differential pairs, and comparators if biased accordingly. These circuits can have one transistor biased below the crossover point in the curve shown in Figure 11.21, while the other could be biased above.

As can be seen in Figure 11.22, the DIBL of halo-implanted transistors continues to decline with channel length even up to 10 μm [drain-induced threshold voltage shift, (DITS)]. Unless this effect is well modeled, it can present some surprises on silicon or when the process is shrunk for logic performance boost.

Figure 11.23 shows how well and polyproximity can affect device V_{th}. Attention must be paid when laying out matched transistors in view of this newly exacerbated proximity effect.

Power Supply Rejection Ratio (PSRR) Scaling of logic transistors has resulted in power supply voltage reduction that has limited the use of certain circuit techniques, such as folded cascode circuits that provide better PSRR. New circuit ideas have emerged that provide better PSRR that still fit within the power supply headroom (see Chapter 4 for details). Other power supply problems include noise coupling through the low-resistance substrate of the epitaxial process. A triple-well process will eventually be needed to isolate digital noise, in a mixed-signal design, from the sensitive analog circuits. It is important to keep the triple-well area small, to reduce capacitive noise coupling [27]. Use of guard rings and strategically placed substrate and well taps causes substrate noise to become common-mode noise, which is rejected in differential circuits.

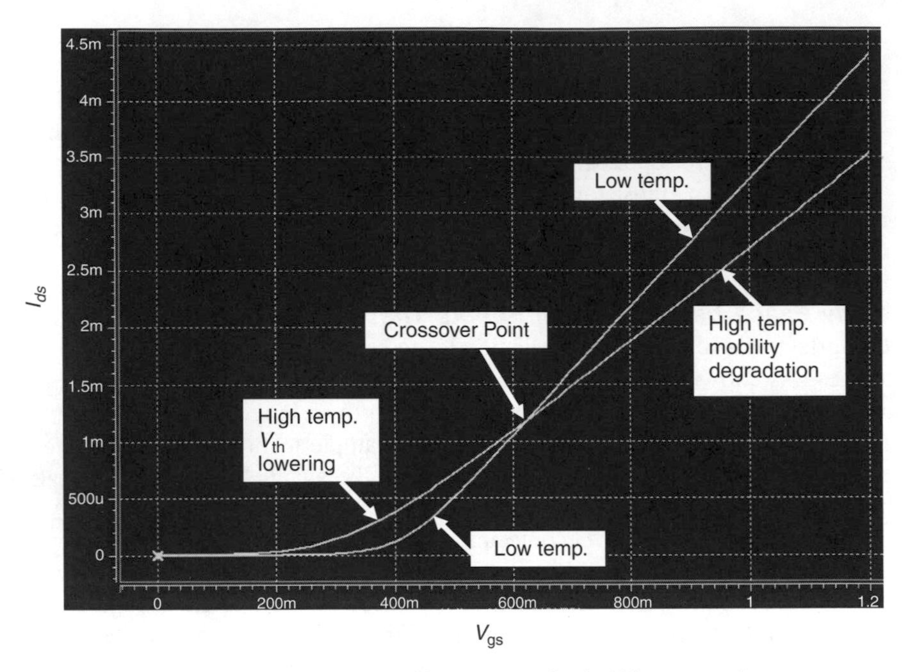

Figure 11.21 Temperature effect on I_{ds} of sub-100-nm transistors.

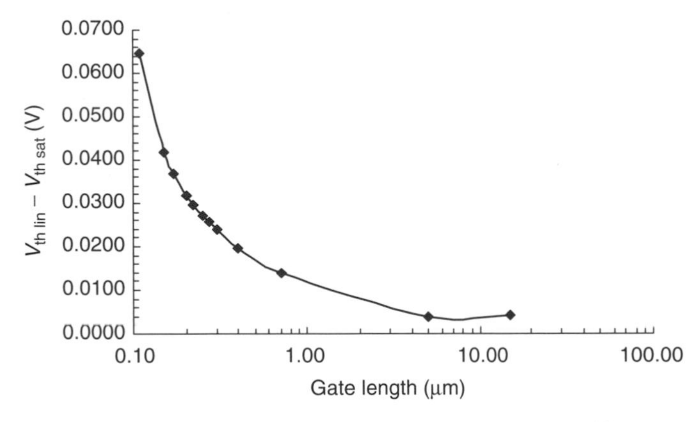

Figure 11.22 Drain-induced threshold voltage shift.

Package and system modeling of the supply impedance is now very important, especially for high-performance chip designs. $L(di/dt)$ is the most significant voltage drop in such designs, and the supply impedance must be designed to keep the $L(di/dt)$ plus *IR* drop to within the design budget, usually 10% of the supply voltage for high-performance processors. This would require on-chip decoupling capacitors, package capacitors, and capacitors on the system board. At a minimum, the decoupling capacitors on the chip should be 10 times the

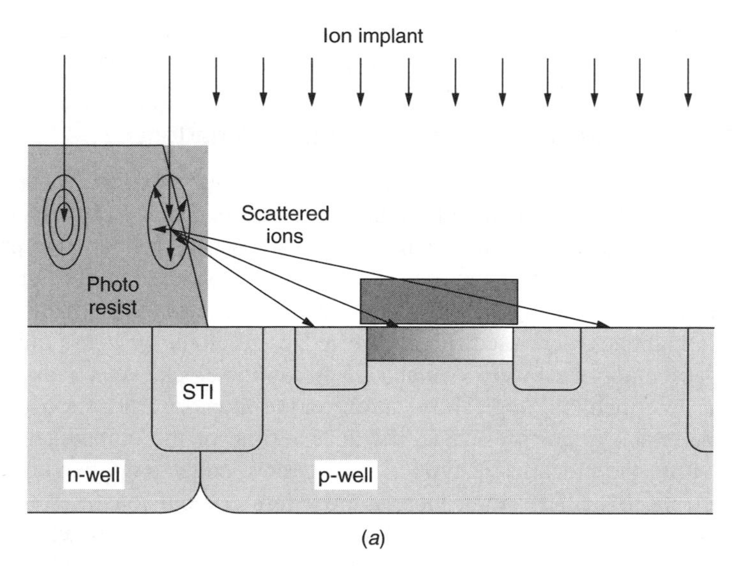

(a)

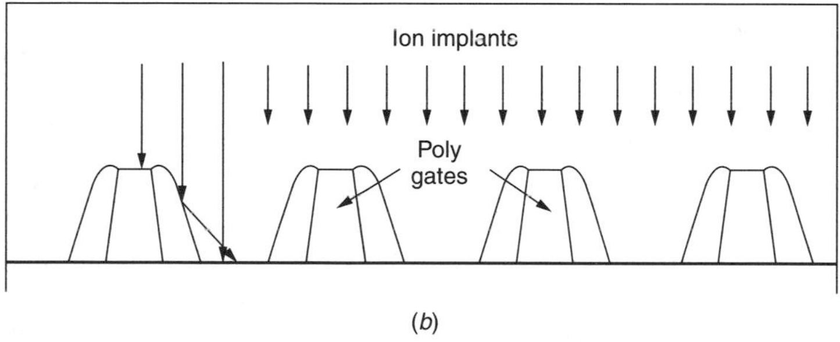

(b)

Figure 11.23 (*a*) Well proximity effects. (*b*) Poly proximity effects.

equivalent switching capacitance (C_{eqv}) of the chip, where

$$C_{\text{eqv}} = \frac{\text{chip power}}{V_{dd}{}^2 \times \text{frequency}}$$

The chip power is determined using the worse-case vector, which means that the C_{eqv} value will be larger and can better maintain the power to the chip during the peak power demand period. Hence, performance degradation during the period of peak power demand due to supply voltage droop will be kept to a minimum due to the well-decoupled supply. Package and system board modeling is a very important part of the design in order to meet the supply impedance goal of a high-performance system and is beyond the scope of this book.

In mixed-signal designs in an epitaxial process, sharing V_{ss} but separating the V_{dd} supply will result in the lowest noise coupling from the digital domain [12]. This also provides an opportunity for using a higher voltage power supply than in digital circuits, to provide the necessary headroom for some designs. Sharing

V_{ss} simplifies the electrostatic discharge (ESD) protection as well (see Chapter 5 for a full discussion on ESD protection in the nano-CMOS regime).

11.2.4 Digital Circuit Strategies to Deal with Variations

Digital circuits are usually more tolerant to process variation; however, some digital circuits, including self-timed circuits and matched delay circuits, can be extremely sensitive to process variation. Self-timing is used primarily in embedded memories such as cache memories. It was used most commonly during the period when clock frequency was low. To reduce the access time of memories, self-timing techniques were used to generate edges to clock the sense amplifiers (SAs), so that memory data were available earlier in the clock cycle. This enabled one-cycle access, including logical operation on the memory data, for better performance. As clock frequency scales, the access time of the embedded SRAM has come within the clock cycle time, so a lot more edges have become available to clock the SAs. Therefore, there is now less compulsion for self-timing to generate edges. The only other need for self-timing is to save power in cases where the SAs are not clocked until an address changes, while the clocked design requires clock gating to reduce clock power. It is still a lot easier and more robust to gate the SA clock than to self-time it. Many schemes are designed to mitigate the impact of variation on design robustness if one must self-time. We discuss next a self-timed scheme used in SRAM.

Self-Timing Strategies Traditional self-timed memory relied on a single SRAM cell to drive the dummy bitline, which is then converted to a full CMOS level and fanout to drive the SA clock line [28]. As can be seen in Figure 11.24, a single-cell self-timing scheme is very sensitive to process variation that causes cell drive variation resulting in higher self-timing delay variation. To avoid failures due to the higher self-timing path delay variation, more margin is needed—at the expense of performance.

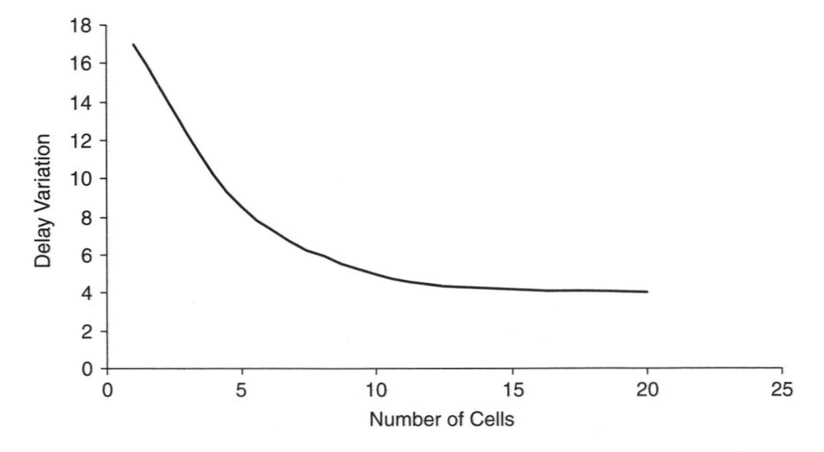

Figure 11.24 Effect of number of cells on self-timed delay variation.

As shown in Figure 11.24, lower self-timed path delay variation can be achieved if more than one cell is used [13]. The use of several cells averages out cell current variations. A multicell self-timed scheme is illustrated in Figure 11.25. To minimize cell drive variation it is important to have another column of dummy cells at the edge of the array next to the self-timed dummy column. In the memory array, metal density is very consistent, making it easier to control ILD variation. Due to the regularity of the metal lines, resistance variation due to CMP chemical–mechanical planarization (CMP) is kept to a minimum, provided that the fabricator optimizes CMP for the metal density as found in the memory array. Most fabricators understand the need to reduce resistance variation in a memory array and will optimize the process around the memory array. Even so, there will still be some resistance variation due to barrier metal thickness and wire width variation.

Self-Timed Margins Figure 11.26 illustrates a typical race condition in self-timed designs. In this illustration, delay in Out1 must be less than the delay in Out2; otherwise, functional failure would result. Due to process, voltage, and

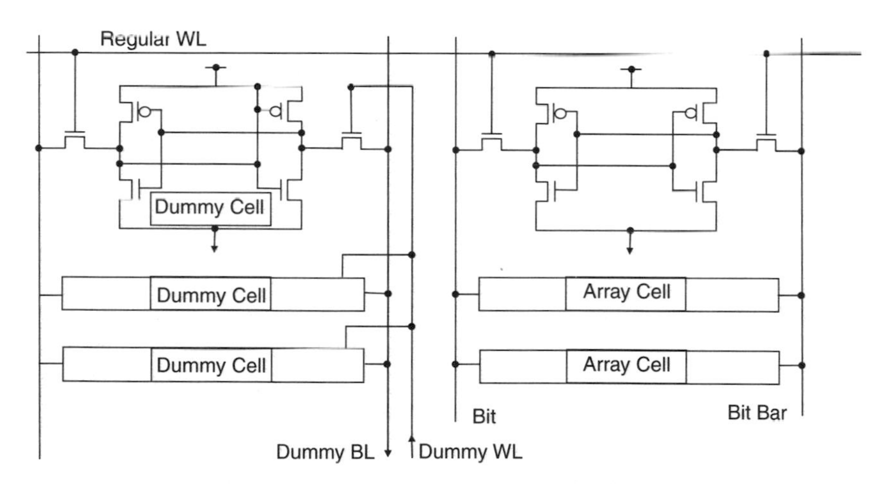

Figure 11.25 Multicell self-timed scheme.

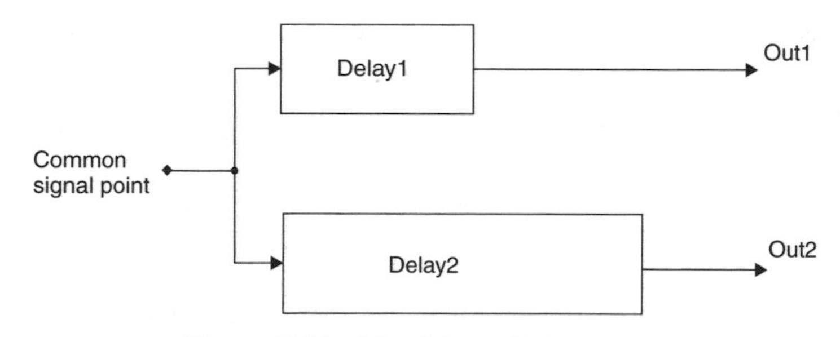

Figure 11.26 Margining self-timed paths.

temperature (PVT) variations and layout differences, Delay2 may become shorter than Delay1 on silicon, due the fact that some local effects are not fully or correctly modeled or anticipated during the design phase. When this happens in a self-timed design, it will result in a functional failure which will not work at any frequency, including at very low frequency, and will require design redo to restore even basic functionality. This is a very serious and costly design failure. To safeguard against such a situation, we add margins to the simulation model to cover for the unanticipated effects, so as to reduce the probability of such a functional failure.

As mentioned earlier, the speed of Delay2 may not match that of Delay1 due either to some unanticipated effect or if the circuit is not fully optimized. The following analysis translates the margin into a physically meaningful parameter that can be used to verify the margin of the self-timed circuit. The self-timed circuit in Figure 11.26 at the verge of failure can be represented as

$$\text{Delay2} \times (1 - M) = \text{Delay1} \times (1 + M) \qquad \text{where } M \text{ is the self-timed margin}$$

Simplifying, we obtain

$$M \times (\text{Delay1} + \text{Delay2}) = \text{Delay2} - \text{Delay1}$$

Hence,

$$M = \frac{\text{Delay2} - \text{Delay1}}{\text{Delay1} + \text{Delay2}}$$

Typically, M is set to 0.25 for prelayout and 0.15 for postlayout extracted simulations over all practical corners. The use of statistical models is highly encouraged for more realistic corner coverage. Further details on statistical modeling are given in Section 11.3. Regardless of the self-timing margin, every self-timed path must have metal programmable options to increase the margin to at least 30% in all practical corners. As mentioned earlier, self-timed race failure is catastrophic for a chip; the addition of metal programming options can lead to a quick loop fix. The metal options must be designed to affect a self-timing margin change in as little as one layer and no more than two layers. This is important, since mask cost is on the rise, especially for nano-CMOS process nodes. If possible, design the programming change at as high a metal level as possible to allow for a quick fabrication turnaround time for the fix in the event that a self-timing margin change is necessary.

Delay Variation Due to Slow Nodes Slow nodes manifest themselves as high-fanout nodes, long unrepeated lines, and signals through pass gates and cascading pass gates. Pass gates present themselves as large resistors to the signal, just like long unrepeated lines. When more then two pass gates (unbuffered) are in a signal's path, the result is a really slow node that must be dealt with. Slow nodes could also be weakly driven nodes, as in the case of signals through cascading pass gates and long, unrepeated signal lines. The weakly driven nodes are more

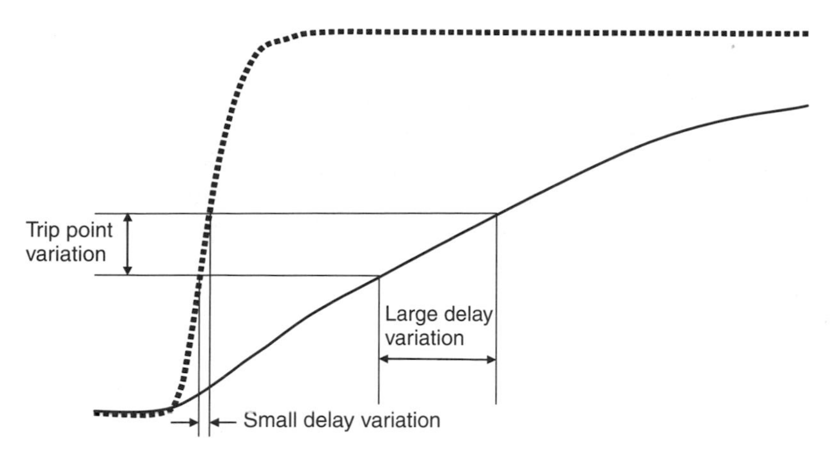

Figure 11.27 Trip point versus delay variation.

susceptible to noise coupling into the far-end node where the receiver resides. There is another hazard that affects all slow nodes, including high-fanout nodes. As shown in Figure 11.27, variation in the input trip point of the receiver will translate into a larger input delay variation due to the gentle slope of the input signal on the slow nodes. Maintaining an input slew rate enables the design to better tolerate P-to-N process skew that affects gate input threshold or trip point.

In some circuits, such as an arithmetic block, there will be pass gates in the data path if pass gate adders are used. In some cases there could be several pass gates in series in the data path unless the designers add buffers between the cascading full adders. This adds delay in the critical path. There are ways to mitigate this by using differential cascode voltage switch (DCVS) logic instead [31][32].

Pulse Flop Clock Generator Design Strategies

Match Trip Points Pulse flop operation and design are not covered in this book; refer to other circuit design texts for a detailed discussion. Full understanding of the pulse flop operation is needed to appreciate the following discussion on process variation issues that affect the pulse generator and operation of the pulse flop. Figure 11.28 shows a typical pulse generator for pulse flops. Inv1 through Inv3

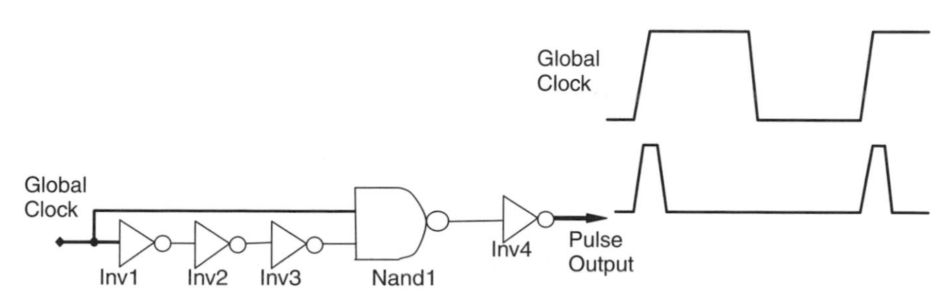

Figure 11.28 Typical pulse flop pulse generator.

form a delay chain that defines the pulse width of a pulse generator. Pulse generator pulse width variation has a serious impact on the hold time of a pulse flop. The input trip point of Nand1 and Inv1 must be matched; otherwise, the pulse width varies with the global clock edge rate variation. A longer pulse output width will result in a longer hold-time requirement but offers a longer transparent time. If the logic cone feeding into the pulse flop is not properly balanced in timing, the longer transparent period due to the wider pulse width can cause a hold-time problem even when there is a maximum time path from the same logic cone.

Let us consider the case where Nand1 has a higher trip point than the input trip point of the inverter chain, starting with Inv1 in Figure 11.28. As the global clock rises, Inv1 trips first and starts the delay chain going, while Nand1 has not quite reacted to the global clock input. This in effect shortens the output pulse width of the pulse generator because the inverter delay chain times out sooner with respect to the rising edge of the pulse generator output. The delay after Inv1 until Nand1 triggers will be the amount of shortening of the pulse generator pulse width. As can be seen in Figure 11.27, the clock rise time change can alter this delay, thus changing the pulse width. The clock rise time can change for several reasons, and the change can affect the hold time of the chip and cause catastrophic failure. The flip condition where the Nand1 trip point is lower than Inv1 will increase the pulse width and hold time requirement of the pulse flops. In cell-based designs where the pulse flops characterization condition assumes that the trip point of Inv1 and Nand1 are matched, hold time failures can result if the trip points of Inv1 and Nand1 are not matched, as that changes the actual hold time requirement of the flops.

Set the input trip point slightly below $V_{dd}/2$ (lower middle third) but not too low; otherwise, ground bounce will be an issue. The reason for this is that the edge placement error is lower at a point low on the clock rising edge. Since the pulse generator only references the rising edge of the clock, this technique ensures more accurate clock reference and lower latency from the clock edge.

Pulse Generator Output Waveform Peak The pulse width must be wide enough to ensure that the pulse reaches V_{dd} under all load conditions that the pulse generator must drive, over all practical corners. This is to make sure that the pulse width is deterministic. If the pulse width reaches V_{dd} under all load conditions, the pulse will always be discharged from the same voltage under the same PVT conditions and will therefore be deterministic. This eliminates pulse width variation beyond what is attributed to the PVT conditions. The other reason for having the clock pulse reach V_{dd} is to make sure that the flops always see the same drive level at its clock input, thereby avoiding varying setup and hold time due to varying gate drive.

Pulse Generator Delay Tracking of Data Path Delay The delay chain formed by Inv1 through Inv2 is by necessity constructed with transistors of minimum size, to keep the power down. This is where we have to trade power consumption for process tracking of the data delay. The devices must be large enough so that the delay is not dominated by parasitics. The parasitics along the delay chain must be minimized as you would on the data path that is optimized for speed. The delay chain speedup ratio must match the data path speedup closely over the practical corners to avoid running into a hold time violation. If the data path speeds up more than the delay chain, especially for dynamic pulse flops, we could end up in a situation when the input data to the dynamic flop change before the pulse resets.

The last element in the delay chain (Inv3) must have the same stack height as the logic flop driven by the pulse generator. If the flop that received the clock pulse from the pulse generator is not a simple flop but a dynamic logic flop, Inv3 in the delay chain must have the same stack height as the dynamic logic that is preceding the flop (see Figure 11.29). This allows the delay chain to track the logic delay over process corners. Figures 11.30 and 11.31 illustrate the need to relax spacing rules as well as poly end-cap coverage to reduce device variation due to processing distortion of drawn polygons.

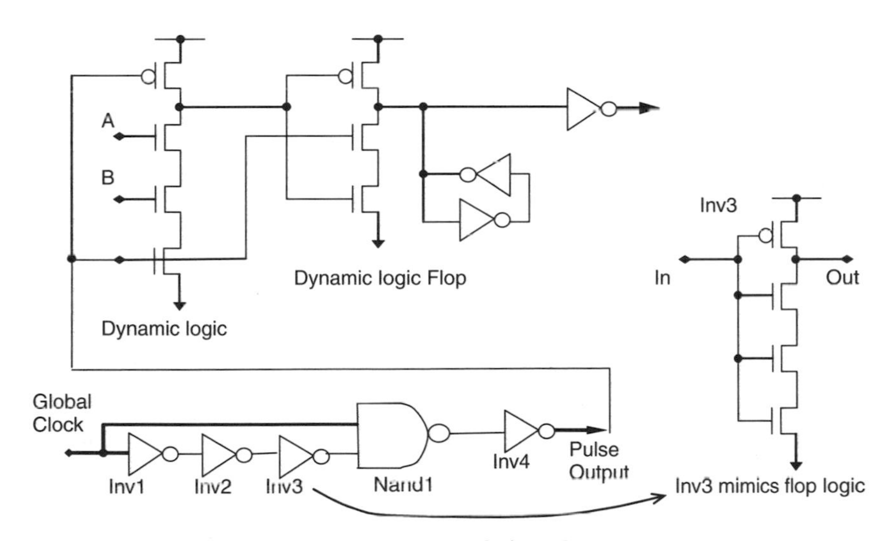

Figure 11.29 Delay tracking technique for pulse generators.

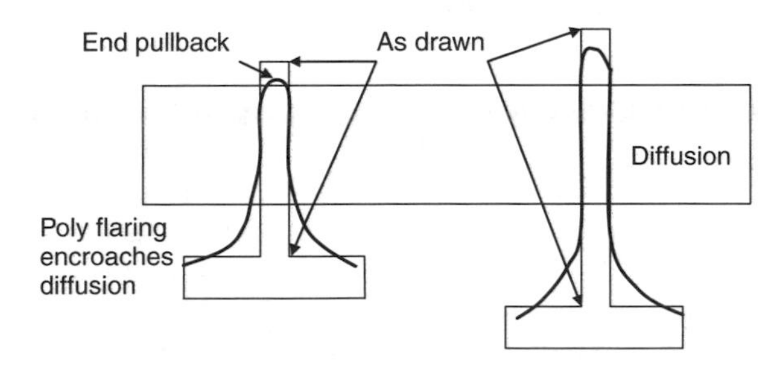

Figure 11.30 Poly flaring.

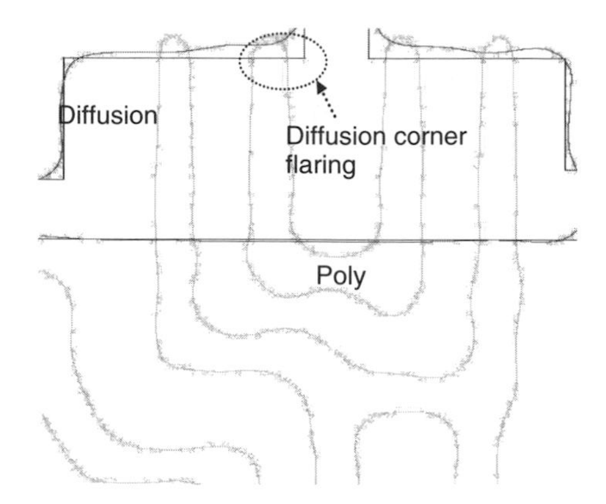

Figure 11.31 Poor end-cap coverage for poly at diffusion corner.

11.3 CORNER MODELING METHODOLOGY FOR NANO-CMOS PROCESSES

SPICE modeling has become the most critical component for enabling designers to determine necessary design margins to meet the stringent requirements of modern IC circuits. With the ever-increasing speed requirements, margins have continued to decrease, forcing designers to rely more heavily on models for an accurate reflection of the process, including its expected variation. The traditional approach for model development has been to use a nominal case adjusted to a foundry process control methodology and then to develop corner models that are worst case for digital logic. The process variance has not scaled equivalently with the critical dimension scaling, which has made this source of error more pronounced, especially on the deep submicron processes. There is now a real need for statistical models for a more accurate representation of the process. Figure 11.32 shows a diagram of the various levels of process variation. Each level in the process flow can add additional variation to the device performance. Understanding the contribution at each stage is important for creating accurate statistical models.

11.3.1 Need for Statistical Models

The process corner model approach creates unrealistic process combinations and leads to overdesign, especially as design margins become smaller. This is illustrated in Figure 11.33, a scatter plot of the NMOS and PMOS $I_{D\ \mathrm{sat}}$ measurements over numerous wafer lots. Here, fast–slow and slow–fast (FS and SF) corners rarely occur. This makes sense from a process standpoint since PMOS and NMOS devices are only partially correlated. For example, if we consider the various parameters that can vary, such as oxide thickness, gate length, gate width, channel doping, and halo implant, some of them (e.g., oxide thickness and channel length) will vary similarly for PMOS and NMOS devices, while others

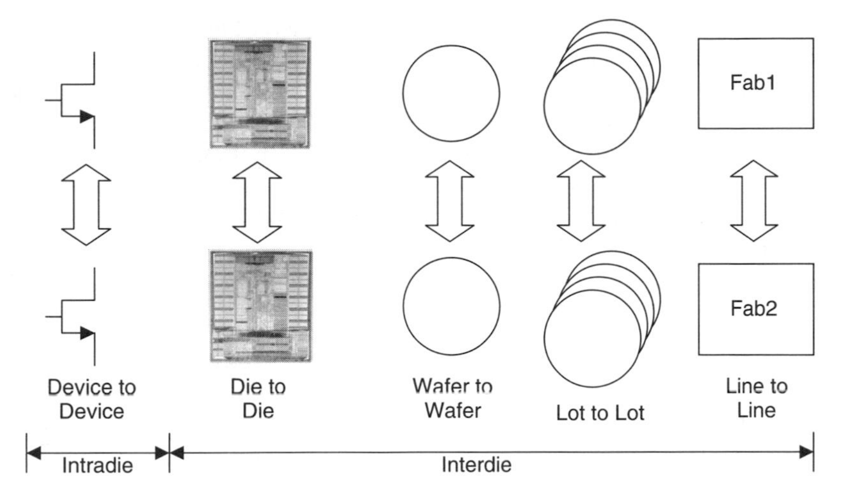

Figure 11.32 Various levels of process variations.

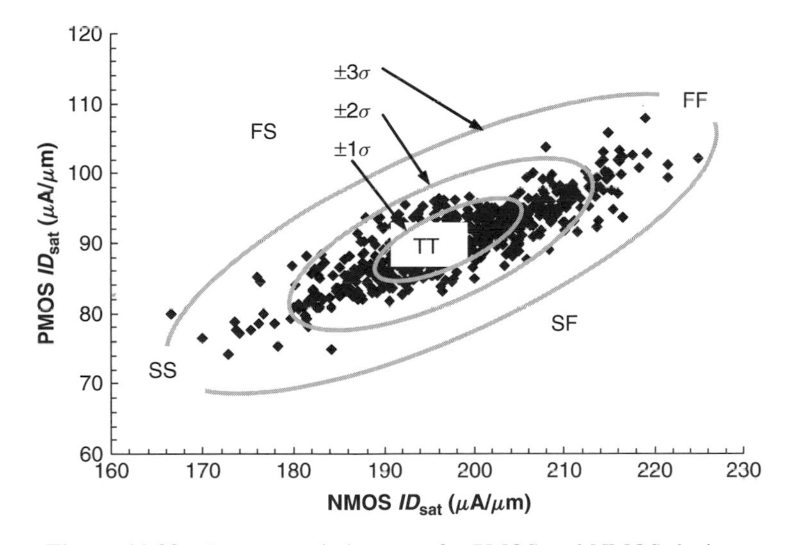

Figure 11.33 Process variation map for PMOS and NMOS devices.

will not be correlated and will vary independently. Additionally, the variance of the process will have both localized and global components. Process corners do not provide this partitioning of the variance, so it is impossible to determine the effect of localized variation between devices based on corner models.

Identifying the worst-case corner for analog circuits becomes difficult. The concept of fast/slow may not be applicable. For an operational amplifier, high gain/low gain may make more sense, but which digital process corner corresponds to the high-gain case for the amplifier may be difficult to say, since it is dependent on the specifics of the amplifier architecture. Identifying what process corner represents the worst-case corner becomes more difficult as subblocks are combined to form more complex systems such as a data converter. The analog circuit may end up being overdesigned if the analog circuit is simulated using the digital process corners, especially given the already limited design space for analog circuits. Overdesign of a circuit can result in increased complexity, larger die size, and potentially, a missed market window and is therefore best avoided if possible. If we consider the variation of several parameters that can vary for a process, the combined variance can be expressed as

$$\sigma_{\text{total}} = \sqrt{\sigma_{t_{\text{ox}}}^2 + \sigma_L^2 + \sigma_W^2 + \sigma_{N_p}^2 + \sigma_{N_n}^2 + \sigma_{\mu_p}^2 + \sigma_{\mu_n}^2 + \cdots} >> 3\sigma$$

Combining the variation in this manner can result in significant overdesign of a circuit if it must meet the performance requirements at these extreme cases.

The use of statistical modeling allows the designer to estimate the functional yield of a given design before it has been fabricated. This information is crucial for making trade-offs during the design cycle rather than postfabrication. The designer can look at subblocks within a design to determine the contribution of each of these components toward the overall system yield, allowing emphasis to be placed on the most critical portions of the design. The designer will also be able to make an assessment of device sizing effects on the functional yield.

11.3.2 Statistical Model Use

Statistical models are based on a first principles approach to measuring the source of variation and translating that variation into SPICE model parameter variation. The first step is to identify the independent factors and capture their long-term variation. An example of this is shown in Figure 11.34, which shows the capacitance equivalent thickness (CET) variation in oxide thickness over a period of time. This information is translated into a histogram, allowing the mean and standard deviation values to be extracted. These values are then entered into a

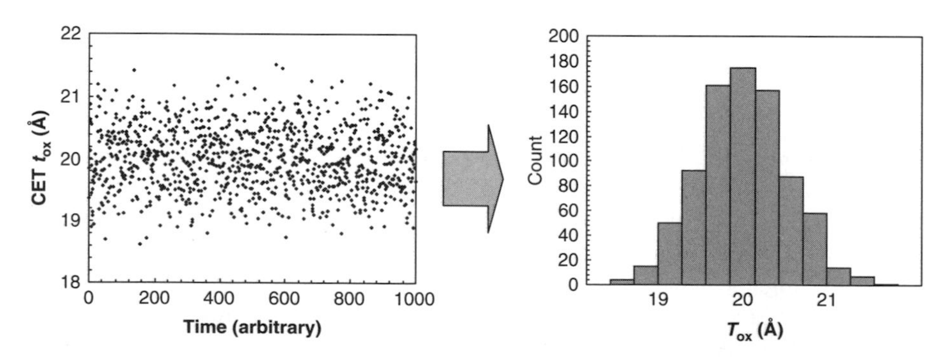

Figure 11.34 Oxide thickness variation over time for a given process.

model such that the independent model parameter is modeled by its nominal value plus the standard deviation variable. Physical parameters that can be considered include doping concentrations, oxide thickness, mobility, gate width, and gate length. It is crucial that the parameters selected be applied correctly to the SPICE models to ensure that their effects are simulated correctly. For example, it is common practice simply to vary the threshold voltage of a device to look at the process variation effects, but this does not capture back gate biasing correctly, so erroneous results will be obtained. This is partially what makes this task so difficult since the SPICE models do not have a physical context entirely.

The next step is model correlation. Normally, a parameter such as threshold voltage, VTH0, is set to a fixed value such as VTH0 = 0.4. This would now become VTH0 = 0.4 + VTH_PVAR where VTH_PVAR is defined to be AGAUSS($\overline{M}$, σ, N), where $\overline{M}$ represents the mean value, σ the variance, and N the number of standard deviations represented by σ. Use of this approach would not capture the threshold voltage dependency on oxide thickness, so it is better to represent it as [36]

$$V_{\text{th0}} = V_{\text{FB}} + 2|\phi_F| + \frac{qN_S x_{t1} + qN_P(X_{\text{dep}} - x_{t1})}{C_{\text{ox}}}$$

where $C_{\text{ox}} = \varepsilon_{\text{ox}}/t_{\text{ox}}$, $t_{\text{ox}} = \bar{t}_{\text{ox}} + \sigma_{t_{\text{ox}}}$, $N_S = \overline{N}_S + \sigma_{N_S}$, $N_P = \overline{N}_P + \sigma_{N_P}$, and $V_{\text{FB}} = \overline{V}_{\text{FB}} + \sigma_{V_{\text{FB}}}$. The parameters are as follows: N_S is the doping density between 0 and x_{t1}, and N_P is the doping density between x_{t1} and the depletion depth X_{dep}. All other terms have the standard meaning already defined. Using this representation for the threshold voltage allows a multitude of process parameters to be accounted for such as the flat-band voltage and channel doping. This also captures the effects of the substrate biasing as well, making the overall simulation more accurate. Once the appropriate parameters are obtained, it is possible to run multiple simulations to obtain a distribution for parameters that can be measured on wafers such as threshold voltage or $I_{D\text{sat}}$. The real-world measurements can be compared to the simulated distribution to validate the distribution generated by the model.

The standard deviation of each of the parameters is typically not the same for both device types. Similarly, there is a significant dependency on the device size as well. This size dependency is greater for the channel length, especially for very small channel lengths. Figure 11.35 shows the localized difference in threshold voltage between two identical NMOS devices placed side by side to provide the maximum degree of matching, with varying size for a deep submicron process. These data do not include device displacement that will add further to the variation. Localized variation may not be too important for digital logic since it tends to average out, especially for deep levels of logic, but it becomes crucial for analog design. This localized variation can be used to determine the optimum device size for critical components such as a differential pair.

Consideration of both the local (intradie) and global (interdie) variation represents a reasonable model for the variation. The process variation can be

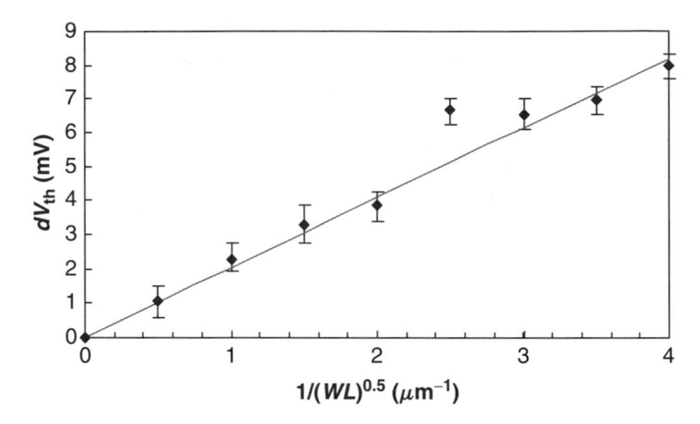

Figure 11.35 Threshold variation as a function of device size.

represented by [34]

$$\sigma^2(\Delta P) = \frac{A^2_{\Delta P}}{WL} + S^2_{\Delta P} D^2$$

where $\sigma(\Delta P)$ is the standard deviation of the process parameters, ΔP. The device channel width and length are represented by W and L. The displacement between devices is represented by D, and the parameters $A_{\Delta P}$ and $S_{\Delta P}$ are process-dependent constants that must be determined by measurements. The first term represents the localized variation, and the second term represents the global variation that is dependent on the physical displacement between devices. In some cases this model may not provide the necessary insight into the process variation [35]. For this reason, it may be best to form the variance in more components to allow great analysis of the various places that variation can be introduced and the overall impact. One may go to the level of detail shown in Figure 11.32, where a variance component is assigned for each level. This approach will allow much more insight into the product yield, but obtaining meaningful information on the additional variation at each level can become difficult.

This approach is applied to a phase-locked-loop charge pump to estimate the degree of current mismatch that can be expected. The results of these simulations are shown in Figure 11.36. Here it is assumed that the design can handle ±6% mismatch of the current resulting in 15 die that are outside that range, or a 97% yield. If this yield is deemed adequate, no further design effort is required. If a higher yield is necessary, the circuit can be redesigned. This redesign may require entirely new charge pump architecture, or simply resizing critical devices to decrease the variability. Figure 11.37 shows how the threshold voltage variation decreases when the device size is increased. The y-axis shows the threshold voltage shift, while the x-axis shows the normalized device size (area) when normalized to a minimum-sized device for a 100-nm process. It is possible to reduce the overall system variation by sizing up critical devices selectively.

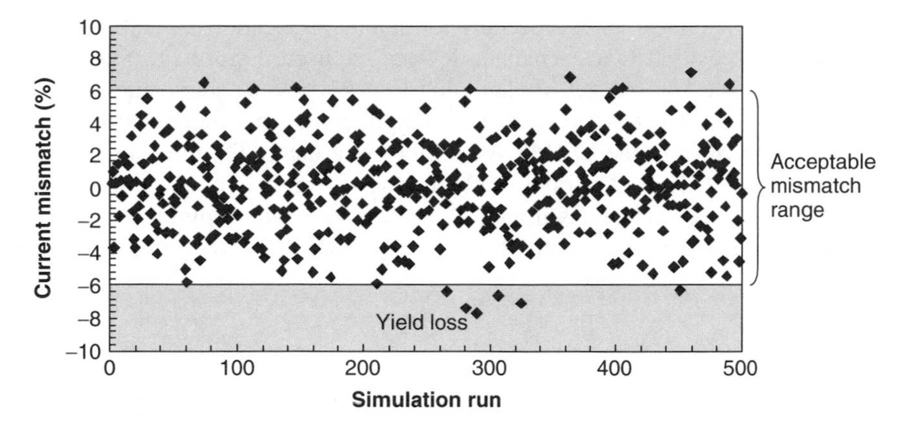

Figure 11.36 Charge pump circuit current mismatch induced by localized and global effects on threshold voltage variation.

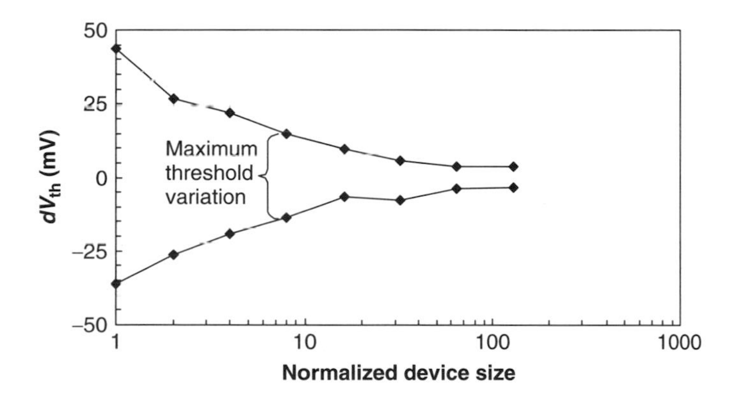

Figure 11.37 Threshold voltage variation as a function of device size.

11.4 NEW FEATURES OF THE BSIM4 MODEL

The implementation of BSIM4 models has allowed a significant improvement in simulation accuracy for the deep-submicron processes. BSIM4 models incorporate several important features previously missing from the BSIM3 models, which include modeling of the halo or pocket implant, gate-induced drain leakage (GIDL), gate direct tunneling, and trench isolation stress effects. Trench isolation stress effects are discussed at length in Chapter 4.

11.4.1 Halo/Pocket Implant

The halo/pocket implant is used to reduce the threshold voltage roll-off for very short channel devices, but this implant results in significant DITSs for longer-channel devices. The halo/pocket implant increases the g_{ds} value in long-channel devices, which is undesirable, especially for analog applications, which is one of the primary places that longer-channel devices are used. Figure 11.38(*a*) shows the location of the halo/pocket implant, and Figure 11.38(*b*) shows the resulting DITS effect for a 100-nm process. This output impedance degradation is not modeled completely in the BSIM3 version because the DITS does not consider the effect of the halo/pocket implant. Modeling of the halo/pocket implant has been achieved by no longer assuming a uniform substrate doping. A limitation still occurs because the DITS output resistance model does not include the body bias effect.

11.4.2 Gate-Induced Drain Leakage and Gate Direct Tunneling

The various components of off-state leakage are shown in Figure 11.39 along with a relative indication of the influence for several process generations. The gate leakage is projected to become a more significant factor at the 90-nm technology node and beyond, but source–drain leakage remains the primary issue. BSIM4 models allow the gate leakage to be modeled, but at a cost of additional simulation

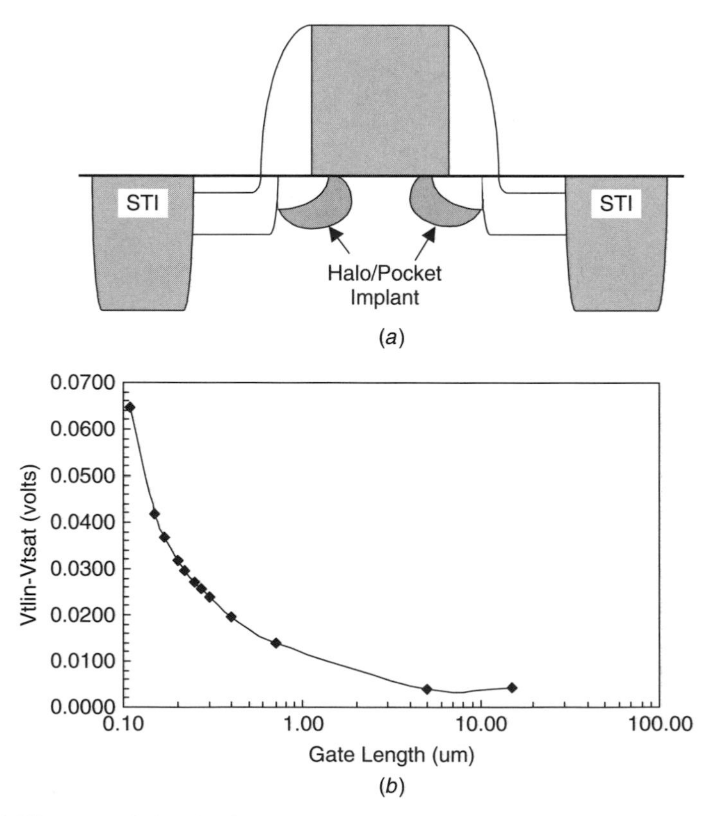

Figure 11.38 (*a*) Halo/pocket implant used on deep submicron processes. (*b*) Resulting simulation of DITS for a 100-nm process.

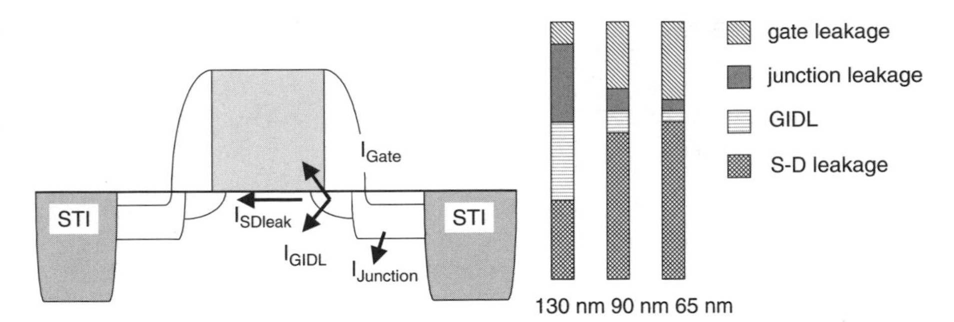

Figure 11.39 Transistor off-state leakage components and the relative scaling with process.

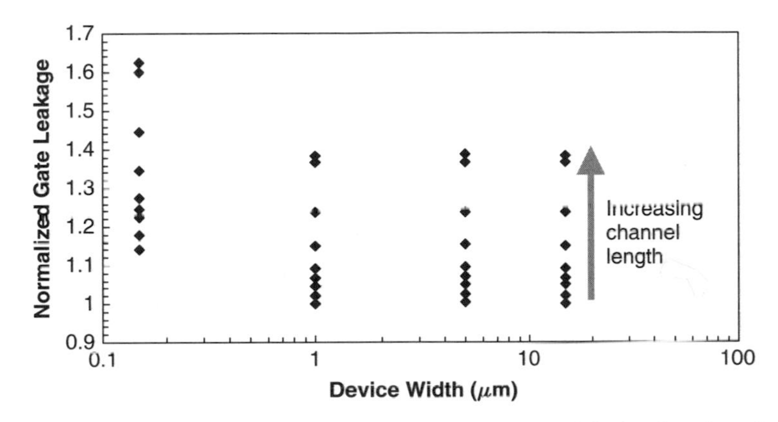

Figure 11.40 Normalized total gate leakage as a function of device length and width.

time since the gate leakage must be evaluated at each gate bias point since it is dependent of the potential across the gate. Figure 11.40 shows the total normalized gate leak current as a function of device width for various gate lengths ranging from 0.2 to 15 μm. Figure 11.41 shows the GIDL effect for a thin oxide device on a 100-nm process. The GIDL current is in the nanoampere range. A weak dependency on the bulk bias can also be observed.

11.4.3 Modeling Challenges

Although BSIM4 represents a significant improvement over BSIM3 models, it still does not account for all factors that can have a pronounced affect on device performance. Many of these effects relate to how the device is laid out and the physical location of adjacent devices: (1) dogbone devices to realize narrow-width devices, (2) well proximity effects, and (3) shallow trench isolation stress effects (these effects can be modeled postlayout). A suggested approach to use is to avoid layouts that aggravate these effects wherever possible since they

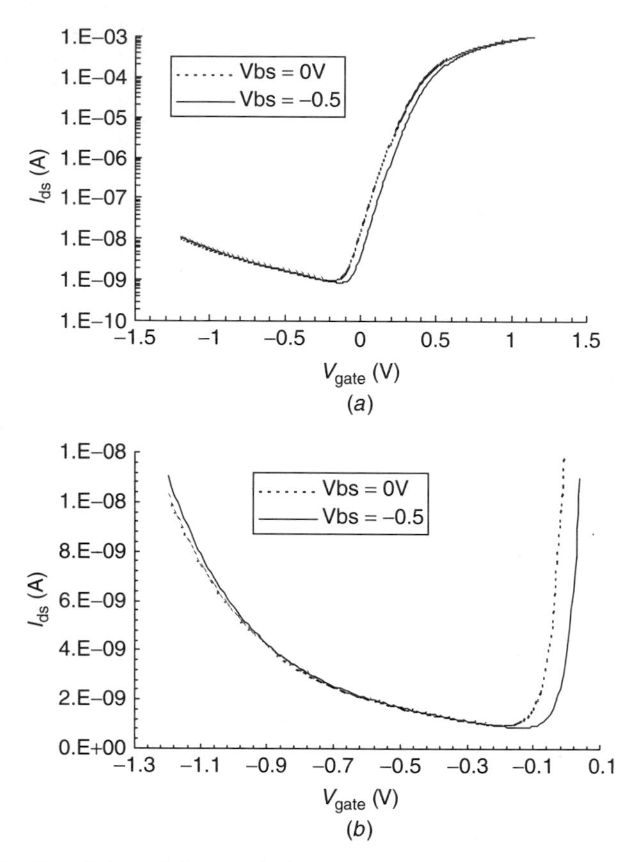

Figure 11.41 Simulation of the gate-induced drain leakage over (*a*) a wide gate voltage range and (*b*) a zoomed area to show the bulk bias influence.

are difficult to model. This approach can lead to a serious constraint with the physical implementations, increasing the overall die size. A second approach is to develop macro models that allow these effects to be modeled. These models can be generated for the most critical circuits within a design, such as an SRAM cell to ensure that the highest level of accuracy is obtained. These macro models should be parameterized to allow maximum flexibility. Correlation between the model and early test chip results is required to ensure that the models are accurate.

11.4.4 Model-Specific Issues

BSIM4 models use nonphysical parameters to have high accuracy for short/narrow devices. The use of nonphysical parameters makes the model parameter extraction procedure much more complicated because of the correlation between short- and long-channel parameters. Insufficiently modeled physical effects such as doping dependent mobility models for the halo/pocket implant technologies are resulting in some discrepancy between the modeled device and the physical device. The reverse short-channel effect (RSCE) needs to be modeled as well to further improve model accuracy. With each progression of BSIM model comes an increase in the number of parameters, giving rise to an increase in the simulation time and memory requirements. It is crucial to balance the number of parameters with the need to have reasonable simulation times.

11.4.5 Model Summary

Modeling of halo/pocket implanted devices has been improved significantly with BSIM4. The much needed gate direct tunneling model required for design on 90 nm and below is also available. The parameter extraction approach has become much more complicated, and the number of parameters has increased significantly. Macro models can be used to allow modeling of some of the layout specific issues, but they must be correlated with actual silicon measurements to confirm their accuracy. There are still quite a few more effects that must be incorporated into the model, but this must be done such that it does not significantly affect the complexity or simulation run time.

11.5 SUMMARY

The principles presented in this chapter can be applied to many other circuit and layout types to minimize the impact of variation on their functionality as well as manufacturability. As we scale the technology well into the nano-CMOS regime, dealing with variation will be part and parcel of all design methodology, including ASIC design. Some designs are more sensitive to variation and would require more care during the design stage to anticipate possible pitfalls so that we can design around or take special precautions so that variation will not adversely affect the circuit functionality and manufacturability. Designers must learn to create variation-insensitive circuits if they are to have high-yielding product that meets the design target as well. The concept of conventional variation has evolved from digital corner methodology to the incorporation of statistical variation of fundamental physical parameters at both the intra- and interdie level. In Chapter 10 we dwelt more on the design for manufacturability aspects of the design and in most cases will be helpful in reducing the impact due to variability.

REFERENCES

[1] International Technology Roadmap for Semiconductors, *http://public.itrs.net*.

[2] K. Bernstein, Design, process, and environmental contributors to CMOS delay variation, tutorial, *IEEE International Solid-State Circuits Conference*, Feb. 2003.

[3] S. Borkar et al., Parameter variations and impact on circuits and microarchitecture, *IEEE Design Automation Conference*, pp. 338–342, 2003.

[4] Berkeley Predictive Technology Models, *http://www-device.eecs.berkeley.edu/~ptm.*

[5] Y. Cao et al., New paradigm of predictive MOSFET and interconnect modeling for early circuit design, *Proceedings of the IEEE Custom Integrated Circuits Conference*, pp. 201–204, June 2000.

[6] Y. Cao et al., Design sensitivities to variability: extrapolations and assessments in nanometer VLSI, *IEEE International ASIC/SoC Conference*, pp. 411–415, Sept. 2002.

[7] S. R. Nassif, Design for variability in DSM technologies, *IEEE International Symposium on Quality Electronic Design*, pp. 451–454, 2000.

[8] C. Visweswariah, Death, taxes and failing chips, *IEEE Design Automation Conference*, pp. 343–347, 2003.

[9] K. A. Bowman, S. G. Duvall, and J. D. Meindl, Impact of die-to-die and within-die parameter fluctuations on the maximum clock frequency distribution, *IEEE International Solid-State Circuits Conference*, pp. 278–279, 2001.

[10] M. Eisele, J. Berthold, D. Schmitt-Landsiedel, and R. Mahnkopf, The impact of intra-die device parameter variations on path delays and on the design for yield of low voltage digital circuits, *IEEE Trans. VLSI Syst.*, Vol. 5, No. 4, pp. 360–368, Dec. 1997.

[11] D. Burnett, K. Erington, C. Subramanian, and K. Baker, Implications of fundamental threshold voltage variations for high-density SRAM and logic circuits, *IEEE Symposium on VLSI Technology*, pp. 15–16, 1994.

[12] Y. Cao et al., Yield optimization with energy-delay constraints in low-power digital circuits, *IEEE Conference on Electron Devices and Solid-State Circuits*, Hong Kong, Dec. 2003.

[13] S. Mukhopadhyay and K. Roy, Modeling and estimation of total leakage current in nano-scaled CMOS devices considering the effect of parameter variation, *IEEE International Symposium on Low Power Electronics and Design*, pp. 172–175, 2003.

[14] A. Srivastava, R. Bai, D. Blaauw, and D. Sylvester, Modeling and analysis of leakage power considering within-die process variations, *IEEE International Symposium on Low Power Electronics and Design*, pp. 64–67, 2002.

[15] H. Q. Dao, K. Nowka, and V. G. Oklobdzija, Analysis of clocked timing elements for dynamic voltage scaling effects over process parameter variation, *IEEE International Symposium on Low Power Electronics and Design*, pp. 56–59, 2001.

[16] S. Lin and C. K. Wong, Process-variation-tolerant clock skew minimization, *International Conference on Computer-Aided Design*, 1994.

[17] B. Gieseke et al., A 600 MHz superscalar RISC microprocessor with out-of-order execution, *IEEE International Solid-State Circuits Conference*, pp. 176–177, Feb. 1997.

[18] H. Ando et al., A 1.3 GHz fifth generation SPARC64 microprocessor, *IEEE International Solid-State Circuits Conference*, Feb. 2003.

[19] M. Bohr, Interconnect scaling: the real limiter to high performance ULSI, *Proceedings of the IEEE International Electron Devices Meeting*, pp. 241–244, Dec. 1995.

[20] K. Bernstein et al., *High Speed CMOS Design Styles*, Kluwer Academic, Norwell, MA, pp. 41–45, 1998.

[21] A. Kahng and M. Sarrafzadeh, Modern physical design: part V, tutorial, *International Conference on Computer-Aided Design*, Nov. 1999.

[22] D. Bailey and B. Benschneider, Clocking design and analysis for a 600-MHz alpha microprocessor, *IEEE J. Solid-State Circuits*, Vol. 33, No. 11, Nov. 1998.

[23] C. Bittlestone, A. Hill, V. Singhal, and N. V. Arvind, Architecting ASIC libraries and flows in nanometer era, *Design Automation Conference*, June 2003.

[24] K. Osada et al., Universal-V_{dd} 0.65–2.0 V 32 kB cache using voltage-adapted timing-generation scheme and a lithographical-symmetric cell, *IEEE International Solid-State Circuits Conference*, pp. 168–169, Feb. 2001.

[25] K. Bernstein, Design, process, and environmental contributors to CMOS delay variation, *SCCS near Limit Scaling Workshop*, 2003.

[26] A. Asenov et al., Increase in the random dopant induced threshold fluctuations and lowering in sub-100 nm MOSFETs due to quantum effects: a 3-D density-gradient simulation study, *IEEE Trans. Electron Devices*, Vol. 48, No. 4, Apr. 2001.

[27] P. Larsson, Measurements and analysis of PLL jitter caused by digital switching noise, *IEEE J. Solid-State Circuits*, Vol. 36, No. 7, July 2001.

[28] K. Osada et al., Universal-V_{dd} 0.65–2.0-V 32-kB cache using a voltage-adapted timing-generation scheme and a lithographically symmetrical cell, *IEEE J. Solid-State Circuits*, Vol. 36, No. 11, Nov. 2001.

[29] M. Yamaoka, K. Osada, and K. Ishibashi, 0.4-V logic library friendly SRAM array using rectangular-diffusion cell and delta-boosted-array-voltage scheme, *IEEE Symposium on VLSI Circuits*, 2002.

[30] D. Harris and M. A. Horowitz, Skew-tolerant domino circuits, *IEEE J. Solid-State Circuits*, Vol. 32, No. 11, Nov. 1997.

[31] G. A. Ruiz, Evaluation of three 32-bit CMOS adders in DCVS logic for self-timed circuits, *IEEE J. Solid-State Circuits*, Vol. 33, No. 4, Apr. 1998.

[32] L. G. Heller and W. R. Griffin, Cascode voltage switch logic: a differential CMOS logic family, *IEEE International Solid-State Circuits Conference*, pp. 16–17, 1984.

[33] K. Okada, Statistical modeling of device characteristics with systematic variability, *IEICE Trans. Fundam.*, Vol. E84-A, No. 2, Feb. 2001.

[34] M. J. M. Pelgrom, C. J. Duinmaijer, and A. P. G. Welbers, Matching properties of MOS transistors, *IEEE J. Solid State Circuits*, Vol. 24, No. 5, pp. 1433–1440, Oct. 1989.

[35] C. Michael and M. Ismail, Statistical modeling of device mismatch for analog MOS integrated circuits, *IEEE J. Solid State Circuits*, Vol. 27, No. 2, pp. 154–166, Feb. 1992.

[36] W. Zhang and Z. Yang, A new threshold voltage model for deep-submicron MOSFETs with nonuniform substrate dopings, *Microelectron. Reliab.*, Vol. 38, pp. 1465–1469, 1998.

INDEX

Nano-CMOS Circuit and Physical Design, by Ban P. Wong, Anurag Mittal, Yu Cao, and Greg Starr
ISBN 0-471-46610-7